ADVANCING TECHNOLOGY:

ITS IMPACT ON SOCIETY

ADVANCING TECHNOLOGY:
ITS IMPACT ON SOCIETY

Donald P. Lauda
Indiana State University

Robert D. Ryan
St. Cloud State College

WM. C. BROWN COMPANY PUBLISHERS
Dubuque, Iowa

HM
221
, L35

To our Wives, Sheila and Barbara

Contents

PART III

An Era of Change

Foreword

Permanence is dead, proclaimed Alvin Toffler in a recent bestseller, *Future Shock.* In pronouncing the victory of transience and dynamic change over the relatively static and stable elements of human history, Toffler emphasized what is becoming increasingly obvious to all students of the past and present: Society is changing at an ever-accelerating pace. Technology is both cause and effect of this rapid social change. And, unless we understand the dynamic interplay between technology and society, we cannot make our own future; instead, we will simply react to the surge of events rather than direct them.

From the very beginning of mankind, technology has been one of the most dynamic components in plotting the course of human destiny. Man's reliance upon his technology goes back to the origins of our species. Indeed, anthropologists seeking the emergence of homo sapiens can separate man from his pre-human ancestors only on the basis of tool-using and tool-making. Some other animals make and use tools, but none became so dependent upon them as man. Possessing only natural endowments of hands, feet, and teeth, our pre-human forebears might have remained but a minor primate species, and perhaps even have become extinct. But, anthropologists tell us, tools made possible anatomical and physiological evolution to homo sapiens. Man could not have become homo sapiens, man the thinker, had he not at the same time been man the maker, homo faber.

Not only was technology basic to the emergence and survival of our species, it also played a major role in the beginnings of civilized society. The development of settled communities, the start of civilization, rests upon a technological innovation: agriculture. Up to that time in pre-history, men had been hunters and food-gatherers; in a sense, they had been parasites upon nature. Once men discovered that they could cooperate with nature in growing their food supply, there arose the possibility of settled and civilized life.

We can trace the impact of technology on society in the first great civilizations which developed along the river valleys in Egypt, Mesopotamia, India, and China. These are known as "irrigation civilizations," because sizeable irrigation works were required if crops were to be grown year after year.

To develop such works required the elaboration of social and political institutions. Some of these institutions seem so evident today that it is difficult to conceive of their having to be born. Yet, as we know from studies of prehistoric man and from today's primitive peoples, such institutions are not "natural" to man. Because of their scope, irrigation works required a high level of social integration among large groups over a wide area, and this in turn brought forth the state. True, there had been some form of government in earlier societies, but this was usually exercised over small kinship groups. Such ties no longer sufficed to regulate the relations among men; an impersonal and abstract government system had to develop. Religion too was affected: the old tribal gods were superseded or supplemented by a super-tribal deity.

Irrigation civilizations also required armies, for they were an attractive target for barbarians outside. Some men had to devote their energies to defense. Above the farmers and soldiers, a directing or organizing class, perhaps originally a priestly group, provided a third social grouping. Further specialization of labor resulted in the emergence of artisans and craftsmen as well as professionals— scribes, lawyers, judges, physicians.

Because these irrigation civilizations produced a surplus of products, trade also developed. Trade required merchants, and also money, credit, commercial law—and, in order to deal with distant traders, the beginnings of international relations.

The knowledge required in constructing and maintaining irrigation works, the complex economic transactions for payment of taxes and exchange of goods, required records; and this of course brought writing. Information had to be organized into learnable and teachable knowledge; thus these civilizations developed the first educational systems, the first schools and the first professional teachers. There also emerged concepts of justice, expression in the arts and poetry, and, eventually, philosophers and the world religions.

The pattern of man's life and work laid down at the very beginning of civilization lasted for some 5,000 years. Only relatively recently, some 200 years ago, was the basic social organism again subjected to an upheaval similar to the wrench from a nomadic, tribal society to a settled, agricultural, politicized unit. Here again technology was a major factor. The Industrial Revolution, beginning in 18th-century Britain, extending throughout the Western world in the 19th and 20th centuries, and still taking place in other areas of the globe, thoroughly transformed society. For millennia the hearth and home had been the center of production: agriculture provided the livelihood for the bulk of mankind, and the ways and places of living and working were rooted in the past. Industrialization changed all that. Men were transplanted from their ancestral dwellings to industrial towns; they were introduced into novel ways of working and making a living; the discipline of the factory replaced the rhythms of farming; and the family, religious institutions, values, and virtually every other aspect of society were revolutionized in this process. Indeed, to describe the social results of the Industrial Revolution is to delineate our contemporary world.

Because technology's impact is so obvious, some scholars claim that technology is the prime determinant of the social process. Indeed, there are some who fear that advancing technology has become an end in itself. To such critics as Jacques Ellul and Herbert Marcuse, technology has become man's master instead of his tool.

To others, however, technology is neutral. Lynn White states, "Technology opens doors, but it does not compel man to enter." This view of technology as an enabling instrument raises further questions: Who determines which doors to open? And, once one passes through, does technology determine the contours of the room into which one enters? If technology is viewed simply as a means, there is a question of who determines the ends, and also the danger of the means themselves becoming ends.

Some social thinkers, such as Lewis Mumford, claim that technology has become an evil mechanism which has prevented man from the realization of human goals. To Mumford, the technology of an earlier time provided a symbiosis of nature and human values, but he fears that today's technology has become too powerful an authoritarian force.

Such problems are not merely historical questions. They are real and present problems, as is evidenced by contemporary concerns over the direction and control of technology. The story of past technological development might itself provide clues to this dilemma. Whether technology is good or bad—and the historical evidence indicates that it can be used for both benign and harmful purposes, to make life easier for man or to destroy human beings, to tame a hostile nature or to spoliate the environment—the fact is that the evolution of technology reveals the power of the human intellect, imagination, creativity, and skills. Technology has enabled man to feed and clothe himself, to conquer space and time, to master his environment and subdue nature, to create structures of great beauty, and to relieve man's muscles and mind of onerous burdens.

By showing us what man can do with his hands, mind, and tools, the story of technology provides us with hope that the great problems facing us today, many of which have been caused by the heedless growth of technology or its use in behalf of narrow and selfish interests, can also be resolved. Man's technical triumphs provide manifest proof of his ability—if he wills it—to cope with the human and social problems of today and tomorrow.

Melvin Kranzberg
Case Western Reserve University

Acknowledgments

Melvin Kranzberg, "Foreword," written for the book.

Robert Theobald, "Dialogue-Focuser: The Impact of Technology." From *Dialogue on Technology*, edited by Robert Theobald, copyright © 1967, by The Bobbs-Merrill Company, Inc., reprinted by permission of the publishers.

Morris C. Leikind, "The History of Technology: Man's Search for Labor-Saving Devices." Alice Mary Hilton (Editor), *The Evolving Society*, New York: Cybercultural Research Press, 1966, pp. 23-33. Used by permission of the editor.

Eric E. Lampard, "The Social Impact of the Industrial Revolution." From *Technology in Western Civilization, Volume I: The Emergence of Modern Industrial Society, Earliest Time to 1900*, edited by Melvin Kranzberg and Carroll W. Pursell, Jr. Copyright © 1967 by The Regents of The University of Wisconsin. Reprinted by permission of Oxford University Press, Inc.

Wilbert E. Moore, "Conditions for Industrialization," *The Impact of Industry*, © 1965. Reprinted by permission of Prentice-Hall, Inc. Englewood Cliffs, New Jersey.

Jacques Ellul, "Technique in Civilization." From *The Technological Society* by Jacques Ellul, translated by John Wilkinson. Copyright © 1964 by Alfred A. Knopf, Inc. Reprinted by permission of the publisher.

Melvin Kranzberg and Carroll W. Pursell, Jr., "The Importance of Technology in Human Affairs." From *Technology in Western Civilization, Volume I: The Emergence of Modern Industrial Society, Earliest Time to 1900*, edited by Melvin Kranzberg and Carroll W. Pursell, Jr. Copyright © 1967 by The Regents of The University of Wisconsin. Reprinted by permission of Oxford University Press, Inc.

Nels Anderson, "The Nature of Work." From *Dimensions of Work*, by Nels Anderson. (New York: David McKay Company, Inc., 1964) Used by permission of the publisher.

Robert Ryan, "Women Work—Why?"

Leonard Nadler, "Helping the Hard-Core Adjust to the World of Work." From *Harvard Business Review* (March-April 1970). Copyright © 1970 by the President and Fellows of Harvard College; all rights reserved.

Richard Armstrong, "Labor 1970: Angry, Aggressive, Acquisitive," *Fortune,* October, 1969, pp. 94-97, 144, 146, 150. Used by permission of the publisher.

Jack T. Conway, "Ideological Obsolescence in Collective Bargaining," An address presented to the Industrial Relations Conference, Institute of Industrial Relations, University of California at Berkeley, 1963. Used by permission.

Stanley L. Englebart, "How It All Started." From Stanley L. Englebart, *Computers,* New York: Pyramid Publications © 1962, Ch. 2. Used by permission of the publisher.

Don Fabun, "Work: The Myth That Became a Monster." From Don Fabun, *The Dynamics of Change,* Englewood Cliffs, New Jersey: Prentice-Hall, Inc., 1967, pp. 8-17. Used by permission of the publisher.

Charles E. Silberman, "The Real News About Automation." From the January 1965 issue of *Fortune* Magazine. This article has appeared in a book, *The Myths of Automation,* published by Harper and Row.

Ben B. Seligman, "The Social Cost of Cybernation." Alice Mary Hilton (Editor), *The Evolving Society*, New York: Cybercultural Research Press, 1966, pp. 159-166. Used by permission of the editor.

James Boggs, "The Negro and Cybernation." Alice Mary Hilton (Editor), *The Evolving Society*, New York: Cybercultural Research Press, 1966, pp. 168-172. Used by permission of the editor.

Charles K. Brightbill, "What is Leisure?" From Charles K. Brightbill, *The Challenge of Leisure*, © 1960. Reprinted by permission of Prentice-Hall, Inc., Englewood Cliffs, New Jersey.

Sebastian de Grazia, "Leisure's Future." From Sebastian de Grazia, *Of Time, Work and Leisure*, The Twentieth Century Fund, New York, 1962.

"83 Billion Dollars for Leisure—Now the Fastest-Growing Business in America." Reprinted from *U.S. News & World Report.* Copyright © 1969, U.S. News & World Report, Inc.

Gilbert Burck, "There'll Be Less Leisure Than You Think," *Fortune,* March, 1970, pp. 86-89, 162, 165-166, 168, 171. Used by permission of the publisher.

Robert Theobald, "The Background of the Guaranteed-Income Concept." From *The Guaranteed Income* by Robert Theobald. Copyright © 1965, 1966 by Doubleday and Company, Inc. Reprinted by permission of the publisher.

Erich Fromm, "The Psychological Aspects of the Guaranteed Income." From *The Guaranteed Income* by Robert Theobald. Copyright © 1965, 1966 by Doubleday and Company, Inc. Reprinted by permission of the publisher.

H.F.W. Perk, "Economic Aspects of the Evolving Society." Alice Mary Hilton (Editor), *The Evolving Society,* New York: Cybercultural Research Press, 1966, pp. 191-200. Used by permission of the editor.

Harvey R. Hamel, "Moonlighting—An Economic Phenomena," *Monthly Labor Review,* October, 1967, pp. 17-22. Used by permission of the U.S. Department of Labor.

Whitney M. Young, Jr., "The Split-Level Challenge," *Saturday Review,* August 23, 1969. Copyright © 1969 Saturday Review, Inc.

Bernard Asbell, "Illiteracy: The Key to Poverty." From *The New Improved American* by Bernard Asbell. Copyright © 1965 by Bernard Asbell. Used by permission of McGraw-Hill Book Company.

Paul O. Flaim and Paul M. Schwab, "Employment and Unemployment Developments in 1969," *Monthly Labor Review,* February, 1970, pp. 40-53. Used by permission of the publisher.

Donald P. Lauda, "The Complications of Change."

Don Fabun, "The World Alters as We Walk in It." From Don Fabun, *The Dynamics of Change,* Englewood Cliffs, New Jersey: Prentice-Hall, Inc., 1967, pp. 2-3, 5. Used by permission of the publisher.

Max Ways, "Era of Radical Change," *Fortune,* May 1964, pp. 113-115, 214-216. Used by permission of the publisher.

Wilbur H. Ferry, "Must We Rewrite the Constitution to Control Technology?" From *Saturday Review,* March 2, 1968, copyright © 1968 Saturday Review, Inc.

Allan J. Topol, "Law and The Nation." From *Saturday Review,* August 3, 1968, copyright © 1968 Saturday Review, Inc.

Eric Hoffer, "Drastic Change." From *The Ordeal of Change* by Eric Hoffer. Copyright © 1954 by Eric Hoffer. Reprinted by permission of Harper & Row, Publishers, Inc.

Philip Hauser, "The Chaotic Society: The Social Morphological Revolution," *Vital Speeches of the Day,* pp. 22, 28, 30, October 15, 1968. Used by permission of the publisher.

Lawrence E. Hinkle, Jr., M.D., "The Growing Importance of Human Ecology," *Bell Telephone Magazine,* November/December, 1967, pp. 8-11. Used by permission of the publisher.

Paul R. Ehrlich and John P. Holdren, "Population and Panaceas: A Technological Perspective," *Bio Science,* December 1969, pp. 1065-1071. Used by permission of the publisher.

Stuart Chase, "The Control of Population." From Stuart Chase, *The Most Probable World,* New York: Harper and Row Publishers, 1968, Chapter 4. Used by permission of the publisher.

Bernard Asbell, "Can We Survive the Madding Crowd?" From Bernard Asbell, "The Danger Signals of Crowding," *Think,* July/August, 1969, pp. 30-33. Used by permission of IBM.

D.W. Brooks, "The Specter of World Hunger," *Inspection News,* The Magazine of Retail Credit Company, May/June, 1969, pp. 3-5, 17-19. Used by permission of the publisher.

Don Fabun, "The Protein Path: Hunger Begins with a Hungry Plant," *Kaiser Aluminum News,* April, 1968, pp. 33-37. Used by permission of Kaiser Aluminum and Chemical Corporation.

Alan F. Westin, "Life, Liberty and the Pursuit of Privacy," *Think,* May/June, 1969, pp. 12-21. Used by permission of IBM.

Myron B. Bloy, Jr., "Technology and Theology." From *Dialogue on Technology,* edited by Robert Theobald, copyright © 1967, by the Bobbs-Merrill Company, reprinted by permission of the publisher.

Stuart Udall, "But Then Came Man," *Vital Speeches of the Day,* July 1, 1967, pp. 569-573. Used by permission of the publisher.

Irwin Hersey, "Are We Beginning to End Pollution?" *Engineering Opportunities,* September, 1969, pp. 9-12, 14, 16, 19-20, 22, 25-26, 28-30, 34-38, 40, 42. Used by permission of the publisher.

Paul Remirez, "Thermal Pollution: Hot Issue for Industry." Reprinted by special permission from *Chemical Engineering,* March 25, 1968, pp. 48-52. Copyright © 1968, by McGraw-Hill, Inc., New York, New York 10036.

Charles A. Schweighauser, "The Garbage Explosion," *The Nation,* September 22, 1969, pp. 282-284. Used by permission of the publisher.

Anthony Hannavy, "Can Engineering Cope with the Debris of Affluence?" *Product Engineering,* October 9, 1967, pp. 37-44. Used by permission of the publisher.

Eric B. Outwater, "The Disposal Crisis: Our Effluent Society." From *National Review,* February 24, 1970, Permission National Review, 150 East 35th St., New York, New York 10016.

Gus Tyler, "Can Anyone Run a City?" From *Saturday Review,* November 8, 1969, copyright © 1969, Saturday Review, Inc.

Joseph Wood Krutch, "Must Technology and Humanity Conflict?" *Bell Telephone Magazine,* March/April, 1969, pp. 2-7. Used by permission of the publisher.

Nigel Calder, "The Control and Use of Technology," *The Nation,* January 4, 1965, pp. 3-5. Used by permission of the publisher.

T.A. Wilson, "A Crucial Time for Technology," *Vital Speeches of the Day,* June 15, 1968, pp. 533-536. Used by permission of the publisher.

Athelstan Spilhaus, "The Next Industrial Revolution," *Science,* Vol. 167, p. 1673, 27 March 1970.

Gilbert Burck, "Will the Computer Outwit Man?" From the October 1964 issue of *Fortune* Magazine. This article has appeared in a book, *The Computer Age,* published by Harper and Row.

E.J. Mishan, "Abyss of Progress," *The Nation,* November 4, 1968, pp. 466-468. Used by permission of the publisher.

R. Buckminster Fuller, "The Year 2000," Architectural Design, 26 Bloomsbury Way, London W.C. 1, February, 1967, pp. 62-63. Used by permission of the publisher.

George A. Miller, "Some Psychological Perspectives on the Year 2000," *Daedalus,* Summer, 1967, pp. 883-896. Used by permission of the publisher.

Arthur Kantrowitz, "The Test," *Technology Review,* May, 1969, pp. 45-50. Used by permission of the publisher.

John McHale, "The Future of the Future," *Architectural Design,* 26 Bloomsbury Way, London W.C. 1 England, February, 1967, pp. 65-66. Used by permission of the publisher.

PART I

Chronology of Technology

INTRODUCTION

Rapid change in recent times has left those caught up in its wake somewhat breathless. So complex are the factors that created the change and the consequent effects that mankind could be thrust into an era where that which once was thought to be good will in fact be problematic.

In the recent past there was time to permit the individual and society to adjust to the pace of advance. This generally is not possible today. In fact, many advances are scuttled on the "drawing board" in favor of new and more advanced concepts.

This acceleration of time, or more explicitly the shortening of time from concept to implementation, can be best illustrated by the statements of historians who tend to agree that there has been more technological advance during the last few decades than in all of history.

If the history of man is to be thought of as an evolution through technological advance, then man must be considered as a tool-developing and tool-using animal. Physically weak when compared with the other species he had to compete with, he used his superior intelligence to develop tools for survival. Other animals were equipped with teeth, claws, speed and other physical characteristics whereby they could either evade or overcome their enemies. Man, to assure his existence and to provide for the necessities of life—food, shelter, clothing and security—developed tools.

As man evolved his life became more complex, as did his tools and his social systems, and the forces that continued to shape his society.

The authors of this section discuss the impact of technological advance from the beginnings of man to recent decades. It is essential that the reader be aware of the conditions for industrialization, the techniques and/or skills necessary, and the effects on society and the individual. A study of history will provide a greater understanding of the present and place our technological development in perspective.

Dialogue-Focuser:
The Impact of Technology

Robert Theobald

This is a new style of document called a dialogue-focuser. Its purpose is to summarize the state of the debate on a particular subject. Dialogue-focusers are not copyrighted, for they attempt to reflect what all of society has discovered about a particular topic: They therefore belong to the total society.

Areas of Essential Agreement

Man today is in the process of achieving the power he has sought so long, the power to remake his environment. Thus, even though we presently apply only a small part of available knowledge, abundance has been achieved in the United States. This means that it is possible, by making appropriate decisions, to call forth additional production to meet needs. The potential for human betterment, however, is not limited merely to the satisfaction of material needs; ongoing research is producing new tools for perception of man as a biological system affected positively and negatively by interaction with an environment he himself helps to create, for improvement of the individual's understanding of himself and his physical and mental perception, for greater inter-personal understanding, and for the comprehension of relationships of human group to human group.

The dramatic increase in man's power stems from four developing realities. First, there is the drive toward the production of effectively unlimited energy, which provides the possibility of carrying through any task defined as worthwhile. Energy makes it possible to turn salt water into fresh water, to mine low-grade ores to obtain necessary metals, and to change the environment.

Second, there is man's increasing skill in manipulating the basic building blocks of nature on the micro-scale. This provides the possibility of designing materials with any desired set of characteristics; it also makes it possible to manipulate the genetic inheritance of man.

Third, there are ever greater numbers of people who have been so educated that they can continue to widen the frontiers of knowledge.

Finally, there is the development of the computer. The first commercial computer was installed in 1950: The number of computers operating in the United States grew to 5,000 by 1960 and is conservatively estimated to reach 70,000 by 1970. During this period there will be an increase in the power of

3

the computer by several orders of magnitude; in addition, time-sharing will allow several people to use the same computer simultaneously. The computer should not be perceived only as a logic machine of rapidly growing capacity and speed. It is indeed quite possible that the effect of the computer itself is less crucial than the change in the pattern of thinking which it is forcing. Fuzzy thinking is incompatible with the use of the computer and complex machine systems; the consequent rigor in patterns of thinking is spreading far beyond those areas where computer applications exist or are probable in any near future.

These increases in man's power to control his environment are leading to fundamental clashes between the goals of the industrial age and the emerging commitment to provide each individual with the maximum possibility to develop himself within an appropriate societal framework. First, present international relations are based on the assumption that each country should be able to defend itself against all potential attackers; it therefore becomes necessary to adopt any weaponry-system, because a potential enemy might otherwise gain an overwhelming advantage. The result is that each country destroys its own security in the process of seeking to obtain it.

Second, our present socioeconomic system requires the use of any technique or technology promising to provide a competitive advantage. A cheaper technique introduced in one firm but not in another, will weaken the position of the second; a country introducing more technological advances than another will strengthen its own position. Such a drive toward greater efficiency, and the minimization of human labor can be favorable, but not when jobs, essential to an individual's sense of himself, are eliminated. In addition, certain types of individuals with low skills and education are unable to compete with machines at any wage providing a decent standard of living.

Third, most people can only support themselves by holding a job; in order to provide jobs for everybody, people in the society must be willing to use all the goods that can be turned out. This constraint in the socioeconomy is increasingly undesirable because:

—people can only absorb and integrate a certain amount of information, and excesses above this level result in sensory overload and consequent blanking of the senses' ability to provide the individual with knowledge of himself; this in turn prevents the growth of the individual's potential;

—the drive to maximize production causes such quantities of waste and pollution that the survival of life on the planet is threatened.

Each dialogue-focuser will necessarily impinge on many others. The nature of the dialogue in other areas must certainly be implied rather than detailed.

Areas of Disagreement

Are present dominant styles of academic analysis adequate for studying technology? Some argue that we have understood the scientific method, that this method is always appropriate, and that there is, therefore, no need to examine the techniques employed. This view is challenged by most of those deeply involved in the debate on technology who argue that new tools of analysis and new methods of communication are required. Some believe a fundamental difference lies between the analytic techniques required to examine improvement in the functioning of an already existing socioeconomic system and the task of imagining or inventing a new one.

The growing consensus about the applicability of new styles of analysis does not imply agreement about the essential nature of these new styles. The most fundamental disagreements are in terms of using subjective or objective techniques. Some authorities are engaged in refining such instruments as input-output analysis, computer based planning, game theory, etc. These techniques assume that there is an objective reality which can be understood and controlled, that the consequent task is the development of better methods of measurement and control. Other authorities are convinced that reality is subjective, that the world tends to become what it is believed to be—particularly given man's power to change his environment. Those advancing the subjective view argue that the primary task must therefore be to discover the type of world in which man would wish to live. A few people are convinced that neither the objective nor the subjective modes, taken *alone,* are adequate, that it is essential that both modes of viewing reality remain in constant tension, and that both can contribute to the task of designing the future. (The parallel to long-standing philosophical disagreements is obvious.)

Even those who see most relevance in objectivity tend to devalue the importance of "facts," i.e., isolated pieces of data relevant for particular moments of time, because they recognize:

—that the methods used in examining and collating data to produce "facts" are such that the "facts" serve to support existing behavior rather than to challenge existing patterns of reality.

> Most "facts" are not single events, but rather they relate to overall patternings of behavior and relationships. The ways in which we perceive patterns of behavior and relationships are themselves heavily structured by the environment in which we live. Thus societally perceived "facts" tend to support the existing socioeconomic order.

—that the pace of change is presently so great; "facts" therefore change with such rapidity that they are, of necessity, inadequate guides to understanding.

At the very least, then, those examining the implications of technology agree that they must be concerned with "trends" in data rather than with "facts"; in other words they must examine not single pieces of data relevant for a particular instant of time but rather the expectable changes in data over a relevant time period. It is increasingly argued, however, that trends are also inadequate as guides to understanding the future because:

—"trends" are "facts" extended in time; the assumptions about social reality which determine our perception of "facts" have the same effect on our understanding of the meaning of "trends."

—the present rate of change cannot persist over an extended period of time because no rapid growth trend can continue for ever within an essentially closed environment. It is therefore inevitable that changes in trend will develop in coming years and decades. It is impossible, however, to determine which existing trends will be limited and which reinforced, without an examination of the desires affecting people, communities, nations and the world in general. The probable direction of change also depends on the values which are affirmed and the values which are actually followed.

Downgrading the importance of "facts" and "trends" is one major factor leading to increasing debate about the nature of the constraints imposed by the environment, the nature of the change process and the nature of man; this latter subject shades into a debate about the nature of the good society. Central to all these discussions is cybernetics, an understanding of the science of communication and control. The central element in cybernetics is the recognition that systems—whether animal, mechanical, economic, social, human, etc. —include feed-back mechanisms and that the existence of appropriate feed-back mechanisms is a requirement for the survival of any system.

> Feed-back means that when an action takes place, the initiator of the action receives information as to what has followed his action so that he/it can be sure that the desired effect is being achieved. No assumption about the nature of "desirability" is made here; the issue of how to decide on the desirable falls outside cybernetic theorizing. Three examples of feed-back may prove helpful:
> 1. when an individual runs a temperature, information about the deviation from the desired temperature level is conveyed to other parts of the body which then try to reduce the deviation from the norm.
> 2. similarly, a thermostat in a room reacts to the temperature around it; if the air is too cold, the thermostat sends a message causing the furnace to start up; when enough heat has been generated the thermostat sends a message to the furnace causing it to turn itself off.
> 3. a functioning free-market mechanism is based on cybernetic principles: rising prices cause additional people and companies to believe it would be worthwhile to produce more of the expensive item; this increase in the supply causes the price to fall. Similarly, when the price falls, people and companies tend to produce less and this brings the price back up.

Cybernetics also proves that change in one part of the system will cause changes, both expected and unexpected, in other parts of the system. This means that any significant innovations will not only affect the area of immediate impact, but many other areas as well. Very complete and sensitive understanding of a particular system is required if understanding the impact of various types of changes is to be possible, in particular if determining those changes which will have limited impact, and those with impact which will continue to develop over time, is to be possible.

Successful directed change is only possible, therefore, if those attempting it understand the current situation and process. In other words, it is always essential to work within the actual situation, to recognize that change will only be acceptable if it appears evolutionary even though it may, in the end, have revolutionary consequences. The requirements for purposive change are therefore:

—that those who wish to bring it about understand the system well enough to perceive the changes which will appear as evolutionary,

—that those who wish to bring about change have a clear idea of the limitations on desirable action imposed by environment and by the "inherent" nature of man.

—that those who wish to bring about change have a clear idea of the type of individual and/in/with society they wish to develop.

But while cybernetics makes clear that the elements of the ecological system interconnect, there is little agreement about the degree of stability or instability in the ecological balance and even less about factors causing a non-reversible change, climatological or other, which would make the earth uninhabitable. There are few, however, who would dismiss the possibility that while man's actions can improve the environment, they can also gravely reduce the carrying capacity of earth.

Similar uncertainty permeates the debate about the nature of man. On one hand, there is general agreement that man is an animal with a relatively long period of maturation during which several stages of behavior tend to appear. On the other, there is general agreement that considerable variation in behavior patterns can develop in order to adapt to the environment. Nevertheless, there is profound disagreement about the range of flexibility in the maturation process and the sets of values possible of achievement.

The extent of the disagreement is increased, and the nature of the disagreement confused, because it is still far from clear what value-sets will be required to live in the era we are entering. Debate on this issue—and the appropriate direction for the socialization process—has only just begun. Indeed, the relevance of debate on this issue is often denied on the grounds that we should not be concerned with requirements for living in the coming era, but rather that we should aim to design the type of world which we desire here and now. At this point, the debate comes full circle, for it is being argued once more that man has the power to create the environment he wishes and that he should therefore be primarily concerned to find ways of deciding on the environment he desires.

A growing number of people on the edge of the debate on technology deny the validity of dichotomy between the necessary and the desirable; they claim that it is essential for man to become honest, responsible, humble and loving if he is to live with the power which he himself has created. This viewpoint suggests that the old tension between intelligent actions and moral actions is disappearing and being replaced by tension between objective and subjective methods of seeing reality.

For some, therefore, the debate on technology ends in an affirmation that it is now essential for man to reach a new stage in his development, that he grow up as a race. This argument has been advanced by scientists, social scientists, theologians and philosophers. Much of the tension in the present dialogue on technology can be traced to the growing belief that man cannot survive in the environment which he himself has created, without fundamental change in his nature, and to the inability of those involved in the debate to perceive how the required changes can be achieved with the necessary rapidity.

The History of Technology:
Man's Search for Labor-Saving Devices

Morris C. Leikind

The road from flint knives and axes to modern computing-machine-directed machine tools and cybernated machine systems has been long and arduous. Among the major achievements of our earliest ancestors were mastery of fire, development of stone tools (knives, spearheads, arrowheads, axes) and domestication of plants and animals.

One of the greatest advances was the invention of writing—an extension of man's memory that made the vast accumulation of information and the recording of knowledge possible. The wide-spread dissemination of information began with the invention of printing and led to the mass production of information and data that, in turn, led to new methods of handling such quantities: computing machines.

Developing sources of power beyond that supplied by man's own muscles began with the taming of animals. It continued with the use of wind and water, and with the discovery of electricity. And now, there is the promise of virtually unlimited power by means of nuclear fusion.

I have been given the difficult assignment of reviewing the *History of Technology,* particularly man's search for labor-saving devices, in twenty-five minutes. Some of you are perhaps familiar with the massive five-volume *History of Technology,* prepared by the late Professor Charles Singer and his colleagues. Attempting to encapsulate the contents of this work for you, I might, thereby, qualify as a *labor-saving device.*

Benjamin Franklin once defined man as a tool-making animal. Perhaps this trait distinguished man from the other members of the animal kingdom. Certainly, the ability to make tools has enabled man to master his environment as no other species has.

It is not possible to give precise dates to man's earliest inventions in attempting to extend his powers, both physical and mental, and thereby to control his environment. Some of the principal developments were, of course, the taming of fire and the ability to make a variety of tools, such as knives, axes, throwing sticks, arrow and spear heads, needles and other devices of flint and basalt, of wood and bone. All of these devices were used by man to fight his fellow men, but, more important, they were tools of the economic process —hunting, fishing, and the domestication and cultivation of plants and animals.

Seldom considered as labor-saving devices, poisonous substances, discovered in certain plants and applied to spear and arrow heads, make hunting much easier, and thus make it possible for one man to kill a large animal— an elephant, a mastodon, or a saber-tooth tiger—where a whole tribe would have been needed to slaughter it before.

During the pre-literate state of human history almost all the important mechanical devices, which extend man's capacity for work, were invented, including the wheel and the lever. Nevertheless, had the human capacity for inventing stopped at this point, mankind might still be in the primitive state of culture represented by the rapidly dwindling survivors of primitive groups today. Fortunately for us, however, an invention made by some one or some group—we shall never know precisely—some 5,000 years ago set mankind on the road we are still traveling today. This is the invention of writing.

Writing is an intellectual rather than a manual development. Its importance simply cannot be overestimated. It provides man with a permanent extension of his memory not only during the lifetime of the individual but, indeed, during the lifetime of the species. To this day, examples of writing which go back to the very beginnings of this art are preserved. Writing made the real beginnings of science and the refinement of technology possible. With its aid, men could, for the first time, think and work in precise, quantitative ways that were impossible when one had to rely on memory and oral communication alone. Writing and record-keeping make the development of such sciences as astronomy, mathematics, arithmetic, geometry, and trigonometry possible. Geometry means, literally, the measurement of the earth, and it is no accident that geometry began in one of the centers where writing was invented —in Egypt, where the annual flooding of the Nile wiped out the boundary markers of the fields and made the re-measurement of boundary lines necessary.

In this same area the colossal monuments of antiquity were erected. The pyramids and sphinxes are more than monuments and tombs to kings; they are monuments to the architects who designed them. They are also monuments to thousands of slaves who labored to build them, and to the administrators who organized these minions to perform almost incredible feats of construction.

We sometimes think that labor-saving devices are the hardware, and we overlook the intellectual advances describable as "organization and administration." A tool of any kind, from an inert stone to a human being, is of little value unless there is a methodology directing its use. An efficient workman with a tool can obviously produce more than an inefficient one; an efficiently directed person obviously has a greater labor output than an inefficiently directed one or one who is not directed at all. In any consideration of labor-saving devices, methodology (including organization and administration) is as important and significant as the tool itself.

The Egyptian pyramids have long been a mystery. They were built without the aid of modern tools by the organized labor of thousands of men. These men had to be assembled in one area. They had to be fed, housed, provided with tools and their labor had to be directed toward one predetermined purpose—conceived and controlled by the will of *one* man. It was a tremendous technological feat to harness the power of a hundred-thousand men at a time when methods for harnessing other natural forces had not yet been devised.

The monuments of the past remind us that by the time men were able to build pyramids, they had already domesticated plants and animals, were beginning to live in urban communities, had discovered division of labor and specialization as labor-saving methods, and had invented various arts and crafts. In short, they had become farmers and shepherds, carpenters, stone-masons, shoemakers, weavers, smiths, metal workers. Herodotus, writing in the fifth century B.C., tells us of the specialists among physicians: "Every physician is for one disease and not for several, and the whole country is full of physicians, for there are physicians of the eyes, others of the head, others of the teeth, others of the belly, others of obscure diseases." He even tells of one specialist known as the "keeper of the anus," who twenty-five centuries later is called a proctologist.

The Greek Miracle

All the advances, of which I speak, had been made, according to available evidence, before the fifth century B.C., on the shores of the Mediterranean— from Asia Minor in the east to the Sicilian islands in the west. On the Greek peninsula and its islands, the Greek miracle, the "mutation in human thinking"—the birth of science—happened. Man began to think abstractly and abstraction became the means of understanding his world. He learned to generalize from particular observations. And this, I submit, was a labor-saving discovery of the highest importance.

The greatest contribution of the Greeks was their very invention of abstraction in pure mathematics. Greek mathematics was applied in astronomy and in such branches of physics as mechanics, hydraulics, and optics. In the biomedical sciences, Hippocrates, Aristotle, and Theophrastus are among those whose magnificent contributions will live forever. We must note, however, that Greek civilization—like all ancient civilizations—was born in a human-powered society and that technological advances we would consider labor-saving were scarce. The Greeks did know the principles of the screw, the inclined plane, the action of pumps, and the use of steam to produce power. Recently, a complex mathematical and astronomical instrument, the *antikythera*, was described by Derek Price. Although this indicates that a more sophisticated technology existed in antiquity than we had hitherto supposed, Greek knowledge was primarily theoretical. Greek science was not harnessed to technology.[1] The Romans, on the other hand, proved their great engineering skill in their roads, aqueducts, and military machines but they seem to have had little interest in science. In both Greek and Roman civilization, the technology was geared to the use of human and animal muscular energy. Hence the size of machines and the objects handled had to be restricted.

In some instances, one technical advance was partly nullified by lack of progress in a related technology. The Romans, as we know, were master road builders. Many of their highways are still in existence. Yet, they were not as effective as they might have been because their horses were not properly harnessed, and, therefore, their horse-drawn vehicles were inefficient.

New Sources of Power

Horse-power was a discovery of the early Middle Ages. The horse was originally used to transport human beings—chiefly in war and hunting—and only rarely as a pack animal. Since antiquity the principal draft animals were oxen. The chief reason for the late use of the horse as a power source was the lack of adequate horseshoes to protect horses' hooves and—even more important—the inefficient harnessing technique used during antiquity. The horseshoe and the horsecollar, which transformed the horse into an economic as well as a military asset, made horse-power a very important factor in the great agricultural revolution of the Middle Ages. The invention of the plough —millennia before—represents the first application of non-human power to agriculture. Dragged by a pair of oxen, this primitive digging stick was suitable for the light, dry soil of the Mediterranean regions. But as agriculture spread northward across Europe, vastly greater power was required to move the heavy, moist soil. Speed was essential because of the shorter season for plant-

---

1.The Greeks thought only the abstract worthy of man. Archimedes had an almost legendary reputation as the inventor of skillful contrivances that were used against the Romans during the siege of Syracuse. Yet, Archimedes considered mechanics and everything directed to use and profit unworthy of being written down. He was a great experimenter, but considered only his deductive (and rigorous) proofs worthy of recording. See Alice Mary Hilton, *Logic, Computing Machines, and Automation* (1963), pages 26-7.

ing. And a well-shod horse with a collar that did not choke him proved far more suitable than the ox. As Lynn White, Jr.[2] has so well said:

> The agricultural revolution of the early Middle Ages was limited to the northern plains where the heavy plough was appropriate to the rich soils, where the summer rains permitted a large spring planting, and where the oats of the summer crop supported the horses to pull the heavy plough. It was on those plains that the distinctive features both of the late medieval and of the modern worlds developed. The increased returns from the labour of the northern peasant raised his standard of living and consequently his ability to buy manufactured goods. It provided surplus food which, from the tenth century on, permitted rapid urbanization. In the new cities there arose a class of skilled artisans and merchants, the burghers who speedily got control of their communities and created a novel and characteristic way of life, democratic capitalism. And in this new environment germinated the dominant feature of the modern world: power technology.

White also points out that:

> ... the later Middle Ages, that is roughly from A.D. 1000 to the close of the fifteenth century, is the period of decisive development in the history of the effort to use the forces of nature mechanically for human purposes. What had been up to that time an empirical groping, was converted with increasing rapidity into a conscious and widespread programme designed to harness and direct the energies observable around us. The labour-saving power technology which has been one of the distinctive characteristics of the Occident in modern times depends not only upon a medieval mutation in men's attitudes toward the exploitation of nature but also, to a great extent, upon specific medieval achievement.

Despite the fact that by the first century B.C. the Hellenistic world was producing elaborate geared mechanisms, there is no evidence that its technicians were attempting to exploit sources of natural power. Hero of Alexandria had devised a steam turbine, but it remained a toy throughout antiquity. Nevertheless, during the period of the Roman conquest of the Near East we get the first glimpses of the harnessing of water power for irrigation and the milling of grain.

It took a thousand years for water power to spread over Europe, and by then men began to look for additional sources. In the eleventh century, tidal power began to be used, and in the twelfth century, the windmill began to appear on the European horizons. By the fourteenth century, Europe had made phenomenal progress toward substituting wind and water power in the basic industries. All over the landscape were mills for fulling cloth, for laundering, sawing, crushing everything from olives to ores, for blowing air into blast furnaces, or grinding and polishing metal for armor, making mash for beer, even for the polishing of jewels.

The success of these power devices led to the exploration of even more sources and now we find men beginning to look to steam as a source of motive power. It was not until the seventeenth century that steam was on the way to

---

2. Lynn White, Jr., *Medieval Technology and Social Change,* Oxford University Press, New York (1962), p. 72.

being finally harnessed; however, at the same time we find experimenters playing with blow guns, air guns and attempting to use the force of expanding vapors and gases.

One of the by-products of these experiments was the rocket. The story of the invention of gunpowder is too confused and complex to discuss here. Nevertheless, I must mention the development of the cannon and other fire-arms which not only revolutionized warfare but had a profound influence on several branches of science. Ballistics is a direct offspring of studies on the theory of the motion of projectiles and the fundamental laws on the conservation of energy that were first suggested by studies on the heat produced by the boring of cannons. We must not forget that the internal combustion engine is merely a cannon with a piston attached to a crankshaft.

The evolution of the crank, which made the translation of reciprocating motion to rotary motion possible, needs to be mentioned. The appearance of the bit and brace in 1420, and the connecting rod in 1430, were the bellwethers of this advance in machine design. But there was more to come; because medieval technicians faced problems of controlling and regulating continuous rotary motion, fly wheels and other forms of mechanical governors had to be invented.

The term *feed-back* was yet to be coined but the mechanical base for the idea was already in existence. Increasing sophistication in machine design led in rapid succession to: the invention of the treadle, used in weaving and in operating lathes; the use of springs as a component of automatic power-machines, such as sawmills; and, lastly, the spinning wheel, adapted to the use of belt transmission of power which made the control of different rates of speed in various parts of the same machine feasible. The culmination of medieval mechanical efforts came with the invention of the mechanical clock around the middle of the fourteenth century. With this, the prototype of the automatic machine was created.

These remarkable inventions were made in the course of about five centuries. Since the earliest known tools go back about five-hundred-thousand years, this was indeed a remarkably rapid revolution.

The Printing Press

But the rate of progress accelerated to an even greater degree when the printing press was invented and modern science began. Just as one of the great landmarks of human history was the invention of writing, during the fifteenth century an equally important invention was made—printing from movable type. The printing press made possible the mass production and democratization of knowledge. Before the invention of the printing press, only a select few could read or write. Manuscripts were rare and exceedingly expensive. Even the rise of universities from the twelfth century had little impact on the production and supply of written, transmittable knowledge.

Teaching was, with few exceptions (as in monasteries), an oral process. The printing press changed all that. In 1543, almost a century after the Gutenberg Bible was issued in 1452, two books were published which were to have a revolutionary impact on humanity and dated the beginning of modern science. Copernicus, in *De Revolutionibus Orbium Caelestium (On the Revolution of the Celestial Orbs)*, gave us the modern concept of the solar universe, and the Belgian anatomist, Andreas Vesalius, in *De Humani Corporis Fabrica (On the Structure of the Human Body)*, gave us the modern concept of the human body. Georgius Agricola produced his most famous work, *De Re Metallica,* on mining, geology, and metallurgy. These, and books that followed, such as William Harvey's discovery of the circulation of the blood, published in 1628, and Galileo's writing on mechanics and physics, pre-saged the information explosion in science that was triggered by the invention of the printing press.

If we define science as cumulative, verifiable, and communicable knowledge, we can see how the invention of printing gave science and technology a tremendous impetus. New ideas could be disseminated widely, and, like seeds falling on fertile soil, they germinated and gave birth to additional new ideas and techniques, *ad infinitum.*

The organization of scientific societies in the seventeenth century, such as the Royal Society in London in 1662, marked the beginning of scientific journals and the creation of scientific libraries all over the world. The production of books and papers increased at such a fast rate that the scientific community soon began to complain that there was too much to read. By 1830 the abstract journal had to be invented as a labor-saving device.

The sixteenth and seventeenth centuries were periods of tremendous intellectual activity which had a profound impact on technology. A number of new sciences were hatched. The rise of anatomy and physiology laid the foundations for modern medicine. Alchemy gave way to chemistry, and physics, through the work and Galileo and Newton, advanced rapidly. The interaction of science and technology became more fruitful as exemplified by the invention of a variety of important scientific and technical instruments such as the microscope, telescope, and vacuum pump, the barometer, the magnetic compass, and a variety of mathematical instruments, including "Napier's Bones" and the mechanical calculators of Pascal and Leibniz. These devices by extending man's sensory powers and his ability to manipulate the objects of scientific study led to even greater advances.

Electricity

One of the subjects which had long attracted considerable attention was electricity. The Greeks had toyed with it and had given the subject its name. Little progress, however, was made until the seventeenth century. Again it was the invention of instruments which conditioned the advances. The vacuum pump of Hooke, von Guercke, and others, and the electric battery of Volta as well as the Leyden jar, opened the door to a whole series of studies, including those of Benjamin Franklin, Galvani, Volta, Ohm, Ampere, Faraday, Henry, and Clerk Maxwell. These discoveries were of significance not only scientifically but also technologically, because they ushered in the age of electrical power in industry.

In the age of electricity the world was transformed. The work of Faraday, Henry, Maxwell, and Edison had direct and profound effects on the economy, with vast labor-saving consequences. The work of these men had scientific consequences of even greater import. Attempts to understand the nature of electricity and electromagnetic forces led in rapid succession to the discovery of the electron, of x-rays and radioactivity, of the structure of the atom, and last, but not least, of techniques for releasing and harnessing the energy locked within the nucleus of the atom.

And now the final discovery remained. Man, as we have seen, has been progressively mastering the various forces of nature to serve his ends. He has harnessed men, animals, the winds, water, coal and oil, electrons, and nuclear energy. For each of these systems he has required and devised methods of control. The newest, and perhaps most important instrument, is the computing machine that now promises to affect our lives in ways which can be only dimly foreseen. In the span of human history, the computing machine is so recent an invention that it has scarcely begun to attract the attention of historians.

To go into the fascinating history of these machines would be beyond the scope of this essay except to indicate that the detailed history of computing machines[3] would require a knowledge not only of the history of mathematics —going back to the invention of the oriental abacus (3000 B.C.), "Napier's Bones" (1617), Pascal's (1642) and Leibniz's (1642) calculating machines, Jacquard's loom (1801), Babbage's analytical engine (1833), and Vannevar Bush's differential analyzer (1931)—but also an account of the invention of the vacuum tube and the transistor. And still, it would be only part of the story. Really to understand the nature and function of computing machines, we must also study a history of physiology, notably the work of Claude Bernard (1878) and Walter Cannon (1932) and their theory of homeostasis (self-regulation). And finally, the rapidly developing field of neurophysiology has made its contributions to the cybercultural revolution. It is interesting to note that two of the greatest contributors to the age of cyberculture, Mathematician Norbert Wiener and Physiologist Walter Cannon lived and worked on the banks of the Charles River in Cambridge, Massachusetts.

3. See Alice Mary Hilton, *Logic Computing Machines, and Automation,* Spartan Books, Washington (1963), distributed by the I.C.R. Press. Paperback edition, Meridian Books (1964). Chapters one, six, and nine.

In ancient Egypt, one man—the Pharaoh—could command the labor of one-hundred-thousand men. Today, one man—who need not be the Pharaoh—can command the direct energy equivalent of one-hundred-thousand men. Are we approaching the time when even that one man can be released?

The Social Impact of the Industrial Revolution

Eric E. Lampard

The Industrial Revolution: Triumph or Catastrophe?

The Industrial Revolution was for long represented by its historians (and by many of its contemporaries) as a "sudden and violent" change which inaugurated an age of economic exploitation and social unrest. The revolution, they say, began in the late 18th century when steam engines were harnessed to new machines in the textile mills of central and northern England. It led inexorably thereafter to the crowding of scores of thousands of men, women, and children into slum quarters of the bleak towns which mushroomed about the factories. There the workers were driven to labor long hours under abominable conditions that were not materially remedied before the mid-19th century through the mounting pressures of trade unions and the passage of benevolent factory acts. "The effects of the Industrial Revolution," said Arnold Toynbee in 1881, "prove that free competition may produce wealth without producing well-being. We all know the horrors that ensued in England before it [the revolution] was restrained by legislation and combination."

For Toynbee and most other critics of the early industrial society, it was the factory-owner who appropriated the economic benefits, while the worker and "the community" bore the costs. When the Industrial Revolution later spread to continental Europe and North America, it appeared that many of the same processes were repeated in circumstances which, however much they might differ in local color and detail, gave rise to similar forms of exploitation and the same kinds of social abuse. In these later industrial revolutions, likewise, time and suffering were required before public-health and factory regulations, backed by a more enlightened and democratic public opinion, could prevail against the factoryman's greed and the cruel indifference of town governments.

During the present century, this almost wholly pessimistic view of early industrialization has come under critical review. As a result, the optimism expressed by some contemporary observers in the late 18th and early 19th centuries found new support in the writings of 20th-century scholars. Historians were able to show that many of the technological and organizational novelties commonly associated with the Industrial Revolution had, in fact, been unfolding over previous centuries on the Continent as well as in the British Isles. This held for the engines utilizing atmospheric pressures and even for the so-called uniformity system of interchangeable-parts manufacture which underlies modern mass production. Likewise, factory organization did not break in upon a stagnant world of industry, but had its prototypes and forerunners dating back at least to the Italian Renaissance. Many economic and social abuses of "the bleak age," moreover, were found to antedate the steam engine and the textile mill.

Social reformers, such as the German businessman and revolutionary propagandist, Friedrich Engels, had clearly romanticized the realities of a bygone rural life and industry in which "workers enjoyed a comfortable and peaceful existence" and where "children grew up in the open air of the countryside . . . and there was no question of an 8 or 12 hour day." Hours of toil in household and workshop had most likely been long and arduous and, according to a contemporary spokesman for the factory interest, William Cooke Taylor, child labor in Lancashire "was at its worst and greatest height before anybody thought of a factory."

The researches of economists have also revealed that, notwithstanding the possibility of dislocation and hardship in areas of industrial decline, the workers who migrated to new areas of industrial revolution received rising real wages and could expect to enjoy a higher level of living. Finally, it was suggested, the shocking squalor of the manufacturing towns—which had alarmed reformers and reactionaries alike at the time—was to be better understood in terms of the circumstances rather than the rank cupidity of the capitalist class.

Neither the original pessimistic view nor the more recently found optimism of the revisionists, however, can be taken as altogether single-minded efforts at historical description and analysis. Nor should they be dismissed out of hand as mere political or class prejudice. Both positions contain the not-uncommon mixture of history with advocacy which marks the treatment of any really *live* issue—from the Reformation to the causes of world war. The bitter indictment of an Engels from "the Left" and the complacent apologetics of a Cooke Taylor from "the Right" refer, in fact, to much the same bodies of historical evidence and rehearse many of the same arguments that have served pamphleteers and scholars down to the present day. Revived interest in the historiography of industrial revolution is, indeed, related to the current phase of "competitive coexistence" in which the leading Cold War protagonists recommend their own particular styles of industrialization and social change to newly developing countries in Asia, Latin America, and Africa.

The pessimistic left focuses its critical attention upon the immediate experience of industrial revolution and dwells upon the "full" costs of the economic and social upheavals that have characterized the early stages of industrialization carried out under capitalist auspices. The optimistic right addresses itself rather to the ultimate outcome and insists that the benefits from economic growth and rising levels of living accrue to the entire population and not to the owners of capital alone. Moreover, they remain unimpressed by the more recent performances of the collectivist alternative. A critical view of the Industrial Revolution is put forward by the socialist and, oddly enough, the old-style conservative, while a favorable judgment is usually rendered by the classical liberal and his latterday descendent, the new-style conservative, who has made himself the ideological custodian of industrial achievement. Whereas the prosecution's frame of reference is generally social responsibility and the idea of "community," that of the defense is liberty of the individual and "free" enterprise in the pursuit of wealth.

Since the historical specialists have so far tended to adopt opposing views on the social impact of the Industrial Revolution, it is not surprising that the general reader has difficulty making up his mind on the subject. When most of the historical evidence is, so to speak, short-run and parochial, concerned only with a few decades in time and with one or two localities, it is not easy to develop any adequate sense of a prolonged revolutionizing process. Also, where the established frameworks of interpretation are either too partial (narrow and biased) or too abstract (theoretical), it is hard to see the longer-run process unfolding in any authentic context of history. The divergent views on the Industrial Revolution outlined above serve to underscore the hazards involved in generalizing about a many-faceted process of social change, the character and effects of which *in the event* were highly diverse. Nevertheless, within the brief scope of this chapter, the next task will be to construct a socio-economic framework in terms of which the historical experience of industrial revolutions can be more fully understood.

The Economic Benefits and Costs

From an economic standpoint, any industrial revolution may be regarded as the initial reorganization of a society's productive resources that gives rise to a rapid and "self-sustaining" increase in per-capita incomes. The metaphor of "revolution" denotes a social departure in which the total output of goods and services begins, virtually for the first time, to grow at a discernibly faster rate than population. From the purely economic standpoint, therefore, industrial revolution is but another term for accelerated economic growth. But it is also more than the phenomenon of economic growth since it comprehends both the departure from some traditional order of society characterized by, among other things, little or no economic growth, *and* the inauguration of a new "industrial" order characterized by, among other things, continuous and cumulative increases in output and income. The long-run tendency for per-capita output and incomes to rise is to be included then among the major consequences of a social movement or transformation called "industrial revolution."

Industrial revolution is thus a particular form of social change. Its occurrence transcends explanation in purely economic terms. For example, changes in technology and social organization—which may occur independently of changes in economic variables such as incomes and investment—are often among its mainsprings. Hence, our concern here is the *social matrix* itself, whose changing contours, taken in conjunction with dependent economic movements, embody the revolution's social impact.

Requirements for economic growth may be baldly stated as continuous increases in the per-capita supply of productive resources, that is, the labor and capital (including raw materials) from which all goods and services derive. In short, growth depends on the *supply capacity* of the economy which governs potential output at any time. The means to achieve rising output per capita are twofold: (1) an absolute increase in the quantity of resources utilized per head of population by the addition of more labor and/or capital (including materials) than heretofore; (2) an increase in the relative *efficiency* with which available resources are utilized by altering productive methods of organization so that any given output per capita is obtained from a smaller input of resources. The labor and/or capital withheld from the production process by more efficient operations may then be utilized in raising output in some other line of activity. Thus a unit of resources spared through greater efficiency (2) is equivalent to a unit of resources added to the per-capita supply (1). By whichever means or combination of means the supply capacity of the economy is enhanced, output is potentially increased. In short, the "problem" of economic growth resolves itself into the problem of developing in the population a style of behavior that will enlarge the society's capacity for production.

The concept of growth as a supply function must be distinguished from the concept of the determination of total product. Total product at any time is normally dependent on total demand, within some upper limit set by the output potential of the economy. The analysis of the determination of total product is thus the analysis of demand. Growth analysis, on the other hand, is concerned precisely with the upper limit of supply potential and, more especially, with the rate at which the upper limit of supply changes over time. Our analysis of industrial revolutions, therefore, is concerned not so much with actual output or total demand as with the changing supply capacity of the economy.

The economic behavior which is most appropriate to enlarging the supply capacity is called saving and investment. The act of diverting a part of the existing resource supply from current consumption and replacement activity is termed *net* saving; the act of converting such saving into the enlargement of labor and capital stocks in the future is *net* investment. To augment productive capacity and supply potential, therefore, requires saving beyond mere replacement levels and investment beyond mere replacement levels. Simply to replace existing stocks of inputs would represent a condition of zero net saving and investment, a condition that Adam Smith in the 18th century already recognized as a "stationary state."

Just how rapidly, and in what measure, a rising rate of net investment will return higher incomes per capita in the future depends on: (1) the proportion of total income that is saved—the saving ratio; (2) the amount of investment required on average to produce a unit of output—the so-called capital-output ratio; and (3) the rate of growth of population. Given the level of saving (1), the growth of product and income will vary inversely with (2) the capital-output ratio. The higher that ratio, the more capital that is used up on average in producing a unit of output, the slower will a unit of investment pay off in added output (and vice versa). Hence the critical relevance in all growth calculations of (3) the population factor, since it represents the denominator that is divided into the output numerator at any time to give output per capita.

This highly abstract and artificial framework of economic and demographic (population) variables permits a closer understanding of the "problem" of industrial revolution. Unless a country can borrow or otherwise obtain *extra* resources from abroad, it must "squeeze" its saving out of resources that are currently available at home. When a large proportion of the resources that are withdrawn from current consumption and replacement are used not so much for the provision of new capital goods (the celebrated textile machines, for example) as for the production of necessary "social overhead" (roads, bridges, canals, harbors, warehouses, hospitals, sewage facilities, business and residential construction, and so on), the incremental yield of goods for personal consumption by workers and other members of the population is likely to be low. The additional income earned by workers in the installation of the social overhead, in fact, will have no corresponding or "off-setting" equivalence in increments of consumer goods. Prices of the necessities that loom large in the budgets of workers' families are likely to rise at least until such time as higher prices serve to attract resources into enlarging the economy's supply capacity for consumer goods. Thus, at the very moment when a society is first preoccupied with expanding its overall supply capacity (and largely because of the preoccupation), prices may rise and *real* incomes per capita fall.

From the standpoint of the mass of working consumers then, the achievement of economic growth under the constraints of industrial revolution may impose considerable hardship for an indefinite period. This would almost certainly be the case if the job opportunities connected with the installation of the social overhead were to encourage couples to marry younger and have larger numbers of children. The increase in population would, in the short run, heighten pressure on the already limited supplies available for consumption.

But one important mitigating circumstance in many low-income pre-industrial situations is the likelihood of a degree of "disguised unemployment," or more accurately, *underemployment* of existing resources. In some activities, for example, the marginal product of additional units of input may be zero or actually negative in the sense that employment of additional units of input would add less to income than to costs. Hence, the successful deployment of resources from such typical lines of underemployment as traditional agriculture into activities where marginal products are likely to be positive, such as commerce or industrial manufactures, would in itself be tantamount to an absolute increase in the society's total resources—simply because labor productivity is higher in the modernizing sectors. An alternative strategy would be to keep existing labor and equipment at work for twelve instead of ten hours a day and, without any evident loss or gain in efficiency, total resource input per capita would nonetheless be increased by the additional man-hours of labor input.

The central question—whether or not industrial revolutions necessarily impose hardship on the mass of the population—probably turns on: (1) the degree of underemployment, and (2) the rate of population increase. If underemployed resources abound, if there is "slack" in the traditional system, the necessary movements of economic variables need impose no severe deprivation in consumption terms. In such an instance, the population variable would be decisive.

The Industrial Revolution and "The Social Question"

It is not inconceivable, therefore, that per-capita incomes might actually rise during the course of an industrial revolution. Yet even in that best of all possible worlds, the broad masses of the population and their spokesmen might still find the social costs of economic change excessive. A society is always more than a market place, and a livelihood is never the whole of life. Almost any of the economic growth possibilities considered above would have entailed profound alterations and disturbances in the inherited pattern of everyday life —regardless of changes in average levels of living.

In social terms, the transition from a traditional or pre-industrial society to an industrial society may be characterized as the disintegration and "break up" of an old order and its gradual replacement by a new order which is organized and integrated along different lines. What might at first appear to be rather commonplace matters (to 20th-century people) of changing jobs, shifting from one type of occupation to another, moving away from home or other familiar scene, getting poorer or richer, would, under conditions of industrial revolution, turn out to be nothing short of the disintegration and displacement of an older way of life.

During the first phases of the Industrial Revolution, the distress and bewilderment of populations in towns and in the country was, if anything, aggravated by the very gradualness of change. Aware that the old order was in flux, few could yet be sure that they or their children would find a place in the new "dis-order," or of what the outcome would be. Certainly the traditional agrarian society, with its high incidence of poverty, hunger, and disease, and its legal and social discriminations, was not Utopia. The idealized image of America held by Europeans from the 16th century was itself a significant rejection, in some ways, of the world that was known. Yet to many caught up in the throes of industrialization, the old order acquired a sufficiency and equity, a benign corporateness, as it receded further into the past. For a country like England with relatively low density of population, commercial vitality, and "ancient" liberties, even the recent past would seem to be "Merrie England"; but in France or Germany the golden age was still further back in the lost order of the medieval world. By the second quarter of the 19th century, contemporaries already spoke of the need for social reconstruction as *the* social question."

The "social question," the "social problem," or the "labor problem" represented efforts to comprehend and cope with the gathering impetus of industrialism. Contemporary diagnoses and prescription, of course, varied with the observer's position and outlook in life, and with the place and date of his observation. Nevertheless, the posing of the question at any time revealed the growing concern among different segments of society that the new and potentially more productive ways of livelihood were contingent upon the acceptance of far-reaching alterations in the conduct and conditions of everyday life.

Under industrialism the focus of the social question changed almost as rapidly as the environment itself. The age-old phenomenon of pauperism and the poor was gradually transmuted into the more pressing "social problem" of a self-conscious working class or "industrial proletariat"; while the response to that issue in turn brought the entire social order into question. Under industrial conditions it was increasingly difficult to accept poverty as part of some inscrutable God-given natural condition; it was truly a social question. With the further unfolding of industrial development in the later 19th century, however, it was apparent that larger numbers were adjusting to the routines and discipline of the new order, reconciled that for them at least, the benefits in income and acquired status, on balance, outweighed the heavy costs. Changes in social organization and individual behavior tended to assume a more regular form such that, underlying the politicized "problems" of adjustment to social change, some observers could discern the formative societal processes: the matrix of industrial society.

It is possible for the historian of the Industrial Revolution to define the essential contours of these underlying movements in the changing context of the social question. The problems identified by those who experienced the revolution in their own lives—problems of population, the factory, the city, class conflict, the right of association, illiteracy, and persistent poverty— furnish important clues to the processes that were actually shaping the industrial order.

Conspicuous among the problems of early industrialization in a number of countries, for example, were those relating to: (1) the recruitment and training of workers for specialized installations of manufacture such as mills and factories; (2) the adjustment of workers to the routine of factory operations and to impersonal "industrial" relations with their employers in matters of work discipline, hours, conditions, and wages; also (3) the adjustment of working populations to the housing, public health, congestion, rules and conditions of changing town life; (4) the political integration of the greatly enlarged working classes into the emerging social structure. This meant achievement of full civil rights, political recognition, and the right of association. Other problems included: (5) the formation and integration of new middle classes of talent and education—the independent professionals, public servants, and higher grades of technical and clerical "white collar" workers—into a social hierarchy with the older middle classes of property-owners, merchants, and industrialists; (6) the adjustment of personal behavior to continuing alterations in the social environment, or the accommodation of inherited rules and mores to new patterns of individual behavior; and (7) development of the means, such as enlargement of communications and other service facilities, provision of educational and training systems, inauguration of forms of "social" insurance against periodic economic distress, physical disability, and old age, and so on, to institutionalize social change.

Among these "typical" problems of early industrialism—freely adapted from a more comprehensive list suggested by the German scholar Wolfram Fischer—item number (7), purposeful and controlled adaptation to industrial change, clearly overlaps and even subsumes some of the others. Indeed, insofar as it raises the larger issue of the overall coherence and direction of industrial change, the question of adaptation becomes virtually "*the* social question" itself. Underlying the many issues of personal and institutional adjustment was the pervasive tendency toward greater specialization and differentiation of functions in almost all aspects of life. As a consequence, the Industrial Revolution involved greater functional interdependence in the economic sense among more specialized groupings of workers and machines in different regions and countries, at the same time as the older hierarchical forms of personal and social interdependence were dissolved. Many of the changes, costs, and benefits, of early industrialism were rooted in this intensified socio-economic division of labor. Inherited relationships and structures that proved to be too deeply embedded in small market towns and traditional agrarian communities disintegrated.

Population Movements

The long-run tendency of industrialism toward greater specialization and differentiation of functions manifested itself in three characteristic population movements: (1) occupational differentiation, involving the shift of workers into full-time, specialized employments, notably in non-agricultural industries but in agriculture also; (2) spatial differentiation, or territorial redistribution of population, especially in the form of urban concentration; and (3) social-structural differentiation, involving upward social mobility and fuller participation in shaping the social order, through the achievement of civil rights for all citizens and expansion of opportunities for attaining "middle class" status. To be sure, the organization and structure of traditional societies are also characterized by varying degrees of functional-structural differentiation. But the effect of the Industrial Revolution was to break up the prevailing order insofar as it was characterized by smaller and more evenly distributed populations, comparatively undifferentiated rural-agricultural occupations, and by the ranking of classes or "estates" in established hierarchies of status.

The Industrial Revolution put mobility in the place of stability. As a consequence of progressive economic specialization, industrial populations tended to become, on average, more skilled and productive, more urbanized in residence and, with rising levels of education and per-capita income, more "bourgeois" in their social orientation and style. In short, the industrial regime required occupational, residential, and social mobility.

The transition from the traditional to the industrial, however, was a long drawn out, uncertain process. At any time institutional and structural rigidities in the old order could, and in some countries did, arrest the course of social renovation and economic progress. In the late 18th and early 19th centuries —the period with which we are presently concerned—the shape of things to come was everywhere unsure. Under the technological and organizational conditions of the Industrial Revolution, moreover, the very same occupational and residential shifts that were a condition of economic progress were also a primary source of dislocation and stress. In conforming to these structural movements individual members of the population felt the social impact of the revolution in the pattern of their own lives.

Conditions for Industrialization

Wilbert E. Moore

The clamor for change is so general in the modern world that it hardly appears credible that most of the world's societies and cultures seemed sleepy and content through the opening four decades of the twentieth century. The contentment, we now have reason to believe, was either a sham or simple apathy in the face of apparently fixed inequities. And the sleepiness was partly an apathetic sham and partly a distorted perception in the eye of the beholder. Yet it is certainly true that for most of the nonindustrial parts of the world —and in numerical terms, that means most of the world—the safest basis for predicting social behavior by the day or by the week or by the year was knowledge about "how it had always been." The winds of change have provoked a reconsideration of the seemingly somnolent societies, and the alacrity with which novelties have been embraced—in frequently bizarre combinations —has provoked scholarly soul-searching over the supposed inviolability of sanctified tradition. Theoretical structures shook, and first-hand observations were degraded into snapshot versions of a richer and more dynamic reality.

Though the discomfiture of the intellectuals provides a mean and temporary solace to prideful fools, the reshaping of science concerns us rather less than the reshaping of social systems. Yet the latter can be viewed only as chaotic or worthy of only journalistic reporting unless we develop new instruments for systematic observation and for ordering observation into meaningful and predictive generalizations.

Among all the manifestations of discontent and disorder, of changes occasionally orderly and often volatile, the ones that engage our primary attention here are those that are somehow related to economic modernization. This may seem rather lacking in daring, and even lacking in a suitable sense for what is important, since many of the events that exemplify the end of quiet continuity present themselves in a political context. Disorder is intrinsically political, but the discontent that underlies it may have little to do with politics. We have urged the view that economic change is both a goal widely shared in underdeveloped areas and an instrument for the achievement of other goals such as education and health and even national power.

What, then, are the requirements for economic modernization or industrialization? They include the positive side of discontent, the quest for improvement. But they also include, ironically, a considerable measure of political order, a condition notably lacking in some of the very areas where the demand for economic improvement is articulate to the point of stridency.

Generalization is both important and hazardous.[1] At a minimum, a discussion of the social conditions for economic development should provide a systematic checklist for particular, local analyses. A somewhat more ambitious aim, which is the one sought here, is to establish types of relationship that will be applicable to any particular area by a process of filling in details. The maximum aim of theory is to establish highly general "laws" of economic growth. The hope for this achievement is, it appears, premature, although suggestive attempts have been made and are by no means fruitless.

1. Much of the balance of this chapter has been adapted, with major changes in organization, from Wilbert E. Moore, "The Social Framework of Economic Development," in Ralph Braibanti and Joseph J. Spengler, eds., *Tradition, Values, and Socio-Economic Development* (Durham, N.C.: Duke University Press, 1961), Chap. 2.

Our objective is to outline a way of analyzing the social framework of economic development and to note at least some generalizations about the way various conditions interrelate. Our procedure will be to start with organization —that is, with action systems necessary for a modernized economy. We shall then consider the institutional order—the web of laws and less formal rules within which economic organization must act. Finally, we shall consider ideological and motivational elements: ultimate values, collective goals and aspirations, and the drives and purposes that animate individuals in the pursuit of economic activities. We shall thus be using a kind of metaphor of concentric circles, comparable to our procedure in subsequent chapters in delineating the consequences of economic transformation.

It must be noted that we are following this procedure for analytical clarity, but the result will not be an order of priorities in terms of either time or urgency. Such priorities must differ according to the circumstances of particular areas. Nor is our procedure a strictly logical one, since goals must be considered as precedent to the means for their attainment. But the order adopted reflects the view that the gross goals of economic transformation are not in serious question. Ideological and motivational conditions almost appear as subtleties, as variables doubtless affecting the pace of change and even some of its detailed order, but often in ways less direct than, say, financial organizations or property laws.

Organization

The organizational requirements of modern economic systems are in one sense true by definition. That is, if industrialization is equated with factory production, all that remains is to specify the social characteristics of the latter. However, it is also true that almost any economic development will require rationalization of organization and the creation of concrete systems of action designed for specific and limited functions. Although forms of rationally constituted productive organization are probably much more common in preindustrial societies than is commonly supposed,[2] they constitute a central and pervasive characteristic of economically advanced societies.

Our concern here is to identify a kind of minimum checklist of organizational requirements for economic modernization, without closely examining their exact forms as social systems. Our question can be put simply: What must be done in an organized way if a traditional economy is to be transformed into a modern one? Note that the underdeveloped economies of the contemporary world are far from homogeneous in their political and economic structures. Some have been colonial areas until very recently and thus have at least a veneer of modernity in their legal systems. Many have had some involvement

2. See Stanley H. Udy Jr., *Organization of Work: A Comparative Analysis of Production Among Nonindustrial Peoples* (New Haven: HRAF Press, 1959).

in international trade and thus have at least a veneer of modernity in their commercial arrangements.

Resource Utilization

Resources have significance for human affairs as means for the achievement of human ends, or as limiting conditions on the possibilities and costs of economic production. The soils that man has learned to use for growing food and fiber or for pasturing livestock, the stones and minerals that man has learned to transform into useful and ornamental objects or into ingredients or machines for fabricating those objects, and the inanimate sources of power that man has learned to harness as a supplement to his energy or that of domesticated animals—these resources are distributed very unevenly over the earth's surface. Any particular piece of the earth's land and water areas that has been marked out as constituting a political unit may be rich or poor in resources, or rich in some and poor in others. Of course, the picture may change through time, with new discoveries as a result of explorations for metals or fossil fuels such as coal and oil, or new techniques such as those that transformed uranium from a metal mainly useful as a ceramic glaze to an awesome source of power. Plans for industrialization must proceed from known resources, however, subject to change as geographical and technical discoveries are made.

The known resources and known techniques for their use need not be exclusively domestic, however. Even the European pioneers in industrialization relied on trade in raw materials as well as in finished products. The United States and later the Soviet Union with their large and richly endowed land areas were more nearly self-sufficient in resources, though not in capital and technology. As their industrial processes and consumer tastes demanded more and more exotic materials, the physical technology of transportation and the social technology of international trade made their economic growth less dependent on their own material endowments.[3]

Though trade and the accumulated technical knowledge that permits choice and substitution of products and processes have reduced the importance of local resources for industrial development, it remains true that unused raw materials make the developmental process easier. Exports of minerals or fossil fuels permit capital imports, and extraction and related activities will normally attract foreign capital.[4] Nonetheless the possibility and rate of economic growth may be less affected by resources than is the exact composition of production. That is, where particular resources are locally available, their

3. See Joseph J. Spengler, "Summary, Synthesis, and Interpretation," in Spengler, ed., *Natural Resources and Economic Growth* (Washington, D.C.: Resources for the Future, 1961), pp. 275-303.

4. See John H. Adler, "Changes in the Role of Resources at Different Stages of Economic Development," in Spengler, ed., *Natural Resources . . .*, pp. 48-70. See also Richard Hartshorne, "Geography and Economic Growth," in Norton Ginsburg, ed., *Essays on Geography and Economic Development* (Chicago: Department of Geography, University of Chicago, 1960), Chap. 1.

fabrication is more likely to be a leading component of industrial output. This is one of the principal reasons that food processing and textile manufacturing figure prominently in early stages of industrialization, since the resources used are primarily agricultural in origin.

Very small countries that are also poor in natural resources will experience unusual difficulties in achieving substantial economic growth. Those that have succeeded—Switzerland, Denmark, Belgium, Netherlands, and, more recently, Hong Kong and Puerto Rico—have done so in part by importing raw materials and exporting finished products. Their success has rested, therefore, on increasing the effectiveness of use of their *human* resources. Small countries in Africa and Latin America may also be able to overcome the limits set by poor resources, but only if they become in a sense part of larger economic units through trade and simultaneously upgrade the qualities of their potential workers.

Financial Organization

Banks and other financial organizations are necessary as a means of assembling and distributing funds and credits. Unless the state is the exclusive investor, banks and similar organizations are needed to tap the savings that commonly exist even in poor countries. Since the assembling of the various "factors of production" is not an instantaneous process and often requires the making of plans and placing of orders months and even years in advance of the intended completion date, a system of debts and credits is essential.

It is banal to observe that the principal economic problem of the underdeveloped areas is the shortage of capital. However, some aspects of the problem are far from elementary. And there is ample theoretical and empirical reason to doubt that large capital supplies would automatically solve the remaining difficulties of economic development. The historical record in the countries now well advanced in industrialization indicates that a high rate of savings (or *capital formation*) does not necessarily produce a high rate of gross output or output per capita. As Kuznets concludes from his review of historical trends in capital formation,

> Capital formation does not matter as much as capital utilization. And utilization depends upon a host of economic and social conditions which sometimes permit attainment of high rates of growth with little capital, but at other times impede the growth-inducing effect of even large amounts of capital.[5]

Substantial savings cannot be expected in tribal or other subsistence economies. Peasant and other agrarian economies, however, may have considerable "frozen" savings not put to productive use. The savings of a family may be

---

5. Simon Kuznets, *Quantitative Aspects of the Economic Growth of Nations: VI. Long-Term Trends in Capital Formation Proportions,* Supplement to *Economic Development and Cultural Change,* 9, 4 (July 1961), 56.

almost entirely represented by gold and precious stones made up into jewelry for women (or, occasionally for men also). Gold and silver coins may be accumulated and hoarded. The assets embodied by jewelry may be used as security for loans (commonly at usurious rates) or even converted to cash by sale in cases of emergency. But these attempts at self-insurance provide little or no benefit to the economy or return to the hoarders. The shortage of developmental capital may thus derive partly from the lack of confidence in the safety of investment and in the stability of currencies, and from the lack of reliable channels of investment for the small investor. Existing investment channels, if available and utilized at all, may be overly conservative or may divert savings to relatively unproductive uses such as loans that help the farm family survive until the next crop is harvested but without improvements in capital or techniques that would increase productivity.

The harnessing of savings for capital expansion is both a condition and consequence of industrialization in something more than a definitional sense. Yet the rates and mechanisms of doing so cannot be generalized through time and space. The historic record is mixed, as just noted, and the contemporary developing areas add further types of variation.

Socialist states can attempt, more or less effectively, to capture and utilize all savings by collectivizing all forms of production and taking all profits either directly or through taxation. Other economies must find alternative ways to induce savers to become investors rather than hoarders. In free enterprise systems, business profits become a principal source of capital accumulation through being reinvested in the same or other activities. Profits, however, may simply be spent on increased current consumption by their recipients. In many underdeveloped countries, increased luxury consumption has negative effects on the economy, since it is likely to involve luxury imports. This provides no effective increased market for domestically produced goods, and uses up scarce foreign exchange needed for capital imports. Foreign investments pose a special problem, for the profits may be neither spent nor invested locally, but instead be repatriated for the benefit of the investors, to spend or invest in the country supplying the original capital.

Even high profits may not be sufficient to divert savings from one economic sector to another: for example, from agriculture to industry. The risks may be judged as disproportionately higher. In addition, unfamiliar organization and technology are involved. These possibly rational considerations are often supported by nonrational ones. This is particularly true where ownership of land involves the principal basis of both security and prestige in the traditional social structure, while newer economic activities at best provide merely money but not aristocratic social standing.

Large-scale industrial production normally requires capital beyond the means of even wealthy families. There are three fundamental alternative ways of getting the necessary capital: from a large foreign corporation, from a domestic corporation that will pool the resources of many investors, or from the government, which uses its taxing or borrowing power to acquire the funds. There are, of course, many varieties and mixtures of the means of assembling savings and directing them toward investments consistent with continuous economic growth. The machinery of investment is a necessary counterpart of the organization of production.

In newly developing areas, the needs for investment to create expanded productive facilities intersect the rising demand for current consumption. Even if there were no population growth—and that is a seriously complicating factor, as we shall see—the poverty that is a major incentive for economic development is also a major deterrent to rapid growth. For the private investor in the oldest industrial economies, savings represent deferred consumption. In the meantime, he receives additional income, in the form of interest, dividends, or trading profits, from investing his savings. The desperately poor are unlikely to afford voluntarily the luxury of postponing consumption, unless this is somehow linked to the quest for economic security and familial obligations toward children.

Transportation and communications systems are also requisite conditions both for national economic integration and for links with the industrial and commercial world. With the rich pool of accumulated technology now available to newly developing areas, their forms of transportation and communication at any given time may not exactly duplicate those of advanced industrial countries, and certainly there is no need to replicate the original sequence of development. Air strips may be built before hard-surfaced roads, and radio stations before telephone lines. Still, the various forms of transportation and communication in current use in advanced countries have somewhat distinct functions, and newly developing countries are likely through time to broaden their range of facilities.

Resource utilization, commercial and financial organization, and a network of transportation and communication comprise the minimum essential organizational conditions for industrialization. A host of other organizational forms are highly correlated with economic modernization but scarcely qualify as prior conditions. Research institutes, public health services, or an electric-power grid may greatly facilitate economic growth, but there are enough examples of alternative arrangements to caution us against viewing them as necessities.

The Institutional Order

It is the general function of institutions, as complexes of norms or rules of conduct, to relate standard patterns of action in a society—often encompassed in concrete social organizations—to the general system of functional requirements and values of that society. This statement, although cryptic, has some implications that are substantive as well as conceptual. The set of prescriptions and expectations comprised by the institution of monogamous marriage, for example, appears, correctly, to have primary relevance to family and household composition. By the same token, however, the institution has relevance for the heterosexual relations of adults generally, the socialization of the young, the mode of distribution of goods and services, the modes of assigning general social status, and—in more attenuated ways—maintenance of order and preservation of values. In other words, assignment of an institutional complex to a particular functional area—familial, economic, political, religious—is always in some measure improper. It is precisely one function of institutions that they be *relational,* that is, that they provide the bonds or cement among particular patterns of social action.

These general comments are prompted by the circumstance that we shall here be considering the significance of institutions that are in the first instance "economic," before moving to other institutional requirements. In no instance, however, is the designation precise, since we shall be dealing always with *degrees* of relevance for the production and distribution of goods and service.

"Economic" Institutions

Traditional economic analysis would have dealt with land, labor, and capital as "factors of production" and later theorists would add "entrepreneurship" or "organization." This conceptual framework has basic difficulties, as there is really little sensible or useful distinction between land and capital, and entrepreneurship may properly be considered a particular kind of skilled labor. It seems preferable, therefore, to adopt a somewhat different conceptual scheme. We shall here deal with three interrelated institutional complexes—property, labor, and exchange. All three seem to define aspects of relations between persons and "things"—their control, transformation, and distribution. However, all also define social relationship—between owners and nonowners, between persons performing complementary tasks, between sellers and buyers. No society is or could be lacking in rules governing such relationships, but their form and content differ and these differences are centrally relevant for the possibilities of economic development.

A *property* relationship is at the minimum triadic: the person or other social unit, the object or locus of scarce values, and the potential challenger to "rights." Property relationships may also be very complex, with various social units holding common or diversified rights in the same locus of value, varying rights of disposal, transfer, use and appropriation of increase.[6]

The important point in the present context is that the property systems of most underdeveloped areas do not favor modern forms of economic enterprise. This is generally true, for example, of the older property laws and practices in Latin America. The application in the Spanish colonies of quasi-

6. See Wilbert E. Moore, "The Emergence of New Property Conceptions in America," *Journal of Legal and Political Sociology,* 1, 3-4 (April 1943), 34-58.

feudal land-tenure conceptions (often through the specific forms of the *repartimiento* and *encomienda,* which "entrusted" indigenous populations and their labor to landlords) encouraged large estates but not necessarily their efficient operation or easy transfer to more efficient producers. The *hacienda* system and many other forms of plantation agriculture have often been wasteful of both labor and capital (land and its fertility). Perhaps most importantly it established the basis of a socioeconomic elite dedicated to traditional economic and social forms.

Modifications of traditional property systems have not necessarily brought them "closer" to the institutional requirements for economic development. Land reforms, for example, may lead to more intensive utilization but often still with archaic techniques and with overcapitalization of equipment that must be held by each cultivator of a small holding. The Mexican *Ejido,* which like most land reforms was undertaken primarily in the interests of social justice rather than economic modernization, has discouraged transferability of property and has probably contributed to hidden unemployment in agriculture.[7] Put in the bluntest possible terms, peasant proprietorship is more likely to impede than to facilitate economic development.

Any property system consistent with modern forms of production tends to "expropriate" the worker from ownership of most tools of production. The avoidance of this by socialist organization or nationalization of capital is only partial and partly fictitious, since essential controls of capital allocation and disposition remain in the hands of relatively limited numbers. The economies of scale characteristic of modern economic enterprise can be achieved only by correlative unity of control, however the ancillary rights and benefits may be distributed. Those same economies of scale are likely to require collective forms of proprietorship, whether through the state or through the private corporation that mobilizes the financial resources of many individual investors.

Labor provides a second major focus for economic institutions. The norms governing productive work are aspects of more general norms that control status and role assignments in a continuing social order. In a nonindustrial or, more properly, in a nonmarket economy the separation of useful activities into "labor" and "other" is likely to be difficult and abstract, except with reference to the production of physical goods.[8] Any society is characterized by considerable specialization of roles, including in this narrow sense, productive roles. Generally, although not entirely,[9] this specialization in nonindustrial societies is determined on grounds of age, sex, kinship, and hereditary social position. Competence is then developed for predetermined positions in the process of socializing the young. Such an institutional structure

---

7. See Wilbert E. Moore, *Industrialization and Labor* (Ithaca, N.Y.: Cornell University Press, 1951), pp. 237-38.

8. See Wilbert E. Moore, "The Exportability of the 'Labor Force' Concept," *American Sociological Review,* 18 (February 1953), 68-72.

9. Udy, *op. cit.*

is likely to function adequately for limited specialization and minimal change. It is scarcely suitable for an economic system that entails extensive specialization and relatively rapid change in occupational structure. The difficulty of institutional adaptation is made more acute by the circumstances that new forms of demand for labor represent a radical shift in social role, and are intrusive in the traditional structure.

Among the many institutional difficulties with respect to labor mobility, several may be especially noted. Strong traditional attachments to the land (with ancillary handicraft production) discourage movement to other modes of production without the pressure of poverty. Moreover, where large-scale agriculture is the norm, various covert forms of peonage or debt servitude are by no means unknown.

The problem of securing enterprisers, managers, and technicians for modern economic production is partly educational—the sheer undersupply of skills. The institutional order in the normative sense is also relevant, however. Traditional forms of division of labor have placed considerable value on public administration and the older professions, less on business administration, risk-taking investments, or various kinds of engineering. It is not clear how acute this institutional barrier to development may be, but there is some suggestion[10] that both managerial innovation and capital formation are hindered by persistent high evaluation of a "leisure class" and the choice of current luxury consumption over reinvestment and productive expansion.

The third institutional complex of primary relevance to the economy is that of *exchange.* Some minimum form of exchange prevails in any society for transfer of products from the specialized producer to the general consumer. However, the elaboration of exchange relationships is likely to be limited by the degree of productive specialization. Typically, too, exchange relations are closely intertwined with other social bonds in preindustrial societies. Even "markets," which are a feature of many agrarian areas, may involve a complex of social relationships and of barter according to traditional terms of trade rather than consideration of current supply and demand.[11]

Economic development depends upon institutional transformation in the direction of "impersonal" markets, not only for goods but also for labor. These are linked by profits, salaries, and wages, all of which are virtually meaningless in the absence of a commodity market. Wherever transportation facilities and communications permit in Africa, Asia, and Latin America, there is evidence of the gradual or abrupt transformation of traditional trading relations into something approximating the economists' model of markets.[12]

10. See Frederick Harbison and Charles A. Myers, *Management in the Industrial World* (New York: McGraw-Hill Book Company, 1959).
11. See Talcott Parsons and Neil Smelser, *Economy and Society* (New York: Free Press of Glencoe, Inc., 1956).
12. See Richard H. Holton, "Changing Demand and Consumption," in Wilbert E. Moore and Arnold S. Feldman, eds., *Labor Commitment and Social Change in Developing Areas* (New York: Social Science Research Council, 1960), Chap. 11.

If one were to attempt a one-word summary of the institutional require-ments of economic development, that word would be *mobility*. Property rights, consumer goods, and laborers must be freed from traditional bonds and restraints, from aristocratic traditions, quasi-feudal arrangements, paternalis-tic and other multibonded relations. Such mobility necessarily creates tensions and readjustments. The orderliness of social life becomes and remains precari-ous, for there is no stable future time unless the modernizing efforts themselves fail.

Order and Change

In addition to those normative complexes that have primary relevance for the production and distribution of goods, there are other systems of values and prescriptions for behavior that modify or affect the course of economic devel-opment. We need to note here especially the problems of maintaining a suffi-cient degree of reliability and persistent predictability in social action, on the one hand, while the system is embarked on programs of rapid change and encountering its side effects.

Because manufacturing, as distinct from many aspects of trade and com-merce, depends heavily on fixed capital installations, *political order* is a prime requisite for industrialization. This requirement is accentuated by the extensive development of a credit structure, often on long-term bases, that characterizes both the capitalization and the trade of industrial societies. The geographical extent of political order is also of some importance, since the "factors of production" for manufacturing must typically be assembled from scattered places that often lie within different national boundaries.

The notable political instability in many underdeveloped countries cannot fail, accordingly, to have an adverse effect on many forms of economic develop-ment. Mere transfers of power, with or without electoral sanction, may have little consequence, of course, but rapid changes in legislation and administra-tion set up difficult if not impossible conditions for long-term planning and commitment of resources.

Though order is essential, it serves as a framework for massive changes. At early stages of industrialization, which is our focus here, a structural revolution in productive processes is involved. But that revolution is not a single event or a mere transition from a stable condition of poverty to a stable condition of wealth. Change becomes a continuing and often accelerating state of the social system.

Change, in fact, becomes institutionalized. This is most obviously the situation with reference to the encouragement of *science and technology*. Initially it is the systematic knowledge of the nonhuman environment that is paramount, but large-scale production and distribution and the persistently troublesome questions of human aspirations also make *social* knowledge rele-vant. The development of exact science and its application to particular problems in the form of various technologies rest upon education systems and research establishments, but these in turn rest upon norms that specify a rational, problem-solving orientation to the natural and social universe.

It is often noted that latecomers in the process of economic development are heirs to the accumulated technology of the older industrial areas, and thus do not need a complete recapitulation of the sequence or the time initially required for that accumulation. It is less often noted that any technology is part of a functional context, beginning with such mundane considerations as parts' suppliers and repair facilities, extending to skilled technicians, and finally to the institutional system itself. When we add the important qualification that many technical developments are "labor-saving," which is not a matter of fundamental economic importance in most underdeveloped areas, we have seriously modified the easy acceptance of the "acceleration principle" in the diffusion of technology.

In most societies in the history of the world, change has been largely unintentional and often simply the response to natural or social crises. It is a special feature of industrial societies that a great deal of social, including economic, change is deliberate. This is closely related to the norms governing science and technology, particularly as the latter are viewed in an appropriately extended sense. Historically it appears that the development of change as a *norm* was slow and characteristic precisely of inventors and entrepreneurs. In the contemporary world the primary source of this institutional transformation is likely to be governmental, only gradually extending to more "private" sectors of social systems. The acceptance and active fostering of the "rational spirit" in all aspects of social behavior,[13] with its attendant subversion of the power of tradition, appears to be a necessary if often unwelcome correlate of economic growth.

Ideology and Motives

Standard anthropological and sociological analyses of societies, cultures, and lesser social systems place great emphasis on the integrative function of *values.* The emphasis is not misplaced, if it stops with the functional importance of ultimate explanations and justifications for specific beliefs, rules, and patterns of action. The emphasis is misplaced, however, if such values are regarded as immutable, and therefore as "permanent" sources of differences in social systems or at least as tremendous barriers to the acceptance of any such social novelty as new forms of economic activity. Precisely because of their pervasive, integrative function, values are likely to be slow to change, and to furnish resistance to innovation in subtle ways. Nevertheless, some value

13. See Wilbert E. Moore, "Measurement of Organizational and Institutional Implications of Changes in Productive Technology," in International Social Science Council, *Social, Economic and Technological Change: A Theoretical Approach* (Paris, 1958), pp. 229-59.

changes do occur, and "traditions" may evolve, become adapted to objective changes in social conditions, and even decline.[14] Crude historical experience also offers ample evidence that ideologies are transferable between social systems. Thus belief systems ranging from the strictly religious (such as Christianity or Islam) to the seemingly secular (such as economic development) have shown remarkable powers of expansion among otherwise diverse cultures.

One of the most commonly noted features of industrial societies is their internal diversification. A complex division of economic function is integrated through the impersonal operation of the market or the quasi-impersonal discipline of administrative organizations. Urbanization and other forms of geographical mobility potentially bring together people of highly diverse backgrounds. Income and status differences result in quite variable styles of life. A multitude of associations vie for members, whether to press economic and political interests or simply to represent expressive and recreational affinities.

Behind such diversity there are three principal common orientations: a minimal cognitive consensus, an acquiescence in if not positive acceptance of a normative order without which coordination could not emerge from specialization, and a minimal consensus on ultimate values.

The importance of formal education in providing common cognitive orientations will be noted in later discussion. Many less formal agencies of socialization operate also—at work, in the market, in urban neighborhoods. Increasingly, also, "mass communication" media are used both for quick dissemination of information and for propaganda and persuasion. The person in transitional situations must learn a multitude of facts and skills, from survival tactics in urban traffic to the arbitrary divisions of life's activities into temporal units.

The normative order, which had our attention in the preceding section of the chapter, needs one point of further emphasis here. It is in the general nature of rules to be specific and therefore to arise in particular action contexts: the family, the work place, the market, the school, the church. It does appear, however, that some more general "normative orientations"—generalized principles of correct conduct—may operate pervasively. Promptness and a rational orientation to decisions are two examples of such generalized norms in industrial systems. The transferability of such norms from one action pattern to another aids individual role-playing: "If in doubt, follow the general rule." Transferability also serves as an indirect, integrative linkage among highly specialized contexts of social behavior.

A minimum value consensus is also a theoretical necessity of a viable social order. Not only is the normative order usually referable to common

14. See Bert F. Hoselitz, "Tradition and Economic Growth," in Braibanti and Spengler, eds., *Tradition, Values, and Socio-Economic Development,* Chap. 3.

values, but such values may be given specific ideological explication and thus serve as direct incentives to appropriate action.[15] Standards of equity and justice; the allocation of wealth, power, and position; the maintenance of institutional balance—these serve as value premises for particular sets of rules. Additionally, political and religious ideologies may provide goals as well as standards of conduct.

Beliefs: Sacred and Profane

The debate over the importance of Protestantism in the rise of capitalism is scarcely relevant to most developing areas, yet the problem of ultimate values is still critical. Collectivist ideologies, for example, may assume religious overtones, promising, among other things, that worldly immortality can be gained by developing the economy for generations yet unborn ("building the socialist fatherland"). Nationalism and patriotism always have religious elements, whether linked to traditional religious beliefs or not.

Subtle questions remain, however, concerning religious beliefs. As we noted briefly when discussing motivation and enterprise, the achievement orientations encouraged in some parts of Protestant Christianity may be either absent or less socially disciplined in other religious systems. Does the otherworldliness of Hinduism or Roman Catholicism, combined with an emphasis on acceptance and adaptation instead of active improvement, preclude economic development? Does the somewhat more hedonistic other-worldliness of Islam, coupled with an authoritarian view of worldly power, have similar negative effects? The questions multiply, but the answers do not.

The questions are, to repeat, subtle, since the gross evidence of a strongly secular goal-orientation toward economic growth is manifested "everywhere." Even in cultures traditionally based on an ideological system that emphasized other-worldliness, the desire for change in this world is constantly increasing. Indeed, the desire for such change has itself become a *spiritual* force of great importance in those areas of the world.[16]

When translated into motivational terms, the goal of economic growth is most widely and enthusiastically endorsed as producing an improvement in the material conditions of life. Yet there are other uses for wealth, including patronage of the arts, construction of monuments and monumental buildings, contributing to cultivated knowledge, and purchasing destructive weapons. It is the utility of economic growth as a kind of "universal means" that mainly accounts for its "ultimate" quality, an end in itself. To achieve that end changes in practices and institutions become the requisite means.

15. See Joseph J. Spengler, "Theory, Ideology, and Non-Economic Values, and Politico-Economic Development," in Braibanti and Spengler, eds., *Tradition, Values, and Socio-Economic Development,* Chap. 1.
16. See Arnold S. Feldman and Wilbert E. Moore, "Commitment of the Industrial Labor Force," in Moore and Feldman, eds., *Labor Commitment . . . ,* Chap. 1.

The ideological goal is a necessary but by no means sufficient condition for social transformation. The "means" turn out to be new patterns of daily existence, and thus in conflict with an intricately interrelated social structure. These patterns of behavior and their normative codes in turn relate to goals and values other than economic development or material well-being. Since material well-being is not the sole goal of any society, and could not be if it is to survive as a viable system, the value conflict is not trivial or simply based on temporary ignorance or misunderstanding.

The undoubted costs (in terms of value sacrifices) entailed in economic development may be partially offset not only by new distributive rewards of one sort or another, but by additional collective ideologies.

It is commonly, and probably correctly, assumed that wherever economic development becomes a matter of public policy (and that is nearly everywhere) the state is likely to play an active role, at least in surmounting barriers. Although the economic activity of the state in the historic "laissez-faire" economies should not be minimized, there is ample reason to assume that the contemporary state will figure more largely as an agent of growth than was true in the past.[17] Questions of political loyalty and participation therefore assume an importance directly relevant to economic development, in addition to the role of the state as the focus of national integration and identity and the ultimate enforcement agency for social codes.

The repeated association between deliberate economic development and extreme nationalism is surely not accidental. Nationalism presents an essentially nonrational unifying force that may ease and rationalize the hardships of personal change. As Smelser has observed,

> Because the existing commitments and methods of integration are deeply rooted in the organization of traditional society, a very generalized and powerful commitment is required to pry individuals from these attachments.

In his view, "xenophobic national aspirations" and political ideologies such as socialism are the functional equivalents of religious values such as Protestantism. In new nations the forms of nationalism include using the former colonial power as a scapegoat for present dissatisfactions and the attempt to establish an older history and continuity of traditions prior to the interregnum of the colonial period. And it is especially in the former colonies that a prior sense of national, or even cultural, identity scarcely existed. When the transition to independence is also accompanied by extensive efforts at economic revolution, various intermediate social structures that shared or captured loyalties in the preindustrial system are undermined. Nationalism—often in the garb of Arab or Indian or African socialism—is offered as a source of identity to substitute for the tribe or village. Success in establishing a

---

17. See Hugh G. J. Aitken, ed., *The State and Economic Growth* (New York: Social Science Research Council, 1959); Karl de Schweinitz Jr., *Industrialization and Democracy: Economic Necessities and Political Possibilities* (New York: Free Press of Glencoe, Inc., 1964).

nationalist ideology provides a rationale for the multitude of changes in way of life, though we shall discuss later the low probability that genuinely democratic institutions will quickly emerge. Failure to achieve national political unity is likely to retard or prevent programs of economic change.

Technique in Civilization

Jacques Ellul

Traditional Techniques and Society

What was the position of technique in the different societies which have preceded ours? Most of these societies resembled one another in their technical aspects. But it is not enough to say that technique was restricted. We must determine the precise characteristics of the limitations, which are four in number.

First, technique was applied only in certain narrow, limited areas. When we attempt to classify techniques throughout history, we find principally techniques of production, of war and hunting, of consumption (clothing, houses, etc.), and, as we have said, magic. This complex of techniques would seem to modern man to represent a rather considerable domain and, indeed, to correspond to the whole of life. What more could there be than producing, consuming, fighting, and practicing magic? But we must look at these things in perspective.

In so-called primitive societies, the whole of life was indeed enclosed in a network of magical techniques. It is their multiplicity that lends them the qualities of rigidity and mechanization. Magic, as we have seen, may even be the origin of techniques; but the primary characteristic of these societies was not a technical but a religious preoccupation. In spite of this totalitarianism of magic, it is not possible to speak of a technical universe. Moreover, the importance of techniques gradually diminishes as we reach historical societies. In these societies, the life of the group was essentially nontechnical. And although certain productive techniques still existed, the magical forms which had given a technique to social relations, to political acts, and to military and judicial life tended to disappear. These areas ceased to respond to techniques and became subject instead to social spontaneities. The law, which had traditionally expressed itself in certain customs, no longer had any character of technical rigor; even the state was nothing but a force which simply manifested itself. These activities depended more on private initiative, short-lived manifestations or ephemeral traditions, than on a persevering technical will and rational improvements.

Even in activities we consider technical, it was not always that aspect which was uppermost. In the achievement of a small economic goal, for example, the technical effort became secondary to the pleasure of gathering together. "Formerly, when a New England family convoked a 'bee' (that is, a meeting for working in common), it was for all concerned one of the most pleasurable times of the year. The work was scarcely more than a pretext for coming together."[1] The activity of sustaining social relations and human contacts predominated over the technical scheme of things and the obligation to work, which were secondary causes.

Society was free of technique. And even on the level of the individual, technique occupied a place much more circumscribed than we generally believe. Because we judge in modern terms, we believe that production and consumption coincided with the whole of life.

For primitive man, and for historical man until a comparatively late date, work was a punishment, not a virtue. It was better not to consume than to have to work hard; the rule was to work only as much as absolutely necessary in order to survive. Man worked as little as possible and was content with a restricted consumption of goods (as, for example, among the Negroes and the Hindus)—a prevalent attitude, which limits both techniques of production and techniques of consumption. Sometimes slavery was the answer: an entire segment of the population did not work at all and depended on the labor of a minority of slaves. In general, the slaves did constitute a minority. We must not be misled by Imperial Rome, Greece under Pericles, or the Antilles in the eighteenth century. In most slaveholding nations, slaves were in a minority.

The time given to the use of techniques was short, compared with the leisure time devoted to sleep, conversation, games, or, best of all, to meditation. As a corollary, technical activities had little place in these societies. Technique functioned only at certain precise and well-defined times; this was the case in all societies before our own. Technique was not part of man's occupation nor a subject for preoccupation.

This limitation of technique is attested to by the fact that in the past technique was not considered nearly as important as it is today. Heretofore, mankind did not bind up its fate with technical progress. Man regarded technical progress more as a relative instrument than as a god. He did not hope for very much from it. Let us take an example from Giedion's admirable book, in which he elucidates the small importance technique had traditionally.

In our day, we are unable to envisage comfort except as part of the technical order of things. Comfort for us means bathrooms, easy chairs, foam-rubber mattresses, air conditioning, washing machines, and so forth. The chief concern is to avoid effort and promote rest and physical euphoria. For us, comfort is closely associated with the material life; it manifests itself in the

1. George C. Homans, quoted by Jerome Scott and R. P. Lynton.

perfection of personal goods and machines. According to Giedion, the men of the Middle Ages also were concerned with comfort, but for them comfort had an entirely different form and content. It represented a feeling of moral and aesthetic order. Space was the primary element in comfort. Man sought open spaces, large rooms, the possibility of moving about, of seeing beyond his nose, of not constantly colliding with other people. These preoccupations are altogether foreign to us.

Moreover, comfort consisted of a certain arrangement of space. In the Middle Ages, a room could be completely "finished," even though it might contain no furniture. Everything depended on proportions, material, form. The goal was not convenience, but rather a certain atmosphere. Comfort was the mark of the man's personality on the place where he lived. This, at least in part, explains the extreme diversity of architectural interiors in the houses of the period. Nor was this the result of mere whim; it represented an adaptation to character; and when it had been realized, the man of the Middle Ages did not care if his rooms were not well heated or his chairs hard.

This concept of comfort, closely bound up with the person, clearly takes death for granted, as did man himself; man's awareness of death likewise profoundly influences his search for an adequate milieu. Giedion's study is convincing. Medieval man did not dream for an instant that technique had any influence at all, even on objects which today we consider completely material and consequently of a technical order.

This limitation of the sphere of action of technique was increased even more by the limitation of the technical means employed in these fields. There was no great variety of means for attaining a desired result, and there was almost no attempt to perfect the means which did exist. It seems, on the contrary, that a conscious Malthusian tendency prevailed. It was expressed, for example, in the regulations of the guilds concerning tools, and in Roman law, by the principle of the economy of forms. Man tended to exploit to the limit such means as he possessed, and took care not to replace them or create other means as long as the old ones were effective. From the judicial point of view, the principle of the economy of forms led to the creation of the fewest possible legal instruments. Laws were few, and so were institutions. Man used the utmost ingenuity to obtain a maximum of results from a minimum of means at the price of fictions, transpositions, applications *a pari* and *a contrario,* and so on. This was also true industrially. Society was not oriented toward the creation of a new instrument in response to a new need. The emphasis was rather on the application of old means, which were constantly extended, refined, and perfected.

The deficiency of the tool was to be compensated for by the skill of the worker. Professional know-how, the expert eye were what counted: man's talents could make his crude tools yield the maximum efficiency. This was a kind of technique, but it had none of the characteristics of instrumental technique. Everything varied from man to man according to his gifts, whereas technique in the modern sense seeks to eliminate such variability. It is understandable that technique in itself played a very feeble role. Everything was done by men who employed the most rudimentary means. The search for the "finished," for perfection in use, for ingenuity of application, took the place of a search for new tools which would have permitted men to simplify their work, but also would have involved giving up the pursuit of real skill.

Here we have two antithetical orders of inquiry. When there is an abundance of instruments that answer all needs, it is impossible for one man to have a perfect knowledge of each or the skill to use each. This knowledge would be useless in any case; the perfection of the instrument is what is required, and not the perfection of the human being. But, until the eighteenth century, all societies were primarily oriented toward improvement in the use of tools and were little concerned with the tools themselves. No clean-cut division can be made between the two orientations. Human skill, having attained a certain degree of perfection in practice, necessarily entails improvement of the tool itself. The question is one of transcending the stage of total utilization of the tool by improving it. There is, therefore, no doubt that the two phenomena do interpenetrate. But traditionally the accent was on the human being who used the tool and not on the tool he used.

The improvement of tools, essentially the result of the practice of a personal art, came about in a completely pragmatic way. For this reason, we can put in the first category all the techniques we have classified with regard to intrinsic characteristics. A small number of techniques, not very efficient: this was the situation in Eastern and Western society from the tenth century B.C. to the tenth century A.D.

The world of technique had still a third characteristic prior to the eighteenth century: it was local. Social groups were very strong and closed to outsiders. There was little communication, materially speaking, and even less from the spiritual point of view. Technique spread slowly. Certain examples of technical propagation are always cited; the introduction of the wheel into Egypt by the Hyksos; the Crusades; and so on. But such events took millennia and were accidental. In the majority of cases, there was little transmission. Imitation took place very slowly and mankind passed from one technical stage to the next with great difficulty. This is true of material techniques, and even more so of non-material techniques.

Greek art remained Greek in industrial projects such as pottery-making, even when imitated by the Romans. Roman law did not extend beyond the Roman borders, whereas the Napoleonic code was adopted by Turkey and Japan. As for magic, that technique remained completely secret.

Every technical phenomenon was isolated from similar movements elsewhere. There was no transmission, only fruitless gropings. Geographically, we can trace the compass of a given technique, follow the zones of its influence, imitation, and extension; in almost every case we find how small was the extent of its radiation.

Why was this so? The explanation is simple: technique was an intrinsic part of civilization. And civilization consisted of numerous and diversified elements—natural elements such as temperament and flora, climate and population; and artificial elements such as art, technique, the political regime, etc. Among all these factors, which mingled with one another, technique was only one. It was inexorably linked with them and depended on them, as they depended on it. It was part of a whole, part of the determinate society, and it developed as a function of the whole and shared its fate.

Just as one society is not interchangeable with another, so technique remained enclosed in its proper framework; no more would it become universal than the society in which it was embedded. Geographically there could be no technical transmission because technique was not some anonymous piece of merchandise but rather bore the stamp of the whole culture. This entails much more than the existence of a simple barrier between social groups. Technique was unable to spread from one social group to another except when the two were in the same stage of evolution and except when civilizations were of the same type. In the past, in other words, technique was not objective, but subjective in relation to its own culture.

It is understandable, therefore, that technique, incorporated in its proper framework, did not evolve autonomously. On the contrary, it depended on a whole ensemble of factors which had to vary with it. It is not accurate to conceive the movement in the oversimplified manner of Marxism, as first the evolution of technique, and subsequently the alignment of the other factors. This view is accurate for the nineteenth century but it is false for history as a whole. Certain important covariations traditionally existed, and these factors, covariant with technique, changed according to the type of civilization. There was, for example, the association of technique and the state among the Egyptians and the Incas; of technique and philosophy in Greece and China. Francastel has shown how technique could be "absorbed and directed by the arts," as happened, say, in the fifteenth century, when it was subordinated to a plastic vision of the world, which imposed on it limits and demands. At that time, there existed a whole "civilization well provided with technical inventions, but which deliberately undertook to use them only to the degree in which these inventions would allow it to realize an imaginative construction." Thereafter, we find a complicated "art technique" and, as elsewhere, we almost never find technique in a pure state.

The consequence was an extreme local diversity of techniques for attaining the same result. No comparison or competition existed yet between these different systems; the formulation: "The one best way in the world" had not yet been made. It was a question of the "best way" in a given locality. Because of this, arms and tools took very different forms, and social organizations were extremely diverse.

It is impossible to speak of slavery as all of a piece. Roman slavery, for example, had nothing to do with Teutonic slavery, or Teutonic slavery with Chaldean. We habitually use one term to cover very different realities. This extreme diversity divested technique of its most crucial characteristic. There was no single means which was judged best and able to eliminate all others by virtue of its efficiency. This diversity has made us believe that there was an epoch of experimentation, when man was groping to find his way. This is a false notion; it springs from our modern prejudice that the stage we find ourselves in today represents the highest level of humanity. In reality, diversity resulted not from various experimental attempts on the part of various peoples, but from the fact that technique was always embedded in a particular culture.

Alongside this spatial limitation of technique, we find a time limitation. Until the eighteenth century, techniques evolved very slowly. Technical work was purely pragmatic, inquiry was empirical, and transmission slow and feeble. Centuries were required for: (a) utilization of an invention (for example, the water mill); (b) transition from a plaything to a useful object (gunpowder, automatons); (c) transition from a magical to an economic operation (breeding of animals); (d) simple perfecting of an instrument (the horse yoke and the transition from the simple stick plow to the train plow). This was even more true for abstract techniques. Abstract techniques, I maintain, are almost non-transmissible in time from a given civilization to its successor. We must be somewhat skeptical, and in any case prudent, when the evolution of techniques is presented as an evolution of inventions; actually this development was never more than potential. There is nothing to prove that true technique existed heretofore, that is, in the sense of generalized application. It is possible to compile a fine catalogue of seventeenth-century inventions, and to deduce from it that a great technical movement was in force at that time. Many writers have fallen into this error—among them, Jean Laloup and Jean Nelis. It is not because Pascal invented a calculating machine and Papin a steam engine that there was a technical evolution; nor was it because a "prototype" of a power loom was built; nor because the process of the dry distillation of coal was discovered. As Gille has very judiciously noted: "The best-described machines in the eighteenth century *Encyclopédie* are possibly better conceived than those of the fifteenth century, but scarcely constitute a revolution." The initial problem was to construct the machine, to make the invented technique actually work. The second consisted in the diffusion of the machine throughout the society; and this second step proceeded very slowly.

This divergence between invention and technique, which is the cause of the time lag we have spoken of, is correctly interpreted by Gille in these words: "There was a discontinuity of technical progress but there was probably a continuity of research." Gille shows clearly that technical progress develops according to a discontinuous rhythm: "It is tied up with demographic or economic rhythms and with certain internal contradictions." This discontinuity still contributes to evolutionary lag today.

Slowness in the evolution of techniques is evident throughout history. Very few variations seem to have occurred in this constant. But it cannot be maintained that this slowness was completely uniform. Yet, even in periods that appear rather fertile, it is clear that evolution was slow. For example, Roman law, which was particularly rich in the classical period, took two centuries to find a perfect form. Moreover, the number of applied inventions was sharply restricted. The fifteenth century, in spite of its importance, produced no more than four or five important technical applications. The natural consequence of this evolutionary slowness was that technique could be adapted to men. Almost unconsciously, men kept abreast of techniques and controlled their use and influence. This resulted not from an adaptation of men to techniques (as in modern times), but rather from the subordination of techniques to men. Technique did not pose the problem of adaptation because it was firmly enmeshed in the framework of life and culture. It developed so slowly that it did not outstrip the slow evolution of man himself. The progress of the two was so evenly matched that man was able to keep pace with his techniques. From the physical point of view, techniques did not intrude into his life; neither his moral evolution nor his psychic life were influenced by them. Techniques enabled man to make individual progress and facilitated certain developments, but they did not influence him directly. Social equilibrium corresponded to the slowness of general evolution.

This evolutionary slowness was accompanied by a great irrational diversification of designs. The evolution of techniques was produced by individual efforts accompanied by a multitude of scattered experiments. Men made incoherent modifications on instruments and institutions which already existed; but these modifications did not constitute adaptations. We are amazed when we inspect, say, a museum of arms or tools, and note the extreme diversity of form of a single instrument in the same place and time. The great sword used by Swiss soldiers in the sixteenth century had at least nine different forms (hooked, racked, double-handed, hexagonal blades, blades shaped like a fleur-de-lis, grooved, etc.). This diversity was evidently due to various modes of fabrication peculiar to the smiths; it cannot be explained as a manifestation of a technical inquiry. The modifications of a given type were not the outcome of calculation or of an exclusively technical will. They resulted from aesthetic considerations. It is important to emphasize that technical operations, like the instruments themselves, almost always depended on aesthetic preoccupations. It was impossible to conceive of a tool that was not beautiful. As for the idea, frequently accepted since the triumph of efficiency, that the beautiful is that which is well adapted to use—assuredly no such notion guided the aesthetic searchings of the past. No such conception of beauty (however true) moved the artisan who carved a Toledo blade or fabricated a harness. On the contrary, aesthetic considerations are gratuitous and permit the introduction of uselessness into an eminently useful and efficient apparatus.

This diversity of forms was manifestly conditioned by vainglory and pleasure—the vainglory of the user, the pleasure of the artisan. Both caused changes in the classic type. And why not include as well that pure fantasy which runs through all the creations of Greece and the Middle Ages?

All this led to a modification of the given type. The search for greater efficiency likewise played a role, but it was one factor among several. The different forms were subject to trial and error, and certain forms were progressively stabilized and imitated, either because of their plastic perfection or because of their usefulness. The final result was the establishment of a new type derived from its predecessor.

This diversity of influences, which operated on all technical mechanisms, explains in part the slow tempo of progress in these areas. To obey a multiplicity of motives and not reason alone seems to be an important keynote of man. When, in the nineteenth century, society began to elaborate an exclusively rational technique which acknowledged only considerations of efficiency, it was felt that not only the traditions but the deepest instincts of humankind had been violated. Men sought to reintroduce indispensable factors of aesthetics and morals. Out of this effort came the unprecedented creation of certain aspects of style in the 1880's: the tool with machine-made embellishments. Sewing machines were decorated with cast-iron flowers, and the first tractors bore engraved bulls' heads. That it was wasteful to supply such embellishments soon became evident; their ugliness doubtless contributed to the realization. Moreover, these flourishes represented a wrong road, technically speaking. The machine can become precise only to the degree that its design is elaborated with mathematical rigor in accordance with use. And an embellishment could increase air resistance, throw a wheel out of balance, alter velocity or precision. There was no room in practical activity for gratuitous aesthetic preoccupations. The two had to be separated. A style then developed based on the idea that the line best adapted to use is the most beautiful.

Abstract techniques and their relation to morals underwent the same evolution. Earlier, economic or political inquiries were inexorably bound with ethical inquiry, and men attempted to maintain this union artificially even after they had recognized the independence of economic technique. Modern society is, in fact, conducted on the basis of purely technical considerations. But when men found themselves going counter to the human factor, they reproduced— and in an absurd way—all manner of moral theories related to the rights of man, the League of Nations, liberty, justice. None of that has any more importance than the ruffled sunshade of McCormick's first reaper. When these moral flourishes barely encumber technical progress, they are discarded— more or less speedily, with more or less ceremony, but with determination nonetheless. This is the state we are in today.

The elimination of these evolutionary factors and of technical diversification has brought about a transformation of the basic process of this evolution. Technical progress today is no longer conditioned by anything other than its own calculus of efficiency. The search is no longer personal, experimental, workmanlike; it is abstract, mathematical, and industrial. This does not mean that the individual no longer participates. On the contrary, progress is made only after innumerable individual experiments. But the individual participates only to the degree that he is subordinate to the search of efficiency, to the degree that he resists all the currents today considered secondary, such as aesthetics, ethics, fantasy. Insofar that the individual represents this abstract tendency, he is permitted to participate in technical creation, which is increasingly independent of him and increasingly linked to its own mathematical law. It was long believed that rational systematization would act to reduce the number of technical types: in the measure that the factors of diversification were eliminated, the result would be fewer and more simple and precise types. Thus, during the latter part of the nineteenth century—in the mechanical, medical, and administrative spheres—exact instruments were available from which fantasy and irrationality had been totally eliminated. The result was fewer instruments. As further progress was made, however, a new element of diversification came into play: in order that an instrument be perfectly efficient, it had to be perfectly adapted. But the most rational instrument possible takes no account of the extreme diversity of the operational environment. This represents an essential characteristic of technique. Every procedure implies a single, specific result. As Porter Gale Perrin puts it: "Just as a word evokes an idea which exactly corresponds to no other word," so a fixed technical procedure generates a fixed result. Technical methods are not multipurposive, or adaptable, or interchangeable. Perrin has demonstrated this in detail with reference to judicial technique, but it also holds for everything else. Take the well-known example, cited by Pierre de Latil, of a machine, brought to the highest possible pitch of perfection, the purpose of which was to produce from cast iron, at a single stroke, cylinder heads for aircraft engines. The machine was 28 meters long and cost $100,000. But the moment the required type of cylinder head was changed, the machine became good for nothing; it was unadaptable to any new operation. A judicial system may function perfectly adequately in France but not in Turkey. For true efficiency, not only must the rational aspect of the machine be taken into account, but also its adaptation to the environment. A military tank will have a different form depending on whether it is to be used in mountainous terrain or in rice paddies. The more an instrument is designed to execute a single operation efficiently and with utmost precision, the less can it be multipurposive. A new diversification of technical apparatus thus appears: today instruments are differentiated as a result of the continually more specialized usage demanded of them.

The field of aviation gives us one of the best examples of this. Aircraft are described by the use to which they are put. We have, correspondingly, extremely precise and more and more diversified types. The list of French military aircraft, consisting at the present of five great categories, is as follows: (1) strategic bombers, (2) tactical bombers, (3) pursuit planes, (4) reconnaissance planes, and (5) transport planes. These five categories are subdivided further; there are altogether thirteen different subtypes, none of which are interchangeable with one another. Each has very different characteristics resulting from more and more refined technical adaptations.

The same extensive differentiation is found in much less important areas. A recent brochure of the world's largest refiner of lubricating oils lists fifteen different kinds of lubricants designed exclusively for automobiles. Each type corresponds to a definite use, each possessing specific qualities, and all equally necessary.

A fourth characteristic of technique, which results from the characteristics just enumerated, is the possibility, reserved to the human being, of choice. Inasmuch as all techniques were geographically and historically limited, societies of many different types were able to exist. For the most part, there was an equilibrium between two major types of civilization—the active and the passive. This distinction is well known. Some societies are oriented toward the exploitation of the earth, toward war, conquest, and expansion in all its forms. Other societies are inwardly oriented; they labor just enough to support themselves, concentrate on themselves, are not concerned with material expansion, and erect solid barriers against anything from without. From the spiritual point of view, these societies are characterized by a mystical attitude, by a desire for self-dissolution and absorption into the divine.

Human societies are variable, however. A group which has hitherto been active might become passive. The Tibetans, for instance, were conquerors and believers in magic until their conversion to Buddhism. Thereafter they became the world's most passive and mystical people. The reverse can also take place.

The two types of society coexisted throughout history; indeed, this seemed necessary to the equilibrium of world and man. Until the nineteenth century, technique had not yet excluded one of them. Moreover, man could isolate himself from the influence of technique by attaching himself to a given group and exerting influence on this group. Of course, other constraints acted on him; the individual was never completely free with respect to his group, but these constraints were not completely decisive or imperative in character.

Whether we are considering unconscious sociological cohesion or the power of the state, we find these forces always necessarily counterbalanced by the existence of other neighboring groups and other loyalties. There was no irrefutable constraint on man, because nothing absolutely good in respect to everything else had been discovered. We have noted the diversity of technical form and the slowness of imitation. But it was always human action which was decisive. When several technical forms came into contact, the individual made his choice on the basis of numerous reasons. Efficiency was only one of them, as Pierre Deffontaines has demonstrated in his work on religious geography.

Although the individual existing in the framework of a civilization of a certain type was always confronted with certain techniques, he was nevertheless free to break with that civilization and to control his own individual destiny. The constraints to which he was subject did not function decisively because they were of a nontechnical nature and could be broken through. In an active civilization, even one with a fairly good technical development, the individual could always break away and lead, say, a mystical and contemplative life. The fact that techniques and man were more or less on the same level permitted the individual to repudiate techniques and get along without them. Choice was a real possibility for him, not only with regard to his inner life, but with regard to the outer form of his life as well. The essential elements of life were safeguarded and provided for, more or less liberally, by the very civilization whose forms he rejected. In the Roman Empire (a technical civilization in a good many respects), it was possible for a man to withdraw and live as a hermit or in the country, apart from the evolution and the principal technical power of the Empire. Roman law was powerless in the face of an individual's decision to evade military service or, to a very great degree, imperial taxes and jurisdiction. Even greater was the possibility of the individual's freedom with respect to material techniques.

There was reserved for the individual an area of free choice at the cost of minimal effort. The choice involved a conscious decision and was possible only because the material burden of technique had not yet become more than a man could shoulder. The existence of choice, a result of characteristics we have already discussed, appears to have been one of the most important historical factors governing technical evolution and revolution. Evolution was not, then, a logic of discovery or an inevitable progression of techniques. It was an interaction of technical effectiveness and effective human decision. Whenever either one of these elements disappeared, social and human stagnation necessarily followed. Such was the case, for example, when effective technique was (or became) rudimentary and inefficacious among the Negroes of Africa. As to the consequences of a lapse in the second element, we are experiencing them today.

The New Characteristics

The characteristics of the relationship of technique, society, and the individual which we have analyzed were, I believe, common to all civilizations up to the eighteenth century. Historically, their existence admits of little discussion. Today, however, the most cursory review enables us to conclude that all these characteristics have disappeared. The relation is not the same; it does not present any of the constants recognizable until now. But that is not sufficient to characterize the technical phenomenon of our own day. This description would situate it in a purely negative perspective, whereas the technical phenomenon is a positive thing; it presents positive characteristics which are peculiar to it. The old characteristics of technique have indeed disappeared; but new ones have taken their place. Today's technical phenomenon, consequently, has almost nothing in common with the technical phenomenon of the past. I shall not insist on demonstrating the negative aspect of the case, the disappearance of the traditional characteristics. To do so would be artificial, didactic, and difficult to defend. I shall point out, then, in a summary fashion, that in our civilization technique is in no way limited. It has been extended to all spheres and encompasses every activity, including human activities. It has led to a multiplication of means without limit. It has perfected indefinitely the instruments available to man, and put at his disposal an almost limitless variety of intermediaries and auxiliaries. Technique has been extended geographically so that it covers the whole earth. It is evolving with a rapidity disconcerting not only to the man in the street but to the technician himself. It poses problems which recur endlessly and every more acutely in human social groups. Moreover, technique has become objective and is transmitted like a physical thing; it leads thereby to a certain unity of civilization, regardless of the environment or the country in which it operates. We are faced with the exact opposite of the traits previously in force. We must, therefore, examine carefully the positive characteristics of the technique of the present.

There are two essential characteristics of today's technical phenomenon which I shall not belabor because of their obviousness. These two, incidentally, are the only ones which, in general, are emphasized by the "best authors."

The first of these obvious characteristics is rationality. In technique, whatever its aspect or the domain in which it is applied, a rational process is present which tends to bring mechanics to bear on all that is spontaneous or irrational. This rationality, best exemplified in systematization, division of labor, creation of standards, production norms, and the like, involves two distinct phases: first, the use of "discourse" in every operation; this excludes spontaneity and personal creativity. Second, there is the reduction of method to its logical dimension alone. Every intervention of technique is, in effect, a reduction of facts, forces, phenomena, means, and instruments to the schema of logic.

The second obvious characteristic of the technical phenomenon is artificiality. Technique is opposed to nature. Art, artifice, artificial: technique as art is the creation of an artificial system. This is not a matter of opinion. The means man has at his disposal as a function of technique are artificial means. For this reason, the comparison proposed by Emmanuel Mounier between the machine and the human body is valueless. The world that is being created by the accumulation of technical means is an artificial world and hence radically different from the natural world.

It destroys, eliminates, or subordinates the natural world, and does not allow this world to restore itself or even to enter into a symbiotic relation with it. The two worlds obey different imperatives, different directives, and different laws which have nothing in common. Just as hydroelectric installations take waterfalls and lead them into conduits, so the technical milieu absorbs the natural. We are rapidly approaching the time when there will be no longer any natural environment at all. When we succeed in producing artificial *aurorae boreales,* night will disappear and perpetual day will reign over the planet.

I have given only brief descriptions of these two well-known characteristics. But I shall analyze the others at greater length; they are technical automatism, self-augmentation, monism, universalism, and autonomy.

The Importance of Technology in Human Affairs

Melvin Kranzberg and Carroll W. Pursell, Jr.

Our Present Technological Age

In the late afternoon of November 9, 1965, a small electrical relay in a power station in Ontario, Canada, failed. Within a few minutes the flow of electric energy throughout much of the northeastern section of the United States and part of Canada had ceased. Some thirty million people, including those in the great metropolitan areas of Boston and New York, were plunged into darkness. Coming as it did, during the evening rush hour when people were on their way home from work, the shutting off of electric power left hundreds of thousands of New Yorkers stranded in subway trains, confined in elevators stalled between floors of towering skyscrapers, or caught in monstrous traffic jams created by the absence of traffic lights. Even when they finally reached home, many of the now-disconcerted city-dwellers found it to be without warmth, without hot food, and without light. Here was a dramatic demonstration of modern man's dependence on the machine.

Disaster was narrowly averted. Emergency generating equipment allowed essential equipment to function in hospitals and institutions, and with a sense of shared adventure, Americans sought to help their neighbors in a surprising display of good humor and humanity. The great urban centers were able to limp along through the night without many of the technological devices and comforts which characterize life in 20th century America. Yet, had the shutdown of power lasted over a much longer period, it is clear that a considerable disaster could have occurred and that much of civilization as we know it would have been seriously disrupted.

For the fact is that we live in a "Technological Age." It is called that, not because all men are engineers, and certainly not because all men understand technology, but because we are becoming increasingly *aware* that technology has become a major disruptive as well as creative force in the 20th century. The "biggest blackout" of November 1965 gave ample proof of the role of technology in determining the conditions of our life and heightened our awareness of our dependence upon machines, tools, vehicles, and processes.

Equally important, the "biggest blackout" also demonstrated the close relationship between man and his machines from another angle. For while the immediate cause of the power failure was apparently the breakdown of a mechanical component—an electrical relay—this failure might not have occurred had prior decisions been taken to provide "backup" systems, nor would it have extended over such a wide area had the man in Ontario monitoring the power switches acted immediately on the information given him by the dials on his control panel (when he saw the power drop in the Canadian system, he could have switched off the American connection and prevented the power loss in the New York system). Once the blackout had occurred another human failing was revealed: the power company serving New York City was unable to restart its plants immediately because no auxiliary equipment had been provided for that purpose, it being incomprehensible to the engineering mind that such an event could occur.

What distinguishes our age from the past is, first, our belated recognition of the significance of technology in human affairs; second, the accelerated pace of technological development that makes it part-and-parcel of our daily living in ever-increasing measure; and, third, the realization that technology is not simply a limited or local factor but encompasses all men everywhere and is interrelated with nearly all human endeavor.

Man has always lived in a "Technological Age," even though we sense that this is particularly true of our own time. The modern tractor-driven plow represents a higher level of technology than the heavy, crooked stick with which primitive man—or, rather, woman—scratched the soil; and the hydrogen bomb is an infinitely more complex and lethal weapon of destruction than the bow and arrow. Nevertheless, the stick-plow and the bow-and-arrow weapon represented the advanced technology of an earlier era. The heavy stick with which our primitive ancestors prepared the soil for planting enormously increased their ability to wrest a living from an inhospitable and unpredictable nature. Similarly the bow and arrow greatly added to their larder when used to kill game for food. And when used upon their own kind, bow-and-arrow weaponry also gave the first possessors a decided advantage over an enemy who still relied upon rocks and clubs and who could be brought down from afar before their close-range weapons could be brought to bear.

What is Technology?

While the influence of technology is both widespread and fundamental, the term cannot be defined with precision. In its simplest terms, technology is man's efforts to cope with his physical environment—both that provided by nature and that created by man's own technological deeds, such as cities—and his attempts to subdue or control that environment by means of his imagination and ingenuity in the use of available resources.

In the popular mind, technology is synonymous with machines of various sorts—the steam engine, the locomotive, and the automobile—as well as such developments as printing, photography, radio, and television. The history of technology is then regarded as simply a chronological narrative of inventors and their devices. Of course, such items form a part of the history of technology just as chronologies of battles, treaties, and elections form a part of military and political history. However, technology and its history encompass much more than the mere technical devices and processes at work.

An encyclopedic five-volume work on the history of technology, edited by the late Dr. Charles Singer, defines its subject as "how things are commonly done or made . . . [and] what things are done or made." Such a definition is so broad and loose that it encompasses many items that scarcely can be considered as technology. For example, the passage of laws is something which is "done," but the history of law certainly is not the history of technology.

An element of purpose is stressed in another definition of technology as "man's rational and ordered attempt to control nature." Here the definition is too tight, for while it would include much of technology, many elements would not fit within its limits. The development of certain kinds of toys, for example, does not constitute an attempt to control nature. Furthermore, not all technology exists for the purpose of control, nor, has all past technological endeavor been rational and systematic.

In addition, much of man's technology is devoted to elements which are part of his physical environment but which are not necessarily part of "nature." The various means that man has devised for purposes of controlling the flow of traffic in congested cities are in response to a highly civilized and urban environment which is not a part of the natural environment. Any definition therefore must be extended to include the man-made as well as the natural environment.

To limit the definition of technology to those things which characterize the technology of our own time, such as machinery and prime movers, would be to do violence to all that went before. Indeed, a good case can even be made for considering magic as a technology, for with it primitive man attempted to control or at least influence his environment—a perfectly straightforward goal of all technology. If we now feel that our ancestors used their magic without much success, let us not fall into the error of equating technology only with *successful* technology. The past abounds with failures—schemes that went awry, machines that wouldn't work, processes that proved inapplicable—yet these failures form part of the story of man's attempts to control his environment. Albeit unsuccessful, many of these failures were necessary preliminaries toward the successes in technology.

Sometimes technology is defined as applied science. Science itself is viewed as an attempt by man to *understand* the physical world; technology is the attempt by man to *control* the physical world. This distinction may be briefly put as the difference between the "know-why" and the "know-how." But technology for much of its history had little relation with science, for men could and did make machines and devices without understanding why they worked or why they turned out as they did. Thus for centuries men produced usable objects of iron without knowing the chemical composition of iron and why the various changes occurred in smelting and working it; indeed, they could successfully make things of iron even when they had false theories and incorrect understanding of metallurgical processes. Even today much technology does not represent an application of science, although in such sophisticated technologies as those involving nuclear science, scientific understanding is closely linked with technical accomplishment.

Technology, then, is much more than tools and artifacts, machines and processes. It deals with *human work,* with man's attempts to satisfy his wants by human action on physical objects.

We must use the term "wants" instead of "needs," for human wants go far beyond human needs, especially those basic needs of food, clothing, and shelter. Technology administers to these, of course, but it also helps man to get what he wants, including play, leisure, and better and more commodious dwellings. He cultivates a taste for more exotic foods than those necessary to still the pangs of hunger. He yearns to achieve faster and more lasting communication with others. He wants to travel abroad and be entertained, and to fill his house and his life with beauty as he sees it.

Emphasis upon the "work" aspect of technology shows that it also involves the organization as well as the purpose of labor. For example, the pyramids of Egypt are monuments to the technology of that early civilization. The pyramids demonstrate even today how much can be done with very little in the way of tools but with much ingenuity and skill in the organization of labor. In our own day the efficiency of new tools and processes can only be maximized by utilizing efficient organization. We are increasingly forced to think in terms of "systems," and even decision-making now can sometimes best be done by machines.

The Comprehensiveness of the History of Technology

The nature of invention itself requires that the history of technology be more than a mere tabulation of inventors and their creations. Invention does not come about simply because a creative person decides that he is going to "build a better mousetrap." Invention is a social activity, much affected by social needs, by economic requirements, by the level of technology at a given time, and by sociocultural and psychological circumstances. The fact that some inventions "come before their time" indicates the importance of the sociocultural milieu, and it raises the whole question of the nature and origin of creativity.

Even if we were to try to limit the history of technology to inventions, we would be forced to deal with many social, political, economic, and cultural aspects of civilization. For example, ever since World War II, which stimulated nationalistic feeling in Russia, Soviet scholars have been publishing reports of "firsts" by Russian inventors and scientists. Although most Americans have shrugged off these Russian attempts to claim priorities in inventions, the facts are that some of the Russian claims are well founded and that individual Russian scientists and inventors during the 19th century were the peers of their counterparts in Western Europe and the United States. Yet even if we were to accept all of the Russian claims, we would still face another question: why did Russia lag behind other European nations in industrialization? The answer to that is not to be found in the mental prowess and inventive capabilities of the Russian people, but rather in the complex of social and political circumstances under which invention and innovation thrive.

It is not enough simply to discover who first had the idea for an invention, nor even who first patented the device; we must also see when, why, and how this invention actually came into use. The answers to these questions involve much more than the purely technical factors, which is why the history of technology is such a comprehensive subject. It covers every aspect of human life and must go back to the very beginnings of the human species.

Technology and the Emergence of Man

Anthropologists seeking the origins of mankind have attempted to differentiate between what constitutes "almost man" and the genus *Homo,* man himself. The chief distinction they have found is that man employed tools, thereby distinguishing him from his almost-human predecessors.

Man, as we know him, surely would not have evolved or survived without tools. He is too weak and puny a creature to compete in the struggle with beasts and the caprices of nature if armed with only his hands and teeth. The lion is stronger, the horse is faster, and the giraffe can reach farther. Man has been able to survive because of his ability to adapt to his environment by improving his equipment for living. As Gordon Childe has pointed out, the specialized equipment man uses differs significantly from that of the animal kingdom. An animal is capable of using only that equipment which he carries around with him as parts of his body. Man has very little specialized equipment of this kind. Moreover, he has discarded some of the organic "tools" with which he started and has relied more on the invention of tools, or extracorporeal organs, that he makes, uses, and discards at will. This invention and use of extracorporeal equipment has enabled man to adapt to nature and to reign supreme among the animals on earth.

Archaeological anthropologists continue to discover older and older fossils of human-like skeletons, almost always surrounded in their graves by primitive tools or implements. It has even been postulated from these findings that technology is perhaps responsible for our standing on two feet and for our being *Homo sapiens,* Man the Thinker. Thus, man began to stand erect so that he might have his forearms free to throw stones; he did not throw stones simply because he was already standing erect. Modern physiology, psychology, evolutionary biology, and anthropology all combine to demonstrate to us that *Homo sapiens* cannot be distinguished from *Homo faber,* Man the Maker. We now realize that man could not have become a thinker had he not at the same time been a maker. Man made tools; but tools made man as well.

Technology and the Advance of Civilization

The very terms by which we measure the progress of civilization—Stone Age, Bronze Age, Iron Age, and even Atomic Age—refer to a developing technological mastery by man of his environment.

One indication of the start of civilization—the development of settled communities—rests upon a technological innovation: agriculture. In the prehistory before that time, men had been nothing more than hunters; in a sense, they had been parasites upon nature. We do not know exactly how or when agriculture began. Recently there has been found evidence of agricultural communities in the Middle East dating as far back as 8000 years ago. Once men discovered that they could co-operate with nature by sowing seeds and waiting for nature to perform the miracle of growing crops, there arose the possibility of settled and civilized life.

Unlike the hunter, the agriculturist could not afford to live in constant conflict on all sides. Rather, he had to learn to co-operate not only with nature but with other human beings. If he spent too much of his time in fighting, he could have neither the time nor the energy for carrying on his agricultural pursuits. Yet if he ran away from his enemies, his crops would go unattended and he would lose his means of livelihood. With the introduction of agriculture, therefore, civilized society began to emerge. This both spawned and depended upon man's dawning awareness that he must live and work together with others if he was to survive. It is a reasonable, though optimistic, extension of this concern to hope that man has, in the many thousands of years since, begun to realize that he is part of a larger community and that there is a need to co-operate with other human beings in order to advance his control over nature. No longer are his actions, thoughts, or aspirations confined to his immediate locale. Rather he must learn to consider all mankind since he has acquired the skill literally to reshape or destroy the world with the technology at his command.

In terms of energy, there has been transition from human muscle power to that of animals, to wind and water, to steam and oil, to rockets and nuclear power. With machines, we have witnessed change from hand tools to powered tools, from craft shops to mass production lines, from the beginnings of job definitions and quality control to computer control of factories.

The advance of material civilization has not been without interruption, and cannot be portrayed on a graph as a straight line climbing constantly upward through time. Instead, periods of great technological progress have sometimes been followed by eras of relative stagnation, during which time very little advance was made in man's control over nature. Moreover, materialistic techniques may progress while cultural activities such as music, art, literature, and philosophy seem to retrogress. There have been times when religious, philosophical and artistic activities achieved great heights while technology seemed to rest on a plateau.

Technology and Western Civilization

Technology and its modern twin, science, are the distinguishing hallmarks of recent Western civilization. The Scientific Revolution of the 17th century was reinforced by a Technological Revolution of the 18th and 19th centuries. These revolutions brought something to our culture that had been unknown to the earlier Western civilization of Greece and Rome or the Eastern civilizations of India and China. Science and technology differentiate our society from all that has gone before in human history and all that has taken place in other parts of the world. While the roots of our Western religious and moral heritage can be found in the Judeo-Christian-Greek tradition, contemporary Western culture is perhaps based more upon science and technology than upon religious and moral considerations.

If we wish to test the hypothesis of the uniqueness and significance of Western civilization, we need merely ask ourselves what "Westernization" means to non-Western societies. To them, it means the acquisition of the products of Western technology, not the political institutions, religious faiths, nor moral attitudes which the West has developed over the centuries. When we speak of the "Westernization" of Japan during the late 19th and early 20th centuries, we refer to the acceptance and the borrowing of Western technology by the Japanese. Similarly, many of the underdeveloped nations of the world want to borrow from the West today. While they often specifically reject Western moral and social attitudes, they want desperately the material advantages which technology can bestow upon them, even though they criticize the West's "materialism." To much of the world, the "American Way of Life" does not mean democracy, much less free enterprise. It means material abundance within the reach of all men; and social and political "isms" become relevant only when they retard or encourage the gaining of that goal.

The attitudes and values of Western man himself have been deeply affected by technological advance. For centuries men thought that it was their lot to earn their living by the sweat of their brows, and there was little hope for material abundance here on earth. In the past, technology was primarily concerned with furnishing the human needs of food, clothing, and shelter. It still serves to fulfill those needs, but now so successfully that modern technology for the first time in history has produced in the United States a society which has not only a surplus of goods but a surplus of leisure as well.

The Human and Social Elements in Technology

There have always been those, especially since the Industrial Revolution, who have seen new technologies as a threat to "human values." In the late 18th century the excesses of the growing industrialism in Great Britain—symbolized in William Blake's description of the "dark Satanic mills"—tragically alienated a large and influential segment of our common humanistic tradition. Many within that tradition—including artists, writers, and philosophers—have to this day continued to deplore the Industrial Revolution and our modern urbanized and industrialized society that has issued from it. This estrangement has led in some cases to a failure of the humanities to perform their functions as prophets of mankind. The tragedy is that this alienation, which leaves us all the poorer, seems so unnecessary.

We have come to think of technology as something mechanical, yet the fact remains that all technical processes and products are the result of the creative imagination and manipulative skills of human effort. The story of how man has utilized technology in mastering his environment is part of the great drama of man fighting against the unknown.

Furthermore, the significance of technology lies not only in the uses of technology by human beings, but in terms of what it does to human beings as well. If we regard the telephone, for example, only as a system of wires through which a tiny current passes from mouthpiece to earphone, it would seem to have little interest, except to technicians and repairmen, and virtually none to historians except for the antiquarian desire to discover who conceived the idea and reduced it to practice. But the greater significance of the telephone lies in the newly found ability to transmit voice communication between persons over long distances. It is the communications function of the telephone that gives it importance. The principal significance of this particular bit of technology— as in the function of every technological item—is its use by human beings.

The essential humanity of technology is nowhere better demonstrated than in the fact that it too, like the noble heroes of Greek tragedy, carries within it a fatal flaw which threatens always to lay it low. It is no longer possible, if indeed it ever was, to believe that progress is either inevitable or uniformly beneficent. Granted that technology has contributed to man's material progress, its social repercussions have not always been a boon to all segments of the population. There have been victims of the rapid social readjustment to industrial growth, notably the factory workers in Britain during the early days of the Industrial Revolution. This has led some critics to claim that technology presents two faces to man: one benign and the other malignant. The latter face is most frequently represented today by the destructive potentialities of intercontinental ballistic missiles armed with nuclear warheads.

Yet it advances our understanding very little to say that technology wears two faces, as though one were comedy and the other were tragedy. Technology, in a sense, is nothing more than the area of interaction between ourselves, as individuals, and our environment, whether material or spiritual, natural or man-made. Being the most fundamental aspect of man's condition, his technology has always had critical implications for the status quo of whatever epoch or era. Changes have always rearranged the relationships of men—or at least of some men—with respect to the world about them. Not a few of the historic outcries against technology (or, more properly, against some changes in technology) have been essentially protests against a rearrangement of the world's goods disadvantageous to those who complain.

Some "defenders" of technology claim that it is neutral, that it can have socially desirable or evil effects, depending on the uses which man makes of it. To deny this and to say that technology is not strictly neutral, that it has inherent tendencies or imposes its own values, is merely to recognize the fact that, as a part of our culture, it has an influence on the way in which we behave and grow. Just as men have always had some form of technology, so has that technology influenced the nature and direction of their development. The process cannot be stopped nor the relationship ended; it can only be understood and, hopefully, directed toward goals worthy of mankind.

PART II

World of Work

INTRODUCTION

The current era demands that the nature of work and man's contribution to society be re-defined. Those who are aware of the changing "world of work" realize that the Protestant Work Ethic may no longer be valid. It may well be that an individual's contribution cannot be equated with punching a time clock or by following the adage of "a day's pay for a day's work."

Those such as Robert Theobald and Erich Fromm advocate the guaranteed income concept. Others, such as De Gracia, propose that leisure may be a way of life before too long. These concepts are based on the possibility that from 2 to 10 percent of the population can produce the goods and services for all. Why is this a possibility? Computers and automation—these are to be the slaves of our society. Just as the Romans and Greeks, centuries ago, built a society on the backs of slaves, we now face the likelihood that we could have a society based on leisure with machines as our slaves.

The possibility of a future leisure society may cause us to fail to place current issues in proper perspective. The authors of this section deal with the elements of the labor force, unemployment, automation, leisure and related topics. Only through fully comprehending the current issues, and solving them, can we move on into the future and realize the potential that is there.

The Nature of Work

Nels Anderson

Definitions of work tell us little about it, and apparently the making of such definitions has never given man much concern. We think about it more than formerly partly because leisure is coming to us, and the readiest way to understand leisure is to describe it in relation to work. We see work as economic activity for a purpose, while leisure activity is usually an end in itself. When we raise and try to answer questions about work, we abstract it into an idea. Then we discover that, whatever the definition we make, we are considering something that is not the same today that it used to be. We also find that, whatever definition we make, it leads us in different directions.

We discover too that the implications of work are diverse and its nature so diffuse and intriguing that it needs to be viewed from different perspectives. The beginning we make in this chapter is very general.

A Brief Look at Meanings

Few men who work do much thinking about its meaning; there is no reason why they should. They may think about what they will get out of it, or if they think about what they are doing at the moment their attention may be on how it relates to what has just been done or what comes next, or how it relates to what someone else is doing. One may give thought to the satisfaction coming from work that goes well. The frustrations coming from work that goes wrong may be something to think about. The work itself shapes the thought. One may also think of the easiest and most efficient way to get work done, but this is not giving thought to the why and wherefore of work as such.

We therefore venture the opinion that most men who work do not concern themselves with the whys and wherefores; that task goes by default to the philosophers, who write about work and the workers. For the most part, they have tried to fit work into some larger and more comprehensive scheme of things, with the object usually of sustaining and often sanctifying that scheme of things and fixing people into the roles they have long played.[1]

Most philosophical explanations of work—at least those that have lived long, spread widely, and taken deep root—have been incorporated into folk thinking and into one religious ethic or another. Among the exceptions would

---

1. Adriano Tilgher, *Work, What it Has Meant to Man through the Ages* (New York, Harcourt, 1930), one of the few to study work in historic perspective, finds that work has been considered in most of the great religions and philosophies, but apparently attention is mainly on what is expected of man.

be the views held by the ancient Greeks and Romans.[2] These were less philosophies of work than attitudes about work: who should do and must do the manual work, and whose privilege it was to do the thinking, the praying, the curing, the fighting, and the ruling. Such distinctions, of course, did not originate with the Greeks, but those who were above the peasants, the artisans, and the slaves made a fine art of nonwork activity. They created the gentleman-citizen.

The gentleman had his obligations to the community, but labor was not one of them. Indeed, he might lose status if he did manual work. Tilling, hewing, grinding, cooking, weaving, sewing, washing (except one's self), bearing and fetching were among the activities performed for the gentleman by lesser people. Things were often so when urban life evolved, but the Greeks set an example in making use of nonwork time as a cultural leisure. Many before them and since have enjoyed the same nonwork privileges, holding both work and workers in low esteem.[3]

Apparently it has always been true that those with no choice but to work also had their various levels of status and, where a kind of work remained changeless for generations, each type of worker or craft, whether he worked with wood, clay, stone, metal, glass, cloth, leather, foods, drinks, or whatever, acquired certain traditional views and lore about his occupation. Motivations for these views and lore were mixed. On the social side, men have always managed to get status from their work. One might recognize, with mental reservations, that his occupation was lower in status than some others, but he took satisfaction in feeling that his work was superior to certain other occupations.

On the economic side, workers in one occupation guarded their secrets against outsiders. They kept alive certain beliefs about the ills that would befall the encroacher on their work. The tailor's needle would turn on the impostor and prick him. The ax of the carpenter would fight back if held by inexpert hands, and whoever took up the trowel of the mason must expect the trowel to trick him. This monopolist attitude toward particular types of work, which found expression in a strict inheritance of occupations as property, evidences what Parsons and others call *ascription-orientation*. The worker and his occupation were closely associated.[4]

2. C. Wright Mills, *White Collar, the American Middle Classes* (New York, Oxford, 1951), p. 215. "To the ancient Greeks, in whose society mechanical labor was done by slaves, work brutalized the mind, made man unfit for the practice of virtue.... The Hebrews also looked upon work as 'painful drudgery,' to which they added, man is condemned by sin. ... In primitive Christianity, work was seen as punishment for sin but also as serving the ulterior ends of charity, health of body and soul, warding off the evil thoughts of idleness. ... The church fathers placed pure meditation on divine matters above even the intellectual work of reading and copying in the monastery."
3. William G. Sumner, *Folkways* (Boston, Ginn, 1940, 1st ed., 1906), p. 162, notes how ancient civilizations held work in low regard, except for that of chiefs, teachers, medicine men, soldiers, etc. In recent times in the West "labor has been recognized as a blessing or, at worst, as a necessity which has great moral and social compensations and which, if rightly understood and wisely used, brings joy and satisfaction."

Under the ascription thesis, work, the occupation, and the worker were as one, whereas in modern work the worker and the occupation are separated, much as work is separated from all nonwork activity. In former times, something about the person, like the bowlegs of the cowboy, symbolized his occupation. Often it was the hat (or cap), the jacket, or even the haircut; whatever the badge, it was proudly displayed. The individual did not change from his father's occupation or, if necessity made a change necessary (the father having several sons), he entered an occupation of equal or higher rank.

In countries where the occupational hierarchy separated work categories by sharp lines, various caste systems developed. These might become so crystallized that new types of work could hardly enter. If new types of work entered, as must be with industrialization, those freest to accept it were often the very lowest castes, the untouchables.

We may say that until very modern times there was no interest in examining the phenomena of work, or speculating about them. We need to study it now because work is bought and sold in measured quantities. We have come to measure and evaluate it according to quality. Let us consider what that means.

The New Division of Labor

There have always been divisions of labor between the old and the young, the male and the female, the strong and the weak. With the rise of crafts in towns and cities the division of labor came more to be associated with tools. The worker was one who used tools, and that was the activity known as work. The philosopher used pen and paper, but these were not tools in the sense of the spade, the hammer, or the trowel. The philosopher did not regard his activity as work or occupation, but merely as a preoccupation. Such a division of labor, associated with tools used, materials worked with, and sorts of articles produced, was made possible by the urban market and an exchange system that enabled a craftsman to invest his labor in something he could sell and with the money buy the essentials of life. This division of labor along craft lines, so integrated in the exchange-money economy and with potentially expanding markets, contained within itself the germ of its own final elimination.

During the several centuries while it lasted, the craft division of labor served well in its own monolithic way as a force for order in town and city. Each craft had its place in the community, and each worker had his place in some craft. The guilds assumed responsibility for their own social and welfare problems. On the one hand, the guilds endeavored to keep competition under

---

4. Harold A. Gould, "Castes, Outcastes, and the Sociology of Stratification," *International Journal of Comparative Sociology,* Vol. 1, No. 2, September 1960, pp. 221-22. The other type of stratification is the *achievement-oriented* which separates person and occupation and is uppermost in the industrial civilization. Where ascription-orientation is strong "occupation is rooted in kinship structures."

control, and on the other they held the line against inventions that might change their ways of work. But competition did not submit to restraint, and inventions came in spite of the guilds. The guild way of working and making money could not suppress the urge to find better and faster ways of getting more money. Then came the Industrial Revolution.

Man under urbanism, using Mumford's expression, had achieved the "capacity to impose work on himself," and this in time not only "gave him greater security and freedom but made possible a more organized type of society."[5] The guild system carried the evolution of work in this direction a few steps but could not, because of its handwork limitations, go farther. But the guilds did this: for the first time in human history they brought a degree of dignity and respectability to work, in that they made the "calling" of the craftsman something to be proud of. The first decades of the Industrial Revolution played havoc with that dignity, but this turned out to be a passing setback.

Industrialism was not a division of labor along craft lines; the crafts themselves were subdivided, in some cases to extinction. In fact, "division of labor" hardly applies to what took place and is still taking place in the industrialization of work. This development is too well known to call for description here. It is, however, pertinent to our theme for this reason: it made the study of work necessary. The machine was introduced into the work process, and ways were found for transferring much human work to machines. We can even think of a division of labor between men and machines, and the balance in this relationship changed with each technological advance. This depended in part on the skill with which a unit of work, a task, could be simplified into its elements.

In this process the craftsman's occupation became fragmented beyond recognition, while its mysteries and lore were brushed aside. Only naked work remained, something that could be described in terms of multiple motions; lifting, pushing, pulling, turning, and so on. But subdivision and simplification called for work organization, which meant functional interdependence. In an elemental way the work of the craftsman had been organized, but in relation to him and his tools. The new organization was outside the worker. It involved, on the one hand, an integrated array of machines spatially and functionally located in the production process and, on the other hand, a distribution of workers manipulating the machines, but co-ordinated one to another in this production process. The performance of each depended on the performance of others. Here was a type of dependency and interdependency never known before, nor even visualized by utopian writers.

5. Lewis Mumford, *The Condition of Man* (New York, Harcourt, 1944), p. 4. He adds, "But in origin work and play have the same common trunk and cannot be detached; every mastery of the economic conditions of life lightens the burdens of servile work and opens up new possibilities for art and play."

The Road to Work Rationalization

What naturally followed, after the simplification and fragmentation of an occupation and the array of separate performances, had to be integration. The products of the subdivisions of the whole process had to be assembled as the work flowed along and the completed product appeared at the end of the process. This has been called the rationalization of work. The disintegration of the old occupation through rational organization of work leads to integrated work results. In the course of this development work finds its meaning in motion that can be measured and skill that can be described with reasonable precision. It becomes something that the manager or engineer can estimate against money values.

Moreover, in this rationalization of work, prediction is possible. Since the types of work can be classified into different categories according to the skills required and other capabilities which the worker must meet, and since the performance of the different categories is known from study and experience, it is possible to estimate, for example, what the output of ten workers in any category would be over a period of any given number of days. This can be done only in countries where such a rationalized conception of work has become incorporated into the habitual ways of community living. It assumes a labor force adapted to the discipline of these new work ways, commonly called industrial, although it does not need to be in a factory.

Any type of industrial work, for which the tasks are separated and the outputs of workers in the co-ordinated process are assembled to make the product, calls for organization. Each type of work organization, whether it makes hats, rubber articles, sport equipment, automobiles, or washing machines, becomes within itself a system of dependent and interdependent relationships. But that is not all. Each such organization is linked in dependent-interdependent relationships with others. In one plant the raw materials may be prefabricated; special parts may be made in another plant; another work organization may provide special services, like repair and replacement. Modern work gets part of its meaning, then, in terms of types of interdependence. This awareness of interdependence and the capacity to behave rationally toward it enter into the discipline of the industrial labor force. But here is one of the glaring lacks in the nonindustrial developing country.

This industrial type of work, with men and machines in combination, with its changing element of interdependence and its unique labor-force discipline, has been evolving without plan or direction into a rational work system mainly during the past century. Much of the understanding of work as it figures in this system is the result of work study during the recent seven or eight decades. Most of this study has been focused on the motion and skill aspects of work in the production process, mainly in the work place. Outside the work place scholars and commentators have given thought to labor as a factor in production, to the labor-machine combination, and to the role of the capitalist-entrepreneur.[6] But the real analyst of work and the changer of its character has been the engineer or the manager having responsibility for technological supervision.

And these analyzers and changers of work who brought about the modern division of labor, which is continually being modified by them, have not been concerned about the social, psychological, or philosophical meanings. Their attention has been focused on performance, output, and unit costs. To serve their ends, skill and performance each had to be reduced to its component elements for purposes of quantification and cost estimates. However, as organizers and manipulators of work in their places of operation—factories, shops, offices, even public administrations—in the course of defining and refining tasks they have profoundly influenced life outside their work places. The engineer is the true revolutionary.[7]

While this thought will be developed in several contexts later, it needs to be mentioned here that modern work acquires some of its meaning because it is time sold. Many kinds of work are traded in the labor market. In the work place the buyer of work must know the number of workers with each type of skill needed to operate the organization. It is to his interest not to use a highly skilled worker to perform a task which can be done by a worker of less skill, which means less cost. It is also in his interest to eliminate any highly skilled operation that can be divided into a number of semiskilled operations, especially if the change increases production. The need to reduce labor costs by reorganizing the work, by the introduction of machines, by the simplification of work itself, leads ever to some change in the division of labor. Work means time for the buyer as for the seller, as it also means money for both.

Work Abstracted and Precisioned

It is for the engineer to find ways of getting the most out of work for the least cost, whether he is considering the work of men or machines. He must find ways to abstract work so that the production of men and machines can be equated. Both must be measured against such other abstractions as time, space, money. It is only as this can be done that estimates can be made regarding the kinds, amounts, and costs of work in future operations, as when the contractor estimates the cost of building a bridge, or when the manufac-

6. Maurice Dodd, "Entrepreneur," *Encyclopaedia of the Social Sciences* (New York, Macmillan, 1948), Vol. V, pp. 558-60. A brief comprehensive statement of the role of the entrepreneur and capitalist in industrial history.
7. Thorstein Veblen was one of the first to write with insight about the evolution of industrial work. See his books, *The Instinct of Workmanship*, 1914, and *The Engineer and the Price System*, 1921, or pertinent extracts from Veblen's works by Max Lerner, ed., *The Portable Veblen* (New York, Viking Press, Inc., 1948).

turer agrees to deliver on or before a certain date a given quantity of goods at a pre-estimated price.[8]

To abstract work in order to subject it to precision reckoning means also to depersonalize it, in about the same sense that urban life is depersonalized. In the ascription-orientation civilization mentioned above, work and worker are associated as one and the occupation is a property of the kinship group. But in the new situation work is achievement-oriented and is rarely associated with the person. The job is built into the organization at the work place, much as machines are built into the structure of the plant. To depersonalize work, however necessary for industrial efficiency, means to detach from it any identification of the worker either with the performance or the product. Brave efforts are being made in some industries, with the aid of scholars who should know better, to reverse the depersonalization trend. On his part, the employer may be no less detached than the worker. It is difficult to see how this depersonalization trend can be avoided or why that should be desired. Work cannot be moved back to its preindustrial ascriptive role except at the cost of productivity.[9]

For the citizen in the old days, to look at his shoes was to think of the man who made them to order, and to look at his coat was to think about his tailor. It is rarely that the citizen today thinks of the workers who made articles he buys. He may be partial to a hat of a certain brand but may never bother to learn where it was made, how it was made, or by whom. In these terms the worker is out of mind as much as the cow is to the city child who associates milk with bottles from the store. This separation of the worker from the product begins at the work place where work is abstracted and where one draws pay for hours of performance. We must recognize that the abstraction which depersonalizes and sets work so completely in a sphere apart from personality (and even from the product) may be a human achievement, for it also sets work outside the sphere of living. Complaints notwithstanding, this new meaning of work may be progress.

In this trend, so it seems, work is coming to be activity to which one gives some of his time, but he need not be responsible for it. Those responsible for it are also selling their time, but not giving their whole personality. Work in

8.   In addition to the engineer, the physicist has also examined work at close quarters. He visualizes the unit of work for measuring purposes as the erg, the energy needed to lift a gram one centimeter. The large calorie, says the chemist, is the amount of heat required to raise the temperature of one kilogram of water one degree centigrade. Heat liberated by the body at work exceeds that while at rest, and the quantities can be measured. A man of average weight while sitting consumes about 100 calories in an hour, against 200 calories if he walks at the rate of 2-1/2 miles per hour. Walking upstairs calls for 1,100 calories per hour. See J. Stanley Gray, *Psychology of Human Affairs* (New York, McGraw, 1946), p. 485.

9.   Glen U. Cleeton, "The Human Factor in Industry," *Annals of the American Academy of Political and Social Science,* Vol. 274, March 1951, pp. 17-24. This article and other writings by Cleeton express well the views of those who hold for the humanizing of industrial work. Efforts should be made to stimulate fellowship in the work place. A companion argument relates to neighborhoods; one is asked to be neighborly, however he himself may feel about it.

this impersonal sense must be amenable to quantification and precision planning. It must be predictable and rational in its administration.

Work Motivations and Rewards

It may be a foolish question to ask why man works, but philosophers and churchmen have been giving their different answers for a long time. The Christian answer, for example, lays it all to the sin of Adam. In these answers people tend to find meaning for working at all. Quite aside from explanations on how things came to be so, there are other motivations for working than the penalty reason given in the Adam and Eve story, and of these Santayana lists three.

> The motives for work which have hitherto prevailed in the world have been want, ambition and love of occupation; in a social democracy, after the first was eliminated, the last alone would remain efficacious. Love of occupation, although it occasionally accompanies and cheers every sort of labour, could never induce man originally to undertake arduous and uninteresting tasks, nor to persevere in them if by chance or waywardness such tasks had been once undertaken. Inclination can never be the general motive for the work now imposed on the masses.[10]

Although it is implied by Santayana that want has been eliminated in the social democracies (and he was apparently thinking of the United States), want is relative, much like poverty. He says nothing of ambition; nor need we, since it diffuses itself into other motivations. Most people would answer that love of occupation as a motivation for work rarely finds expression in most modern work which, as we note above, is "imposed on the masses." It is perhaps true that one does not incline naturally to most of the work offered to most people. The motivation may be a qualified type of want. One may not be pressed by hunger unless unemployment comes his way, and this, of course, he must avoid.

Perhaps a more universal motivation, one that does much to give meaning to work, is the essentially social fact that most of us know work is expected. If everyone else works except us, we don't feel that we belong. He who works is rewarded with social approval—usually with greater approval if he works well and advances himself. If he is not regular in his working, if he endeavors to avoid work, social sanctions in the way of criticism may fall upon him, and his friends may turn from him. With work, as with other social expectations, one learns to do, and often gets satisfaction in doing, what is generally expected. In this situation the incentive is social, and so is the reward.

Since one is expected to work and usually does, work in return brings social as well as material rewards. While the social expectations make proper

---

10. George Santayana, *Atoms of Thoughts,* Ira D. Cardiff, ed. (New York, Philosophical Library, 1950), p. 34.

allowance for rewards, they normally do not press the individual to strive for more than sufficient for his needs. Here we meet a striking difference between work expectations in developing and advanced industrial societies. In the former it is not uncommon for one to be envied or suspected if he works beyond the usual measure and endeavors to accumulate surpluses. In the advanced countries expectations go beyond the limit of sufficiency. The man in the developing land, if he works harder to accumulate surplus, will be confronted with family demands. He belongs to a kinship group, and he may not aspire to individual success and be honored for it as in the Western world where, quoting Gray, after he has met his physical needs he can strive "to gain mastery over things and people."[11]

The most satisfying rewards of work, which give it meaning in terms of individual well-being, inhere in the material gains beyond subsistence. Then it is, declares Mumford, that the "social meaning of work derives from the acts of creation it makes possible."[12] Self-sufficiency is more easily achieved once subsistence sufficiency has been met. While this does not apply to Mumford, some writers overemphasize creativity, or deplore too much the lack of opportunity for "creative expression" in most modern work. They forget how little creativity there was when work was several hundred years in the hands of the guilds. We will return to this in later chapters. It is our introduction to another thought.

We have already mentioned that a degree of self-respect derives from the fact that one works. He may not like his job, but he prefers work to idleness outside the fellowship of the occupied. He may have low status in his work, but without it he would have next to no status. Also a satisfaction is realized from knowing one's job, which is conducive to attitudes of integrity toward work. After admitting much that is said about the depersonalization of modern work, the fact remains that the worker on the job has an infinite number of opportunities to demonstrate these attitudes of integrity. Were this not the case, the productivity of the total labor force would not have increased steadily as it has over the past several decades. Work is a challenge to the performer at any level of training, and how workers feel about it is articulated in such American expressions as "the job is the boss," "botchers get caught up with," "you can't hide bad work." This discipline is not imposed by foremen, supervisors, or managers.

---

11. Gray, *op cit.,* pp. 515-16. Regarding the psychology of rewards for work, these observations are added: the promise of reward must be fairly certain, it must not be too long postponed, and it must be attainable.

12. Mumford, *op. cit.,* p. 5. From the same page: "The role of work is to make man a master of the conditions of life; hence its constant discipline is essential to his grasp of the real world. The function of work is to provide man with a living; not for the purpose of enlarging his capacities to consume, but for liberating his capacities to create."

Women Work — Why?

Robert D. Ryan

The United States has recently witnessed a dramatic shift in the role of women in the labor force. The turning point occurred during World War II. It would not be unrealistic to state that even as the last bombs were falling on Pearl Harbor on December 7, 1941, the women of the United States were preparing to put down their housewifely aprons and take up the welder's masks and riveting guns. During the war, the United States had approximately 10 million men in uniform. Therefore, it was necessary that the women of the United States be called upon to fill this vast void created by the men who were in the services and the additional jobs created by the war effort. It was expected that when the war ended, the women would return to their homes, and once again become "only" homemakers. Contrary to these expectations, the majority of the women remained in industry, or had developed a new attitude toward "working" and realized that they too could contribute to their society in ways other than in the home.

But why, when the war ended and the men returned to the civilian labor force, did "Rosey" not return to the kitchen? Why, as a point of fact, has the number of women in the labor force continued to increase? Why have women accounted for the major share of the growth of the labor force during the last decade? Why, when automation and increased manufacturing efficiency allows a single worker to produce as much as several did previously, does the number of women in the labor force continue to rise? Finally, why are women even busier today preparing themselves for work outside the home?

The answers to these questions are multi-faceted and may be found in demographic, economic, social, technological and personal factors. There are those who say that it is primarily attitudinal on the part of both the husband and employing agencies. Others state that it is the recognition of the equality of rights, and still others attribute the trend to strictly technological advancements where labor-saving equipment, prepared foods, and virtually maintenance-free materials shorten the time required for domestic chores.

Today women comprise about 35 percent of the labor force, or about one out of every three workers. However, in 1900, women comprised only about 18 percent of the labor force. By 1920 it had increased to 20 percent; in 1930 to 22 percent; 1940 to 24 percent; 1950 to 27 percent; 1960 to 33 percent; and 1970 to 35 percent. Another revealing picture of women in the labor force is the fact that fifty years ago the average woman worker was single and 28 years old. Today she is married and 41 years of age.

If analyzed, it appears that there is no single cause, but rather a combination of all recognized or accepted reasons why women are entering the labor force in ever-increasing numbers.

When women are queried on why they work, their statements should be interpreted with caution. In many cases, the overt statement may be only a concealing reason to cover up a perceived feeling of need for working, or the response could result from guilt feelings, because they still possess the concept that the woman's place is in the home. Generally, women respond that they work because of financial needs. It is obvious that women vary widely in their reasons for working, and that financial needs are legitimate and real. Although the significance of financial motivation cannot be minimized, there are other motives that cause women to enter the labor force. In formulating reasons, it is apparent that financial need must be discussed first.

I. *Financial*

Into this category are placed, first of all, those women who must work because they are heads of household. These are the single, widowed, divorced, separated, or special cases. There are many married women who must work in order to support the family unit. These women may be part of a family unit where the husband is unable to provide adequate support because he is disabled, unable to command a position which will ensure adequate support for his family, and other circumstances. However, there are women who are members of family units where the primary wage-earner, the husband, is earning a sufficient income to maintain a satisfactory standard of living. This then leads to perceived financial needs which are the result of a personal value system. These come in the form of larger homes, more elegant homes, appliances, second cars, college education for their children, or for most any item which can be considered to have a social status value. Although the significance of financial motivation cannot be minimized, there are other motives present to cause women to enter the labor force.

II. *Advertising and the Cashless Society*

Advertising through various media has scored an unbelievable success in the United States, to the point where Americans are purchasing long before their ability to pay. American business enterprises have made it extremely simple to charge. You have all heard it, "Nothing down and a small amount per month," even for small purchases. Spending before earning, by means of charge accounts and easy-pay plans, has in many cases made a second wage-earner in the family a necessity. The American family unit, desirous of material things that American industries can provide, has created a situation where another wage-earner must be acquired. The most likely additional wage-earner is usually "mother."

III. *Attitudinal*

Not too many years ago it was considered unacceptable for a man's wife to work. The traditional attitude which has prevailed until recent years is that when a woman marries, she accepts the role of motherhood and homemaker. Today a dichotomy exists; for example, especially in young married couples, the wife is expected to work to earn money for a down payment on a home or for other necessities to get the couple off and running. However, this generally leads only to continued work. The couple very quickly adjusts to two incomes and the thought of existing on one income sends fear to their minds. Or very realistically, they now have time payments, fixed expenses which they could not meet on one check. It appears that working wives, especially in young couples, now have become an American way of life. Today the American public almost tends to look down on the young married woman, especially if she does not have any children, who does not work.

IV. *Increased Educational Opportunities*

The last several decades have witnessed an increasing number of women attending colleges, business schools, and other institutions to prepare themselves for a variety of occupations and professions. Upon completion of their education, these women seek work for a period of time before marriage and rearing a family. This training and work experience gives the woman a feeling that her time is worth money, that she can contribute to her society, and her time should not be spent only at home. The general attitude of the American public today is causing nonworking women, especially those who have special skills, talents, or education, to feel guilty. They're not contributing to their society and are wasting a very important segment of our total labor force.

V. *Maintain Competence*

Much has been written concerning men in our labor force, specifically men in scientific and technical areas, in regard to the difficulties they encounter keeping pace with technological developments and advancements. They must engage in self-study, in-plant training, or courses offered at nearby educational institutions. This is an increasing trend for women in the work force. They also have specific skills or talents which will become obsolete if they do not stay in the labor force. Therefore, many women choose to remain in the labor force or return to their specific occupation or profession periodically in order to upgrade or maintain their knowledge in their specific field. As job requirements continue to rise, it would seem that this motive will be increasingly important in the years to come.

VI. *Interaction with Peers*

There are those women, especially in the professions, who choose to continue to work to avoid an intellectual death. Challenging interaction with peers is frequently stated as one of the reasons that women enter the labor force and continue to work.

VII. *Social Contact*

Over and over it is stated that women feel they must get out of the home to converse with adults. The husband is constantly being faced with the argument that he gets out to the office every day and converses with adults and has intellectual stimulus, whereas the wife stays home and talks primarily with the children, with the exception of occasional coffee klatches and phone calls. It can be readily understood that this feeling is paramount and lack of adult social contact would cause many women to seek work outside the home.

VIII. *Family Planning*

Current medical advances have made it possible for the modern woman to effectively plan the size of her family. The result has been an increase in smaller family units. Today about one half of the women marry by age 21 and have their last child at about age 30. By the time her youngest child is in school, the mother still has about 30–35 years of active work life ahead of her, if she so chooses.

IX. *Industry Wants Women to Work*

Business and industry are constantly encouraging women to enter the labor force. They are being asked to work for a variety of reasons.

    a. They generally work for less wage.

    b. Dexterity—they have nimble fingers. Electronics industries are hiring women almost exclusively for their assembly work, data processing, key punching, and other operations. For example, the assembly activity of electronics industries is almost 99 percent performed by women.

    c. Women are extremely patient. They are not as easily bored as men.

    d. Women have been more content, and are not constantly striving for increased pay, promotions, and fringe benefits.

Points a and d could be due, at least in part, to the fact that many of these workers are married, and therefore are not the primary wage earner of the family unit. If a woman is the primary income source, these reasons are generally not true cause, because the income of the family unit depends upon her occupation. In any case, the points a and d are not in the best interests of American women. There are organizations today, such as women's liberation, who are working for equality of women workers in all areas, and as many have said, the salary, position and promotion should be determined by the services rendered, and not by sex.

X. *Bored with Homemaking*

Many women today do not find a challenge or feel rewarded from being only a housewife. They quickly become bored with the routine duties of the home, such as washing and ironing, dishes, preparing meals, and other daily chores. Others find children nerve-racking and choose to hire a housekeeper, and find employment outside the home.

XI. *Self-Enrichment*

With the changing attitude of the American public, business, and industry toward women in the labor force, it is apparent that personal achievement and contribution are becoming increasingly important in the reasons why women work. There is no question that women are now desirous of making their full contribution to the family, labor force, and society.

XII. *Alimony*

A new dimension has been added to the family unit and that is second marriages. The divorce rate is increasing every year in the United States, and accordingly second marriages are increasing. The husband is required to pay alimony to his first wife and support payments for the children of his first marriage. The woman who chooses to marry a divorced man may be faced with the reality that in order to make her marriage a success, she will have to enter the labor force for added income to support the family unit, because a high percentage of the husband's income will be required for alimony and support payments.

It is apparent that there are many reasons for women working, especially married women. When asked, as was stated earlier, most women will say they work because of financial need. However, if analyzed further, and a true motivation uncovered, it is thought that this statement would not be entirely true.

There appears to be little doubt that life styles, family life and responsibility in the United States are undergoing drastic change, which is due, at least in part, to technological developments. The trend of working mothers is on an upward spiral, irrespective of the age of the woman or her children.

The entire picture of the labor force fairly indicates that women must take advantage of all the education and training available to them, and must develop their talents and abilities to the fullest extent possible. In this era of rising demand for more skilled workers and of accelerated technological advance, women must be positive and flexible in their attitudes. They must be willing to adjust, learn, and make the necessary changes. They must be alert to new job opportunities, and to new education and training programs. Only if they are fully prepared by adequate education and training, and a willingness to learn anew, will they be ready for the challenges, demands, and promises of tomorrow's society.

In summation, it is recognized that American women have always contributed their full share to the economic status of the country. However, the pattern of contribution has drastically changed. Not too many years ago, the contribution was made primarily through skills and talents which women developed and utilized almost exclusively within the home. As the United States became more industrialized and specialized, and the concomitant business and office requirements developed, the responsibilities of women were subsequently transferred from the home to factories, stores, and other agencies which requested and utilized her services. Today American women are increasingly seeking the means by which they can make their full contribution to their family and society.

Helping the Hard-Core Adjust to the World of Work

Leonard Nadler

The employer must install a "support system" to change the new employee's behavior and his expectation that he will fail.

Foreword

An ambitious attempt to turn large numbers of the urban poor into regular members of the work force is a phenomenon of the current U.S. business scene. Laudable as the attempt is, the author contends that it will fail unless the corporations involved accompany their job training efforts with carefully devised and coordinated systems to transform the typical attitude of the hard-core unemployed—an attitude compounded of suspicion of business's motives, isolation from our "white" society, lack of concern over what is expected of an employee, and preoccupation with personal problems. The author reports on his HBR-sponsored study of the support systems which five companies have built to supplement their hiring and training structures. He also offers some suggestions based on these observations and other experience.

Mr. Nadler is Associate Professor of Adult Education and Employee Development in the School of Education at George Washington University. An original member of the President's Task Force on the War Against Poverty (1964), he has consulted with many companies on employment of the urban poor, and he has designed and conducted training programs for several Office of Economic Opportunity projects.

Hiring the so-called hard-core unemployed is acknowledged to be a crucial part of what U.S. business can do in the national effort to solve the problems of our great urban areas. Business is increasingly shouldering this task, and those companies that subscribe to the goals of the National Alliance of Businessmen (NAB) have pledged to sign up several hundred thousand of the hard-core before this decade is very old.

Putting these persons into jobs where they can be productive, gain self-respect, and have the hope of advancement is a new experience for most companies. Still newer to business are the complexities of training and guiding them so they can perform their jobs adequately and stay in the ranks of the employed.

Merely following the usual training practices is inadequate, for these men and women have been long thought of as unemployable. They are typically poorly educated and without skills; and, more serious, their attitude, characterized by tardiness and absenteeism, betrays a great lack of understanding of what is expected of them in the world of work.

So what is necessary is not only training that gives the person skills, but company procedures that help produce new behavior patterns and "support" those patterns while he is undergoing training and also later while he is becoming adjusted to his job.

It is safe to say that companies which fail to install effective support systems will fail in their programs to hire and absorb the urban poor into their work forces (beyond token numbers).

Since support systems have been installed by only a few companies, they have not been studied much with a view toward assisting neophyte companies about to undertake such a task. One program (that of Lockheed Aircraft Corporation) has previously been analyzed in the *Harvard Business Review*.[1] Another recent HBR article was instructive in its vivid picture of the grim backgrounds of more-or-less typical trainees (at Westinghouse Electric Corporation's East Pittsburgh, Pennsylvania plant).[2]

With the idea that the subject should be examined further, HBR commissioned a study of five companies' ongoing programs, which train men and women for such entry-level positions as mechanic's helper, sweeper, wire drawer, and clerk. This article is a report on that study.

The five companies were chosen because they had several things in common. Each was a multi-site organization (in three cases I studied operations at more than one site); thus they were large enough so that, without good controls, management decisions could easily be misinterpreted or even lost as they filtered down from headquarters. Each company also had a very competent, experienced training director, who was situated in such a way that he could provide guidance and leadership for his company's program from the corporate level.

1. James D. Hodgson and Marshall H. Brenner, "Successful Experience: Training Hard-Core Unemployed," September-October 1968, p. 148.
2. Theodore V. Purcell and Rosalind Webster, "Window on the Hard-Core World," July-August 1969, p. 118.

These are the facilities that participated:
○ Boeing Company (Renton, Washington)
○ Eastman Kodak Company (Kodak Park Division, Rochester,
    New York)
○ Eastman Kodak (General Office, Rochester)
○ Westinghouse Electric Corporation (East Pittsburgh)
○ Westinghouse (Bloomfield, New Jersey)
○ UAL, Inc. (25 in 11 cities)
○ Bankers Trust Company (New York City)

I shall not try to compare the programs at these companies. They are in a large sense not comparable, since an effective support system must be based on factors peculiar to the organization and the situation in which it operates. After describing the programs, however, I shall offer some guidelines based on my study and other experience in this field.

Before I discuss the support systems themselves, I shall say something about the training practices around which the systems are built.

Training Programs

The training of new hires usually has at least three identifiable components: attitude changing, remedial education, and job skills development. Here are the details:

1. *Attitude changing*—Many of the recruits have never known regular employment or lived in environments where steady employment was part of the norm. They must be helped to rid themselves of their expectation of failure and to develop a self-image of themselves as regular members of the work force. The training tries to enhance their self-esteem. Absenteeism and tardiness are manifestations of their inability to adjust to their new environment. This aspect of their behavior is a serious concern of the training program, and much attention is paid to it.

An additional element of attitude changing is related to appropriate dress and behavior on the job. The purpose is not to force conformity but rather to help the new employee understand the permissible range of dress on the job. In many cases, safety regulations dictate what is permitted.

How this element of training is approached varies, of course, with each company. Boeing provides a spectrum of instruction, ranging from getting along with others and black history (where appropriate) to how to pay taxes and buy a home—with the aim of bolstering self-sufficiency and pride in being a contributing member of society. Westinghouse/East Pittsburgh involves its trainees in intensive group discussions about themselves, their environment, and their heredity, while Westinghouse/Bloomfield conducts sessions on the world of work and the individual in relation to his community. Bankers Trust holds informal discussions and conducts tours of such places as the employees' dining room—all designed to make the new hires feel at home.

2. *Remedial education*—"Functional" literacy (being literate enough to function in our society as an economic unit) is generally considered to have been achieved on completion of the fifth grade. Many of the urban poor have not reached this level, and so they need remedial education, particularly in arithmetic and English.

The Boeing program, which usually lasts 14 weeks, allots three hours a day for remedial work. Westinghouse/Bloomfield adds material on consumer education to the curriculum. Bankers Trust uses outside resources—the American Institute of Banking and the Board of Fundamental Education—which carry on such training for many New York City area banks. EK/Kodak Park has a contract with the Board of Fundamental Education for this work. English for employees is provided by the Adult Basic Education branch of the Rochester school system.

3. *Job skills development*—The range of job skills training naturally is directly related to the kind of entry-level position for which the recruit is being prepared.

At Boeing, five hours in the eight-hour day are devoted to preparing trainees to qualify specifically for one or more of the 32 different job titles. The emphasis is on performance close to what will be expected of them on the shop floor. At UAL, the program duration ranges from two weeks for janitors to eight weeks for ramp servicemen.

One of the more elaborate programs is EK/Kodak Park's Hands On center, which provides up to 12 weeks of training designed to make the new hire eligible as a helper-skilled trades, which is a basic entry-level position. But for those who demonstrate ability there is a higher level program of vocational training leading to apprenticeship.

The location of training facilities varies greatly. Boeing uses a facility central to its various plants, which cover a wide area around Seattle. Westinghouse/East Pittsburgh located its facility right in the plant, where it shares a building with the vocational training unit. Westinghouse/Bloomfield, however, established a training center in the black ghetto area of nearby Newark. UAL, which has operated programs in 11 cities, relies almost exclusively on in-house facilities and trainers. Bankers Trust uses quarters that are provided by the American Institute of Banking.

Before I examine the support systems, I should mention that the federal government is trying to encourage greater corporate involvement in training the disadvantaged through its subsidy program, by which a company can recover a significant portion of its training expenses. The procedures have been simplified and are supposed to be simplified even more this year.

In discussions with Manpower Administration officials during late 1969, I was advised that efforts are being made to eliminate as much paperwork and red tape as possible from the proposal and funding procedures.

Examining Support Systems

Corporate support systems for absorbing the hard-core unemployed into the work force have five main elements: (a) organizational involvement, (b) pretraining preparation, (c) training support, (d) job linkage, and (e) follow-up procedures.

I shall discuss them at some length in turn, and for quick reference I have capsulized them in *Exhibit I* and illustrated them with examples from my study.

Organizational Involvement

A company's program may fail despite the wholehearted backing of top management and involvement of affected supervisors, but it certainly will fail without it.

The company's commitment must be communicated from the highest echelons to the lowest, and frequently. This includes, of course, informing unaffected employees about the program and its goals. As part of its communications effort, Eastman Kodak shows to all employees a 15-minute film in which the chairman of the board talks of management's total commitment to helping solve urban problems.

A challenge for highly decentralized companies is communicating the corporate goals so that far-flung units will keep them in mind. UAL has made a national commitment to employ the urban poor and has quotas in the cities where it operates and where an NAB program exists. UAL is decentralized, but it stresses that its objectives are companywide, not the goals of individual personnel and training departments.

Commitment leads a company into many practices which it would not undertake in the normal course of business. For example, during the early work experience of the new hire at Boeing, his supervisor is allowed a 7-1/2 percent differential in performance. Even if the employee's performance falls short of standard by as much as 7-1/2 percent, his supervisor is not penalized for the lower output.

Unreserved organizational involvement calls for an educational effort beyond mere communication of goals; training of many employees other than the new hires becomes necessary. This is particularly true of supervisors, who, as the point of confrontation between the new employee and the organization, are key men in the program.

Though training of supervisors is sometimes carried on as a special arrangement, more often it is part of the regular training procedure. For some time Bankers Trust has operated a three-day "critical issues" program for supervisory personnel. Part of it is a one-day session concerned with the relationship of the supervisor with the disadvantaged employee.

Exhibit I
Summary of Support Systems Observed in Five Companies

Elements	Aspects	Examples
Organizational involvement	Training of supervisors	Part of three-day "critical issues" program
	Plant-wide education	From plant manager to foreman
	Observable commitment	7½% differential in production for recruits
Pretraining preparation	Recruit made aware company wants him	Put on payroll of future department
	Counselor assigned	Helps recruit through personal as well as job-related problems
Training support	Realistic environment	Actual production for company units
	Personal help	Legal assistance
	Relationship with staff	Future foreman visits new employee
Job linkage	Bridge between training and job	Employee included in decision on where he will work
	Interdepartmental coordination	Counselor, foreman, and others consult often
Follow-up procedures	Reinforcement of new behavior patterns	Frequent, informal contact with counselor
	Extra job instruction	16-week period of intensified supervision

Each major Eastman Kodak division has its own supervisory training program. All foremen, however, attend these three sessions common to every one: "Basic Supervisory Program" (for new appointees), "The Supervisor and His People," and "Talking With People." Each emphasizes the "human relations" aspects of being a foreman, and a segment of each is devoted to aspects of working with the new hires. Selected supervisors also attend half-day "understanding" sessions conducted by a community action organization, Rochester Jobs, Inc.

At Boeing, a supervisor scheduled to receive a new hire undergoes four hours of special training (if he has not had it before). When possible, this instruction is completed in the last week before the trainee reports to his supervisor.

Provision must also be made to educate segments of the organization other than the supervisors, to establish a more receptive climate. At Westinghouse/Bloomfield sensitivity training is given to the plant manager, the superintendent of each section, and the personnel department, as well as all general foremen. Each divisional training department at Eastman Kodak has developed supplemental case studies and discussion materials, based on actual incidents and situations, for use throughout the division.

Pretraining Preparation

As I indicated earlier, it is essential for the employee's self-respect that the company convince him that he was hired to make a contribution, for which he is being paid. If he is made to feel that he is receiving a handout, his training will be a waste of time and effort and his chances of becoming a self-supporting citizen will be seriously hurt.

So it is crucial to have a job waiting for him at the end of the training period and to make him aware of it. At Boeing, for example, he is told when hired that one of the 32 entry-level categories is open for him; that the position (usually as some kind of assembly mechanic) can lead to a better one with more pay; that the training he will undergo is part of the necessary preparation for the job; and that when he completes this first assignment, he will be transferred to a job site.

The new employee at Westinghouse/Bloomfield is told that a specific job has been set aside for him. For the men it is wire drawing (the highest unskilled position at the plant), and for the women it is wire drawing, inspector-packer, or final inspector.

At EK/Kodak Park, the new hire is introduced to his foreman before he reports for training. At that time he is actually placed on the payroll of his future department, and the check he receives comes from its payroll rather than from a training payroll.

UAL also hires men for specific jobs and trains them accordingly. At the time of writing, the positions included ramp serviceman, airplane cleaner, dining service helper, kitchen helper, general clerk, and janitor.

Westinghouse/East Pittsburgh prepares men for the entry-level job called sweeper. (The title is something of a misnomer; a sweeper also performs many other tasks like bringing materials to work stations and taking away finished work.) Later he can move up to a machine operator position or other semiskilled work. The exact departmental assignment is not made at the time he is hired.

Help from counselors: Another necessary element of this phase is assignment of a counselor to the new employee. The counselor is designated as the person the employee should turn to whenever he has a problem at work or in his personal life.

In programs where the staff is small—and in some others where it is not so small, as at Westinghouse/East Pittsburgh—the trainer and the counselor may be the same person. At Bankers Trust, the person guiding the training was once part of the training staff, but is now in the personnel department. In this case the early training is conducted outside the installation, and so he is not administering his own program.

A question that often arises is: Should the counselor be of the same ethnic or minority background as the new hire? There is no single answer.

Boeing selects its counselors on the basis of ability to understand the trainee's problems and identify with him, as well as give him a role model for upward mobility in the work force. Naturally, a counselor may be better able to identify with the trainee if he is of the same background, but that is not the primary consideration.

EK/Kodak Park makes no attempt at ethnic correlation; its program is not large enough to support many staff members. The counselor is white, while the training group is predominantly black. As far as I could determine, this has worked out well.

Training Support

The more closely the instruction simulates the actual job conditions, the more easily the trainee will be assimilated into the employee ranks.

For the health of the organization as a whole it is beneficial to try to establish a relationship between the training program and the organization's other elements. This serves to dispel the mystery that often surrounds an unusual company activity.

EK/Kodak Park hit on a contractual relationship between its Hands On training center and other departments whereby the center arranges with other Kodak Park units to produce certain parts for stock. These parts—junction boxes, conduits, and other materials used mostly for maintenance—previously were purchased from outside sources.

The center also contracts for repair work which is within the recruits' capabilities. On one visit I observed trainees doing welding who two weeks earlier had never handled a welding torch. One was welding a locker; to preserve its appearance when painted, he had to weld with a fine bead. Another was welding a large frame for holding tubing to be returned to the production floor. His weld had to meet standards, but it did not require a finished bead.

An additional tie with the rest of Kodak Park is the availability of trainees to work elsewhere in the plant in the event of an operational breakdown. Under Kodak policy, a unit in the chemical process may draw on manpower from other parts of the organization in order to restore operation quickly. The trainees are available to join the work crews for which they have been hired if those crews are tapped for emergency repairs.

One small, but significant, support element at the Hands On center, which is familiar to vocational and industrial arts teachers, is the practice of having each trainee make his own tool box. The box, which will accompany him later when he moves from one assignment to the next, is a tangible sign of accomplishment. One can imagine the care that is lavished on its manufacture.

Personal help: It is essential that the employer have available at the outset the full range of personal services which the newly employed urban poor typically need. Although these services are not directly job related, they are critical because personal problems can ruin the preparation of the recruit for his job. For example:

☐ Working mothers who have preschool children need provision for placing the children in responsible hands, so many companies have established day-care centers.

☐ Transportation is a big problem for new hires. Boeing faced up to this by providing buses which pick up the trainees at their door and bring them to the Kent training center in Renton. They provide more than just transportation, however; to each bus, holding some 11 recruits, a counselor is assigned. He uses the travel time for discussion of the trainees' common problems and concerns.

The buses are also a symbol reinforcing the aims of punctuality and regular attendance. If the trainee does not appear for the bus, the counselor may ring his doorbell. If he is not to be found, the counselor notifies the center to have a second, unscheduled bus come later. In the meantime, the counselor tries to locate the recruit and determine the reason for his absence.

Westinghouse/East Pittsburgh has organized a car pool system and has arranged with the local transportation officials to change some bus schedules and routes to make it easier for trainees to travel to and from the plant.

☐ Legal difficulties are a recurring problem with new hires. Westinghouse/East Pittsburgh has a young lawyer on a retainer basis who is on call to go "downtown" and negotiate with city authorities whenever a trainee is involved with the law. His absence because he is in jail or in court retards his training progress. Timely provision of legal services minimizes the loss.

Involvement with staff: Success of the training program hinges largely on the relationship of the recruit to his supervisor and counselor during this period. At Boeing, the ratio of trainer to trainee is one to seven—about half the company norm in its instruction program—which promotes close individual attention. EK/Kodak Park has one supervisor, who is a journeyman, for every four to six new hires.

Every Friday afternoon during Bankers Trust's six-week program the recruits report to their counselors to discuss their progress and their relationship to the positions they are preparing for. The counselor also tries to draw out the new employees on any personal problems they face which might affect their employment.

As I have mentioned, those companies that put the recruit on a department payroll as soon as he is hired usually have him meet his future foreman at that time. EK/Kodak Park not only does this, but also has the foreman visit the trainee while he is at the Hands On center.

At the same time the foreman meets with the faculty to discuss the training program itself, if desired, as well as the progress of "his" trainee. This includes a review of the "Level of Accomplishment of Training," a list of 95 tasks which the new employee must perform satisfactorily before he graduates.

Some companies draw from the ranks of foremen when staffing their training centers. EK/Kodak Park's faculty consists of four men who handle a hundred recruits annually. Two of the four are rotated every six months. The other two are permanent faculty, but they also have had line experience.

Boeing has two lead instructors, one in the area of remedial work and one in trade training. The latter has a staff assigned to the work from the shop area. The former has many years of experience in adult basic education.

Job Linkage

To transfer the new hire smoothly from the training center into the position he will hold, management must establish a comprehensive procedure to bridge the gap between the two situations.

At Boeing, during the last week of instruction, a meeting is held, attended by the trainee's counselor, his immediate supervisor, and the personnel representative in the shop where he will be assigned. The purpose of this is to inform the shop foreman and the personnel representative about any personal problems of the trainee and about the assistance that will still be available from the training center.

During the last week of the program at Bankers Trust, an evaluation procedure begins for determining what the employee's assignment will be. The trainee, his counselor, and his supervisor review the possibilities. During his last week, the trainee finds out where he will be working and starts instruction on the skills he will learn in on-the-job training after he reports to his department.

Under this program, Bankers Trust has not assigned any new hires to blue-collar jobs, such as janitor or bank guard. They are being prepared only for white-collar positions in the various departments.

At Westinghouse/Bloomfield the job has been previously determined. Now a decision must be made as to the shift and section desired by the employee and available in the plant. As with any new employee, the personnel office tries to place him according to his desires, while taking care to avoid a segregated work shift. (The receiving foreman participates in the final decision on accepting a new employee.) A close relationship exists between the trainers at Newark and the employment interviewers at Bloomfield, which facilitates effective job placement.

Follow-Up Procedures

Even after the new hire is on the job, some continuity with the training experience must be maintained.

The most serious problem after training continues to be lateness and absenteeism. However prompt and reliable in attendance the employee may have been during the training period, once he is on the job his attendance tends to become erratic.

Most members of minority groups have had experiences where employment opportunities evaporated and well-meant promises were not kept. So the new worker emerging from the ghetto probably is "testing" his employer, not, as might be thought, trying to find out how much he can "get away with." He wants to know whether he is really wanted or is just another statistic to make somebody's NAB pledge look good or to fill some government-imposed quota.

A key element is providing sufficient reinforcement of the new employee so he realizes that he is wanted and that he has a bona fide job. When he becomes convinced that he is not merely being used to make a quota, his absence and lateness record almost always improves.

This reinforcement is best provided by continuous, informal contact with the counselor who was part of his previous successful experience in training. The counselor helps him to adjust to a world that has many strange elements and confronts him with a range of completely new problems.

It is not only easier, but more beneficial, for the counselor to visit the employee, rather than vice versa, because this also enables the counselor to talk to the foreman about the employee.

For the foreman, the counselor is an extra resource in training the new hire.

Absenteeism usually calls for home visits, and the foreman, of course, cannot accept responsibility for this task. Although the personnel or medical departments may become involved, the counselor should be called on for some of this work because of the trusting relationship he has established with the employee. But he must avoid the appearance of spying or invading privacy; if the new employee has been on welfare, he has had enough of that.

Those of us who have lived by the clock from our earliest days take the alarm clock for granted as essential in getting us to work on time. But in many inner-city households this appliance is not part of the standard equipment, for it never has been needed. The counselor should see to it that every new hire has one. At Westinghouse/East Pittsburgh, after the first paychecks have been distributed, the counselor accompanies the new employee to the nearest store to make sure he gets an alarm clock.

Extra job instruction: Even in cases where there is intensive training before assignment to a position, extra instruction may be necessary later. Boeing, for example, has a follow-up plan, coordinated by counselors and foremen, which continues for 16 weeks after the job placement.

Where the foreman and counselor have jointly identified an additional training need of a new hire, the foreman assigns one of his hourly men to provide the instruction. The hourly man receives an extra 20 cents per hour while instructing, which is charged to the training budget, not to the operating department.

The program is endorsed by and coordinated with the union having jurisdiction. Moreover, the union has agreed not to consider the temporary increment a raise in pay which must continue after the instruction period ends.

Bankers Trust has developed the position of job coach. He is a supervisor or assistant supervisor who is assigned to give the new employee on-the-job instruction when he reports to the department. The job coach makes out a weekly report on the employee's work and attendance for the counselor.

In some situations the job coach is supposed to work through the counselor rather than directly with the employee. For instance, he calls the counselor immediately when an employee is absent or late. Because the counselor has been working with the new hire since his initial contact with the bank and is familiar with the individual and his record, he is in a better position than the job coach to determine what to do.

The job coach also reports on the employee's level of work. These reports are reviewed frequently by the counselor. The job coach can recommend several salary increments for the new employee early in his tenure on the job. The approach used by Bankers Trust has been such a success that the bank is planning to extend it to all new employees in the organization.

Organizational Perspectives

Once a company has committed itself to the kind of program I have been discussing, it will find that it must change its hiring and training methods in order to meet this different kind of challenge.

The experience of the Ford Motor Company changed the pattern of hiring in Detroit. Up to 1967 a person had to travel out to the gate of a Ford plant to seek employment. After a tedious, time-consuming, expensive trip, what did he face—an employment interviewer who asked him to fill out an application (which he may not have been able to read) and then gave him the standard response, "We'll let you know." How different when Ford went into the ghetto of Detroit, found thousands eagerly seeking work, and quickly signed them up.

The company may discover that it must try to alter attitudes at all echelons of the organization. Whitney Young, Jr., Director of the National Urban League, tells of a company that spent vast amounts of time, energy, and money to improve its image and encourage the recruitment of employees. Then one day a young black came to the front office (not the personnel office) seeking employment. The receptionist looked up and said, "What can I do for you, *boy?*"

Organizations are finding that involvement in these programs encourages organizational renewal. Suddenly they are forced to look at traditional patterns, and usually the need for change becomes obvious.[3]

A company that is not willing to build the necessary support system should not undertake the program. Without supports, a program will likely cause more problems than it can solve. With an effective support system, the benefits to the organization are many.

Pointers on Programs

At this juncture I shall offer some observations on the operation of support systems, including those of the five companies I studied and others in my experience.

Support systems: In my investigations I found it difficult to identify the point at which the support system terminates; it seems just to fade away. This is dangerous, for it can lead to empire building and conflict of responsibilities.

There should be an agreed-on terminal point at which the new hire becomes part of the regular work force. The point may vary with the person, the part of the company, or the nature of the job to be performed. But its identification is crucial, and the lack of such specificity is a weakness in most programs with which I am familiar.

Although many of the companies I have observed have included job linkage as part of their systems, it has not been as strong as it should be. The trainers still seem hesitant about coordination with the rest of the organization. Involvement of the foreman during training is an important part of job linkage, but to be effective it must be made part of standard procedure.

Few companies as yet are involved in work-study programs. In these, the company contracts with a local organization to provide students with work experience closely related to the courses they are taking from that organization. The company does not necessarily hire the graduates. Under the Vocational Education Act of 1968 there is a push to involve more corporations in such programs. When they do get involved, they will find that the student-worker must be integrated into a support system at some point.

Departments specially set up to absorb employees on completion of their training can be helpful or be a problem. EK/General Office, which contracted with a Rochester organization for a work-study program, set up a vestibule department to absorb into the company those graduates whom it had hired. After a suitable time, and when they had met company standards, these employees were routinely assigned to other departments.

Westinghouse/Bloomfield, on the other hand, had to abandon its special group. All the graduates of the Newark training center were sent to the

---

3. For more detail on this, see my article, "Minority Group Employment: Unforeseen Benefits of Specialized Supervisory Training," *Personnel,* May-June 1969, p. 17.

vestibule and a segregated department was thus taking shape. It could easily have developed into a temporary station on the way out of the company. Safeguards must be built in so that the department gains prestige and status in the company's organization.

The buddy system is one of the most researched aspects of support procedures, but it was not a significant part of the systems I have observed. Perhaps it is so much advertised because it is highly visible. Companies that have programs for training buddies may object, but I believe buddies must "develop" naturally rather than be assigned.

Another question is: Should buddies be of the same ethnic group? There are proponents on both sides of this question, but I found no evidence to indicate that ethnic matching is either necessary or desirable.

Counselors: Guidelines on selecting counselors are difficult to formulate, since the position in corporations is so new. Few persons have had the experience, and perhaps none has had the training.

The counselor must of course be an honest, interested, and mature individual. He must be able to relate to the new hire, but this does not mean that he must have come from a ghetto, or even be an ethnic or racial match-up with the recruit. The counselor should be carefully selected, carefully watched, and immediately transferred if he is not proving helpful to the new hire and the new hire's supervisor.

During the study I was often told that the supervisor is the person to whom the new hire relates and who helps him. But this is not the perception of the counselor or the new hire: the bond is between the counselor and the new employee.

And here lies a problem: it may be very difficult for the counselor to disengage from the new employee. As a result, the latter often becomes so dependent on his counselor that even after he has been accepted as a regular employee he tends to bring his problems to his counselor rather than his foreman.

This dependency is a problem for the supervisor as well; he wants to be able to run his own shop. The program restricts his prerogatives of hiring and firing. The counselor keeps coming around to check on things—not on the supervisor's initiative, but on his own or the new employee's. The situation is more serious if it continues after the support system has been withdrawn. Now, the supervisor feels, the new employee should turn to him instead of to the counselor.

When, during my visits to companies, I mentioned this problem to anybody, I was berated for not taking the counselor's side. I could leave; pity the poor supervisor who cannot withdraw from such an attack!

Training facilities: Since a goal of the training program is quick adjustment once the employee enters the job situation, the training facility should provide a close simulation of reality. It should be at the work site or at least in a location that is highly work oriented.

Larger companies are finding that using one facility to serve several work sites can be advantageous. It provides flexibility to deal with fluctuating job openings and unforeseen cutbacks. But where the facility is removed from the work site, job linkage must be given special attention. A feeder operation is insufficient; giving the new hire a card and telling him to report does not bridge the gap between training and the job.

Eastman Kodak's Hands On center is a model of a good vestibule training facility. It is a real work situation, with provision for learning and the necessary possibility of "failing successfully." The trainee can try out his new skills, can learn to work with others in an industrial situation, and has reinforcement from experienced trainers. The center is located at the actual work site.

Some companies contract out certain parts of their training programs. Bankers Trust, by contracting out to the American Institute of Banking, avoids having to provide a special classroom facility for the remedial education portion of its program.

Training staffs: A company intending to conduct its own training must recognize that the trainers will need skills that are not usually part of their background. They will be dealing with a different kind of learner.

Attitude training is an absolute necessity in such a program. Management has to ask itself, "How competent are our trainers in dealing with attitudes when the middle-class background factor is absent?" Too much of our attitudinal training is based on a middle-class set of values. So a company must have trainers who are familiar with other sets of values and know how to accommodate to them.

Remedial education—literacy training—is best conducted by experts. And we have too few of them. The market is flooded with material, but almost all of it is unevaluated. A company's regular training staff is not equipped to make quality judgments of it. The local board of education is a resource often overlooked, and a few companies and at least one foundation (the Mott Foundation of Flint, Michigan) not only provide good materials but will also contract for the remedial work itself.

Therefore, a company does need a specialized training staff, but it is possible to develop one within the existing group. The trainers should be offered some incentive to become involved in an endeavor that is not part of their normal work.

Concluding Note

Whether to get involved in employing the hard-core disadvantaged is a decision that is becoming a restricted choice for many companies.

Some have been forced to by government edict, by community pressure, or the feared escape of markets. Some have turned to this labor source because of manpower problems; the banks of New York City, to cite one example, have a reported average of 1,000 vacancies a week, and some banks have been obliged to turn to Harlem and other previously untapped areas of the city.

Many companies, of course, have undertaken this kind of program for more altruistic reasons.

Whatever the motives, more corporations in this decade will be discovering that merely hiring and training these "unemployable" men and women is not enough. These suspicious, nearly defeated persons must be helped to achieve a significant change in attitude before they can take their places as self-respecting, full-fledged members of the work force.

So these companies will be obliged to embark on a frustrating, time-consuming undertaking which, if carried out painstakingly, can yield rich rewards. In trying to build support systems in their organizations, they have little corporate experience to guide them. In this article I have indicated the approaches which some companies are taking.

This is not a one-year or two-year program, but one lasting several years. By the end of this decade I hope we can look back on a series of successes by industry in providing an avenue for the hard-core disadvantaged to enter the world of work.

Labor 1970: Angry, Aggressive, Acquisitive

Richard Armstrong

Deep forces are gathering that could make the coming year a time of epic battle between management and labor. Some see a time of testing for the economy, a shaking and trying of the structure of collective bargaining, as in strike-riddled 1946, or 1952. Beginning with the electrical manufacturers this month, and on through trucking, meatpacking, rubber, and automobiles next year, five great industries must reach new three-year contracts. Just as the companies begin to prepare for an expected slowing in the economy, and an expected crunch on profits, they will be met with wage demands that are rocketing upward along with inflation. Such a combination of economic events is unusual, and dangerous.

Union leaders are free with strike predictions. The Rubber Workers, whose contracts expire next April, have built up a strike fund of $18 million, three times the sum that was at hand to mount a 101-day strike three years ago. "We think of it as a deterrent, like Nixon's ABM," says a Rubber

Workers' official, "but we'll use it if we have to." "Inflation is like an infection in the air," says a Teamster leader. "The members want all the good things now. They don't want to be rational where their pocketbooks are concerned." The erosion of wages by inflation is real, but at this point in the spiral, wage increases pump directly into still higher prices. A round of trade-union "victories" could present the President and the country with all the worst of two worlds: an economic downturn in which inflation persisted, stamped in by three-year contracts.

There is no assurance that even record offers will avert major strikes. Labor's top leaders have been buffeted recently by the general failure of an ambitious campaign to impose union coalitions on giant companies and whole industries, and this year the leadership split into two rival organizations. These men badly need to win a few.

Even where a contract is successfully negotiated, it may be rejected by the rank-and-file union members, who are in an acquisitive and rebellious frame of mind. The blue-collar worker is in the crosscurrent of social change, disgruntled about his bosses and "the system," and sensitive to the black-power revolution within the ranks of labor. One expression of all this is an unprecedented number of contract rejections in recent years. "A strike is not just economic, but political as well," says an economist at the U.S. Bureau of Labor Statistics. "Union leaders do not find it advisable to say so publicly, but perhaps they need a round of strikes to blow pressure and pull the membership together."

A recent round of outrageous settlements in the building trades has had an ominous impact all across organized labor. One contract that made newspaper front pages everywhere was that of the Brotherhood of Painters in Kansas City, which won a three-year deal that will increase wages and benefits by 67 percent, to $8 an hour. "I have been shaken in my time," says Louis Seaton, vice president for personnel at General Motors and a thirty-year veteran of collective bargaining, "but that Kansas City contract was a doozy." The victory of a few Kansas City painters impinges on the world of Louis Seaton because a mammoth settlement in Kansas City inflates expectations by the union rank and file everywhere. "Our guys don't just read the funnies," said a Teamster official. "They read about these big settlements, and then they're unhappy no matter how much we get them. Frankly, it's one of our main problems."

Where will the ceiling be found in the big industrial contracts over the next year? "Eight percent," predicts an economist at the National Industrial Conference Board. "Six is beginning to look like model restraint. These big industries can't, and won't, go as high as 10 percent." Union economists can get up past 9 percent at the flick of a pencil, merely by adding the long-term annual increase in productivity per man-hour (3.2 percent) and the current annual rate of price inflation (6.4 percent).

The size of all contract packages has been increasing steadily and this trend shows no sign of leveling off. The median raise in wages and benefits during 1967 was 5.2 percent a year, when averaged over the life of the contract. That figure rose to 6 percent in 1968 and to 7.1 percent in the first half of 1969. In the building trades, the median was a staggering 15 percent in the first half, which, with compounding, means a total increase by the end of three years of more than 50 percent. When contracts reach this size, the compounding effect is worth noting. In the last year of such a contract, the increase on the *increases* of the first two years would come to 4.8 percent, which is higher than the median for all union contracts three years ago. The Machinists Union recently won a settlement of 32 percent from the airline industry, a clearly inflationary package and the largest in the history of the union. Even so, Machinists at United and Continental voted the contracts down by large margins and held out for more.

These alarming signals bear with the greatest immediacy on the men at a long table in a conference room on the tenth floor of the General Electric Building on Lexington Avenue in Manhattan. General Electric, which sets the pattern for the industry, is negotiating with representatives from a twelve-union coalition headed by the International Union of Electrical Workers, with a deadline of October 26. G.E. has a long history of skillful and extremely tough bargaining, but this year it is feeling the force of inflation and of a tight labor market that is pushing wage rates upward almost everywhere. At the same time, G.E. is trying to foresee conditions a year from now, when the economy will probably have turned down and the competition for sales will be fierce.

There is speculation that the final offer may be in the range of 18 percent over three years. This would be a fat settlement for G.E., but would it be enough? "For the past ten years, G.E. has played the unions that it deals with against each other and used this whiplash to defeat collective bargaining," says Irving Abramson, general counsel for the I.U.E. "We have fallen far behind other industries, and in companies where we have real bargaining power, we have to fight like hell to break the G.E. pattern. This is the year of the showdown." I.U.E. President Paul Jennings warns: "This may be a knock-down, drag-out fight." I.U.E. officials are predicting a strike.

Where the Raises Went

To understand the intensity of rank-and-file sentiment, it is important to understand the economic case behind it. After an unparalleled economic boom, the average U.S. worker in private industry, by some calculations, is slightly worse off today than he was four years ago. According to the Bureau of Labor Statistics, this average worker with three dependents is grossing $115.44 a week now, compared to $95.80 four years ago. But in 1957–59 dollars, after federal tax and social security deductions, he is taking home only $78.49, compared to $78.88 in 1965. Most union members have done somewhat better, but not all of them. An average production-line worker at G.E.'s turbine plant in Schenectady, assuming that he has three dependents, is making $93.96 today in after-tax dollars of 1957–59, compared to $101.26 in 1965.[1]

One major reason that the union man's lot has not improved is that the leadership, in what by hindsight must be counted a major tactical error, gave up cost-of-living escalators in a number of contracts during recent years. "We had them and we bargained them away for other things," says one A.F.L.-C.I.O. official. "Now we're scrambling to get them back." "The cost of living had been relatively stable for some time, and we did not see this inflation in our crystal ball," says Secretary-Treasurer Cleveland Robinson of Distributive Workers District 65 in New York. The number of workers covered by escalator clauses declined from four million in 1959 to 2,500,000 last year. In a number of unions, including the electrical industry and autos, unions accepted what they call a "cap," or limit on the escalator. Under the cap in the G.E. contract, workers have had cost-of-living raises since 1966 totaling 2.75 percent, while the cost-of-living index has risen nearly 12 percent. The cap in the automobile industry contracts has recently been saving the companies more than $100,000 every hour the assembly lines run. Restoration of a full escalator clause is on a lot of union shopping lists.

"Millions of Moonlighters"

Some caveats are in order. There is reason to believe that the B.L.S. figures on worker income somewhat overstate the plight of the average worker's family. The Federal Reserve Bank of Chicago points out that per capita real disposable income rose by 3 percent in 1967 and again in 1968, well above the average of 2.3 percent over the past twenty years. Part of the difference in the two sets of figures can be explained by an increase in income from interest and dividends, which does not help the average worker much, and by the fact that the per capita income figure includes the imputed rental income of owner-occupied housing, which continues to rise. But there is also the fact that the proportion of married women holding jobs has increased steadily for a decade. These women add to family income but tend to depress the B.L.S. "average worker" statistics, since theirs tend to be lower-paying jobs. Moonlighting apparently has also increased dramatically and helps account for the disparity

---

1. Many government workers are faring better, which says something about government's failure to practice what it preaches. Federal employees have had raises totaling 21 percent since 1965, including a 9 percent raise in July. As labor leaders are fond of pointing out, the "average" Congressman's pay went up 42 percent this year, and the salary of the President of the United States is up 100 percent.

in the statistics. "There are millions of guys hustling at some part-time job," says Nicholas Kisburg of Teamsters Joint Council 16 in New York. "It's the only way they can make it."

The B.L.S. figures do not take fringe benefits into account, and therefore understate total worker compensation. But, as Nathaniel Goldfinger, research director for the A.F.L.-C.I.O., points out, they do not take state and local taxes into account either, and these have been rising sharply. Some economists say these factors roughly balance each other. In Goldfinger's view, the beneficial result of the Kennedy-Johnson years was not improvement in workers' real spendable earnings, which was modest, but a massive increase in total employment. "The non-supervisory employee has been getting the short end of the stick all through the 1960's," he says, "and now the pressure from the membership is on."

Collective bargaining was never designed, of course, to protect union members from tax increases. And the federal surtax was expressly devised to remove money from worker pockets and thereby to ease inflationary pressures. Paradoxically, the higher tax rate and the increase in the social-security tax that went into effect last January 1 fuel inflationary psychology, since they contribute to the virtually universal feeling among union members that they have been standing still or falling behind. "Our history has been one of real wages moving ahead faster than prices," says Herbert Bienstock, New York regional director of the Bureau of Labor Statistics. "For the first time over such an extended period, there has been no gain in real income, despite the fact that the past four years have been a time of sharply higher money wages."

Labor rejects, with all the anger of the righteous, the theory that wage increases have been a major cause of the inflation. "Prices and profits were already on an upward climb before unit labor costs went up at all," says Carrol Coburn, research director for the United Auto Workers. As late as the beginning of 1966, unit labor costs in manufacturing were holding down near the 1957-59 level, while prices were beginning to rise. Most economists agree that in its initial phase the inflation was "demand-pull" and was the result of the Johnson Administration's fiscal and monetary policies, which caused the money supply to increase faster than the nation's output of goods and services, rather than labor-management wage actions. A number of economists, including the University of Chicago's Milton Friedman, argue that this is almost invariably the case: that wage increases are a result of inflation and not its cause.

Speeding up Inflation

"Certainly you cannot make out any case that what has happened to wage income is responsible for inflation," Secretary of Labor George Shultz told FORTUNE recently. "Fiscal and monetary forces produced the inflation—

government policy—and it can only be corrected by government policy. When you look at the figures on real spendable income and see that there has been no rise, you can't get too excited about preaching to people that they ought to restrain themselves. This is one of the problems. What the unions say is true. But is it relevant to the economic situation that will exist next year?"

The present problem is that, whatever started the inflationary spiral in the first place, wage increases have been *adding* to the spiral and increasing its rate of climb. And such extraordinary increases as those achieved by the building trades under monopolistic duress have added to its velocity. Paul McCracken, chairman of the Council of Economic Advisers, notes that last year, while productivity increased by about 3 percent in the private, non-farm economy,

The Dilemmas in Recent Wage Trends

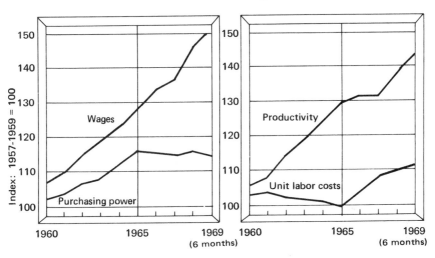

The union case for higher wages can be traced in the chart at left, which shows what inflation has done to purchasing power. All through the 1960's, the average weekly wage in manufacturing (a figure that includes both blue- and white-collar workers) has continued to climb steadily, from $89.72 in 1960 to $127.31 in the first half of 1969. But since 1965 this increase has been illusory, as the trend line for purchasing power reveals. The take-home pay of this average worker with three dependents (here charted in constant 1957-59 dollars) has actually declined slightly, from $88.06 in 1960 to $87.21 today. Part of the wage increase has been going for higher federal tax and social-security payments, and the rest has been swallowed by inflation. (Fringe benefits are not taken into account in these figures, but neither are state and local taxes. Some economists say the two factors would about balance off.) The chart at right presents the problem as viewed from the management side. The productivity index, which in this chart is a measure of output per man-hour in manufacturing, with 1957-59 as the base, has increased by 36 percent since 1960. During the early 1960's productivity increased faster than wages, and the labor cost of a given unit of output actually declined, reaching a low in 1965 of 99.9 percent of the 1957-59 base. This is as it should be, if price increases are to be avoided. After 1965 wages went up faster than productivity, and until labor costs began moving up, adding fuel to the inflationary spiral.

labor costs per man-hour went up by 7.5 percent. Unit labor costs are soaring now because wage increases have been accelerating at the same time that productivity per man-hour in the private economy as a whole has actually been declining. This unhappy trend in productivity is apparently the result of such cyclical factors as the employment of marginal workers.

"You apply the pressure, fiscal and monetary restraint, and you see the change," says Shultz, a former professor of industrial relations and dean of the University of Chicago's Graduate School of Business. "But things happen with lags, and while the economy rearranges itself the picture looks confused. You see the change first in real G.N.P., where there is a flattening trend, a rate of increase that is now less than the long-term sustainable rate. As the pressure is felt in the marketplace, companies try to fight costs, and with a further lag this will lead to a rearrangement in the wage picture. Wages lagged on the upturn of inflation and will probably lag on the downturn." In the meantime, Shultz warns against "these 15 percent contracts floated off into the future . . . Just as it's possible to price goods out of the market, it's possible to price labor out of the market too, if you get the rates up too high and give a tremendous incentive to new technology, new forms of organization, and new ways of doing a particular job."

Much depends on the outcome of the Nixon economic policies. Profits have already begun to soften, as many companies find they can no longer pass along all of their increased costs in higher prices. Viewed from the standpoint of classical economic forecasting, there are some grounds for hope. If real G.N.P. holds flat through next spring and unemployment moves up from the present 3.6 percent to 4.5 percent, as *Fortune's* Business Roundup has projected, labor attitudes could change quite rapidly. "Even a few layoffs can have a dramatic effect," says Horst Brand, of the Bureau of Labor Statistics. "Rank-and-file militancy declines. There is a greater tendency to trust the leadership's judgment as to the best policy through the downturn." If the economy tempers itself in the manner prescribed, then the unhappy results of inflation might at least be spread through all sectors: lower profits, less than optimum settlements by unions, and a continuing but lessening rate of increase in prices. The unpleasant fact is that the country has never worked its way out of inflation without a recession in the postwar years, so nobody can be absolutely certain that it can be done.

Wages Come Out of the Market

Theoretically, at least, it can. "Higher wages are primarily the result of tightness in the labor market, rather than of higher prices," says Shultz. "Don't forget that 75 percent of the labor force is nonunion. You can't simply point to certain big union contracts and say that is the course the economy is going to take." With these few sentences, Shultz is summing up a great deal of recent economic study, which holds that while price increases may frame labor demands, the market for manpower determines generally what labor can get.

A number of economists have attempted to demonstrate the strong relationship between unemployment and wage rates with various models of the "Phillips curve," from A. W. Phillips' study of the British economy from 1861 to 1957. One such mathematical model, constructed by George L. Perry of the University of Minnesota, indicates that price increases, of themselves, account for only one-third of the increase in wage rates during inflationary times. The rest of the wage increase is a factor of low unemployment and relatively high profit margins—in other words, of the working of supply and demand on the labor market. The wage-price spiral, says Perry in his book, *Unemployment, Money Wage Rates and Inflation,* "is one that disappears rather than becoming explosive or indefinitely self-perpetuating."

In the 1958–60 period, Perry found, a 5 percent unemployment rate had about the same effect on wages and prices that a 4 percent rate would be expected to from his model, indicating the persistence of hard-core unemployment, then as now, and the difficulty of solving it without overheating the entire economy. This certainly agrees with the perceived facts in cities such as Detroit, where the recent 3.5 percent national unemployment rate seems like overemployment. "The companies are hiring anybody they can find, guys that haven't had a job in years, including men with criminal records," says a U.A.W. official. This is socially useful. What is distressing is that this hard core seems to exist so far outside the normal labor market that it is not pulled in until the market is extremely tight and an inflationary spiral has begun. Perry also found that a relatively stable economy is less inflationary than a fluctuating one, partly because prices tend to ratchet at the top of the cycle. This would indicate that a sharp recession to curb wage demands would be self-defeating in important ways and that Nixon's aim of "disinflation without deflation" is the wiser course.

As a feasible interim target, economists sometimes talk of the "four-two trade-off," meaning 4 percent unemployment and a 2 percent rate of price inflation. Perry finds this sort of trade-off related mathematically to wage increases no higher than the increase in productivity and to an 11.6 percent average profit on invested capital, which is just about the profit level that actually prevailed last year. One problem is the considerable time lag that Shultz mentions before the economy can rearrange itself. Roundup has projected an actual trade-off of 4.5-3 by late next year. The other problem is that a mathematical model can be knocked askew, at least in the short run, by human passions; and a successful trade-union crusade for three-year contracts at the 25 to 30 percent level could take the economy years to work off.

Management does not plan lightly to go "floating off," in Shultz's phrase, into this kind of future. But business leaders have been "strangely silent," says

Brand, "partly because they are waiting to see what the economy is going to do. They haven't decided where to draw the line, where to stand and fight." Shultz notes that "this is an extremely difficult time for negotiations, such as G.E.'s, when the economic outlook is in the process of change."

Secretary Shultz sees some merit in the front-end load, a type of contract that has been popular this year, in which the bulk of the wage increase comes at once, with smaller installments in the second and third years. "The front-end load can reflect the wage situation that actually exists, the price increases that have already occurred, without presuming that inflation will continue at the same rate into the future." There is some talk at G.E. of a one-year contract offer that would have this effect.

Revolt in the Ranks

Both management and the leaders of labor will need maximum ingenuity in the coming year not only to arrive at contracts but also to sell them to the membership. The rank and file is being swept by a tide of angry revolt, against management, against its own leadership, and in important ways against society itself. Shultz warns that rejection of contracts by the rank and file "may lead to a great deal of difficulty," and he counts on the leadership to serve as "a stabilizing force." The question raised by some company officials is whether present union leadership is vigorous and effective enough to do so.

The only available statistics for contract rejection are for negotiations in which the Federal Mediation and Conciliation Service takes a hand. In these negotiations, contract rejection rose from 8.7 percent of the cases in 1964, to 10 percent in 1965, to a striking 14.2 percent in 1967, tapering off since then to about 12 percent. The most commonly stated cause of rejection, according to a survey taken by the federal mediators, was dissatisfaction because of higher settlements reached elsewhere—those building-trades contracts again.

"Our society rewards groups that put the pressure on," says Ronald Haughton, professor at Wayne State University and a federal mediator. "The rank and file have learned this. Usually, when they reject a contract, the company offers something better." Professor Thomas Kennedy of the Harvard Business School fears that the rejections are becoming "part of the bargaining, part of the theatrics of labor negotiations." A number of companies, he says, now hold back part of their package to offer as a sweetener after the contract is voted down. "The unions have let power pass to the locals and now they are stuck with the results," says an auto-company official. "The stubborn guy, the irresponsible local leader, has the power of the whole organization behind him."

The revolt is extremely trying to society, and to the men on each side of the bargaining table. "We have officials dropping dead with heart attacks at forty-one and fifty-two," said a Teamster leader. "Every executive committee meeting begins with one minute of silent prayer for somebody or other. Look at this stack of local demands for the new master freight agreement next March." The stack was two feet high. "The rank and file have shown limitless desires, and when a local president doesn't deliver they kick him out."

"You can sign a decent contract with the international and still have half your operation closed down by local strikes," said Seaton of General Motors. "In 1967 we began negotiations in July, signed a contract in December, and didn't clear up all the local problems until the following July. We were in the trenches for a year." Local demands, which totaled 11,600 in 1958, grew to "a mountain" of 31,000 in 1967. This year, in reorganizing some of its assembly plants, General Motors was hit with a series of local strikes that cut sharply into sales and profits. "The 1970 negotiations? I don't know," said Seaton with a sly smile. "We don't know what the most pressing human problem of our time is because Walter hasn't told us yet."

At the Point of Wrenching Change

To some degree, rank-and-file militancy is the product not just of inflation but of a deep and generalized anger and frustration among blue-collar Americans at recent social trends—black demands, the welfare rolls, the tax burden, rioting students. "Nobody has paid any attention to this guy," says Kisburg of the Teamsters. "We've all been too busy studying the black worker." "By and large it is the blue-collar worker who is physically at the point of social change in this country," said Goldfinger of the A.F.L.-C.I.O. "Integration of schools, jobs, neighborhoods, poses problems for him that the elitist, managerial crowd knows nothing about." Kisburg predicts that these deep and sometimes unspoken feelings may explode in "more strikes, more runaway locals" this year.

Some rank-and-file unrest stems from a theoretically laudable source: there has simply been an increase in trade-union democracy in the ten years since the Landrum-Griffin Act established the secret ballot and fairer campaign practices. "The same management types who were screaming about 'the overlords of labor' back during the McClellan hearings," says Donald Wasserman of the State, County, and Municipal Employees, "are the ones screaming now about labor anarchy. If you let union members vote, you are going to have to expect them once in a while to vote no." "The members, God bless them," says Kisburg, "just will get worked up now and then and pass some pretty hair-raising resolutions."

The youth explosion is shaking up labor, and this of course is a force that will build instead of wane. President Peter Bommarito of the Rubber Workers says that more than a third of his 208,000 members are under twenty-five. About 40 percent of the hourly employees in the automobile industry have less than five years' service, and according to one union official, "they couldn't care less what happened to Reuther at the Overpass in 1937." So far, at least, polls indicate that young white workers have little interest in the radical ideas popular on campuses. A G.M. official says that the young employees are remarkably well-educated and undisciplined. Casual absenteeism has doubled at G.M. plants in the last five years. Union leaders complain that the young take past union victories for granted and want fatter paychecks, now.

The black revolution is an even more explosive force within the unions. "Organized labor and the black community throughout America are on a collision course," warns Herbert Hill, labor director of the National Association for the Advancement of Colored People. Hill is gradually building up a new body of case law, knocking down racial barriers in building-trades locals and segregated seniority lines in industries such as steel. But he says that "the legal departments of many A.F.L.-C.I.O. unions are behaving like the legal departments of southern school boards, introducing a tangled web of procedural legal questions, forcing us to contest plant by plant, local by local." The top rank of union leadership is virtually lily-white, and militant "black caucuses" have formed in a number of unions, including the Steelworkers and the Garment Workers. The "Dodge Revolutionary Union Movement" (DRUM), which led wildcat strikes at a Dodge plant in Hamtramck, Michigan, last year, has raised high hopes in the New Left with its Marxist rhetoric and has inspired imitators at several other auto plants. Black leaders warn that they will attempt to shut down construction all over the country, as they were able to do recently in Pittsburgh, until they get a fair share of these high-paying jobs. Black workers will inevitably win trade-union power more in keeping with their growing weight within the unions. Hill warns that there will be "more conflict, more interruptions of production" until that happens.

Split at the Top

Have the aging leaders of labor lost their grip? They will gather in convention this month to take the sea breezes, in a familiar scene at Atlantic City. George Meany, still the supple bureaucrat, is getting to be something of a geriatric marvel as well. He was seventy-five recently and told reporters: "There is no such thing as retirement in my plans."

The two largest labor unions in the U.S. will not, however, be represented at Atlantic City. The 2,100,000 Teamsters were expelled from the A.F.L.-C.I.O. in 1957, and last year Walter Reuther quit, pulling the 1,800,000 Auto Workers out of the federation he helped create and had dreamed of one day leading. Reuther was never popular with the labor brass, and he was no match for Meany at internecine intrigues. His support within the A.F.L.-C.I.O.'s twenty-nine-man executive council had dwindled to about five members, while his bitter personal feud with Meany intensified. "Walter was getting a little shook," according to the perhaps unkind assessment of an A.F.L.-C.I.O. union official. "He will have to retire at sixty-five, in three more years, if he sticks to the U.A.W. constitution. He felt like doing something, even if it's wrong."

Last May the Teamsters and Auto Workers called together a club of their own, the Alliance for Labor Action. Its program, approved by 600 delegates in the main ballroom of the Washington Hilton Hotel, was all that Reuther's activist heart could wish: community unions in the slums, a private antipoverty program, including job training and low-cost housing, and a massive organizing campaign in low-wage industries, especially in the South. The A.L.A. also pledged itself to "repair the alienation" of youth and the intellectuals in order to build "a new alliance of progressive forces." The A.L.A. perhaps is helpful to Teamster General Vice President Frank Fitzsimmons, who is trying to establish his own, respectable image now that Jimmy Hoffa is off in federal prison in Lewisburg, Pennsylvania, under two sentences totaling thirteen years. What new opportunity the A.L.A. would open for Reuther is difficult to see. Meany points out that Reuther was chairman of the A.F.L.-C.I.O. organizing committee for years and thus is now criticizing his own record. When asked what the A.L.A. has been up to since the inaugural convention, a U.A.W. official reflects for a moment and says: "Well, it's not going to be a *splashy* organization, you know."

Reuther's departure will weaken labor's drive toward what is known as coalition bargaining—multi-union negotiations designed to impose a standard contract and expiration date on a large, diversified company, or across an entire industry. Unions have looked to this drive as their next great campaign, their answer to the conglomerate. Companies where coalition bargaining has been tried have fought it tenaciously, and so far with some success. After a nine-month strike in 1967–68, unions seeking to impose a coalition on the copper industry gained only a narrowing in the spread of contract expiration dates. The strike exhausted the Steelworkers, who led the coalition, and demonstrated industry's determination not to give unions this massive new bargaining power. Union Carbide and several other companies have defeated coalitions by taking expensive strikes.

Much of the impetus behind the coalition approach—researching possible targets, getting the unions together—came from the A.F.L.-C.I.O.'s Industrial Union Department, which was headed by an able Reuther lieutenant, Jack Conway, who has now departed to become president of the Center for Community Change, a Washington foundation. The U.A.W. joined in a coalition against G.E. in 1966 that won partial success when the National Labor Relations Board ruled that G.E. had to admit all the unions to the negotiations, even though only one contract, that of the I.U.E., was actually under discussion. The U.A.W. is back in the G.E. coalition this year. But with labor split at the top, new coalitions will obviously be more difficult to form. Simply because the economic stakes in joint bargaining are so enormous, this may be the most significant result of the new division in labor ranks.

Room for Leadership

George Meany leaves no doubt about the bad blood between the two groups. He has threatened suspension or expulsion proceedings against any union that joins the A.L.A. A Teamster official in Washington finds that old friends in the A.F.L.-C.I.O. have been "cutting me dead" since the formation of A.L.A. The existence of two labor fronts may well increase competition for contract victories in the coming year. Certainly it will absorb a lot of energy that could be used on the day-to-day problems of constructive leadership.

One veteran federal mediator, Walter Maggiolo, says that in a significant number of cases labor unrest can be traced to the fact that union negotiators get so caught up in the big battle, the size of the total settlement, that they ignore genuine local grievances over working conditions that are as important to the men as wages. He advises negotiators to pay more attention to these local problems. He also advises that whenever possible the final settlement be presented publicly as management's agreement to final union demands, rather than vice versa, so that the union can claim "victory"; that employers insist on unanimous agreement from union negotiators; and that they use their employee communication networks with maximum effectiveness to help unions with the increasingly difficult job of selling contracts to the rank and file. Such skillful procedure could well lessen, although certainly not eliminate, the twin danger of strikes and inflationary settlements in the year ahead.

Ideological Obsolescence in Collective Bargaining

Jack T. Conway

The question I have undertaken to answer in this paper—Can Collective Bargaining Do the Job?—can only be answered prophetically. I cannot cut the future to the measure of my predictions. I can only try.

When it was first proposed that I discuss this question of the relevance of collective bargaining, the New York newspaper strike was still in progress. Major strikes on the docks and the railroads were in suspension at the moment. Editorials in the newspapers throughout the country were suggesting these crises indicated that collective bargaining and unions were obsolete, without recalling that before these activities became obsolete in their judgment, they

An address to the Industrial Relations Conference, Institute of Industrial Relations, University of California, Berkeley, 1963. Used by permission.
Mr. Conway has held several union and government positions. At present he is with the Industrial Union Department, AFL-CIO.

had routinely offered other reasons for doing away with them. The *Wall Street Journal* had, over a period of time, published articles, news analyses, and stories whose central theme was that technological change in many industries, especially the newspaper industry, but not excluding transportation and manufacture, had gone so far and so fast, that the traditional forms of enterprise were obsolete, the traditional skills were obsolescent where they are not already obsolete, and that in a normal lifetime men and women could naturally expect that they would have to acquire two or three competences to keep current with developments, if they could keep pace at all. It was suggested that unions had already passed their time, and that it was now necessary to devise other institutions and methods for performing the service which unions once provided.

The *Wall Street Journal* is too useful a newspaper to be singled out for pillorying in this connection. Actually it only accurately summed up the spate of views and observations which are still appearing in magazines and books, that are skeptical not only of unions, but of the government, and of people. There are suggestions, for example, that human beings are also superfluous and that the task now is to reorganize the society of man to exclude people.

These judgments, which come from the ringside and not from the ring, while fashionable, are not profound, and in my experience, embody no great wisdom. They are manifestations not of institutional obsolescence, or human obsolescence, but of a combination of ideological obsolescence and personal senescence. In the case of unions, for example, the people, frequently relics of the 1930s, who are publishing articles which purport to describe the demise of the labor movement are using exhausted concepts and illusions which blind them to what is actually happening today.

What has been written by these people about my own recent career in a sense is an anecdotal illustration of the inadequacy and irrelevance of this sideline and uninformed interpretation of the evolution of the labor movement. When I left the UAW two years ago to work in the Administration in Washington, it was said, falsely, that I was disillusioned, that I was part of a parade of people leaving the labor movement because it had come to a deadend where there was no longer any function for collective bargaining. Recently, after two very satisfying years in a Federal Agency, when I returned to work in the labor movement, the same people said, "Aha, he is disillusioned with the government."

What this incident, and altogether too many other comments on current events, reveal about these professional intellectuals who proclaim their concern about society is a shallow cynicism in connection with the very agencies and instruments, governmental and nongovernmental, that are operating most effectively to deal with injustice, inequality, and insecurity.

Government, parties, and the labor movement are decried for going too far, not going far enough, for being timid, or arrogant, for overriding individual rights, or for paralysis, for activism, or inactivism.

Moreover these denigrations characteristically carry an emotional charge and almost invariably embody a moral judgment. Not only is it said that collective bargaining is inadequate to some purpose that is never defined, it is also implied that the people who engage in collective bargaining are committing an undefined wrong, or are engaging in something that is unspecifically immoral, and are betraying some ideal ethical principle which once distinguished the labor movement in better, happier times.

These expressions of outrage, of course, mistake the essential nature of collective bargaining, which is a neutral operative activity that is used socially as an instrument for arriving at particular kinds of decisions. A complex apparatus which has evolved in this country over one hundred and fifty years, it incorporates the experience of at least six generations of American employers and employees and works rather well when it is operated and maintained competently. Like any piece of machinery, whether it is a court or a car, the performance quality depends upon its operators. Collective bargaining is neither magic, or automatic. For the best results, it demands intelligence, integrity, courage, and work on both sides of the bargaining table, and a reasonable regard for the traffic rules which regulate economic flow.

Judgments of the success or the failure of bargaining, and there are comparative failures from time to time, as there are successes, imply a measurement against some standard, which most often is called public interest. Unfortunately, the public interest is nowhere precisely defined. Indeed, as the history of labor legislation in the United States demonstrates, it has been impossible to define, at least until now.

Even if there were an adequate, limiting, and accurate description of the public interest, the specification would not be a criterion for judging the effectiveness of collective bargaining. What is better or worse in a particular situation depends not on comparison with an ideal solution but on the contrast with the actual alternatives. It is for this reason, that when a negotiation is over, the people on both sides of the bargaining table cannot guarantee that the agreement reached was an ideal solution. All they can reply to critics is that they came up with the best solution they were able to reach. These imperfect decisions have served and continue to serve the country rather well, in spite of occasional crises which have their own uses and value.

At this moment, collective bargaining has a larger claim to vitality and utility than ever before in the history of the American society. Neither side of the American bargaining table is haunted by an uneasiness over what has been called the end of ideology or the exhaustion of the uses of collective bargaining. In the American unions, it has been estimated that there are about 250,000 men and women in leadership roles and there is no credible indication that these people believe that the labor movement has entered into a decline or that it has become universally infected by cynicism or corruption. Similarly, it is obvious that the extent to which people—wage earners and employers—in every community in the nation are daily engaged in some form of collective bargaining and are by their efforts giving direction to the economy represents a total refutation of the silliness that this movement of some sixteen or seventeen million American wage earners is barren of leadership.

Taking into account the fact that bargaining also comprehends the people on the employer side of the negotiations, it is not an exaggeration to say that bargaining is actually the most widely pervasive democratic activity in the nation and is more widely accepted than ever before in American history. This development alone indicates that it is nonsense to suggest there really was an inspiring labor tradition in some glorious past which somehow has been ground to dust and lost to this generation through the operation of an iron law of bureaucracy.

Most of these capsule summaries of the situation in the American labor movement are captious, capricious, and, in reality, anxiety projections by people who are describing something within themselves rather than the exciting—even though disturbing—variety of developments in the factories and workplaces as well as the neighborhoods and the legislatures of the nation. They recall the Kentucky politician whose constituent conceded that the candidate had gotten the man a job, had arranged for a variety of services for a long list of relatives, but still asked what had been done for him lately. Actually, the labor movement has been performing and is providing an astonishing assortment of essential services to the American people not only lately, but now.

But the actuality should not be exaggerated either. No one in the labor movement would insist that the American unions are without fault, that union leaders all qualify for the Sir Galahad medal, or that the unions are the front ranks of a marching band entering paradise. But what organization or movement in the United States does meet this specification?

If you examine what the American labor movement has accomplished in the last fifteen years, it is impossible to talk of the end of collective bargaining and to speak intelligently. Paid holidays, improvement in vacations, pensions, the establishment of bargaining as an operative feature of the economy, the extension of health and sickness insurance to a majority of union wage earners, the virtual end of violence on the industrial relations scene, the supplementary unemployment program, the adoption, however imperfect, of ethical practices standards for the labor movement, the genuine advances being made against discrimination; very lately, the provisions dealing with automation, the new sabbatical vacation plan set forth in the Steel agreement, the cost-sharing plan between the union and Kaiser Steel, the American Motors Progress Sharing Plan with the UAW, whose innovating features have not yet been properly understood; and most recently, the inauguration of what could be a profound alteration in the collective bargaining process itself in steel and in the automobile industry. There is hope that a year or two from today it will be possible to add to this list the realization and the mastery of new organizing methods appropriate to the demands of the changed industrial topography. Nationally, unions, utilizing a variant form of collective bargaining have had some role in the enactment of certain features of the Trade Expansion Act, of the Area Redevelopment program, and of the Manpower Retraining Act. Except for the initiative of the labor movement, the Fair Employment Practices effort, however far it is from where it should be, would not be operating as well as it is, and the civil rights movement would be more nearly a whisper in the distance, instead of the most challenging development in the nation today.

So, it can be expected that people will rejoin to this recital, sure you have done all these things, but how come you have not established equality in American life, ended nuclear testing, and restored democratic government to Cuba? Nor is this anticipation facetious, for actually, it is clear that most of the criticism of the American labor movement today is based on what has been called the Free Tom Mooney fallacy. During the historic depression it was reported that a tenants' committee met with a landlord over a list of demands and that the landlord finally said he would paint the halls, reduce the rents, turn the heat on at seven o'clock instead of eight, and provide new gas stoves, but, "Tell me," he concluded, "how can I free Tom Mooney?"

When the alienated and rather well-provided intellectuals who today choose the American labor movement for their target (this is not a reference to the people in the civil rights movement who are making proper, if uncomfortable, demands on unions for far more speed and far less deliberation), when these critics aim their spitballs at the unions, essentially they are looking at bargaining and other union activities through unfocused glasses, under a number of disabling misconceptions.

Actually the blur and the badly aimed spitballing are due to the multiple roles unions play in our society:

Their primary activity as bargaining representatives of the people in a particular workplace; their activity as a moral spokesman in the community for wage and salary workers, and the dispossessed, and the alienated, and the persecuted, and the mistreated, and the victims of whatever automatic social trend happens to be running; and their activity, as a political institution, not a party, rather an effectively organized, but by no means omnipotent, interest group, but nevertheless a big one.

While in practice, unions as collective bargaining agents can be influential factors in the determination of wages, hours, working conditions—conditions of employment in the enterprises where they have been standing—unions in fact have no effective control, and not very much influence at any particular time, over the operation of the total economy or the broad society. Unions are never in a position to Free Tom Mooney, about the most they can do is to join other people on a petition.

Unions as bargaining agents, for that matter, have little control over the context of the bargaining situation, that is, the movement of the economy, demand changes, alterations in the society, which can and do change the nature of bargaining, and have in the past in some cases simply eroded the basis for bargaining.

Critics of unions, failing to distinguish among the various functions of unions and the varying powers of unions:

1. make demands on unions they cannot possibly fulfill, and

2. criticize unions for failures in the society or in the economy that are no more the responsibility of the unions than they are of the church, or the Congress, or the President, or the companies, or history itself.

Critics thus ask of unions what they intend to do about:

a. poverty in the United States, that is, the 20 per cent of the people who live outside the national economy, and another 20 per cent who, while they live in the community, live largely as the deprived and not as participants,

b. discrimination in the United States one hundred years after the Emancipation Proclamation,

c. automation and the accelerating technological displacement which is discharging workers from the production process at one end with the speed that doughnuts, or engines, or soft drinks, or cans are released at the other end, and

d. the disappearance of traditional democratic activities in favor of institutional and bureaucratic procedures as the pressure of size and technology move the nation toward computer, vending machine, and information storage methods.

Nowhere in the society is there more concern with these developments than in the labor movement. In most of these areas, the labor movement has been primarily responsible for making the original demands and proposals for dealing with these failures in the society. But what unions are asked to do in connection with these problems is far beyond the competence and the powers of the labor movement, and in some cases beyond the competence and powers of any society or government that is known. Here, it should be acknowledged the demand on the unions to be more effective and to exert powers they do not have also comes from union members, for understandable reasons.

Yet, while unions are taunted for failing to undertake tasks which would require far greater powers, responsibilities, and bureaucracies than anyone has ever proposed for what are essentially voluntary organizations, simultaneously they are targets of a contradictory charge. Ignoring the changes taking place in the big society, which have compelled unions to institutionalize and organize their activities, people who cherish memories from a time when unions were very often only protest groups, insist that the labor movement return to a primitive state of informal grace that survives nowhere else in the community.

Unions, like any other major human activity, need criticism, from inside and outside; the more important they are, the more criticism they need. But some of the criticism that is misdirected at the unions tends to interfere with specific indictments which should be aimed at other institutions. It is easier, for example, to direct fire against unions for their participation in discrimination, than against the companies responsible, or the cities, or the political agencies. Similarly, it is easier to ask what are unions doing about automation than to call attention specifically to the failure of companies in some instances to incorporate plans for the human use of human beings in their long-range programs, or to the failure of the government to implement the public policy set forth in the Employment Act of 1946.

All this having been said, it should be acknowledged that unions, as bargaining agents, as quasipolitical activities, as institutional centers for agitating on moral grounds, are only now emerging from one of their less creative and less responsive periods. In partial extenuation, reference can be made to the difficulties of accommodating to federation unity and to drastic economic and social changes in the society; union leaders, unfortunately, often tend to share the smugness and the complacency of our business-oriented society itself. But no one should fail to take into account the possibility that if the labor movement should slip back into unimaginativeness, into a lack of resilience (which is not likely because of political pressures which will not tolerate immobility), it is indeed conceivable that it will play a less important role in the future.

But even in this circumstance what the labor movement now does—successfully sometimes, less successfully at other times—needs to be done, and will be done, in one institutional form or the other.

Actually, however, the labor movement is beginning to make the adaptive changes that will enable it to contend with those problems with which it is competent to deal and to associate itself with other groups in the society to deal with those problems with which it is concerned but not capable of solving by itself.

No discussion of this question should fail to enumerate the very many different forms that collective bargaining assumes today, involving it in concerns, activities, and decisions which nowhere have been described adequately, certainly not in the texts and treatises dealing with collective bargaining. Indeed what actually happens is complex and varied to a degree that falsifies the conception most people hold of union negotiations.

For example, as the result of the advances made during the last two decades, bargaining proceeds less and less frequently (as in the newspapers) in a hotel meeting room, at midnight, two minutes before a strike deadline, over a new contract. Bargaining goes on continuously in the joint committees which administer the pension programs, the supplementary unemployment benefit funds, and the medical, hospital, and sickness funds. During recent years unions have, it has seemed, been negotiating on a five-day-a-week basis with the Blue Cross-Blue Shield state organizations and other suppliers of health services. Union committees have been compelled to negotiate on actuarial and related problems with the insurance companies. Funds created to deal with personnel changes arising from automation have provided a new bargaining place, have presented opportunities for unprecedented new research into the possibilities of the bargaining process, and have created the need for union leadership training of a kind never attempted before in the labor movement.

But although bargaining is an expanding activity, it certainly cannot do the job, if what is understood to be the job is the provision of full employment, job opportunities for young people, and true security for wage earners throughout their lives. These are public policy questions which require action *outside* the essentially bilateral bargaining process. However, within the limits of what can be done by dealing directly with employers or with associations of employers, something significant and enduring will be established.

Unquestionably, within the next few years collective bargaining will routinely begin to concern itself with the investment of the enormous sums of money which are accumulating in the pension funds, in the SUB funds, and, most recently, in the allocation of money from the progress-sharing funds of the American Motors Corporation or from the economies realized under the Kaiser contract.

This very hasty survey of where collective bargaining is today is intended only to indicate that collective bargaining is not a bankrupt enterprise, but that it has evolved into one of the critical decision-making activities in the economy. Moreover the simpler, less evolved forms of face-to-face collective bargaining will also undergo decisive changes that will be dramatic even if they do not produce picket lines, strike violence, and shrieking headlines. Here, of course, I refer to the move in steel and autos—and it will not stop in these industries —to take bargaining out from under the headline and to make it a cooperative inquiry into economic advance and adaptation. Already the table pounding, the violent language, the masculine vocabularies which were a hallowed if slightly ridiculous feature of the bargaining process of yesteryear have gone by the way.

But beyond the bargaining within plant limits and with specific companies and groups of companies, unions and employers must inevitably—and soon— begin to engage in a three-sided negotiation over the operation of manpower training programs, or tariff problems, and eventually investment and fiscal policy as they relate to jobs and full employment. Moreover, when full employment is an established and continuing feature of our national economy as it must inevitably become, bargaining will necessarily continue within plants and with companies in the customary dialogue, but the crucial decisions will probably be made regionally and nationally as is beginning to be the case in England, in France, and less obviously, but with far greater effect, in Sweden.

Ultimately and inevitably under full employment conditions in the United States, the traditional demands of the union movement will undergo transformations. Already in the highly developed European countries—as in the United States, for that matter—union members with union responsibilities do not conform to the public picture of them as men walking up and down in front of a strikebound shop or factory, with or without picket signs, crying unfair. No one has counted the many thousands of workers in factories and shops and offices whose union duties require them to meet with city, state, and national governments over complicated technical problems which bear vitally on the operation of the society, employment, health administration, housing, city planning, traffic, investment, education; the list of items on the agenda includes every community concern. What should be kept in mind is that the men and women who are now engaged in making these decisions democratically in the community and the broad society only a generation ago were hired hands, people across the railroad tracks with no rights in the plant, no voice in the community, and no meaningful vote in the large social questions which shaped their lives.

Out of this almost invisible but dramatic process will come new institutions in our society, which, far from isolating and alienating human beings, will restore them to the governing process and give more and more of them increasing responsibilities. The new agencies will generate new politics whose issues will be alternative programs for full employment, alternate ways of investing to maintain the dynamics of a growing economy, alternate devices and agencies for facilitating the movement from a manual, largely uneducated working class to an educated society of wage earners. Here one statistic invites examination: in 1930, for all the jobs available in the economy, 32 per cent were unskilled and 25 per cent were either semiskilled or service jobs. Thus 57 per cent of all jobs required relatively little education. By 1970, it has been predicted that instead of a demand category of 32 per cent for unskilled labor, only 5 per cent of available jobs will be unskilled; from 25 per cent, the proportion of semiskilled and service jobs will have shrunk to 21 per cent. Seventy-five per cent of all jobs available will, we are told, require a rather high degree of education. These developments have a significance for the society which is by no means confined to their impact on collective bargaining or the structure of the trade union movement. Trade unionists, however, cannot help but wonder at the influence these developments will have on the structure of unions, local and national, and on the types of personality possessed by those who will become union leaders in the future. In any event, the union structure, the union processes, and union activities, which have already changed more than is generally acknowledged, will inevitably undergo a further dramatic metamorphosis.

For we are, above everything else, a dynamic, mobile, and protean society. The rate of change in the technology, however much it confounds the people caught in the process, does produce equilibrating accommodations. Even when the intellectual reaction to change is apparently out of date and seems to be an effort to solve today's problems with yesterday's answers, the community, making use of a common experience, does adjust and emerge, painful though the adjustment may seem to be.

When one describes the growing participation of the union in the industrial and social economy and the evolution of collective bargaining into an increasingly bureaucratic activity (for the very reason that the society itself is increasingly organized rationally), one runs the risk that critics will be encouraged to cry out, "I told you so, unions *are* becoming one vast bureaucracy."

But the administering bureaucracy that has necessarily evolved in the union is not the bureaucracy critics talk about. They complain, and their complaints would be valid if their description were accurate, of a deadening, self-serving, insensitive bureaucracy lodged comfortably in affluent recumbency on the backs of wage earners. What actually operates in the unions, however, is a rational reconstruction of some features of the economic process to admit more democracy, more flexibility—so that if there must be crises they will not come from blundering, from stupidity, or from inefficiency, or because the proper briefs, or forms, or petitions have not been filed. In contemporary society, where measurements are in millions, hundreds of millions, and billions, there is no escaping the use of efficient administrative machinery. This is bureaucracy, but it is a creative procedural process that organizes and facilitates democratic procedures.

Admittedly, there are dangers from this development, which have not altogether been evaded in Sweden or Israel, two places where the evolution is well advanced. But if some bureaucratization in its invidious sense cannot be escaped, it can also be said that even tribal societies have their flyspecks. The movement will not be from an ideal society to an imperfect one, but rather from one kind of less efficient society to a more efficient but still imperfect endeavor. Bureaucracy will be easier to bear as an imperfection than unemployment or insecurity.

It has been said that politics is the art of the possible. This implies exclusion of the impossible but not of the improbable. Historically, the labor movement—both its political and economic wings—has on occasion played both "practical" and highly improbable politics; it counterpointed obviously possible demands against demands that were not obviously possible. It can be taken for granted that unions, which inevitably carry on their work under a number of countervailing political pressures, will not abandon the element of unreasonableness in their demands (even in the new society) which often make them so exasperating to employers, to governments, and to right-thinking academicians who know perfectly well that what the unions are demanding is impossible, or impractical until the demands are ultimately won. The agitator element in the makeup of the union movement will, as it has in the past, continue to preserve the morality of the labor movement, which is fundamentally not only the justification of unions but also a specific antidote against rigidity in the society. What seems unreasonable to employers may be the basis of moral responsibility to the rank and file.

Can Collective Bargaining Do the Job? If it can't, something else will have to be invented to do what collective bargaining is doing. When it is invented, what will come off the assembly line will not be a computer, but collective bargaining. Even employers who have sought to escape it, at great cost to themselves and the community, have discovered that.

More than anything else, now there is needed, not wailing about the bankruptcy of collective bargaining, or the exhaustion of vitality in the unions, or the end of ideology, but careful, closely focused examination of function and responsibility and potentiality—an approach that is scientific in its integrity, moral in its responsibility to human needs, and practical without complacency in its realism.

How It All Started

Stanley L. Englebart

In the beginning—to borrow a phrase—there was the abacus. This little device came into being some 2,000 years ago and still is the most widely used calculator on earth. In vast areas of Asia it is the only known counting device. Here in the United States it is occasionally seen at work in Chinese hand laundries—but more often only as a decorative item on baby playpens.

The word "calculation" itself comes from the earliest form of abacus which consisted of lines drawn on the ground, with small limestone pebbles to represent numbers. The Latin word *calcis* means lime or limestone and the Latin word *calculus,* which grew out of it, first meant a small piece of limestone and later was expanded to mean any pebble used in counting. Today such words as "chalk," "calcite," "calcium," "calculate" and "calculus" (which, in medicine, refers to a kidney stone while in mathematics means an algebraic technique) all stem from the stones used for counting thousands of years ago.

The abacus (from *abax,* an ancient Greek word for slab) was a direct result of early efforts to count. When primitive man satisfied his needs for food and shelter, he began seeking ways of expressing himself. His first "writings" were the rudimentary drawings we find on cave walls. The earliest were simply representations of what he found in nature—the sun, the moon, the animals he hunted. Soon, however, he wanted to express "how many" animals he had killed in a hunt, "how many" children he had, and so forth. Thus, the development of symbols to indicate "one," "several," and "many."

The next step was a big one—devising symbols to express specific quantities. The first two were, quite naturally, a "two" and a "five": "two" because man had two hands and "five" because he had five fingers on each hand. And by combining the symbols for hand and fingers, he could express many different specific quantities.

The abacus makes use of this two-five or *biquinary notation* system. The Chinese abacus or *suan-pan,* for example, consists of a series of rods and wires on which beads are strung. There are seven beads on each wire—separated by a divider into a set of five on the bottom and two on top. Thus, the number "seven" can be expressed by one bead from the top (equalling five) and two from the bottom.

In the hands of a skilled practitioner, the abacus is an amazingly fast and versatile calculating device. Shortly after World War II, a Japanese bank clerk in Tokyo was pitted against a G.I. clerk in an abacus-versus-desk calculator contest. A host of addition, subtraction, multiplication and division problems were thrown at the two men. But the speeding fingers of the Japanese won out every time.

But the abacus has its shortcomings—otherwise we'd all still be using it. It cannot carry-over tens from one line to another and, as man constantly expanded his mathematical horizons, this became an increasingly vexatious problem.

It wasn't until the 17th Century, however, that the next clear-cut advance came along. Blaise Pascal, a 19-year-old genius tired of adding long columns of tax figures at his father's office in Rouen, France, invented a gear-driven machine the size of a shoe-box on which numbers could be counted through a series of notched wheels. The machine could perform addition, subtraction, and had automatic tens-carrying capability. Results were shown through small openings or windows. And while it wasn't the most efficient device ever invented, it did herald a major step in the evolution of present-day computers.

Over the next half-century there were several new calculating machines invented, but they were primarily modifications or refinements of Pascal's original design. The next important development, in fact, had nothing to do at all with calculating but was extremely important to the advancement of the science. In 1780, a Frenchman named Joseph Marie Jacquard developed an automatic weaving loom which operated from instructions punched into cards or paper tape. The machine was so radical that few people would try it. But Jacquard exhibited his automatic loom at the Paris World's Fair and, as a result, sold many hundreds over the next few years.

The Jacquard punched card loom led directly into what is one of the most unusual and dramatic stories in the history of calculating devices. The year was

1822. Charles Babbage, a moody and often disagreeable young Englishman, asked the British government for funds to build what he called a "difference engine." In effect, Babbage had drawn up plans for the first digital computer. His projected machine could do complex calculations and print out results. There was to be a "memory" made up of the same sort of punched cards used in Jacquard's loom. Cards also would be used for input to the machine and control of successive operations. The machine itself was to have an arithmetic unit, called a "mill," in which to store data; and it was to be able to set up its own results in type, thus avoiding transcription errors.

Babbage was literally 130 years ahead of himself. He built a small working model of the machine but never was able to complete a full-sized device. The reason had nothing to do with his concept, however: the machine couldn't be built simply because the technology of the times was unable to construct the parts he needed. Babbage kept running into financial difficulties and, eventually, died a failure.

A few years ago a Dr. B. V. Bowden, writing in *Think* magazine, described his efforts to track down the Babbage story. Aside from the discovery of papers which proved again the genius of the rebellious Englishman, the trail led to the Baroness of Wentworth, granddaughter of Lady Lovelace who, in turn, was the daughter of famed poet Lord Byron. It seems that Lady Lovelace was as much a genius in mathematics as her father was in poetry. She devised, among other things, a form of binary arithmetic and an "infallible" system for predicting horse race winners. The binary system has stood the test of time and is today used in electronic digital computers. The betting system cleaned out the family fortune and, after her husband abandoned the system and her, forced her to pawn the family jewels.

Somewhere along the line Lady Lovelace met Charles Babbage. Apparently their mutual interest in mathematics was the spark. As Dr. Bowden wrote: "Lady Lovelace often visited Babbage while he was making his machines, and he would explain to her how they were constructed and used. As one of her contemporaries recalled: 'While the rest of the party gazed at this beautiful instrument with the same sort of expression that some savages are said to have shown on first seeing a looking glass or hearing a gun, [she] understood its working and saw the great beauty of the invention.' She worked out some very complicated programs and would have been able to use any of the modern machines. She wrote sketches for several papers, but published only her notes on Babbage, and they were anonymous."

Babbage, Lady Lovelace, Jacquard, Pascal and countless others all set the stage for the introduction of data processing techniques. But it took one man to crystallize this background into the practical devices that are at work today throughout the world. He was Dr. Herman Hollerith, a statistician from Buffalo, New York.

Unlike some of his predecessors, Dr. Hollerith was right in step with the times. He arrived on the scene as the U.S. government was staggering its way through the once-a-decade census count required by the Constitution. He recognized a need and solved it with the world's first data processing installation.

The basic problem was simple: the government had required seven full years to complete the 1880 census. All data had to be handwritten on cards, the cards broken down into various categories, and then each classification counted and re-sorted. It was a backbreaking chore involving millions of cards and by the time the count was finished—it was outdated. With the 1890 census staring them in the face and immigration swelling population ranks by the day, the Census Bureau could envision eight, nine or ten full years to take the next tabulation.

Dr. Hollerith worked out a mechanical system for recording, compiling and tabulating census facts. His system consisted of recording data crosswise on a long strip of paper. Information was indicated by punching holes in the strip in a planned pattern. Then a special machine was used to examine the holes and electrically tabulate the facts.

Aside from his inventive genius, Dr. Hollerith had a splash of Madison Avenue salesmanship. After a few months he abandoned the long strips of paper in favor of cards. The change was for certain technical reasons—but the dimensions and shape of the card approximated the old dollar bill for "standard size and ease of handling." This, in effect, made possible the unit record concept whereby each card represented a separate and distinct record of information. And once information was recorded it could be used time and again for any number of purposes.

Dr. Hollerith's system made possible the 1890 census of 62 million people in one-third the time required for the 1880 census of 50 million. Data was placed in the cards in the form of holes made with a conductor's hand punch. The cards were then positioned one by one over mercury-filled cups. At the touch of a lever, rows of telescoping pins descended on the card's surface. Where there was a hole, the pin simply dropped through into the mercury, made an electrical circuit, and caused a pointer to move one position on a dial. In short, the job of counting was mechanized.

The success of Dr. Hollerith's technique spread rapidly. Before 1900 he set up a similar installation to handle the New York Central's car accounting system. Insurance companies adopted the machines for their actuarial work. Marshall Field in Chicago used punched cards for sales analysis. Dr. Hollerith himself went to Russia to help the government with their census. Punched card techniques were used for cost accumulation at the Penn Steel Company and for sales analysis at Western Electric.

All of this happened as the United States embarked on an amazing technological binge. From about 1880 through 1930 our commerce expanded at a prodigious rate. The railroads pushed west, north and south to open up new markets where there had been wilderness just a few years before. Industry captured mass production techniques to keep goods pouring out by the ton.

The automobile, radio, rayon and brokerage industries grew into giants. Interchangeable parts ensured that a product purchased in one part of the country could be serviced in another. American ingenuity produced 1,330,000 patents in the first third of the 20th century alone.

It was an ideal setting for the introduction of data processing machines. Every bit of this industrial growth created accounting and record-keeping problems. Dr. Hollerith formed a company to develop his machinery and, in 1911, merged it with two other firms to become the Computing-Tabulating-Recording Company. C.T.R.—which eventually changed its name to IBM— had four basic business machines in operation by 1914: a key punch for punching holes in cards; a hand-operated gang punch for coding repetitive data into several cards at the same time; a vertical sorter for arranging cards in selected groups; and a tabulating machine used for adding the data punched into cards.

Over the next half-century the roster of punched card processing machines, sometimes called unit record equipment, increased tremendously. Devices for sorting, punching, verifying, merging, collating, reproducing, printing, tabulating and calculating were developed. They became faster, more efficient, more economical. And a look at the financial statements of Remington Rand and IBM during this period will illustrate graphically how hungry industry and government was for these machines.

But the punched card machine is an electro-mechanical device. That is, electricity is used to drive certain mechanical components and there is a definite limit to the speed and flexibility of the equipment. The physical manipulation of cards inside the machines takes time. Furthermore, since each machine is specialized—used for just one job at a time—it is necessary to move the punched cards from one device to another in order to perform a complete processing chore.

The next step was to coordinate these various operations into a single, multi-purpose machine. As far back as 1915, James Bryce, a consultant to the C.T.R. Company, envisioned vacuum tubes for data processing purposes. But it wasn't until 1937 that the first concrete advance was made. Professor Howard Aiken of Harvard University, working in conjunction with IBM, put together a so-called "computer" made up of 78 adding machines and desk calculators controlled by a piano player type roll of perforated paper. Thus he utilized at once both Jacquard's perforated paper tape idea and incorporated many of the principles advanced by Babbage.

This first "computer" was known as the Automatic Sequence Controlled Calculator (A.S.C.C.). During the 1940s it was used extensively by the U.S. Navy—but one computational chore completed by the machine stands out above all others. Toward the end of World War II, Allied intelligence agents learned that the Nazis were working on an electrically powered cannon. Meanwhile, our primary scientific effort was being directed towards the still theoretical Manhattan Project.

Time was of the essence. If the Germans had latched onto a workable idea, then we'd have to divert some of our brains to a similar project. The intelligence information was broken down into a series of complex formulae and fed into the A.S.C.C. Far faster than even a small army of mathematicians working with slide rule and desk calculator could come up with the answer, the computer reported that the cannon wouldn't work. So while the enemy went on beating a dead horse, we developed the atom bomb.

The scientific demands of World War II proved to be a tremendous stimulus to the development of computers. Most of the momentum came from the urgent needs of engineers and scientists engaged in aircraft and ordnance development. Their schedules were pegged at the bare minimum and they constantly raced the clock to complete vital calculations. Furthermore, the learned men working on the atom bomb project were suddenly coming up against calculations of a magnitude never encountered before.

Such machines as the A.S.C.C. could and did come up with the answers. But they were relatively slow devices depending on switches and relays for movement of internal data. Scientists wanted something even faster.

The first authentic *electronic* digital computer—that is, a machine using electronic tubes rather than electrical relays—was the Electronic Numerical Integrator and Computer, commonly called the ENIAC. ENIAC was developed at the Moore School of Engineering of the University of Pennsylvania, under the auspices of the Ordnance Department of the U.S. Army. It was a monstrous affair containing 18,800 vacuum tubes and its inventors, Drs. J. Presper Eckert Jr. and John W. Mauchly of the Moore School, spent two-and-a-half years just soldering the 500,000 connections required for these tubes.

In operation, the ENIAC could perform 5,000 additions per second and all internal operations were conducted entirely by electrical pulses, generated at the rate of 100,000 per second. And while this is slow compared to present day computers, it was a tremendous advance over the previous electro-mechanical machines.

The ENIAC was used for solving special problems in ballistics at the Aberdeen Proving Ground in Maryland, and saw service from 1947 to 1955. During this period Dr. Eckert and Dr. Mauchly developed another machine called the BINAC which, later, became the forerunner of Remington Rand's famous UNIVAC I.

The ENIAC, however, lacked one important element that was to break wide open the development of these new machines. It did not have a true stored program memory. Instructions for the machine were recorded by removable plug wires—just as instructions for punched card equipment are programmed on interchangeable control panels, cards or paper tapes. As a result, once data was put into the computer it had to progress according to these preset devices. Only in an extremely limited way could the computer depart from the fixed sequence of its instructions.

In 1948, IBM introduced the S.S.E.C.—Selective Sequence Electronic Calculator—which contained 21,400 electrical relays and 12,500 vacuum tubes. In size, speed and capacity it was smaller than the ENIAC. But, more important, it was the first electronic machine to employ a stored program. And with this internal storage system, the computer could modify its own instructions as dictated by the developing stages of work.

In 1950, the Los Alamos atomic energy laboratory presented the S.S.E.C. with "Problem Hippo"—a massive mathematical operation calling for nine million arithmetic solutions. The computer worked for 150 hours to produce an answer. It would have taken a mathematician 1500 years to do the same job.

In effect, the stored program computer took electronic machines out of the "single job" class and made them into "general purpose" devices. With instructions stored internally, they had greater latitude to take on almost any job found in a laboratory and, later, in an office. They became faster, more flexible, more easily controlled. The age of electronic data processing had arrived full scale.

Work:
The Myth That Became a Monster

Don Fabun

"It seems that there were two patriotic Americans who met on the street one day," recounts R. L. Cunningham in *The Redefinition of Work.*

"What's this I hear about you?" demanded one, "That you say you do not believe in the Monroe Doctrine?"

The reply was instant and indignant: "It's a lie. I never said I didn't believe in the Monroe Doctrine. I do believe in it. It's the palladium of our liberties. I would die for the Monroe Doctrine. All I said was that I don't know what it *means!*"

When it comes to considering the meaning of work in our society, we are much like the patriotic gentleman; we are quite willing to die for the right of other people to work, but we don't really know just what it is that we mean when we talk about work.

It's no good looking it up in the dictionary; the best of them all, The Oxford, gives nine pages of small type to the way in which English speaking people have used the word; it has become, to that extent, virtually meaningless in modern times. Actually, what appears to have happened is that somewhere along the way, in our society at least, we have confused work and income, so that the two somehow became related in our minds. When people say, "We want work," they usually mean they want income. The two are seldom related in any proportional sense, yet when we think they are we try to build a social structure that assumes work will somehow solve our problems, when what we really mean is that income will solve our problems, or at least some of them.

How did all this come about? Where and when did we begin to feel that work—in and for itself—as a human activity, was somehow valuable?

There's a long, long trail awinding when one tries to follow the concept of work back to its origins, but there does appear to be some agreement about some aspects of it. Almost any culture you dig into does not like it. The Paradises and Golden Ages of our ancient pasts were primarily places where one did not have to work. And most of the concepts of heaven in our various religions have as their chief characteristic the fact that one does not have to work in them.

It is doubtful that ancient man had any concept of "work" at all, and such primitive societies as still exist frequently have no vocabulary that distinguishes between "work" and "free time."

"In many low-energy societies," says Fred Cottrell in *The Sources of Free-Time,* "the concepts of work time and free time hardly exist. A man does what is expected of him, which we westerners may refer to as the performance of ritual or ceremony, domestic duties, production, military service, etc. What is expected may also include the occupation of time in conversation, sleep, recreation, singing and dancing, or what-not."

"A man in such a culture may feel as constrained by necessity to do one as the other. It is only when we classify his time into categories meaningful to us that work becomes defined. But if we say that he is working only when he is gaining sustenance, then many 'primitive' men had far more work-free time than we have."

As the world moved out of what we now consider primitivism and into more organized social structure, there grew up the myth of some "Golden Age" which was, perhaps, founded on some lingering memory of the good old days when the concept of work did not exist. All the ancient voices, whose thin dry sounds have come down their thousands of years to become the conscience of us all, considered work at best a necessary evil. "In the Socrates of *Xenophon,* work is an expedient," writes Sebastian de Grazia in his definitive book, *Of Time, Work and Leisure,*" in Virgil's *Georgics* it is a necessity and a mock heroism. In Hesiod's *Works and Days* it is a necessity, too, and, worse yet, a curse.

"To the authors of the Bible also, work is necessary because of a divine curse. Through Adam's fall the world became a workhouse. Paradise was where there was no toil. This is the feeling about work one encounters in most of history's years."

The legend of the ancient earthly paradise, where one did not have to work, is preserved in the amber of Lucretius' words. "Earth first," he says, "spontaneously of herself produced for mortals goodly corn crops and joyous vineyards; of herself gave sweet fruits and glad pastures, which nowadays scarce attain any size even when furthered by our labor; we exhaust the oxen and the strength of the husbandmen; we wear out our iron, scarcely fed, after all, by the tilled fields."

When that great classifier and assigner of categories, Aristotle, took a look at work, he could assign it no very high value, except as a way to achieve leisure, or to *not* work. "Nature," he wrote, "requires that we should be able, not only to work well, but to use leisure well. Leisure is the first principle of all action and so leisure is better than work and is its end. As play, and with it rest, are for the sake of work, so work, in turn, is for the sake of leisure."

The legend of a golden age without work was still a lively one even as late as Rousseau, who described it in this way: "The produce of the earth furnished man with all he needed, and instinct told him how to use it, so that singing and dancing, the true offspring of love and leisure, became the amusement, or rather the occupation, of men and women assembled together with nothing else to do."

One may question the logic of the last part of that statement, but there is little doubt that Rousseau did not feel kindly toward work.

Going back a bit, it was the practical-minded Romans who began, a little, to believe there was virtue in work beyond mere sustenance. "In the morning when thou risest unwilling," the emperor Marcus Aurelius tells himself, "let this thought be present—I am rising to the work of a human being. Why, then, am I dissatisfied if I am going to do the things for which I exist and for which I was brought into the world?" But, of course, the only "work" that Marcus had to do was to tell others what to do—not such a bad day to get out of bed for.

The early Christians did not look upon work as beneficial so much as a penance. There was no work in the Garden of Eden. But when Adam sinned, the Lord God said unto him, "Cursed is the ground for thy sake; in toil shalt thou eat of it all the days of thy life. . . . In the sweat of thy face, shalt thou eat bread, till thou return to the ground."

The idea then that work, to be called work, must be something that we do not want to do, or that is at least unpleasant, was imprinted quite early in our western culture. "That work should be painful belongs to its very essence," says the chapter on Labor in the *Syntopicon.* "Otherwise it would not serve as a penalty or penance. But in the Christian view, labor also contributes to such happiness as man can enjoy on earth. The distinction between temporal and eternal happiness is a distinction between a life of work on earth and the activity of contemplation in Heaven. This does not mean the elimination of leisure and enjoyment from earthly life, but it does make labor their antecedent and indispensable condition.

"In all these conceptions of a better life," continues the *Syntopicon,* "labor is eliminated or reduced. The implication seems to be that the labor required for the maintenance of all historic societies is an affliction, a drudgery, a crushing burden which deforms the lives of many, if not all. The pains of toil do not belong to human life by any necessity of human labor, but rather through the accident of external circumstances which might be other than they are. . . . Man might have realized his nature more surely and richly if, like the lilies of the field, he neither toiled nor spun.

"The contrary view would maintain that work is not a curse but a blessing, filling man's hours usefully, turning to service energies which would otherwise be wasted or misspent in idleness or mischief. . . . It is even suggested that useful occupations save men from a boredom they fear more then the pain of labor, as evidenced by the variety of amusements and diversions they invent or frantically pursue to occupy themselves when work is finished. The satisfactions of labor are as peculiarly human as its burdens. Not merely to keep alive, but to keep his self respect, man is obliged to work."

Thus, at least three ways of viewing work are easily discernible. In primitive, survivalist societies, work is the condition of life, but it is not thought of particularly as work. Then there is work as a curse, something we would all avoid if we could. And finally, work as the normal outlet for man's energies, because play is not enough. There is one more main-line idea in our Western culture concerning work—it claims that work in and for itself is of intrinsic value.

This idea may have had its first concrete expression in the Sixth Century when St. Benedict at his monastery at Monte Cassino posted rules for the monks. "Idleness is the enemy of the soul," begins Rule XLVIII. "And therefore, at fixed times, the brothers ought to be occupied in manual labor, and, again at fixed times, in sacred reading."

For the first time not only work as such, but work for a stipulated time, became integral to western thought. In later years we were to confuse the two, so that "putting in the time" became more important than the work. But what was new here at the beginning, with the monks of Monte Cassino, was that work was good for the soul. This was the myth that has become a monster in our times; it drives even the rich to maintain the illusion that they are working, and those who do not work into an incessant apologia for being alive.

"Today, the American without a job is a misfit," comments Sebastian de Grazia. "To hold a job means to have status, to belong in the way of life. Between the ages of twenty-five and fifty-five, that is, after school age and before retirement age, nearly 95 percent of all males work and about 35 percent of all females. Various studies have portrayed the unemployed man as confused, panicky, prone to suicide, mayhem and revolt. Totalitarian regimes seem to know what unemployment can mean; they never permit it."

The "work" monster gained a certain substance from the idea that the progress of a society or a culture is something like the natural progress of the life of a man; as he grows older and works harder, he accumulates more wisdom and more material things.

"Augustine was therefore able to insist," says George Hildebrand in *The Idea of Progress*, "that mankind could well be regarded as a single man, whose earthly experience constituted a gradual advance through education, effecting a slow transition from ignorance to knowledge and finally to faith. . . . In the 17th Century this analogy became fundamental to the formulation of the modern idea of progress . . .

"Progress came to be looked upon as the normal tendency in human affairs, the gradual and inevitable development of the human race."

It is probably no accident that the idea of social progress and the sanctity of work as a means to achieve it grew into a now virtually unexamined ethic at the same time that the Industrial Revolution began to need more "workers." This kind of work was not like the work that had gone on before; it was specially oriented in space (in the factory or foundry) and structured in time (the necessity for the worker to be in a certain place, at certain times, performing certain prescribed activities).

The dangers of this concept of work were seen long ago by Adam Smith, who pointed out, "In the progress of the division of labor, the employment of the far greater part of those who live by labor . . . comes to be confined to a very few simple operations, frequently one or two. . . . The man whose life is spent performing a few simple operations . . . has no occasion to exert his understanding or to exercise his invention. . . . He naturally loses, therefore, the habit of such exertion, and generally becomes as stupid and ignorant as it is possible for a human creature to become."

Later thinkers have come up with slightly different interpretations of what is "wrong" with "work." Niall Brennan, in *The Making of a Moron* finds a somewhat different reason: "The unpleasantness of a job has nothing to do with whether it is repetitive or not. It depends solely on how many of the parts of man are used and how well they are being used. Acting is monotonous in the literal sense of the word; but few occupations use the whole man so intensely. . . . If only a part of a man is being used, the salvation of his sanity depends on what he himself does with the unwanted parts.

"But if ostensibly the whole man is bought by the employer, and only a part is used, or parts wrongly used, and the worker himself is denied right use of his own parts, then his sanity, in the sense of the fullness of his personality, is in danger. Either the unwanted parts atrophy for sheer lack of use, or they are mutilated by misuse, and he ceases to be a whole man."

This same view of industrial labor—which is the symbol of what most of us mean by work—is reflected in Alexander Heron's, *Why Men Work*. He says, "There are few jobs in the industrial world which are inherently interesting; there are reports that even wine samplers, selectors of beautiful models, and private detectives become bored with their tasks."

Again, it gets back to the partial use of the whole man: work in our society may be dull or worrisome or unpleasant, not because it is work, but because so much of the human being is left out of it. Artists of all persuasions, including, of course, mathematicians and physicists, seldom really think of what they do as work (though they may, out of convention, call it that) because so much of them is used up in the process of what they are doing. It is the paper shufflers and the ditch diggers, if, indeed, there are any of the latter left, that find work boring.

We must not, however, believe that any single view of the industrial worker is wholly valid. There are many people to whom work is simply hours to be gotten through, in order to receive sustenance. They accept it as a part of their role in life, unquestioned and unexamined. "There are," says Eli Ginzberg in *The Study of Human Resources,* "further defects in an approach which sees the typical worker as an automaton, frustrated from the time he starts work until the quitting whistle blows. Most men have a realistic opinion of their strengths and weaknesses . . . they know how hard their fathers had to work to support their families, so if their own lot is easier, they are likely to be reconciled to it. . . . But they are not industrial slaves, as Marx called them, nor are they company serfs, forced to do what the employer wants . . ."

Even so, the idea that one should show up at a certain place at a certain time and perform some prescribed activity for a certain number of hours, whether the activity is meaningful or not, remains today what we usually think of as work. We have constructed a society in which participation in this activity almost becomes the goal of life itself.

We need at this point to make a little more precise definition of the two kinds of work. "Object-oriented" work is activity directed toward transforming some natural resource into some object useful to man's needs or wants. Ore transformed into metal, chemicals into plastics, petroleum into gasoline and lubricating oils, water, solar radiation and carbon dioxide into food—these are "object-oriented" types of work. At most levels, this kind of work uses only a small part of a man's total ability and consists of relatively simple, repetitive actions, performed over a prescribed period of time.

"People-oriented" work is directed toward service to others; providing transportation, distributing goods, teaching, social welfare work, government service, providing entertainment or recreation for others. This, too, is repetitive action. But it involves much more of the person who is doing it, and it is constantly refreshed by the new human contacts that occur during the working period.

It is the "object-oriented" work in our society that is being replaced by mechanization, automation and cybernation. In the years ahead, it is the "people-oriented" type of work that is likely to increase. "The Work of the world remains to be done," says Gerard Piel. And the work of the world for the next twenty years would appear to be more and more that which has to do with people rather than objects. Let the machines produce objects; let people become more concerned with people.

The modern work syllogism seems to run something like this: it is natural for a society to "progress," and at the base of progress is the use of human energy in the form of work. Therefore the more people in our society who work, the more will we progress. People who do not contribute to progress by working should not share equally in its fruits with those who do work.

This is a sociological imprint we have inherited from another time and place, namely, the Industrial Revolution of the 18th and 19th Centuries; we continue to act as if our world was the same as then—when, in fact, it is quite different.

"In the United States we produced, in the year 1850, 440 horsepower hours of energy per person in the population," says Fred Cottrell in *The Sources of Free Time*. "In 1900, the figure soared to 1,030, and during 1950, 4,470 horsepower hours were produced. In 1958, about 5,100 was the output. The sources of energy were of course altered. In 1850, human beings produced 13 percent of the energy used, animals 52 percent, and inanimate sources 35 percent. In 1950 humans produced less than one percent, as did animals, and 98 percent was derived from inanimate sources."

As Cottrell points out elsewhere, "A new order of predictable choice—a new hierarchy of values may thus emerge out of a changed flow."

We are beginning to see the consequences of that changed energy flow, and with it the demand for new ways of looking at work and the role it plays in our society. What has happened and is happening is that the necessity for individual human labor as a means toward progress has begun to change over to the management of inanimate energy sources. This new type of "work" precludes the active participation of large segments of our society because the effect of each individual's labor is magnified many times by the inanimate energy he controls.

It is quite possible that much of our current educational system is engaged in preparing young people for "jobs" that simply will not exist in our society by the time these students come into the marketplace. It is equally possible that a goodly segment of the "jobs" being performed in our economy today are simply atavisms of 19th Century concepts of work which have little economic, social, or even personal value today, and will have even less value in the future as our developing technology changes the nature of the use of human energy.

Even today, with most of our technological advances yet to be felt, in the United States only 38 percent of the population is "employed," in the Bureau of Labor Statistics' definition of the word, and during their working years, from 16 to 65, they spend less than 23 percent of their total time engaged in work, even if "fully employed" for all those years.

Of those now fully employed, most are in occupations likely to shrink under the onslaught of an automated technology. "This is still," says Ben J. Wattenberg in his monumental study of the 1960 census, *This U.S.A.,* "predominantly a non-white-collar nation—a nation of makers. Most people do not go to work in a business suit or an office dress. Most (56% in 1964) work in factories and garages, drive trucks, dig ditches, cut hair, clean houses and grow food. . . . But it is equally important to realize this; there are more and more Americans concerned with jobs traditionally associated with the paper-shuffling world of words, figures and abstract thought normally associated with the phrase 'white collar.' The enormous growth of the white-collar occupations has lured many into believing that it has become the majority way of life in America. It has not. Not yet!"

Not yet. But perhaps sooner than most of us want to think. What happens when, as some have predicted, two percent of the American population is employed in producing the necessities of life, and 98 percent is not? How, indeed, can we hope to live meaningful lives in an "economy of abundance?" The tragedy is not, as some seem to believe, that this way of life may come about well within our lifetimes; the tragedy is that, knowing this, we are doing little or nothing to prepare ourselves or the younger generation to cope with it. When and where will we begin to chip away at the antiquated work ethic and come up with new systems and institutions in which leisure, and not work, is the desirable and socially acceptable goal of man?

One scans the horizon of two decades ahead and sees the almost inevitable collision of two great forces—exploding population and exploding technology —and one of the results of that collision will almost certainly be a society in which some other ethic than the sanctity of work will have to be found.

The Real News About Automation

Charles E. Silberman

One of the most sensational pieces of news about the performance of the United States economy in this era of radical change and dire prediction is contained in a statistic that has been ignored by all but a handful of Labor Department economists. The statistic is this: Employment of manufacturing production workers increased by one million in the three and a half years from the first quarter of 1961 to the third quarter of 1964 and by another 700,000 by the fourth quarter of 1965. This increase dramatically reversed the trend of the preceding five years, when 1,700,000 production-worker jobs were eliminated and "the work of the hands" appeared to be going out of style. Such work is very much in style now.

This turnaround in blue-collar employment raises fundamental questions about the speed with which machines are replacing men. It was the large decline in blue-collar employment in manufacturing during the late 1950s and early 1960s, more than the persistence of high over-all unemployment rates, that persuaded so many people that automation was rapidly taking hold, condemning the unskilled and the poorly educated to a vast human slag heap. "The moment of truth on automation is coming—a lot sooner than most people realize," the Research Institute of America warned its businessmen-subscribers in December 1963. "Cybernation," said the sociologist-physicist Donald N. Michael, who coined the term (to refer to the marriage of computers with automatic machinery), "means an end to full employment." The Ad Hoc Committee on the Triple Revolution, a diverse but influential group of private citizens, went even further: in its view "cybernation" means an end to *all* employment, or almost all. Unless "radically new strategies" are employed, the committee warned President Johnson in March 1964, "the nation will be thrown into unprecedented economic and social disorder." It argued that "cybernation" makes a mockery of any attempt to provide jobs for either white or Negro workers. Technological change—the most crucial of the three revolutions the committee is concerned about (the other two are in civil rights and military weapons)—is creating an economy in which "potentially unlimited output can be achieved by systems of machines" requiring "little cooperation from human beings."

Nothing of the sort is happening. Two years of field research and economic and statistical analysis by *Fortune* make it clear that automation has made substantially less headway in the United States economy than the literature on the subject suggests. Fifteen years after the concepts of "feedback" and "closed-loop control" became widespread, and ten years after computers started coming into common use, *no fully automated process exists for any major product in any industry in the United States.* Nor is any in prospect in the immediate future. Furthermore, the extent and growth of several partially automated processes have been wildly exaggerated by most students of the economy. There is, in fact, no technological barrier to full employment.

This is not to deny that technology is changing; clearly it is. Nor is it to deny that such change displaces substantial numbers of workers. Technological innovation is always doing that, and it is always painful to the individuals directly affected—the blacksmiths, harness makers, and coachmen whose jobs were destroyed by the automobile, and the insurance-company clerks now being displaced by computers. But the question raised by Michael, the Research Institute, and the self-appointed Committee on the Triple Revolution is not whether innovation causes displacement of labor. It is, rather, whether technological displacement is occurring at a substantially faster rate than in the past—at a rate so fast, in fact, as to threaten a crisis of mass unemployment similar to that of the 1930s.

The View From Brooklyn

The answer is no. Automation, in any meaningful sense of the term, is only a minor cause of unemployment. The auto workers in South Bend who lost their jobs when Studebaker shut down, the packinghouse workers thrown out of work in Chicago and Kansas City when the meatpackers decentralized their slaughtering operations, the coal miners in Appalachia made idle by the loss of coal's biggest markets, the shipyard workers whose jobs ended when the Navy closed its yard in Brooklyn—all are in trouble and many are in need of help. But they are no more the victims of automation than were the New England textile workers of the 1920s, made idle when the cotton mills first started moving south, or the southern cotton pickers of the 1930s and 1940s, thrown on relief when the mechanical cotton picker came into use.

To the men in question, of course, this fine distinction may seem brutal and irrelevant: what matters to them is not the particular reason they are idle but the poverty—of the spirit even more than of the body—that idleness causes. But to those concerned with relieving and preventing unemployment, the causes *must* be central. There can be no greater disservice to the unemployed—indeed, no greater act of contempt—than to substitute easy slogans for ruthless honesty in analyzing the causes of their joblessness, and thus to fail to ease their plight.

There are two kinds of pitfalls in trying to understand an age like ours. One is to take comfort in some such platitude as "the more it changes, the more it stays the same," thereby underestimating or ignoring altogether what might be termed "the cosmic changes" going on. The other is to become so enamored of the cosmic—to focus so completely on all the possibilities of contemporary science and technology—that one loses sight of the realities of the present. Too many of the people writing about automation and "cybernation" have fallen into the latter trap: they have grossly exaggerated the economic impact of automation. At the same time, curiously enough, some of them may have underestimated or simply ignored its psychological and cultural impact. For technology—*any* technology—has a logic of its own that affects people more or less independently of the purpose for which the technology may be used. The assembly line, for example, dictates a particular organization of work and a particular set of relations among workers, and between workers and managers, whether the line is turning out automobiles, breakfast cereals, or insurance-company records.

The Importance of Imitation

A good deal of the confusion over what's happening stems from a failure to distinguish between what is scientifically possible and what is economically feasible. For technological change is *not* purely a matter of invention, of scientific or technical capability—a fact most defense contractors have had to learn the hard way. On the contrary, as mathematical logician Albert Wohlstetter of the University of Chicago puts it, technological change "has to do with such grubby matters as costs, and uses, and competing purposes: in short, with politics, sociology, economics, and military strategy." Indeed, the great economist Joseph Schumpeter used to argue that invention per se played a relatively minor role in technological change. What was crucial, he insisted, was "innovation"—the process of finding economic applications for the inventions—and "imitation," his term for the process by which innovation is diffused throughout the economy. The time lag between these three steps may have been truncated by the growth of industrial research and development and by the growing recognition that knowledge is the most important form of capital. But a time lag does remain, and it can be substantial, as the disappointingly small civilian fallout from military and space research and development activities attests.

We have misunderstood what is happening, moreover, because discussion of the future of science and technology have turned into "a competition in ominousness," as Wohlstetter describes it. In their eagerness to demonstrate that the apocalypse is at hand, the new technocratic Jeremiahs seem to feel that any example will do; they show a remarkable lack of interest in getting the details straight and so have constructed elaborate theories on surprisingly shaky foundations.

To explain how automation is revolutionizing the structure of production and of employment, for example, Professor Charles C. Killingsworth of Michigan State University told Senator Joseph S. Clark's Employment and Manpower Subcommittee about Texaco's computerized petroleum refining unit in Port Arthur, Texas, which, Killingsworth said, processes "several million gallons of raw material daily." He picked a poor example; the installation actually demonstrates how small the bite of automation is, and how large and numerous are the obstacles to its rapid spread. The unit in question produces about 80,000 gallons a day, 0.6 percent of the refinery's total throughput. The computer installation has been successful, to be sure. But the payoff has not come in reduced employment—the number of workers remained at three per shift for the first two years, at which time productivity gains unrelated to the computer enabled Texaco to cut the number to 2.5 men per shift, a net loss of two men. (Killingsworth told the Clark committee that it was his "guess" that the computer had "replaced a half dozen men in the control room.") The payoff has come, rather, in the greater efficiency with which the unit converts gases into polymers. More to the point, this particular process was picked for computerization because it was "relatively simple when compared with other refinery processes" and because it was one of the few processes for which good historical data (essential if control is to be shifted from people to computers) was available.

Since the first installation Texaco has put a second computer in control of some of the operations in a large catalytic cracker; the number of workers on this process has remained unchanged.

Killingsworth's mistakes about the capacity of the original computerized refinery unit and his exaggeration of its significance typify most of the literature on automation. Donald Michael's influential essay, "Cybernation: The Silent Conquest," published in 1962 by the Fund for the Republic, contains several other such exaggerations. One of them, having to do with the TransfeRobot manufactured by U.S. Industries, is discussed on pages 144-147. Michael's other examples also tend to dissolve under close scrutiny. Two examples:

Michael goes to some lengths to show how "cybernation permits much greater rationalization of managerial activities," e.g., "The computers can produce information about what is happening now . . . built-in feedback monitors the developing situation and deals with routine changes, errors, and needs with little or no intervention by human beings. This frees management for attention to more basic duties." Michael instances "an automatic lathe . . . which gauges each part as it is produced and automatically resets the cutting tools to compensate for tool wear." The lathe "can be operated for 5 to 8 hours without attention, except for an occasional check to make sure that parts are being delivered to the loading mechanism." This description of the lathe came from 1955 testimony before a congressional committee by Walter Reuther, who in turn was quoting from *American Machinist* magazine. Michael's reference to the lathe is almost a complete non sequitur. For one thing, whatever its impact on the machinists in the shop, it did not affect managerial activities at all. For another, no computer was involved. The machine was simply an improved version of a standard automatic lathe that machine-tool manufacturers had been making for several decades.

To show how cybernation permits rationalization of management—in this instance by "combining built-in feedback with a display capability"—Michael also cites the Grayson-Robinson apparel chain's use of a computer to handle "the complete merchandise and inventory-control function." Actually, the chain had neither "feedback" nor "display capability." It did use a computer to give management a weekly report of sales and inventory, but Grayson-Robinson merchandise men did all the buying and reordering. Perhaps the chain would have been better off if the computer *had* handled "the complete merchandise function," for it started a bankruptcy proceeding in August 1962 and was declared bankrupt in November 1964.

The Anonymous Ghosts

Not to be outdone by Michael's invention of the word "cybernation," the consultant, Alice Mary Hilton, president and founder of something called the Institute for Cybercultural Research, has coined the term "cyberculture" to describe the new "age of abundance and leisure" that computer-run factories will soon be forcing everyone to enjoy. Writing about the "cybercultural

revolution" in the fall 1964 issue of the *Michigan Quarterly Review,* Miss
Hilton offered a few original examples of her own:

"In Texas and New Jersey, in the oil refineries—the silent, lifeless ghost
towns of this century—crude oil is processed into different grades of gasoline
and various byproducts—the proportions determined automatically and flexi-
bly as consumers' demands vary. Crude oil is piped in—gasoline and by-
products emerge, hour after hour, day after day, without pause for sleep or rest
or play, without coffee breaks or vacations, sick leaves or strikes. *There are no
workers, no supervisors, no executives; just a few highly trained engineers
standing by in the central control room,* watching their brainchild fend for
itself." (Emphasis added.) Unfortunately, Miss Hilton is closely guarding the
identity of these refinery "ghost towns." In Port Arthur, Texas, however, the
Texaco refinery alone employs 5,000 people, the Gulf refinery 4,000.

Computers are producing fuel for human consumption as well as for
machines, according to Miss Hilton. "In a Chicago suburb, in a bakery as large
as a football field," she wrote, "bread and rolls and cakes and cookies are
produced for millions of households throughout the country by a team of
machines, called a system ... *All the blue-collar and white-collar workers—
of all levels—have been replaced by a silent machine system that labors twenty-
four hours every day* ... The bakery runs itself; the system even maintains
itself," her awe-struck report continued. *"The few human beings still inside
the 'black box' are only nursing the infant cybernation to maturity."* (Emphasis added.) Miss Hilton presumably was describing the Chicago bakery of Sara
Lee, a subsidiary of Consolidated Foods. An elaborate computer system does
indeed control the blending of the ingredients and part of the baking process
to ensure uniformity of product, but people are still required to perform a wide
range of activities—e.g., braiding Danish pastry dough and spreading chocolate icing. And, "the few human beings still inside the 'black box' " come to
about 450 per shift—300 of them in direct labor.

Walter Reuther, from whom Michael took his example of the automatic
lathe, has provided other instances of the devastating impact of automation.
Appearing before Senator Clark's subcommittee in May 1963, Reuther compared the production methods used when he went to work for Ford in 1927
with those now in use. Once it took three weeks to machine the engine block
for a Model T; in Ford's "automated" engine plant in Cleveland, which
Reuther had visited after its opening in the early 1950s, the machining took
only 14.6 minutes, because "the technology in that plant is built around
computers." He added, "The thing we need to understand in order to grasp
this revolutionary impact of such technology is that this automated engine line
... is already obsolete."

Indeed it is. The engine line Reuther described was taken out three years
before he gave his testimony—but not, as he suggested, because the computers

it was "built around" had already become obsolete. On the contrary, the factory had no computers at all. The engine line was taken out because it was just too inflexible; when Ford redesigned its engines in 1959–60 it had to rebuild almost the entire Cleveland factory.

Caught on the Horn of Prophecy

The view that computers are causing mass unemployment has gained currency largely because of a historical coincidence: the computer happened to come into widespread use in a period of sluggish economic growth and high unemployment. Thus it was natural that some who were not looking too closely at the evidence would attribute the unemployment to computers and automation and would assume that a lot more automation must mean a lot more unemployment.

One of the first to push the panic button was W. H. Ferry, vice president of the Fund for the Republic. In a widely publicized essay entitled "Caught on the Horn of Plenty," published in January 1962 and still being distributed, Ferry stated flatly that "The United States is advancing rapidly into a national economy in which there will not be enough jobs of the conventional kind to go around." He then proposed a test: "The next three years ought to suffice to determine whether a liberated margin [his term for the 'technologically displaced'] is in fact in the making. If by 1964 the unemployment rate is close to 10 percent, despite the use of all conventional medications, we may be ready to agree that once again, as in the Thirties, the nation is in a radical dilemma, a dilemma of abundance."

The three years have passed. The unemployment rate did not rise to 10 percent, as Ferry believed it would; instead it declined from its high of 7.1 percent in May 1961 to 4.9 percent in November 1964 and then it proceeded to drop still more—to 3.7 percent in the spring of 1966. In part, of course, the decline in unemployment in 1965 and 1966 was a side effect of the rise in defense spending associated with the war in Vietnam, and unemployment could rise again if defense spending were cut. But to compare our situation with that of the 1930s, when 13 million people—25 percent of the labor force —were unemployed, is to indulge in reckless distortion. The persistence of high over-all unemployment and the changes in the occupational structure of the labor force during the 1950s and early 1960s were due less to automation or technological changes than to a combination of quite different (and in some instances, quite temporary and reversible) economic forces. The remainder of this reading will examine what in fact *is* happening to productivity and consider why automation has proceeded so much more slowly and has had so much less impact on employment than so many had expected.

The Sources of Productivity Growth

If technology were in fact threatening mass unemployment, the threat should be reflected in an acceleration of the rate at which productivity—i.e., output per man-hour—is increasing. What *has* been happening to productivity? Do the statistics of output per man-hour show any evidence of revolutionary changes in the structure of production? In general, the rate of productivity increase tends to accelerate over time, for productivity feeds upon itself. For the last 115 years, however—about as long a period as we can measure—the acceleration has been gradual, averaging about 0.2 percent per decade. Gross private (nongovernmental) output per man-hour grew at an average rate of only 1.3 percent a year between 1850 and 1889.[1] The average jumped to 2 percent during the next thirty years (1889 to 1919), and then to 2.5 percent between the two world wars. Between 1947 and 1960 productivity gains averaged 3.3 percent a year; since 1960 the annual increase has averaged 3.6 percent.

The productivity gains these last five years, it should be pointed out, are running ahead of *Fortune's* 1959 forecast of a 3.4 percent average annual increase for the 1960s as a whole—a forecast that most academic and business students of productivity dismissed at the time as extravagantly optimistic. But note several things about these large productivity gains: Even larger gains have been registered over comparable or longer periods a number of times in the past—for example, from 1920 to 1924 and 1947 to 1953. More important, such figures reflect a lot more than technological change. Firms increase productivity, for example, by making their employees or their machines work harder; by hiring employees with more education or training, or by giving employees on-the-job training; by changing the way in which work is organized (e.g., using a typing pool instead of individual secretaries); or by changing the product "mix" (making more high-priced products).

It is inherent in the way productivity is measured, moreover, that above-average gains are recorded during periods of rising output and below-average gains during periods of stable or falling ouput. Productivity had been held down from 1955 to 1960, partly because output itself had been very sluggish, partly because the huge capital expenditures of 1955–57 and the enormous buildup in R. and D. activities had both, so to speak, diverted manpower from current to future production. When the boom finally got under way in 1961, therefore, business had a reservoir of new products and new production processes to draw upon. More important, almost every industry had some excess capacity, which meant that until mid-1964 or thereabouts, firms could fill incoming orders using only their newest and most efficient facilities. They could also concentrate their capital expenditures on equipment designed to

1. Gross *private* output (rather than gross national product) per man-hour is used because it is impossible to provide a meaningful figure for productivity of government employees.

Chart 1.

Long-term Productivity Growth in the Private Economy

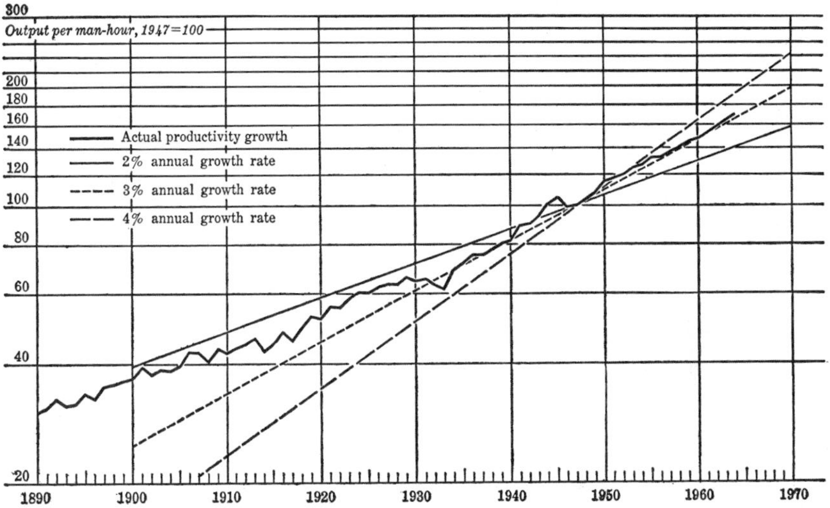

An acceleration in the growth rate of productivity is no new phenomenon, as so many enthusiasts of automation seem to think. On the contrary, the rate has been accelerating for a hundred years or more. (See Chart 1.) Between 1850 and 1889 (not shown), gross private (i.e., nongovernmental) output per man-hour grew at an average of 1.3 percent a year. The rate jumped to an average of 2 percent a year between 1889 and 1919 (thick line), and then again to 2.5 percent between the two world wars. Since 1947 productivity gains have averaged 3.4 percent a year, and just since 1960 3.6 percent a year. The straight lines show what the productivity index would have been had the rate averaged 2, 3, or 4 percent a year.

The chart also helps to place the recent spurt in productivity in historical perspective. It has always grown irregularly, with many spurts and lags. The spurts, of course, are concentrated in years of sharply rising output; the secular gains are more apt to reflect changes in technology, industrial organization, and education. Thus, increases of the sort realized during the first half of the Sixties have been experienced before and have usually been followed by periods of productivity lag. Between 1920 and 1924, for example, output per man-hour shot up by 4 percent a year as the assembly-line and other mass-production techniques gained wide acceptance. In the next five years, however, the rate dropped by more than half. Later, in the six years following World War II, productivity advanced by 3.7 percent a year, but then slowed to an average of only 2.5 percent in the following seven years. Obviously, it is too soon to conclude that the recent 3.6 percent a year rate of productivity growth will last.

increase the efficiency of existing facilities rather than spending to increase capacity. In addition, high unemployment rates made it possible for employers to be very choosy about whom they hired. Taking everything together, there are good reasons for viewing the productivity gains of the past few years as not entirely sustainable for the rest of the decade. In fact, the productivity rise did slow down in 1965 to 2.8 percent, from 3.6 percent the year before.

If the Triple Revolutionists were right—if we were in fact developing "a system of almost unlimited productive capacity"—it would follow that business firms could realize substantial increases in production without having to spend very much on new plants and equipment. But the current business expansion has triggered the greatest capital-goods boom in history—considerably more than twice the size and duration of the capital-goods boom of the mid-Fifties.

Moreover, if a "new era of production" had really begun, it would have to show up in manufacturing, the most critical sector of the economy. Yet productivity has been growing a bit more *slowly* in manufacturing than in the economy as a whole. In the entire postwar period, manufacturing productivity has increased by 2.9 percent a year, vs. 3.4 percent for the private economy. There was an acceleration in the 1960–65 period, to be sure. But these recent gains in manufacturing were smaller than the gains realized in the decade following World War I, when technology was being revolutionized by the assembly line and the endless-chain drive. Between 1919 and 1929, output per man-hour in manufacturing increased by 5.6 percent a year. The acceleration in *over-all* productivity growth since the 1920s has come about because mechanization and rationalization have been applied elsewhere in the economy—e.g., in finance, insurance, retail and wholesale trade.

The Brewmaster's Nose

If there is not now any "cybernetic revolution" in manufacturing, perhaps there will be one soon. How likely is any such event? The answer seems to be —not very. Consider some obstacles that have been encountered in trying to automate continuous-process industries like oil refining, chemicals, and paper.

These industries looked like sitting ducks to manufacturers and designers of computer systems. They are characterized by enormous capital investment in processes in which relatively small increases in efficiency can yield large improvements in profit. By the mid-1950s they had actually gone pretty far toward automation. All that had to be done to achieve complete automation —or so it seemed at the time—was to substitute computers for the human beings monitoring the instruments and controlling the variables of the production process.

The next step was never taken; there are now about 400 process-control computers in use in the United States, but the essential control of the production process remains in human hands, and minds. Refineries and chemical plants, with their miles of pipes and tubes, and paper mills, with their gigantic machines dwarfing the handful of attendants, may *look* as though they are controlled by machines, but they're not. What is literally meant by "automation" or "cybernation," i.e., a process in which a computer or some other machine controls all aspects of the process from injection of the raw material to the emergence of the final product, determining the proper mix and flow of materials, sensing deviations from the desired operating conditions and correcting these deviations as they occur, or before they occur—we are a long way from all this. In the electric-power industry, for example, which has more than

Chart 2.

Postwar Productivity Growth by Industry

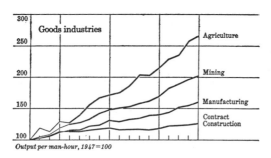

Output per man-hour, 1947=100

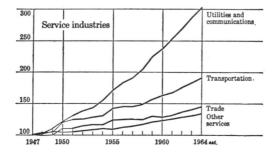

Talk of the over-all rates often tends to blur the widely divergent trends from one industry to another. For all the talk about automation, output per man-hour has increased more slowly in manufacturing than in the economy as a whole since 1947—i.e., by 2.9 percent a year vs. 3.4 percent for the economy. These recent gains in manufacturing are barely half as large as those realized during the 1920s, when a real revolution in manufacturing technology was taking place.

a third of all installations, computers are used mostly as data loggers, recording what happens in the process for the engineers to analyze and study. So far no more than a half dozen or so plants are using computers to control the elaborate sequence of events involved in starting and shutting down a generating station, and only a few plants are using computers to determine the optimum distribution of power throughout the system. In pulp and paper mills, computers serve primarily as data loggers.

Most of these industries have realized large gains in productivity in recent years. But the gains have come less from computers than from better material handling and from installation of larger and more efficient machinery of the conventional sort: for example, bigger electrical turbines and generators, bigger cat crackers in oil refineries, larger-diameter oil and gas pipelines, bigger and faster paper machines. The computers themselves have displaced very few people, if any at all, partly because computers are being used to perform functions that had not existed before; partly because the number of employees involved in controlling the process had already been reduced to the bare minimum needed to take care of emergencies. Computer manufacturers today try to justify their systems by pointing, not to savings in manpower, but to reduction in raw-material costs and increased efficiency in operation.

Full automation is far in the future because, as Peter F. Drucker has observed, "There's no substitute for the brewmaster's nose." The productivity of a paper mill, for example, hinges in large part on such things as a machine operator's ability to establish the proper "freeness," and this he does by watching the water "go along the wire." (The operator watches a mixture of water and fiber going past him; if he feels that the water is traveling too far before draining out, or not far enough, basic operating adjustments have to be made.) There is no *scientific* reason why such operations cannot be automated; in principle, the brewmaster's nose can be too, and in time it probably will be. But the costs of doing so are inordinately large, and the time inordinately long: no industry understands its production process well enough to automate without huge investments of time and capital.

Consider, for example, the basic oxygen process for making steel. The system designers discovered very rapidly that they didn't have data precise enough to enable them to set up a mathematical model describing what actually goes on in an oxygen converter—the first step in designing a computer-controlled system. The data were too crude because the instruments being used were too crude. And the instruments were crude because steelmen didn't know what quantitative information they needed—not because they were uninterested, but because before computers came along they had no use for more refined data. Hence the computer manufacturer must study the process long enough to determine what information is needed, then find instruments sensitive enough to yield that information in quantitative form, then hook up these instruments to a computer to monitor the process long enough to determine what happens and to identify the critical variables. Only then can it try to develop a mathematical model of what goes on in the process.

And setting up the model can be the most intractable job of all. It turns out, for example, that the mathematics of controlling an oxygen furnace, an oil refinery, or a paper mill are in some ways more complicated than the mathematics of controlling a missile or a satellite. The only mathematical technique now available for handling as many variables as are found in most industrial processes is "linear programing." Unfortunately, as Dr. Thomas M. Stout, a process-control engineer and consultant, says, "practically no relationships in nature are linear." Thus, designers of computer process-control systems have had to develop their own mathematical techniques as they went along.

Each new computer installation, of course, makes the next one easier; in time, computers will be able to control more and more production variables. The point is that the change will be gradual and that it will not lead to peopleless plants. The most fully automated refineries, paper mills, and generating stations now imaginable still require a work force something like the present one. Even if computers could handle all the operating variables in a paper mill, for example, the number of operators probably would not be reduced below the present seven per machine. As one production man explains it, "at least fourteen hands are needed immediately to re-thread the paper when it breaks"—and the paper breaks an average of twice a day.

The Elephants and the Mahouts

The obstacles are multiplied several times over when firms try to automate the production of more complicated products that have to be assembled from a large number of parts. Automation, like mechanization in general, proceeds in two ways—either by taking over functions that men perform, e.g., substituting the automobile or the plane for man's feet, substituting the lever, the wheelbarrow, or the power-driven machine for man's arm and shoulder muscles; or by eliminating some of the functions that have to be performed, e.g., eliminating the setting of type through the use of punched tape, eliminating the thousands of operations that are needed to assemble an electronic circuit through the use of printed circuits. Changes that involve the elimination of functions may have great impact when they come—but they come very infrequently.

Most technological change, therefore, involves the mechanization of existing functions—what the brilliant Canadian student of technology, Marshall McLuhan, calls "extensions of man." Three broad kinds of function can be distinguished: muscle power or sheer physical strength; sensory-manipulative operations such as picking things up and moving them elsewhere or guiding a shovel to the right spot with the right amount of force (as distinguished from the exertion of that force); and problem solving, using the brain to analyze a problem, select and process the necessary information, and reach a solution. What distinguishes the computer from most previous technological innovations—what makes it so awesome—is that it can tackle the second and third of these functions, not just the first. This enormous potential of the computer is the kernel of truth—a very large kernel——that the Triple Revolutionists have got hold of, and that gives their arguments so much surface plausibility.

But when we try to apply this newest extension of man to the process of physical production, we start running into difficulty. Many kinds of gross physical activity have already been mechanized out of existence, or soon will be, by simple and relatively inexpensive means, such as conveyer belts, lift trucks, and overhead cranes. The odor of perspiration has largely disappeared from the factory and the construction site. Most of the people left in the production process are involved in sensory-manipulative operations like assembling automobiles or directing a steam shovel. And these tasks—relatively unskilled and uncomplicated as they may appear—are the hardest operations of all to automate.

In addressing a meeting some years ago on the theme "The Corporation: Will It Be Managed by Machines?" Professor Herbert Simon of Carnegie Institute of Technology, one of those working on the furthest frontiers of computer utilization, reflected on a tableau that had been enacted outside his office window the week before, when the foundations for a new building were laid. "After some preliminary skirmishing by men equipped with surveying instruments and sledges for driving pegs," Professor Simon observed, "most of the work [was] done by various species of mechanical elephant and their mahouts. Two kinds of elephants dug out the earth (one with its forelegs, the other with its trunk) and loaded it in trucks (pack elephants, I suppose). Then, after an interlude during which another group of men carefully fitted some boards into place as forms, a new kind of elephant appeared, its belly full of concrete, which it disgorged back into the forms. It was assisted by two men with wheelbarrrows—plain old-fashioned man-handled wheelbarrows—and two or three other men who fussily tamped the poured concrete with metal rods. Twice during this whole period a shovel appeared—on one occasion it was used by a man to remove dirt that had been dropped on a sidewalk; on another occasion it was used to clean a trough down which the concrete slid." Simon concluded, "Here, before me, was a sample of automated, or semi-automated, production."

What the sample suggested was that automation is not, and cannot be, a system of machines operating without men; it can be only a symbiosis of the two. The construction site demonstrated another important fact: we may be further from displacing the eyes, hands, and legs than we are from displacing the brain. The theoretical physicist, the physician, the corporate vice president, the accountant, and the clerk, Simon suggests, may be replaced before the steam-shovel operator or the man on the assembly line.

Our Versatile Children

The reasons are partly technical, partly economic. The technical have to do with man's present superiority over machines in dealing with what Simon calls "rough terrain"—the uneven ground of a construction site, the variations in materials assembled in manufacturing, or the irregularities in the shapes of letters, the sound of words, and the syntax of sentences. Man's versatility in handling rough terrain was never really appreciated until engineers and scientists tried to teach computers to read handwriting, recognize colors, translate foreign languages, or respond to vocal commands. The human brain turns out to be, as Herbert Simon puts it, a remarkably "flexible general-purpose problem-solving device." An adult can recognize over a million variations of the color blue. The merest child can recognize an "e" in upper or lower case, in italics or upright, in boldface or regular, in print or handwriting, in manuscript or cursive, and so on almost ad infinitum—and all of these in an almost infinite range of sizes, colors, thicknesses of line, etc. And he can catch the meaning of words spoken by a voice that is masculine or feminine, high-pitched or low, loud or soft, pronounced with an enormous variety of regional and foreign accents. In short, the central nervous system is an incredibly versatile machine.

Its versatility is equally great—perhaps greater—in dealing with activities involving the coordination of eyes, ears, hands, and feet. "Manipulation is a much more complex activity than it appears to be," Ralph S. Mosher of General Electric wrote in the October 1964 *Scientific American*—even the seemingly simple operation of opening a door. "One grasps the doorknob and swings the door in an arc of a circle with the hinge axis at its center," Mosher explained. "The hand pulling the door must follow an arc lying in a plane at the level of the knob parallel to the plane of the floor, and it must conform to the circumference of the circle defined by the distance from the knob to the hinge axis. In doing this the hand, assisted by the human nervous system, is guided by the door's resistance to being pulled along any other path. In other words, the human motor system responds to a feedback of forces that must be interpreted. A strong robot, lacking any means of such interpretation and free to pull in any direction, might easily pull the door off its hinges instead of swinging it open."

The complexity of the economic considerations that determine what to automate, and when, is shown in International Harvester's construction-equipment factory outside Chicago. One of the plant's three engine-block lines is "automated"—i.e., it employs numerically controlled machine tools to machine engine blocks for enormous earth-moving machines, which are produced in relatively small volume. (For any product produced in quantity, conventional machine tools are cheaper.) But the blocks are moved from one automatic machine station to the next by men, using a simple overhead crane. Right next to this automated line is a conventional machine-tool line turning out vast numbers of engine blocks for tractors and trucks. On this conventional line, first installed in the late Thirties, the engine blocks are moved from station to station by conveyers and other transfer devices. The reason is simple: the volume handled on the "automated" line doesn't justify the cost of installing and operating transfer machinery for a conveyer belt; it's cheaper to move the blocks by hand.

New Jobs, Old Skills

Because it may actually be easier to mechanize or automate clerical, managerial, and professional work than the kinds of blue-collar work that still remain, current discussions of the labor market may be exaggerating the future demand for professional and technical workers and underestimating the future demand for blue-collar workers. The discussions almost certainly overestimate the tendency for automation to upgrade the skill requirements of the labor force. "It is not true," Professor James R. Bright of Harvard Business School, perhaps the most careful academic student of automation in the United States, has written, "that automaticity—automation, advanced mechanization, or whatever we call it—*inevitably* means lack of opportunity for the unskilled worker and/or tremendous retraining problems." In some instances skills are up-graded; in some they are reduced.

Over-all, however, what evidence is available (and there's painfully little) suggests that automation does not radically alter the existing distribution of skills. *Jobs* change, all right, but not the level of skill, particularly as firms gain more experience with automatic equipment. When business computers first came into use, for example, it was generally assumed that computer programmers needed at least a college degree. Today most computer users find a high school education adequate, and even this can occasionally be dispensed with. There is an enormous amount of repetitive work, moreover, under automation. The work may involve a different kind of rote, but it is still rote; it's hard to imagine a much more monotonous job than that of key-punch operator.

The crucial point is that we don't have enough experience with automation to make any firm generalizations about how technology will change the structure of occupations. On the one hand, automation may tend to *increase* the proportion of the population working as mahouts and wheelbarrow pushers in Herbert Simon's metaphor and to decrease the proportion working as scientists, engineers, technicians, and managers, because it may prove easier to displace people at these latter jobs than at the former. On the other hand, rising incomes will tend to increase the demand for services, in which jobs typically involve ill-structured problems and "rough terrain"; the demand for teachers, psychiatrists, journalists, and government officials, for example, is likely to expand faster than the demand for ditchdiggers or light-bulb changers. (The large increase in employment of clerical and professional and technical workers that has already occurred has been less the result of technological change per se than of the fact that industries employing relatively large numbers of such workers, e.g., insurance, education, medical care, have increased their output much more rapidly than industries employing relatively few. There has been relatively little change in the proportion of professional and clerical workers *within* individual industries.

A Question of Costs

Sooner or later, of course, we shall have the technical capability to substitute machines for men in most of the functions men now perform. But the decision to automate would still be an investment decision—not a scientific decision. At any one point in time, businessmen may choose between a wide variety of combinations of capital and labor. Their choice is affected very strongly by the relative costs of capital and labor—illustrated quite clearly, for example, in the fact that International Harvester finds it cheaper to use men than conveyers to move the engine blocks from station to station on that "automated" engine-block line.

In the last analysis, men will not be replaced by machines because widespread substitution of machines for men would tend to reduce the price of the latter and increase the price of the former, thereby creating a new optimum combination of the two. At any given moment business firms will use capital, i.e., machinery, instead of labor in those operations where machinery's advantage over labor is the greatest, and they will continue to use men in operations where the machine's advantage is the least. For the last 150 years of constant technological change, with only rare exceptions, such as in the 1930s, capital and labor have managed to combine in the United States so as to keep 95 percent or more of the labor force employed. This has been a remarkable record, one that has made the United States economy the envy of the world. It would be premature to conclude that this record cannot continue indefinitely.

Meanwhile, if one focuses on a single industry, or a single plant, or a single process within a plant, of course, he can *always* find rather frightening evidence of technological unemployment. Consider this description of technological change in the steel industry, written by an English journalist after touring an up-to-date mill in the United States. "The thing that struck me first was how few men were about," he reported. "To watch the way in which ingots were gripped from the furnaces, laid on rollers, carried on to be pressed, rolled out with steel fingers automatically putting them into position, you would have thought the machines were human." Even more impressive, the Englishman found, was the process for manufacturing steel rails: "From the moment the ore is pitched into the furnace until the rail [is] finished, everything is done by machinery, and *no man has a direct hand in the work.*" (Emphasis added.) The plant was the old Carnegie works at Homestead, Pennsylvania; the time of the visit was 1902.

U.S. Industries' marvelous mythmaking machine[2]

"To tell the truth about automation is not an easy task. Too many people are willing to accept too many myths," the late John I. Snyder, Jr., president and chairman of U.S. Industries and a self-styled "pioneer" in the design and production of automation equipment, complained.

Indeed they are—and one reason is that Snyder himself, through his testimony before congressional committees, his frequent press releases, and his generous assistance to anyone interested in automation, contributed substantially to the cultivation of these myths. One contribution has been that darling of the science fiction writers, the robot—specifically, U.S. Industries' TransfeRobot, an inexpensive ($2,500) automaton capable of replacing people on almost any assembly-line task.

Or so a number of enthusiastic or frightened writers have reported. A close look at both the reports and the record helps explain why "the truth about automation" is so hard to determine.

In 1959 the New York *Times* reported the development of the TransfeRobot 200—"a new concept in automatic machinery designed to eliminate the dull, repetitive tasks of employees in small as well as large plants."

In 1960, Snyder told the congressional Joint Economic Committee that "we stand on the threshold of a new era in automation. As the earlier phase was dominated by the introduction of giant automated industrial complexes, such as refineries or engine plants," he amplified, "the era which we are now entering will have as its distinguishing factor the introduction of small automation units into existing factories. In the past the industrial producer has had to go to automation. In the era we are entering automation will come to

---

2. This case study of one widely publicized piece of "automation equipment" provides a clear insight into the way in which at least some of the myths of automation are generated.

the producer." This new era, Snyder modestly explained, was made possible by U.S. Industries' development of the TransfeRobot.

In June 1961, the New York *Times* reported that U.S. Industries had developed "the first general-purpose automation machine available to manufacturers as standard 'off-the-shelf' hardware"; the report quoted Edwin F. Shelley, then a U.S.I. vice president, to the effect that previous automation devices had been for special purposes or custom-made. "The new machine, called a TransfeRobot," the *Times* continued—apparently forgetting that it had announced the development of the TransfeRobot in 1959—"has been in use some time under test conditions. Now it will be marketed for sale or rent."

In October 1961, the *Times* once again introduced its readers to the TransfeRobot. The occasion this time was the dedication of U.S. Industries' Silver Spring, Maryland, "Automation Center." "The star of the show," the *Times* reported, "was a self-perpetuating device, said to be the first of its kind in industry, called TransfeRobot 200, a registered trademark. It contains its own electronic guiding device and is said to be the first low-price, flexible, automation machine to be produced in quantity for use on any production line." At the ceremonies Shelley announced that "more than fifty TransfeRobot 200's have already been purchased by major companies for use on their assembly lines." This favorable response, he added, "indicates to us that within five years the probable annual market for such devices will total some $100 million."

In January 1962, Donald N. Michael published his widely publicized "Cybernation: The Silent Conquest," described by the *Times* as "a gloomy report of an automated world, in which people will be consigned to the junk heap." As evidence of the versatility of automatic machinery, Michael pointed, among other things, to the TransfeRobot 200, quoting that June 1961 *Times* article. Michael's quotation of the *Times* article was then reproduced in *Computers and Automation* magazine, and a reproduction of *that* reproduction turned up a few months later in a U.S. Industries promotional booklet. The U.S.I. press release that seems to have inspired that 1961 *Times* article had—to use a term that recurs in discussions of the miracles of automation—"closed the loop."

In a July 1963 article entitled "Automation: Its Impact Suddenly Shakes Up the Whole U.S.," *Life* reported that "almost anything that hands of woman can do" the TransfeRobot "can do better, faster, more cheaply." "So far," Snyder was quoted as saying, "we have not been able to find any material or any shape or any size it can't handle."

The other thing Snyder was unable to find for the TransfeRobot was a real market. "I think you're on the wrong track," James H. Cassell, Jr., then U.S.I. public-relations vice president, finally told a *Fortune* researcher in September 1964 after she had tried for months to find out how many TransfeRobots had been sold. "You're pursuing something obsolete." The TransfeRobot, he admitted, had not been a good piece of equipment. Pressed again for sales figures, he guessed they had come to "twenty-four or twenty-five robots"—less than half the number claimed in 1961. Asked which corporations had bought the robots, Cassell said he would have to check and call back. An assistant made the call, explaining that the TransfeRobot 200 had been "marketed"—i.e., lent to potential customers on a trial basis—but that none had actually been sold in the sense in which that term is usually used.

Further digging by *Fortune,* however, turned up six customers that had bought a total of eleven TransfeRobot 200's. But none of the five firms willing to talk to *Fortune* have been able to use the TransfeRobot 200's successfully. One company never used the robot at all; in trial runs, the machine broke down almost daily. Another firm retired the TransfeRobot after six months, and a third is still tinkering with its robots in an effort to get them to perform up to specifications. *Fortune* did find a satisfied customer for the TransfeRobot 210, a later model. This robot, custom designed at a cost of about $10,000, displaced one man.

In any case, U.S. Industries, as still another of its public-relations spokesmen explained, is no longer "pushing the individual TransfeRobot." ("If someone asked for one," he added hopefully, "we would sell it to him.") "The TransfeRobot as such is not functional," Cassell elaborated, "but what we learned from it is being used in our automation systems."

Sales figures on the "automation systems" also proved elusive. "We're concentrating 100 percent in the candy industry," Cassell explained in December 1964. As nearly as *Fortune* was able to determine, two or three such systems were on order, and one was actually in full operation—in the Smiles 'n Chuckles candy factory in Kitchener, Ontario. The $250,000 system handles about 30 percent of the factory's packaging. All told, Smiles 'n Chuckles employs, on the average, 385 people in the Kitchener plant, 85 of them still in packaging; U.S. Industries' system displaced 12 people.

The Social Cost of Cybernation

Ben B. Seligman

The social costs of cybernation may be measured by the net loss of jobs, approximating perhaps two billion dollars per year, but even more importantly, by the alterations in business organization and the creation of social rigidities.

There may be no money value on the latter two. Their cost can be reckoned in terms of their impact on human existence. The promise of creating new jobs stemming from cybernation is an empty one in the light of private industry's inability to provide as many as five per cent of the new jobs opened since 1957.

The most serious social cost involved in the new technology is the imposition of rigid mechanistic modes of behavior on society.

Problems stemming from the introduction of cybernation are much too vast to be discussed adequately in a quarter of an hour. Nevertheless, I shall try, even if the results represent but a sketch of an outline for an adequate analysis. One can think of at least three broad rubrics: (1) irreversible structural changes in the work force; (2) the alterations enforced on business organization; (3) the rigidities that may be imposed on the social structure itself.

Structural Change

Let us proceed to the first, a problem area that the practitioners of conventional wisdom in economics insist is non-existent. But unlike the emperor's clothes, the problem is quite real and there are enough displaced persons—displaced by cybernation or automation—now seeking other jobs who can testify eloquently to its reality. The United States Department of Labor has told us that two-hundred-thousand manufacturing jobs a year will be lost by 1972 because of the advanced technology brought on by the computing machine. From 1953 to 1959, within a short span of six years, eighty per cent of the decline in factory openings could be traced to automation. Moreover, this report, given originally at an Arden House conclave about a year ago, said nothing of those file clerks and accountants whose positions evaporate everytime a fresh piece of data-processing equipment is installed in an office. Other observers are even gloomier—estimates ranging up to forty-thousand lost jobs a week are attributed to the computing machine.

At this rate, simple arithmetical calculation tells us that there would be no work force by 1999 A.D. Of course, such an estimate fails to take into account whatever job creation might stem from the new industrial frontier or from increased demand. Hence, let me offer my own (somewhat more) conservative estimate of *net* job destruction.

If one assumes frictional unemployment of two per cent, as it used to be done (one wonders where and when the current four-per-cent figure got into the act) then, with a work force of seventy-two million and total unemployment of over four million, there would appear to be almost two-and-three-quarter million persons out of work for reasons other than ordinary economic adjustment. Such a presumption is sustained by the fact that unemployment increased seventy-eight per cent in the sixteen years between 1947 and 1963, as contrasted with a twenty-per-cent increase in the work force. One may hazard the guess that perhaps half the unemployment above frictional levels is structural. If this is the case, then almost twelve-thousand jobs a month have been destroyed after all upgrading, new hires, and rehires. The figure, modest as it is, suggests the magnitude of impact.

In a decade, a million-and-a-half jobs have been irrevocably lost. In terms of wages and investment that might have stemmed from such purchasing power, the loss would appear to have been in the order of almost two billion dollars each year.

Cybernation and Jobs

Sometimes we hear that cybernation creates new jobs. If it does, they are not visible. True, there was a 4.3-million increase in the employed work force between 1957 and 1963. But this increase came mainly from federal, state, and local governments: sixty-five per cent of the increase consisted of direct employment by these jurisdictions and of procurement programs on their behalf. Non-profit institutions accounted for sixteen per cent; part-time jobs generated by private demand for fourteen per cent; and full-time jobs created by industry's own effort for five per cent. There is small consolation for the factory worker in the expansion of government employment, since he does not possess the transferable skills. Furthermore, recent reports suggest some doubt about any continued increase in federal jobs. A recent study of employment prospects, issued jointly by the Labor and Commerce Departments, predicted a decline in employment in eighteen major industries, mainly as a result of technology. And in fourteen other industries—including transport, electronics, and trade—only increased demand, it was said, would overcome the effects of spreading labor-saving devices.

The fundamental economic relationships require that, as productivity increases, there be an increase in output or else jobs go down the drain. There is no doubt that productivity, stemming from the new technology, has increased. From 1909 to 1947, the average annual rise in productivity was two per cent; from 1947 to 1960, about three per cent; and from 1960 to 1963, it was 3.6 per cent each year. With sixty million persons in private employment and a three-and-one-half-per-cent rise in productivity each year, there would have to be enough activity in the economy to create well over two million jobs just to keep unemployment from rising, to say nothing of employing those just coming into the work force at the rate of more than 1.5 million each year.

While output has been rising, the pace has not been fast enough to overcome the enhanced productivity stemming from automation and cybernation. It is all too easy to say "Let's produce more!" There is no way of encouraging greater output, at least in the private sector, when there is no greater market, and there is no greater market when jobs are lagging. One solution, which is indeed on the verge of adoption by default, is to recreate Disraeli's "Two Nations" and simply dump the dispossessed onto a social slag heap.

Cybernation and Middle-Level Management

Much has been written already about the capability of cybernated machines. They learn and "perceive." They analyze stock market conditions, establish rocket flight patterns before the shot is fired into space, write television scripts that compare favorably with available scripts written by human television writers, compose music, translate, and play games. They combine high technical competence with just enough I.Q. to keep them tractable. They do precisely the kind of work to which junior executives and semi-skilled employees are usually assigned.

No slur is intended here, for in addition to the ordinary worker, the middle manager, the backbone of the average corporation, will be affected most by cybernation. He has a bleak future indeed, when computing machines relay information to one another, do all the scheduling, and control manufacturing from inception to the point of packaging and rolling into a box car. It is the industrial archon who ultimately wins out; for with the elimination of both plant and office staff, this man at the very top gains even tighter control over the decision-making process.

The sort of organizational looseness that prevailed before the advent of the computing machine is eliminated, and corporate structure becomes more formal, more "integrated," because with the computing machine there must be greater "coöperation." The number of links in the chain of command is reduced drastically; vice-presidents are soon out of a job. No less an authority than Herbert A. Simon of the Carnegie Institute of Technology has said that by 1985 machines will be able to dispense with all middle echelons in business. Production planning is handed over to the digital demon, while both the middle manager and the displaced worker drive taxicabs.

The sociologist may very well ask, whither the American dream of status and success?

Computing Machine Empires

Quite often, the computing-machine engineer tries to build his own empire within the corporation. Fresh to the ways of business life, he unabashedly plays havoc with established relations. He and his programmers cut across all divisions and ignore and undermine the authority of department heads and vice-presidents. Sometimes the new elite does lose out. It has not been unknown for a computing-machine installation to be yanked as a result of corporate internecine warfare.

Usually though, archon and engineers are in complete accord. With the computing machine creating certain expectations, the firm must operate through a series of highly rigid sequences. Flexibility has been dispensed with, for the whole plant is now a single technical structure in which total performance must be "optimized." The engineer examines each step in the process solely in terms of efficiency—industrial logic of the most unremitting kind takes primacy. Under cybernation, the engineer or mathematician is the skilled man in the plant, while workers, those who remain and those who do not, are expected to adjust with equanimity to a situation for which they cannot take responsibility. In fact, the engineer's attitude quite often is tough and hard— too much so for the ordinary man.

What the worker doesn't know, says the engineer, won't hurt him. The "scientists" appreciate only "facts." The human problems of an industrial system frequently have little meaning for them. Unlike the organization men of the 1950's they are usually "inner-directed" disturbers of the corporate peace, freebooters in pursuit of the idols of efficiency. And since efficiency is measured by high profit and low cost, such scientific ruthlessness meets the approval of the archon, who may not know what the scientist is doing.

Top management merely voices a faith based on payoff. Thus the programmer, who often assumes the aspect of a medieval alchemist, runs his own show, designing projects, cutting corporate red tape with abandon, and advising the industrial relations department that labor displacement is none of their business. At best, the engineer can parrot the conventional economic wisdom by repeating that automation and cybernation create new demand and new jobs, up-grade the worker and inspire everyone with its challenge. There must be a certain glory in the marvels of cybernation, but the men who once worked in the chemical plants, the oil refineries, and the steel mills are now out of sight and out of mind.

Distortions of the Value System

Perhaps the most serious social cost stems from distortions of our value systems imposed by cybernation. For the philosophic preconceptions in computing-machine technology are thoroughly mechanistic. The presumption is that all behavior is thoroughly objective and that reality can be compressed into mathematical equations. Once this is done, prediction comes easily. And so does control of the human being and his society.

Certitude, such as was never known before, can be provided, particularly with a high-speed digital machine. Man has no soul, says the psychologist, and his emotions and irrationality are mere outputs emanating from a "black box" whose electronic characteristics will soon be revealed. Introspection is utterly useless, for behavior can be explained in the relatively simpler terms of stimulus and response. The actions of intelligent human beings are thus understood as a product of complicated, but finite and determinate, "laws." It is this philosophical outlook that underpins rote memory experiments, the generation of visual displays on the computing machine, studies in mechanical perception and artificial concept formation. It is confidentially argued that all this will eventually demonstrate how images are transformed into ideas and action by the brain.

An astounding intellectual arrogance infuses the thinking of these specialists. One went so far as to assert that there is too much irrational reverence for human intelligence and that, in fact, there is nothing special about either intelligence or creativity. Intelligence, as ordinarily conceived, is deemed to be but "an aesthetic question, or one of a sense of dignity," not a technical matter. It is really a complex of performances which we may respect but not necessarily comprehend. Consequently, it would be just as easy for these creators of androids to simulate large-scale organizations that replicate the behavior of individuals. Artificial social structures and processes would be made to develop according to certain rules, as the game of philosopher-king continues unabated.

Thus, it is presumed that the machine will itself generate the "time stream of a decision-making process." However, the difficulty is the supposed credibility that it lends to whatever theory the computing-machine operator happens to have in mind. In the social sciences particularly, there are no absolute theories. Everything is "perhaps." Computing-machine experts seldom take the trouble to check their computations against ordinary observation. They expect the common man to accept what they say virtually on faith, simply because complicated equations have been stuffed through a computing machine. In a sense, mechanical or electronic craftsmanship has been substituted for a deeper meaning, and, as Robert Solo remarked, formalism has replaced human thought. It is seldom asked why the choice of the computing machine is necessarily superior to that of judgment rooted in human experience. But —so we are told—the imitation is better than the real thing. And reality is created out of the illusions engendered by the machine.

Disturbing about the efforts to cybernate existence is the fact that conditions are being created that make the manipulation of people too easy. An equation once formulated is presumed to encompass all pertinent factors. As Jacques Barzun said, the individual is to be given a number, stripped of differences, and turned into a manufactured object for analysis and abstraction. The sentient man is replaced by the Compleat Robot, and man's hope and his fate become irrelevant. But complex social and psychological situations cast into a framework suitable for the computing machines are often oversimplified, and the intricacies and subtleties of human response are ignored. Such callousness and expediency fosters and sustains a mass society in which individual uniqueness cannot survive.

The Negro and Cybernation

James Boggs

> To visualize the future rôle of Negroes in a cybernated society, one must review, if only briefly, their past rôle in American society and what this means at the present stage of industrial development.
> Historically, the rôle of the Negro inside this society has been like the rôle of a scavenger. That is, the Negro has been entitled to the leavings, the cast-offs of the whites: jobs which the whites did not want any more or refused to do at all; housing that the whites had moved out of; neighborhoods that whites no longer considered good enough for them; schools that they had abandoned.

In each industry where machinery played a vital rôle, the Negro played a special rôle—that of being the last to be recruited and usually only on an emergency basis, e.g., war.

To visualize the future rôle of Negroes in a cybercultural society, one must review, if only briefly, their past rôle in American society and what this means at the present stage of industrial development.

Historically the rôle of the Negro in this society has been the rôle of a scavenger. The Negro has been entitled to the leavings, the castoffs of the whites: jobs which the whites did not want any more or refused to do at all; housing that the whites had moved out of; neighborhoods that whites no longer considered good enough for them; schools that they had abandoned.

With the Negroes to take over the leavings, it has been possible for white Americans to continue to graduate or progress upwards as the country developed, in very real terms, getting better jobs, better housing, better schools and neighborhoods with every year and always at the expense of the Negroes. The country itself was able to develop because there was the Negro inside the society to fill the void left by those moving upwards, so that there was no waste of obsolete homes, jobs, schools. This is the process that has been taking place ever since the early economic development of the country, in agriculture and in industry.

In the eighteenth century, when tobacco was the main cash crop in the South, Negroes were becoming obsolete as crop-tenders, and in fact, there was widespread consideration of discontinuing the import and the use of slaves and even talk of sending them back to Africa. At this juncture, however, a machine was invented which was to establish the pattern for all future utilization of labor up to the present. This machine was the cotton gin, invented by Eli Whitney, which made it possible for mass labor to be used in the cotton fields. This led to a tremendous demand for slaves.

Ever since, each invention or improvement in machinery has created a need for more manpower. This need for more labor, side by side with the development of new machinery, has played a key role in the rapid economic advancement of the country, attracting wave after wave of immigrants to the labor force who, in turn, became the source of increased capital for future investment.

In each industry where machinery played this vital rôle, the Negro played a special rôle—that of being the last to be recruited and usually only on an emergency basis, e.g., war. Only in agriculture was he the first mass force. Elsewhere he has been relegated to the most menial, manual labor under the worst conditions, e.g., foundry, pick and shovel in the mines, rolling mills, furnace rooms, janitors, material handlers. Thus, in the work process, within American society, the rôle of the Negro has been that of the scavenger. He got the jobs which white Americans would not do, which they considered beneath their dignity, which they had abandoned, or in dying industries.

No More "Negro Jobs"

Now, however, cybernation—i.e., automation with nerve centers operated not by man but by computing machines—is eliminating the "Negro jobs." Thus it is also destroying the process, the ladder, by means of which white workers moved up, leaving the dregs behind to the Negroes. Thus, in the last four years, the employment of Negroes expanded primarily in the civil service and social service arenas—jobs in local, state, and federal government in teaching and social work. Meanwhile, very few Negroes have been hired into industry, mainly because few of them have been taking up in college the kind of technical courses needed in these highly developed industries. Rather, Negroes have been taking, for the most part, courses in those fields which have been open to them—e.g., civil and social service and teaching. However, even here, except for teaching, higher standards have been set for them than those in practice when these jobs were "white jobs." For example, two years of college has become the standard for a Negro clerk typist in civil service work, whereas, years ago when whites dominated this field, a high school education was the standard.

Today, very few college-trained whites are going into these civil and social service jobs. Instead they are going into scientific research and development —for industry, for government—on the basis of the new and highly developed technological level of modern industry. Meanwhile, few Negroes are being admitted into these fields except on a token basis.

Because the difference between Negro and white employment today is so flagrant, many Negroes are demanding that their children get the technological education that will equip them to do what whites are doing. This thinking among Negroes compares to what is taking place in the civil rights field in general, where the great majority of Negroes still believe that all they have to do is have the same thing whites have, and be like whites, and that this will solve their problem.

The Obsolete Occupational Rôle

Meanwhile, however, even whites are going to find out increasingly that despite their technological education, cybernation is going to make their occupational rôle obsolete. The computing machines itself is going to take over more and more of the work that now requires a highly technical education.

On the other hand, the Negro, who has been going more and more into the fields of education, social and civil service, is going into the fields where the important decisions will be made about the structure of society at the stage of cybernation. These fields have been vacated by whites primarily because material accumulation and industrial development have been regarded as the apex of our society while politics and social work have been looked down upon.

This means that Negroes, who can never become a major factor in technology from the vocational aspect, will become a major factor in determining the final disposition of the results of technology, *i.e.,* in the revolutionary and political arena where decisions will be made that govern the use of things rather than how they are to be produced. And these are the most important decisions in a cybercultural society.

Suppose, for example, that General Dynamics or another development corporation is contemplating the production of the kind of electronic equipment that would displace the vast majority of General Motors workers on all levels. The important question would not be whether this *could* be done technically. Rather, it would be what is going to happen to all these former General Motors employees, and how they are to be able to exist without their employment. In other words, the main question would be a political question, not a technical question.

Thus the rôle Negroes are going to play in a cybercultural society is not determined by what they consciously want, *i.e.,* to be employed technically like whites, but by the pace and the manner in which cybernation is taking over. At this time, after all these years of systematic relegation to the rôle of scavenger, it is impossible for Negroes to be integrated into the industrial structure on an equal basis. Integration, at best, will be only a token not a solution for our old problems. Whereas, in past periods, there was a fairly wide range of jobs at the bottom of society which Negroes could fill as scavengers, today there is no bottom left *inside* the industrial structure. Negroes are still at the bottom—but on the *outside* rather than on the *inside*. And being on this bottom, the Negroes are going to be forced, more and more, to struggle in the political arena for the right to share what is achieved inside the economic arena.

The Blessings of Technology

Technology brings with it one major blessing: when it reaches the stage of cybernation it does leave an arena at the bottom for people. Inside production, exploitation becomes the exploitation of machines rather than of men. But when the exploitation of people *inside* the production process is no longer possible, the exploitation of the consumer in the market place begins. Those who control the productive machinery cannot exercise the same controls over the consuming have-nots that they have had over those who labored inside the production process. Being outside the production process, the have-nots can choose to exercise their political powers.

Thus, because the Negroes have been, and still are, the scavengers in the economic arena—"the last hired and the first fired"—they are in the best position to break with the economic tradition that has dominated the United States and that is becoming outmoded. The strategic position of the Negroes is on two levels: (1) whenever they are gaining employment, they do so in civil and social service and in educational fields; and (2) when they are losing employment and are thrown outside any employment possibility, they constitute a mass force with the most concentrated political needs. And, because they have the least to gain from this society, they also have the least responsibility to this society.

It is absolutely absurd to think that the Negroes, having been economically deprived for so long, will catch up economically with the whites, and achieve equality with them on a vocational basis. It is equally absurd to think that the whites, having entrenched themselves both physically and emotionally so deep within the economic structure, will be able to catch up with the Negroes in terms of political orientation or the concentration on human relations. Thus the Negroes, by virtue of their past experience, are better prepared for life and leadership in the new cybercultural society than the whites.

What is Leisure?

Charles K. Brightbill

> *We are unleisurely in order to have leisure.*
> (Aristotle)

DEFINITION

The term *leisure* derives from the Latin *licere,* meaning "to be permitted," and is defined in the modern dictionary as "freedom from occupation, employment, or engagement." Even the term for our revered center of learning, the *school,* derives from the Greek *skole* and the Latin *schola,* which mean not "school" but "leisure." The Greeks believed that the purpose of work was to attain leisure, without which there could be no culture.

Leisure has meant different things in different cultures, and today, unfortunately, there is too little agreement as to what it does mean, and what it implies. There are those who insist that leisure involves so many implications and shades of meaning that it defies definition, or even intelligent discussion, except in terms of values, norms, and cultural orientation in relation to the behavior of particular class, ethnic, and regional groups. Leisure is seen by many as freedom from work. Others view it as an instrument for social control, a status symbol, an organic necessity, a state of calm, quiet, contemplative dignity, or a spiritual, aesthetic, cultural condition. A few insist that leisure is something quite different from free time, or that there just isn't any leisure, or that only negligible amounts of it are present in contemporary society.

In fact, it might be a good thing if a new word could be found for "leisure." Not only is it interpreted in too many conflicting ways, but it is also too often found in the wrong company. The economist George Soule had this in mind when he said that the word "leisure" was undesirable because it brought with it old associations and needed to be justified and defended. He thought that the puritanical concept which made *work* the divine discipline was too much of a handicap. He suggested using "unpaid time" instead of "leisure."

That leisure may bring all or any of these things to mind is true enough. But no matter how one tries to modify the concept of leisure, *time* is its essence. Leisure can no more be divorced from the element of *time* than it can be completely separated from the function of *work.* Ultimately, leisure must be identified with the *when* quite as much as, if not more than, the *how.* That is not to deny, however, that it is the endless ways in which leisure can be used, for good or bad, which make it significant.

TIME

Leisure, then, is a block of unoccupied time, spare time, or free time when we are free to rest or do what we choose. Leisure is time beyond that which is required for *existence,* the things which we must do, biologically, to stay alive (that is, eat, sleep, eliminate, medicate, and so on): and *subsistence,* the things we must do to make a living as in work, or prepare to make a living as in school, or pay for what we want done if we do not do it ourselves. Leisure is time in which our feelings of compulsion should be minimal. It is *discretionary* time, the time to be used according to our own judgment or choice.

Although these three types of time are used in vastly different ways, they do have common characteristics. For example, each time pattern is highly flexible and may be increased or decreased, depending upon various circumstances and conditions. Also, none of these can be defined in terms of what is *good* or what is *bad* use. We may have bad habits in our leisure, just as in our work, or in our eating.

Just as there are different kinds of time, there are two molds of leisure— true leisure and enforced leisure. *True* leisure is not imposed upon us. *Enforced* leisure is the leisure we do not seek—it is the time the victim of confining illness has on his hands, it is the "time off" which grandpa gets when the company says he has reached the retirement age, even though he is fit and wants to continue. Who wants a vacation without a job to which to return? And working does not always mean that we are paid for our efforts.

WORK

To speak highly of leisure is not to disparage or ignore the importance of work. Of all the claims which can be made for its attractiveness, leisure as a substitute for work is not among them. Work is a symbol of growth which challenges and brings a renewal of motives. Industry may often be another word for conscience—sometimes too much so. It is difficult to imagine anyone's being happy without work. We need work quite as much as we need food, to say nothing of most of us having to work if we expect to eat. Work carries with it the feelings of purposefulness and usefulness which are so indispensable to our self-respect. It has its own built-in incentives. Even persons who think they hate work are not so sure when they are out of work. Some people do not discover what work has meant in their lives until they have raised their families and have retired. That we cannot escape the feeling of compulsion while working does not make work any less desirable to us: the oceans of enforced leisure brought by the depression years of the 1930's contributed to a low ebb of human morale.

Yet, if we have to choose between the two threats of *too little regard for work* or *too little regard for leisure,* which shall it be? Robert Louis Stevenson might have answered:

Extreme *busyness, whether at school or college, kirk or market, is a symptom of deficient vitality. . . . As if man's soul were not too small to begin with . . . (some people) have dwarfed and narrowed theirs by a life of all work and no play. . . . It is not by any means certain that a man's business is the most important thing he has to do.* (An Apology for Idlers.)

But as George Bernard Shaw observed, "a perpetual holiday is a good working definition of hell."

Admittedly, we humans would be badly off without the inducements, accomplishments, and satisfactions which come from labor. On the other hand, we should not be too much tied to work. Nor should work be equated with Godliness or treated as the only test of success. When work becomes an end in itself, when one cannot enjoy the things for which he has worked because he is too weary or doesn't have the time or the desire to enjoy them, work is clearly a liability.

Work is the sharpest hazard of all if it causes emphasis on material possessions. Marcus Aurelius cautioned us to remember "that very little was needed to make a happy life." We often work too hard to buy more automobiles, more home appliances, and all kinds of luxuries. An astonishingly large number of persons engage in "moonlighting"—the practice of holding two jobs at the same time. Building our material empire does bring with it more refrigerators, more television sets, more expensive furs, and more high-priced cars; unfortunately, it also creates the need for more aspirin and results in more insomnia, more nervousness, more high blood pressure, and more boredom. A life that follows this kind of work pattern goes in circles. It is a path which always brings you back to where you started—work to buy labor-saving devices to release time for more work to purchase more labor-saving devices. Nobody ever went anywhere on a merry-go-round. In such a hectic existence, the more you accelerate your pace, the more is demanded of you. If you slow down, you get in your own way. It is like being allergic to yourself.

Moreover, our attitudes toward leisure and what we do with it are related to our job attitudes, as has been shown in a study of the off-duty habits of airmen. In fact, the United States Air Force found that what the men did when they were not on duty had a direct correlation to their attitudes toward their jobs. Yet even though leisure may help restore man for work, it does not exist *for* the state of work. Leisure has a much larger and higher role than this. Although those who worship at the shrine of toil would have us believe that we *live to work,* a far more sensible view is that we *work to live.* To look upon leisure *only* as a respite from work is never to discover its full potential. Leisure is the foundation of culture beyond the utilitarian world. It is man's eternal opportunity to overcome his inner impoverishments, although it constitutes no guarantee that he will do so.

Our daily lives can be divided into three parts—sleep, work, and leisure. Part of that time not given to sleep and work is used in performing those duties which are biologically necessary to sustain ourselves. These include eating, cleansing our bodies, and resting (which may idle both the mind and the body). The time that remains is the *true* leisure, that period of time which becomes so significant to us and which causes us to give serious thought to how it is used. Aside from using this *true* leisure to worship, or reconverting it into an opportunity for sleep, rest, and the like, people by and large use it for *play* and *recreation.*

PLAY

Play, despite many attempts to analyze it, is not too well understood as a phase of the living process. Play is the free, pleasurable, immediate, and *natural* expression of animals, particularly the young. It has no observable utilitarian result but expresses itself in a way characteristic of the species. In the higher forms of animal life, play is more diversified and occurs over a longer span of time. The role of play as an adaptive device remains uncertain.

We cannot identify play by its form because there is no such thing as a standard human being. Can we predict with accuracy what kind of play will interest a boy of twelve? Well, if we have a large enough number of healthy, twelve-year-old boys who have been raised in the United States, many, but not all, of them may want to play baseball. We cannot even be certain that they are all "twelve" years old. A youngster may be twelve chronologically but sixteen intellectually, ten physically, and eight emotionally. Play is not stereotyped, mainly because people are different.

Durant says:

Our first great happiness is at our Mother's breast; but our second is in the ecstasy of play. What purpose moves these children to their wild activity? —what secret desire sustains their energy? None: the play is the thing, and these games are their own reward. Children are happy because they find their pleasure in the immediate action; their movements are not means to distant ends; their eyes are upon the things they do, not vainly on the stars; they fall, but seldom into wells. (Will Durant, Mansions of Philosophy.)

RECREATION

Of more significance than play to a society in which a vast amount of off-the-job-time is at the center rather than on the fringe of life is the worthwhile application of spare time. The recreative uses of leisure time may involve amusement, entertainment, participation in games or sports, or engaging in the more frivolous pursuits of life, but also those actions and attitudes which connote relaxation, the potential which leisure has for enriching and developing personality, and the opportunities it presents for the release of our creative powers. Because the recreative use of leisure deals almost exclusively with the enthusiasms of mankind, it is impossible to set limitations upon it! Spontaneous creative resourcefulness can often be found in recreation.

Recreation always invites *activity* of some kind. It may be the vigorous activity of playing tennis or of climbing a mountain, or the less tangible effort of reading a book or listening to music. But some kind of physical, mental, or emotional action, even if not visible on the surface, is necessary. It is *action* as distinguished from *rest.*

Recreation, like play, also has *no single form.* Its content is infinite, because the interests, needs, and wishes of people differ; what is enjoyable to one may be detestable to another. Thus, an action can be *recreation* to one man and *work* to another. We can climb a mountain or read a book, take a swim or listen to a symphony, attend a party or contemplate the passing scene. Here is the chance to be ourselves. My interests, needs, and wishes may differ from yours.

If there is a single word which identifies the recreative pursuit, it is *attitude.* My wife likes bridge; I detest it. The range of interests is infinite. Walking may be fun to the Sunday hiker but not to the postman. Some men fish for a livelihood; many more fish for fun. A role in the little theater may be ecstasy for a housewife, but just plain hard work for the weary Broadway actress. Depending upon one's attitude, something can be both pleasure and drudgery within a matter of moments. Carrying your skis up the mountain, despite the huffing and puffing, can be an experience full of joyful anticipation. But if upon arriving at the summit you discover the ski run is closed, lugging your skis back down the hill can be distasteful work. Carrying this a step further, many writers, painters, actors, musicians, photographers, engineers, and others find the origins of their life vocations in their play or recreation. But once they exchange the primary motivation of personal enjoyment in itself for the basic motivation of monetary compensation, what was once recreation becomes work, with its ever-present element of compulsion hovering over them.

Whether or not something is recreation depends upon the *motive* or incentive of the doer. If the motive is enjoyment and personal satisfaction and the doing of it has its own appeal, it is *recreation.* Rewards, such as improving your health, meeting new friends, and gaining new knowledge may be the result, just as there may be penalties, such as dissipating your energies, losing your savings, or having a hangover. But the primary reason for engaging in the activity is the personal enjoyment and satisfaction that can be found in it.

Another hallmark of recreation is that it *takes place during leisure.* It might be argued that there is *no* time when man is completely free from obligation. There are always some kinds of obligation—obligation to society, family, and the like. Nevertheless, spare or leisure time, in which obligatory pressures are few, is the time in which we are free to do what we please, how we please, with whom we please. It is during *this* period that recreation occurs. Recreation cannot take place during work time, with its charge of compulsion. This is not to say, however, that work cannot be personally satisfying or enjoyable.

The rewarding use of leisure assumes that there is something from which to choose and that we are capable of making a choice. Engagement must be voluntary. It cannot be cast upon the individual with the admonition that he *will* be happy—or else! Compulsion hinders rather than nurtures, self-discovery, self-expression, creativity, intellectual curiosity, self-satisfaction, or personal enjoyment. Opportunities in leisure can be planned and provided but they cannot be ordered, imposed, or forced—and still be called "recreation."

We must, under any circumstances, try to deal intelligently with the problems of leisure. If we do learn how to use leisure to cultivate our minds, hands, and hearts, we shall preserve and strengthen human values as well as make leisure contribute to the order, rather than the disorder, of life.

Leisure's Future

Sebastian de Grazia

What is practical to the free and to the unfree is different. The question of action plays no part. If you can decide who was free—Thales of Miletus or the little maid who laughed to see him fall—you can tell that the free are not against action or practice itself. They see people all about them who merely think they are doing something. Unimportant practice is simply uninteresting. They are not fleeing real life but making their way toward it. On the way they may trip over their own feet, not recognize their neighbor, be ignorant of feminine wiles or where the shopping center is, or what the boys in the back room are saying about machines, but what they know would make them all dizzy, all those whose time is not free for learning, including the maids of Thrace. Action, then is not the point. Contemplation is the sheerest of acts. To try to avoid all action would be unnatural and futile. Reducing intent on the world does seem to reduce activity to a minimum, but only because present standards push it to an artificial maximum.

To justify the life of leisure to the state, any state, not simply the democratic state, is a formidable task. An eminent economist had something like this in mind in saying that it is almost impossible for the scholar to be a true patriot and to have the reputation of being one in his own time.[1] The man of leisure cultivates the mind and may find the truth in his furrow, but he cannot say, when or if he finds it, or some of it, that it will be good for the state today, or next year, or in five hundred years. Besides, the state may appreciate it never. Furthermore, if he finds it, who knows whether he will communicate it? Epicurus wrote three hundred works, of which forty-one were said to be excellent; but he might just have decided to write none. Someone less con-

---

1. The economist who made the observation about scholarship and patriotism is Alfred Marshall. See Arthur C. Pigou, ed., *Memorials of Alfred Marshall,* Macmillan, London, 1925.

cerned than he to save a few good souls might not have written them. Nothing says that whatever one learns in leisure must be communicated to others.

The state may ask why men of leisure engage in talking and writing at all. In writing books or teaching they may be accused of resorting to persuasion, rhetoric, recruiting, or publicity. There are, it seems to me, acceptable explanations. A man's writing apart from letters can be considered a mnemonic device, or a way of reining in runaway thoughts, or else a celebration, an expression of immanence, an overflow of well-being, a richness, a song. "Verily—may the Lord shield me!—Well do I write under the greenwood." At all events it is solely the Epicurean version of the ideal that forswears politics. For the others politics is the only field important enough to pull a man away from his leisurely activities.

Such we saw to be the attitude of the Founding Fathers, as well as that of Plato. There is no doubt in them. The benefits of their life are for their country, if ever it needs them. They would communicate what they know, they would teach or write what they have learned; they would fight, too. Remember that while Aristotle does not admit work into a life of leisure, he will admit war. His logic is simple. Of course war takes you away from leisure and contemplation, but wars are fought to have peace, and peace is good because it offers leisure. Violent action, it seems, is acceptable so long as it is for the purpose of leisure, and not, as the Spartans had it, for training in the interval between wars. If the enemy won the war, leisure would be lost. Aristotle never questions this. It seems to be part of the belief that only Greeks are civilized, and so only they are capable of understanding what kind of state one needs. That state would be best that could create the life of leisure. It could have no higher purpose. Even Epicurus admits that the state's laws might be of some service by keeping wise men from being treated unjustly.

With this as background, perhaps we can further reason about the good state. The desire for tranquillity in the life of leisure leads toward a politics of peace both internally and externally. The convenience of tolerance leads toward a program of politics without crisis or great national efforts, perhaps a form of conservatism. The choice of stability as a criterion for truth and pleasure leads also to a preference for hereditary forms in government and property holding.

This, perhaps, would be the state that could reap most of the benefits leisure has to offer. Under other circumstances a country may consider the ideal of leisure dangerous and actually subversive. A regime whose support is not too steady may fear that truth seeking for itself may take a political turn. The seeker himself admits he does not know where his search will take him. A regime may also take the line that the ideal, especially its nonpolitical version, sets a bad example. By saying that patriotism is a hindrance, it affects loyalty. A bad example might not affect the many, but the few is already too much. Or if a regime is committed to great economic or military enterprises, the vaunting of a life of leisure in the country is a hostile doctrine. Teachings and writings do exist, and if they happen to persuade large numbers of persons to adopt the contemplative life, are they not political tracts?

Perhaps the life of leisure could be made easier everywhere if only the modern state could see something positive in it. Seneca in *De otio* said, "In brief, this is what we can expect of a man: that he be useful to other men; to many of them if he can; to a few, if he can but a little; and if he can but still less, to those nearest him; and if he cannot to others, to himself." Today this being useful by being useful only to oneself cannot be widely understood because the wise man is not a model, nor does heaven smile upon his protection. Otherwise his usefulness could be seen in his living his daily life.

Isn't it easier to hear a contemporary biologist or sociologist or even a political theorist asking, What function do such persons serve in the organism, in society, or in the state? Where is the benefit? On the contrary, there is harm. In this contemporary view the life of leisure is antidemocratic, antisocial, against organization, opposed to work and to most of the things men work for, and indifferent to home, mother, and perhaps even country. One can see why the man of leisure is a rare bird. Not all lands are as amiable as Ionia.[2]

There will be those who retort that though many people hold the above opinion of leisure, yet we must not forget that the advocacy of leisure is permitted nonetheless; for opinion in a democracy is kept free. Anyone can think, write, and speak as he wishes, even if one has nothing but bad to say. The state withholds its interference from any but illegal acts, and to think, write, and speak are not such acts. Still the question arises, is inactivity illegal? If a man does not want to work, but to contemplate, is he a madman? If, to be free from the Lilliputian threads of society, he wants to beg for his bread, should there be a law against it? If he leaves his wife, is he a deserter? By refusing to join the armed forces, is he a criminal? Should there be laws refusing him entrance to certain parts of the country because he has no visible means of support? What do the laws of vagrancy aim at? What is loitering? Omission can sin more than commission. Opinion may be free, yet the laws conspire against a way of life.

There is another defense of the life of leisure that seems to have a special attractiveness today. It has recently become commonplace to look for benefits from those whom the shoe of society habitually pinches, in particular artists and scientists. I have mentioned it here as the benefit of creativeness. As pointed out previously, nonfiction books on the future don't spend many pages on art in the future except to say that with more free time will come more art. Perhaps the lover of leisure can be tolerated or even appreciated for the benefit he brings to society in the long run through his creativeness in science, literature, or art. The argument is by no means absurd. Some will say, though, that it follows that the life of leisure should somehow be worked into workaday life.

2. For an example of the biological perspective, see Innes H. Pease and Lucy H. Crocker, *The Peckham Experiment,* Allen and Unwin, London, 1943.

Then everybody could be creative. An alternating of work and leisure is the usual proposal. At most their solution falls into the formula the Romans expressed as *otium/negotium*. It splits up time, thus again becoming the wrong formula for leisure. For contemporary use it translates well in the duality, free time/work time. As far as leisure goes, irrelevant.

The life of leisure cannot be justified to the state and perhaps also it has exaggerated its independence of the state. It still makes little difference. If it could be justified in terms of the state, then we could speak of its function. If we could do this, the life of leisure would no longer be free. It would have a determined relation to the state. It would become a state functionary. The same subservience would strain leisure were it to be justified in terms of any other society. Should it be possible to confine philosophers and artists in a retreat and say to them, Produce? The life of leisure may accomplish many things; it can promise nothing. Freedom, truth, and beauty is its religion. Let who will go whoring after commodities, and money, fame, wars, and power, too.

We shall go back a little now to bring some of the things we have already discussed in these chapters to bear on the matter of the benefits leisure offers. The phrase "the leisure problem" crops up often these days. Though many use it, they use it with several different views in mind. Many people, indeed most of those who use the phrase, worry that there is too much free time nowadays or that there will soon be too much. This, for the present and for the near future, is wrong. There is not much free time, nor likely to be much.

Many other persons not only believe there is too much free time but also that it is badly spent. In their ranks we should not only include those critics of culture who wish for an uplift in the common man's pleasures, but we should also include the many men, common or not, who are dissatisfied, perhaps without knowing why, with the way their life goes.

Little can be done for the improvers and the culture critics and for all those who, like them, see the problem as too much free time, badly spent. The ideas of work and equality block their way. Seen in this light, things can't get better until they get worse. If things get worse, it will not be their fault as much as that of uneasy people. The worn-out fringe of the Cockaigne dream shows up in the farewell salute familiar in American cities, "Take it easy," and its return, and equivalent of "O.K., I'll try to," or "Yeah, you do the same," both given with the ritual or cynical or resigned but unbeaten air of never, ever, making it. Life in happyland can be improved for them, but not by changing their activities. Work being what it is, the time machine also, and the make-up of most of us being what it is, free-time devices cannot move to a different plane. They can rely less on commodities and purchases, however; they can slow down to a walk in a less hurried setting; they can achieve better taste. But these changes will not come in a vacuum. Changes in political beliefs will take place, too.

Change may be calm or fierce, gradual or swift. Just as it is impossible to keep men from knowing there is widespread unemployment, no matter how dictatorial the press control, no more can people be kept from learning that they don't have lots of free time, don't have a chance of getting it so long as they run after shiny lures into the horizon, and won't have a good life if their cities and towns are the ugliest in the world. It may take a while for these facts to register; it may come suddenly like the click of a door—so suddenly that the words *production* and *efficiency* sour even as we say them.

The third group of people, a handful, believe we need a new texture to our life and that the only hope for it is leisure in the classical manner. They think we need more than processed food and tricky machines and cleverly designed products. They think we need ideas, that all of us need to live in beauty. Without doubt, leisure at times has woven a new texture. It doesn't always succeed, though. There is no way men can say that light and beauty will come to them. Certainly if they try by working harder and harder they only get farther and farther away. Leisure prepares the ground; the rest comes from somewhere beyond man.

If we cannot justify the followers of leisure, more's the pity: we stand to lose more than they. The hope (not the function) that the leisure kind offer politics is that they can learn (and that what they learn they may reveal) about men and politics and their relation to the cosmos. They also offer leisure as the ideal of freedom. All this the state can hope for. A man can hope for more, that through leisure he may realize his ties to the natural world and so free his mind to rise to divine reaches. Man's recognition of himself and his place in the universe is essentially a religious discovery. As such it transcends politics.

The leisure kind themselves are not much interested in justifying their life, and why should we press them? Truth is their justification. There is, they say, something beyond the state that comes to a man in a life of leisure that can come in no other way. The detachment I have already spoken of, and the perspective that a man gains from being able to pull back into himself and, yes, the inspiration. What such a man can do is produce himself. Cultivation of the mind distinguishes him. He shows his godlike nature. Inasmuch as the state too should know its place, it is a good state if it fosters leisure. Just as a good church will try to see that everyone comes to know what contemplation is and will help not only the few who are born to it but everyone to experience it by encouraging retreat and meditation with beautiful stillness—something that architects should remember—so the good state in its support of the life of leisure will try to see that all come to benefit from incorporating some of that life's maxims into their daily world.[3]

3. For a book on architecture and city building that is well aware of some of the problems mentioned here, such as meditation, industry, quiet, promenades, locomotion, and squares, see Percival and Paul Goodman, *Communitas,* Vintage Books, New York, 1947.

The Emperor Justinian built a garrison city on Mount Sinai to protect hermits and desert saints. That is one way. Another is to conceive of leisure as a way of life that is healthier and more satisfying, as an *ars vitae*. The state can expect numerous minor good results from leisure. They are by-products. It can expect bad results too. Out of the situation, traditions, and social arrangements of different countries will come a variety of ways of approaching leisure. For leisure is an ideal. One can only try to get as close to it as possible. The closer to the ideal, the purer the air.

Every great discovery that man makes of his relation to God, to the universe, to his fellow humans, to himself, is so wonderful it calls for celebration. Religion marks it with a holiday. The state too has its holidays. A holiday is universal. It is celebrated not by the discoverers alone, but by all who share the wonders it reveals. It joins the two classes, the great majority and the leisure kind, and all those who have strayed and separated into society's many crannies. The holiday heals whatever rifts exist and reminds men that they are bound together by the one equality with which they came into this world and with which they bow out. So for the few who love leisure and for the many who need free time, the holiday is a day to celebrate the wonder of life.[4]

Leisure, given its proper political setting, benefits, gladdens, and beautifies the lives of all. It lifts up all heads from practical workaday life to look at the whole high world with refreshened wonder. The urge to celebrate is there.

Felicity, happiness, blessedness. Certainly the life of leisure is the life for thinkers, artists, and musicians. Many of the great ones, though seldom attaining it, have throughout their lives given signs of their passion for it. Throughout their biographies runs an attempt to get more free air than the surrounding atmosphere held for them. For that matter, though none of us may ever have been able to live this way, most of us too, perhaps, have had moments when we felt close enough to get glimpses of a truth—that could we have more of the way of life, we would also have more of the truth.

The unwrought leisure kind struggles on. No one can turn them away easily. Their kind of life is too tough. Yet signs appear even now that their day may come, that the United States is moving toward a life where leisure may be possible, that there is a great hunger throughout the land, after so long a life in poverty of spirit. The man invented by the economists has disappeared; *Homo politicus* is bored with precinct politics; the consumer, alone in the furrow scratched by advertising and its allies, noses ahead.

---

4. Strabo in his *Geography* (X, 3, 9) comments on the holiday in the ancient world. "A custom common both to Greeks and Barbarians is to celebrate religious rites in connection with the relaxation of a festival. . . . This is in accordance with the dictates of nature, because the relaxation draws the mind away from human occupations *(ascholemata)* and turns what is truly the mind towards the divine . . . Although it has been said that mortals act most in imitation of the gods when they are doing good to others, it could better be said, that they do so when they are happy, which means when they are rejoicing, celebrating festivals, pursuing philosophy, and joining in music."

The last fact is what confounds observers of the work ethic. They saw off-work time in the nineteenth century condemned as idleness, they saw workers seeking in free time their rest and, in vain, their old pleasures. They saw work become a calling and then lose its meaning. Free time is no longer idleness, and to play on Saturday and Sunday is no longer a sin. Yet men work as much as before; they must somehow. Work still remains in the lead. Free time is still but the parenthesis. At the moment both are marking time. The question is, will the next step be back to work or forward? And if forward, where to? To the circuses, of course, but to anywhere else?

A great fracture in the ethos has taken place. The resultant fault will bring work and time under survey. The American will have to question his identity and ask about his destiny. Why does he work and rush? For bread? To stay alive? Why stay alive? Because he is like all other animals? Because he was taught to stay alive? Anyway, does it matter? What really does matter? A great shifting of substrata is going on, a whole pattern of duties and pleasures seeking to come to rest on something new. Work and time's displacement will bring a fresh inclination. Were our tradition of leisure stronger, we could be more confident that it would settle us where we should have been long ago—in the second stage of political community, the living of a life of good quality. Perhaps this long siege under garrison conditions will enable us to dream better, and dreaming better to build with art and intelligence.

We of the twentieth century may not be here to enjoy the fruits of this questioning and to savor many of the changes. We shall be here now to encourage whoever wants to, to build beautiful things; and to know that in so far as those few succeed, tomorrow's city, slightly mad, not too neat, human, will become a place to stroll, to buy and sell and talk of many things, to eat and drink well, to see beauty and light around.

Those who do not like the way this future looks can "do ye nexte thynge," in whatever way they may choose, to block it. Work, we know, may make a man stoop-shouldered or rich. It may even ennoble him. Leisure perfects him. In this lies its future. For those who do like that future, the next thing is to lean back under a tree, put your arms behind your head, wonder at the pass we've come to, smile, and remember that the beginnings and ends of man's every great enterprise are untidy.

83 Billion Dollars for Leisure—Now the Fastest-Growing Business in America

U.S. News & World Report

Affluent Americans, with more time on their hands and money to spend than ever before, have boomed leisure into an 83-billion-dollar business this year.

That figure tops the current annual outlays for national defense. The money going into travel, sports equipment, campers, boats, summer houses and a host of related items reaches into almost every aspect of the nation's economy.

Manufacturers of everything from croquet sets to cabin cruisers are finding business profits in the leisure-time market. Today, quite literally, pleasure *is* business. And it's the fastest-growing business in the land.

Where the money goes. Largest item in the leisure budget is recreational equipment—boats, camping vehicles, color-television sets, motor bikes and the like.

Total up the bill for these and add to it the money for admissions to sporting events and you get a whopping 38 billion dollars. By comparison, this is 11.4 billion more than the U.S. spent on the same things in 1965.

Growth in such sales is attributed in part to rising demand for such relatively new items as color TV's, campers, snowmobiles and surfboards.

The surge to the great outdoors has put sales of vacation vehicles on a skyrocket course. Travel trailers, motor homes, truck campers and camping trailers grossed nearly a billion dollars in 1968. According to the Recreational Vehicle Institute, the total of such units in service today is about 2.5 million.

Travel trailers—which attach to and are pulled by automobiles—hold a commanding lead over the others. Production in 1968 reached 161,530 units —about 40 per cent of the recreation-vehicle market.

Self-powered motor homes rate as the mansions of the highways. These are built directly on a truck or bus chassis. They offer comfort, convenience and ease of handling—a real home on wheels. They measure up to 36 feet in length and vary in cost from $5,000 to $20,000—depending on the owner's taste and pocketbook.

Motor homes became big business in 1968 with a sales volume of $157,-148,000. That represented an increase of 158 per cent over 1967.

The production of motor homes this year is expected to climb to about 50,000 units.

Rise of snow buggies. Another emerging giant is the snowmobile. Industry officials predict that a million snowmobiles will be in use in the U.S. and

Canada by winter. This all-purpose snow vehicle can carry two persons at speeds up to 50 miles an hour—and tow a sled behind. Average cost is $1,000, although some models can be bought for as little as $650.

The snowmobile has many uses. Ski lodges maintain fleets of them for nonskiers. Businessmen and housewives commute by snowmobile to and from work and shopping when snow closes roads to autos.

Racing with these ski-track, motor-powered vehicles is fast developing an avid following.

Commercial trappers use them to get into remote areas. So do the National Park Service and the Royal Canadian Mounted Police.

Industry spokesmen make this forecast on snowmobile sales for the 1970 fiscal year: 350,000 units with a total retail value of 380 million dollars.

Winter harvest. The rapid success of the snowmobile is explained by Willard E. Frazer of Billings, Mont.—

"Too long has our crop of winter gone unharvested. . . . The development of the snowmobile, with its capacity to open up areas that in the past were beyond the ready reach of mankind during the winter months . . . may well be as significant to Montana's future as was the Whitney cotton gin to the pre-Civil War South. . . ."

Another sport enjoying an upsurge this summer is tennis. Nearly 9 million Americans play tennis—and hand over more than 27 million dollars a year on rackets, balls, and accessories.

Americans pay out 28 million annually on water skiing. Surfers are getting big in the market, too. They bought 9 million dollars' worth of surfboards last year.

Snow skiers—now 4 million in number—rank as lavish spenders. They plank down around 900 million dollars each season getting to the ski slopes, and for equipment, lodging and entertainment costs.

Highly popular with an increasing number of Americans is the pleasure boat. Americans now own more than 8 million boats, which tie up at approximately 5,500 marinas and docks across the nation.

Lure of fairways. There are some 12 million golfers in the U.S., playing regularly on about 10,000 courses.

When you consider the greens fees, club memberships, sales of golf equipment, rental of electric carts and spin-off sales of sprinklers, fairway mowers and ball retrievers, golf stands out as a king-size business.

In addition, there are the tours which offer golf privileges along with rooms and meals in fashionable hotels. It's not unusual for a vacationing couple to spend $50 to $100 a day on a luxury golfing vacation.

More modest are the family float trips down broad rivers, the bargain travel clubs which offer tours to exotic places. And there are always the packed national parks, with more than 40 million visitors this year and the attendant motel rentals.

Fast-rising participation. The Interior Department reports an impressive increase in all outdoor recreation.

National parks are jammed to the crisis point, with a fourfold rise in visitors since 1950.

Sporting "occasions"—the Department defines an "occasion" as a single participation in any sport by one person in a calendar year—are running in the vicinity of 7 billion annually. By 1980, the Department says, the figure will be 10 billion—and it is very likely to hit 17 billion in the year 2000.

Other millions of fishermen, hunters, archers, mountain climbers and joggers contribute to this burgeoning total. Clearly in sight, it seems, is the year when gross sales in the businesses which supply leisure-time equipment climb into the 50-billion-dollar class.

Second-home vogue. The vacation mania is spreading to the housing market, too. A total of 1.7 million American families now own second homes. This number goes up by about 150,000 a year. Sales of recreational housing are expected to hit 1.5 billions in 1969.

These vacation homes are of many kinds: A-frames, condominium apartments, townhouses, factory-made "prefab" units and standard, year-round models.

THE PLEASURE EXPLOSION— AND ITS DOLLAR POWER

	1965	1969 (est.) (billions of $)	Increase
Spending for recreation-sports equipment, reading matter, sporting events, other "personal consumption" products and activities	26.8	38.2	43%
Vacations and recreation trips in U. S.	25.0	35.0	40%
Travel abroad	3.8	5.2	37%
Second homes	0.9	1.5	67%
Swimming pools	1.1	1.4	27%
Vacation land and lots	0.7	1.3	86%
TOTAL	**58.3**	**82.6**	**42%**

Pleasure industries have been growing at an average rate of nearly $6 billion a year since 1965, with no limit in sight.

Sources: American Automobile Assn.; U.S. Dept. of Commerce; Recreational Vehicle Institute; International Snowmobile Industry Assn.; National Swimming Pool Institute; U.S. Dept. of Housing and Urban Development. 1969 estimates—USN&WR Economic Unit.

A typical second home, according to a U.S. Government pamphlet, "Second Homes in the United States," is a single-story structure with four rooms, valued at $7,800.

Nine out of 10 of these dwellings have electricity; 6 have running water and inside lavatories; 2 have central heating.

The average owner of the vacation house has an income in excess of $10,000 a year. He is also dedicated to the idea of getting away as often as possible for the long week-end.

"Fun out of living." One of these owners, a Washington, D.C., public-relations man, says—

"The kids are grown now and finished with college. It's about time I did something with my extra money so that my wife and I can get a little fun out of living."

THE TRAVEL-FOR-FUN INDUSTRY

AT HOME

How Americans will spend $35 billion on vacation and pleasure travel this year—

	(billions of $)
Food	9.5
Lodging	9.5
Transportation	8.0
Entertainment, Other Expenses	8.0
Total	35.0

Source: American Automobile Association

ABROAD

Where 3.9 million Americans spent $4.7 billion abroad last year—

	(millions of $)
Europe, Mediterranean	993
Canada	820
Mexico	630
West Indies, Central America	325
South America	87
Asia, Other Places	167

Plus: $1.7 billion for the cost of getting there.

And: This year, 4.2 million Americans are expected to go abroad, spend $5.2 billion on their junkets.

Source: U.S. Dept. of Commerce

Fun, in this case, is an $18,000 summer home at Charnita, Pa. That's within the national average of $10,000 to $20,000 spent on second homes. Some analysts say that by 1970, one in every five housing starts will be a vacation home.

The drive to get away for the long week-end has set off a corollary boom in vacation-land sales. There are about 900 land-development projects in the U.S., according to Government figures. Because they deal in interstate commerce, these must be registered. The law provides that the prospective buyer be furnished with a property report similar to a securities prospectus.

Many vacation lots start at about $1,500, and the cost can run as high as $30,000 for an exclusive shoreline site. Developers usually build a clubhouse, golf course, beaches, tennis courts and swimming pool for community use.

Yearning to go. The closest rival to outdoor recreation in total expenditures is travel.

Americans spend 35 billions a year in travel, according to the American Automobile Association. This includes vacations, overnight trips and pleasure jaunts of more than 100 miles.

The AAA estimates that Americans will drive 225 billion miles this year, just getting to and from vacation areas.

That is 27 per cent of the total estimated mileage for privately owned vehicles in 1969. Ninety per cent of all domestic pleasure travel involves the automobile.

Americans like short trips. About 63 per cent of all auto travel for leisure is for distances of 200 miles or less. Only 15 per cent are 1,000 miles or more.

Each year, foreign travel becomes a bigger item in the leisure budget. The Department of Commerce says that 4.2 million Americans will go abroad this year and spend 5.2 billion dollars. In 1968, 1.9 million Americans visited

Besides Travel—

10 WAYS AMERICANS POUR OUT THEIR LEISURE MONEY

	1965	1969 (est.)		1965	1969 (est.)
	(billions of $)			(billions of $)	
1. Airplanes, athletic gear, bicycles, boats, campers, motor scooters, snowmobiles and other recreation equipment	6.8	11.2	6. Garden materials	1.0	1.3
			7. Radio-TV repairs	1.0	1.3
2. Radios, TV's, records, musical instruments	6.0	9.0	8. Clubs and fraternal organizations	0.9	1.1
			9. Race-track receipts	0.7	0.9
3. Books, magazines, newspapers	4.9	6.3	10. Other "personal consumption" activities	2.0	3.0
4. Admissions to movies, games, other events	1.8	2.3	Total	$26.8	$38.2
5. Camping, fishing, golf, "participant" amusements	1.5	1.8			

Note: Categories do not add to totals because of rounding.

Source: 1965—U.S. Department of Commerce; 1969—Estimates by USN&WR Economic Unit

Europe and the Mediterranean and spent a billion dollars. Other areas overseas that are highly popular with Americans are the West Indies and Central America.

The urge to soar. Flying for pleasure, in the past considered a sport for the very wealthy, is making significant headway with the general public.

"Discover Flying" a promotion sponsored by some 1,350 dealers and airports across the nation, is proving highly successful.

An important inducement to get beginners off the ground is a coupon which, with $5, purchases an introductory flight lesson.

Veteran pilots say that once a person gets his first taste of handling the controls of a light plane, he becomes a solid flying prospect. From that point, lessons to turn him into a qualified pilot cost anywhere from $750 to $1,000.

Thereafter, if the new pilot wants to buy his own plane, he can get a single-engine job for as little as $7,500—or as much as $40,000.

Flying small planes is particularly popular with American women. Typical of this new breed is Dottie Sager, a junior at the University of Maryland, who learned to fly at the Montgomery County Airpark at Gaithersburg, Md.

Miss Sager, the daughter of a former pilot, completed her first cross-country flight in mid-August. She says this:

"It costs me about $15 an hour to fly this two-seat trainer. It gives me a real sense of freedom. I'm studying interior design at Maryland and I hope to have my own plane some day to fly important clients around. It's really not as expensive as a lot of people think."

In the U.S., there are about 750,000 licensed pilots and 250,000 student pilots. More join the ranks daily.

Support for the players. What it all adds up to—the fliers, the boaters, the campers, the travelers, the tourists and the week-enders—is an astonishing picture of America at play.

Behind the scenes, serving the ever-increasing demand for the trappings of leisure, are the muscle and sinew of American industry.

There'll Be Less Leisure Than You Think

Gilbert Burck

Before long, we keep hearing, work will practically wither away, leaving us all with an abundance of everything, notably free time. No such luck.

Nothing is easier to take for granted in the U.S. than long-term economic growth, and a good many people accordingly take it for granted. The prophets of Automatic Abundance assure us that the economy of the 1970's will grow as effortlessly as crabgrass in a lawn, that technology has solved the classic problem of scarce resources. The big tasks of the 1970's, the A.A.'s aver, will be to distribute production equitably, to improve the physical and spiritual quality of life, and to gain more leisure. They do not go so far as Marx, who predicted that "money-commodity" relationships would sooner or later be abolished—i.e., that things would become so abundant they would be handed out free, "to each according to his need." But many A.A.'s believe that the day is near when people will no longer be condemned to long hours on life's treadmills, and that ambitious labor leaders who are warbling about the four-day and even three-day week are only anticipating the inevitable.

Unfortunately, most of this is nonsense or illusion, or both. The word "affluent," so often used to describe the U.S., is both euphuistic and inaccurate. Granted that the American economy is an engine of production that seems bound to justify the optimistic projections of its expansion, the nation is still far from wealthy. Median family income is about $9,000. Although this represents high living indeed to a subject of the Soviet Union, it is not enough to buy an American family a decent living. As for the "redistribution" about which young revolutionaries talk darkly, it is nonsense double-distilled. If all personal income over $30,000 a year could be redistributed without paralyzing incentives, each family would enjoy only a few hundred dollars more a year. The unhappy fact is that not everybody can yet buy everything he needs, to say nothing of everything he wants. The U.S. is and will remain a "scarcity" economy—one that allocates its limited resources efficiently through the natural feedback system embodied in the profit motive and the market.

Now that improving the quality of life has become national policy, productivity growth is all the more necessary. Controlling pollution, reviving mass transit, rebuilding cities, reducing crime, and providing ample medical care and education will put stupendous additional demands on the nation's resources. Only if our productivity, or output per man-hour, keeps rising at least as fast as it has been, can we do all that we want to do without sacrificing something desirable and important.

The catch is that large and rapid shifts in employment patterns may soon begin to depress the rate of productivity growth. Prices of services will rise inexorably, producing new inflationary stresses. Contrary to all the predictions that automation will throw millions out of work, the scarcest of all resources will be manpower. By 1980 the economy will be able to draw on some 200 billion man-hours a year, up from 165 billion today. But 200 billion man-hours will suffice only if they are employed with increasing efficiency. Meantime the prospect of greatly reducing the hours on life's treadmills remains mainly a prospect. For a long time we'll probably have to work as hard as ever.

An Appetite for Time

The basic reason why carefree abundance and leisure are not likely to fall into our laps like ripe fruit may be put very simply. The more time we save in making goods, the more time we spend providing services. The nation's total output can be conveniently divided into the production of goods (manufacturing, mining, farming, and construction), the provision of services (government, trade, finance, and personal services), and "TUC" (transportation, utilities, and communications). During the past twenty years, output of goods has more than doubled, but productivity of the goods industries rose so much that the number of people producing the goods increased only from 28 to 29 million. In the same years the output of TUC much more than doubled, but the number of people rose only a few hundred thousand, to 4,500,000. But behold the services. The number of people providing them increased by no less than 70 percent, from 28 million to nearly 48 million. Thus the services have accounted for nearly all the increase in total employment since 1950.

The trend seems bound to continue. By 1980, when total U.S. output will have increased by at least two-thirds, employment will have increased by nearly 25 percent. The number of people employed in producing goods will very likely rise a couple of million, 6 percent or so above its present level, but it will then be only some 11 percent greater than in 1948. TUC employment will probably rise a few hundred thousand. So service jobs will again account for the overwhelming bulk of the increase in total employment. They will mount to around 65 million (more than 67 million including the armed forces), or about two-thirds of *all* jobs, and nearly as much as total employment in 1958. By 1990 the services may well account for more than 70 percent of all jobs. As if obeying a law of compensation that dooms men to eternal toil, the services are expanding enough to eat up not only the time that, so to speak, is saved in goods production, but also nearly all the time embodied in the yearly additions to the labor force.

The very nature of the services makes them greedy for time. Goods production depends on a wide variety of services—on trade to distribute products, research organizations to help innovate, legal and financial advice to help make policy, government surveys to help gauge markets. As these services become more complex, more and more of them are being supplied by outside specialists. The very efficiency of goods production, moreover, has played a big part in generating a need for government services such as education, highways, and pollution control.

Perhaps the most important reason services devour time so voraciously is that many of them—legal, financial, and medical advice, for example—depend on live performance and personal contact between buyer and seller. Most of the rest, including research, advertising, and management consulting, are valuable precisely to the extent that they embody a lot of specialized, time-consuming personal effort. Unlike power plants, factories, and refineries, which enormously increase their output per employee with expensive laborsaving capital equipment, most services cannot substitute capital for labor on a large scale.

Finally, service workers are probably less efficient than goods workers. According to Victor R. Fuchs of the National Bureau of Economic Research, who is the country's No. 1 expert on services, service workers put in fewer hours a week than goods workers and receive somewhat lower wages per hour. And their quality, measured by such things as level of schooling, has not been improving as much as that of goods workers.

A Task for a Foolhardy Calculator

As a result of all this, productivity in the services is increasing, on the average, no more than half as fast as in the rest of the economy. A year and a half ago, in "The Still-Bright Promise of Productivity" (October, 1968), *Fortune* estimated that the productivity of the *private* economy had been rising at a little better than 3 percent. The productivity of goods and TUC combined had been rising at around 4 percent, but the over-all average was brought down by the bare 2 percent for the services. Since then government agencies and others have made detailed projections of the economy for the decade of the Seventies. It now appears that the huge and continuous shift of employment to low-efficiency services will by 1980 result in a small but perceptible falling off in the growth rate of productivity in the private economy. Everything else being equal, that growth rate will decline from more than 3 percent to 2.8 percent; the difference is equal to some $40 billion worth of output a year (at today's prices).

Those official figures, moreover, grossly overestimate productivity growth because they omit government employment from the total. It is a standard assumption that government activity cannot be measured because it lacks a marketable output. But the time has come to reckon with government productivity, if not to try to measure it. For government jobs in 1968 amounted to 12 million or 15 percent of all employment (including the armed forces, more than 15 million and 18 percent), and may climb above 20 percent by 1980.

Nearly 70 percent of the federal government's three million employees work in the Defense Department or the Post Office, and it would be a foolhardy calculator indeed who would essay to demonstrate that these employees are improving their output by anything at all. As for the other federal employees, they are probably doing only a little better. But state and local governments now employ 9,400,000, or more than three times as many as the federal government. Nearly five million of these 9,400,000 are educators of one kind or another; their number, up from about 1,500,000 in 1947, has increased ten times as fast as the population and three times as fast as the number of pupils and students. Measured quantitatively, productivity of education has obviously declined. Since there are now on the average a good many fewer students per teacher, the quality of education may be rising. But if so, few experts are yet prepared to argue that it has. And a long time may elapse before any noticeable improvement will be discernible.

The Handsome-Psychoanalyst Effect

Taking one opinion with another, it appears that the productivity of government as a whole is rising by no more than 1 percent a year, and probably less. If total employment is redefined to include government jobs, and if the productivity of these jobs is rising all of 1 percent a year, national productivity is increasing not by 3 percent or more but by only 2.7 percent a year, and by 1980 it will be increasing at only 2.5 percent. Merely revising the figures, of course, will not change projections for the private economy, but revising the figures to include government is realistic and salutary. It verifies statistically the extent to which low-productivity service employment is eating up more and more of total working time. And it also tells us we are deluding ourselves when we suppose the economy is growing as fast as the figures say it is.

Quite possibly, the government is overestimating the real value of some service production. Statisticians in the Labor and Commerce departments measure the production of an industry in terms of its contribution to gross national product, or the value of its production minus the value of materials and services it buys from others. Then they divide the industry's contribution to G.N.P. by the man-hours consumed by the industry, and so arrive at its output per man-hour. Because the official man-hour figures are accurate and production figures are carefully adjusted for rising prices, changes in output per man-hour probably show up pretty accurately in the statistics.

But estimates of service *production* are often tautological. The contributions of medical and business services to G.N.P., for example, are derived from income figures. This means that a handsome dog of a psychoanalyst, specializing in wealthy matrons whose chief ailment is that they have nothing worthwhile to do, may contribute ten times as much to G.N.P. as a hard-working psychologist in a clinic. Or a brilliant management consultant with the bearing of a Churchill may set his tariffs twice as high as an equally brilliant consultant with a stutter. If both men are increasing their productivity at the same rate, then the former is contributing, on paper, more to the economy's growth than the latter. A man or company in what is in effect a monopoly position, in other words, can produce less and yet contribute more to G.N.P. and national growth than a man or company competing in the market.

Even if the productivity of services is rising at 2 percent a year—an exceedingly generous estimate—the time appropriated by them is increasing, and will continue to increase, at a very rapid rate. Since time is money, services obviously cost more and more. What may not be immediately obvious is that their costs and prices must rise even faster than those of goods. For wages in service industries are going up at least as fast as in goods industries. As a matter of fact, organized teachers and hospital employees, despite their slight or nonexistent gains in productivity, have made up for previously depressed rates by negotiating even bigger percentage increases than employees in goods industries. Low-efficiency service industries thus find themselves burdened with permanently rising costs. Unlike industries wherein productivity is increasing briskly, they cannot offset rising pay per worker with rising output per worker. Because service industries account for more than half the U.S. employment and must pass on all or most cost increases as price increases, their inferior productivity performance is one of the nation's prime inflationary forces.

A Kind of Monopoly Position

When the cost of anything rises inordinately and disproportionately, the demand for it tends to soften or decline. But as measured by the Department of Commerce, output of services (and demand for them) is rising at least as fast as output of goods. If services are growing so infernally expensive, why is the country buying more and more of them? The obvious answer is that it needs or wants them badly enough to pay the price. As the economists put it, the demand for them is price-inelastic.

The prices paid for services are also inelastic: when they move, they usually move upward. One reason is that services are not very competitive. To a considerable extent, it is true, services are provided by small owner-managed firms with little market power. But about a third of service employment is accounted for by government and private nonprofit organizations. All told, producers accounting for more than half of total service employment enjoy some kind of monopoly position that encourages them to increase their output but does not pressure them to improve their efficiency. Thus they tend to expand existing services or introduce new ones—as the federal government does when it enlarges a bureau or establishes a new one. Or they try to improve the quality of their output—as departments of education do when they raise the ratio of teachers to pupils. Sometimes the result is better output, sometimes not; often there is no way of telling. When a producer or supplier improves "quality" in this special economic sense, quality means more *input* per unit of output; it does not, alas, necessarily mean better output.

Even competitive services are obliged to improve quality at the cost of higher prices. Motels that used to offer bare lodging, for example, can now meet the competition only by providing wall-to-wall carpeting, color TV, and swimming pools. In one way or another, most services seem driven to expand their output more than twice as fast as their productivity, and thus to keep enlarging their payrolls, their costs, and their prices.

Manufacturing industries provide an edifying contrast. Most manufactured goods are produced by large corporations that are supposed to wield a lot of market power. All would certainly like to get more for their production. Failing that, they would like to justify higher prices by improving the quality of their goods. Auto manufacturers, for example, would love to put so much car into each car that the cheapest models would fetch $3,000. But for all their legendary market power, the auto makers keep getting hauled up by competitors both at home and abroad. So with appliance makers, food processors, and on down the long list of manufacturing industries. Unlike so many suppliers of services, they add relatively few people to their payrolls even while tripling and quadrupling their production.

The Government as Monopolist

All the negative characteristics of most services—low productivity growth, rising comparative costs, lack of market discipline, limited consumer sovereignty, and a pervasive compulsion to expand—are combined in government. In a general way, people demand schools, highways, hospitals, adequate armed forces and police, medical care, and so on. But the size and disposition of government programs have only the most tenuous connections with consumer demand properly describable as such. The consumer's sovereignty over government spending, never very strong, seems to be growing weaker as government spending mounts. As a supplier of services, the government finds itself in the role of a monopolist aiming to meet demand and improve quality without worrying much about costs. Even if government chooses the right services for the public, its immunity to anything resembling market discipline ensures that they will be costly. And no sector of the economy is growing so fast as government. In 1948 federal, state, and local governments employed nearly 5,700,000. The figure more than doubled by 1968 and may exceed 20 million by 1980.

The leisure society is a myth because more and more man-hours will be needed to provide ever-expanding services. The chart at the right breaks down U.S. employment by industries. Between 1948 and 1980, total employment will have increased by about 40 million, from 61 million to more than 101 million. But the number of people employed in goods production will probably have increased by no more than two million, and the number in transportation, utilities, and communications by little more than 500,000. All the other additional people in the labor force, nearly 37,500,000, will have in effect found jobs in government, trade, and other services. In 1980 the services alone will provide jobs for nearly as many people as the entire economy did in 1958. The employment figures for 1948-68 come from the Bureau of Labor Statistics; they include self-employed, household, and unpaid family workers (as in stores). The 1980 figures are FORTUNE estimates. Not included in the chart are the armed forces, which numbered 1,400,000 in 1948, 2,600,000 in 1958, 3,400,000 in 1968.

101,700

18,000-20,000

81,216

2,737   4,600

9,465

3,716   21,000

65,532

2,191

Government ─ ─ ─ ─ ─ ─
    Federal   1,863   5,648

15,058

State and local   3,787   2,827

Services ─ ─ ─ ─ ─ ─ ─
Finance, insurance, and   2,054
    real estate

Business, personal,   8,519   11,086
    professional

20,500

16,659

Trade   11,613

13,589

Transportation, utilities,
communications ─ ─ ─ ─ ─
    4,392   4,563   4,900

Goods ─ ─ ─ ─ ─ ─ ─ ─
    Construction   3,164   4,189   4,065   5,500
        3,522   3,200

Agriculture   8,392   5,352   4,164

Manufacturing and   22,000
    mining   17,074   17,128   20,769

61,058

Employment (in thousands)   1948   1958   1968   1980

About 50 percent of all civilian government employment is accounted for by education, which provides a splendid example of how increased quantity and quality will use up more time. Although education's share of employment may decline a trifle by 1980, the number of people on education payrolls will probably rise from nearly 5 million in 1968 to 7,500,000 in 1980. Enrollment in public elementary and secondary schools, reflecting slower growth in the numbers of school-age children, will increase only 1.2 percent a year between now and 1980. But this low rate will be more than offset by enrollment in higher education, which will increase enormously. The emphasis, moreover, will be on quality. For the ratio of students to teachers will decline, bigger and presumably better schools will be built, and new programs will be inaugurated for disadvantaged youths. The National Planning Association has estimated that 58 cents of every additional dollar spent on education between 1968 and 1980 will represent quality improvement.

In a way not generally understood, the low productivity of government services imposes a heavy burden on the cities. Professor William Baumol of Princeton, writing in *The American Economic Review,* has described the basic problem of the cities as one of supplying services in which productivity is rising very little if at all and whose costs are therefore rising cumulatively and endlessly: police, schools, social services, hospitals, subways, and buses. Most also suffer from what might be called the Quill-Guinan effect. Both the late Mike Quill and Matthew Guinan, past and present presidents of New York City's Transport Workers Union, pursued a policy of demanding work-rule concessions calculated to reduce productivity. At the same time they demanded wage increases that raised transit labor costs at least as fast as manufacturing labor costs. Faced with a similar situation, U.S. railroads and steamship lines got out, or want to get out, of the passenger business. The cities cannot get out of passenger transportation, and hardly anything else. Inexorably and cumulatively, inflation or not, municipal budgets will therefore mount. Aside from standing up to the unions, there is only one way out. Since raising property-tax rates will quickly produce negative returns as taxpayers leave the cities, Baumol argued, the federal government must supply the resources to prevent the crisis.

Probably the most flagrant example of how a nongovernment service can waste time and money is provided by medical and health services. Here employment more than doubled between 1950 and 1967, jumping from 1,400,000 to 3,500,000, and by 1980 will probably increase another two-thirds, or to 5,300,000. Outlays for medical care in current dollars rose from $10 billion in 1947 to $63 billion in 1969 (in constant 1969 dollars, from $24 billion to nearly $63 billion). By 1980 this spending may well rise to $200 billion. Statistical and conceptual difficulties have prevented satisfactory measurement of the output and productivity of the medical-care industry. But Victor Fuchs of the National Bureau of Economic Research has interpreted official industry figures on expenditures, price, and employment to show that real output per man declined between 1947 and 1956, and racked up only a slight gain between 1956 and 1965.

Quality improvements, moreover, have often increased the amount of time it takes medical personnel to do a given job. These days the average doctor is equipped with myriad devices to increase his proficiency and productivity, but precisely because he has so many gadgets and techniques on tap, he takes longer to perform a thorough physical examination than he ever did. "The modern patient usually wants proof from some expensive machine that he's O.K.," says Dr. John Knowles, director of Massachusetts General Hospital. "That takes time and money."

Fireworks in FIRE

Because many of the services that could behave like competitors have neglected to do so, the Antitrust Division of the Department of Justice is beginning to move in on them. There has been much more inflation in personal services than in commodity prices, and a major cause of this inflation, says Richard W. McLaren, head of Antitrust, is price fixing in the services. Some of the cases doubtless will be found in what the Department of Commerce calls "miscellaneous business services," a swiftly growing group that includes business consultants, advertising agencies, janitorial services, and research and testing laboratories. All provide services that business needs and obviously is willing to pay well for. Except perhaps for advertising agencies, they do not have to compete very strenuously; and in any event their fees—and accordingly their measured output and contribution to G.N.P.—keep rising inordinately.

Other targets for Antitrust are "educational services" and "nonprofit organizations." The former category includes commercial and trade schools, libraries, and establishments offering academic and technical courses, nearly all of them nonprofit. They employed fewer than 700,000 in 1958, but will probably employ more than two million by 1980. Nonprofit "membership" organizations, such as trade associations, labor unions, and charitable and religious organizations, now provide about 1,600,000 jobs, and will provide perhaps two million by 1980.

Finally, Antitrust appears ready to set off some fireworks in the service category sometimes labeled F.I.R.E. (finance, insurance, and real estate), whose total employment rose from slightly more than two million in 1948 to 3,700,000 in 1968, and will probably top 4,600,000 in 1980. Early this year Antitrust fired a preliminary salvo against a Maryland association of real-estate brokers, charging it with setting minimum commissions on the sale of real estate and denying listing privileges to noncooperating brokers.

The Elusive Luxuries

Some personal services are trying to survive with few forces, natural or artificial, working for them. They are the products, so to speak, of theatres, concert halls, opera houses, bars, restaurants, bespoke tailoring shops, barbershops, ateliers of various kinds, and hotels and lodgings. Their labor costs are rising as fast as other industries' labor costs, they benefit from no appreciable productivity increase, and most get no subsidy, at least not yet. Some do possess monopoly power; after all, a recital hall featuring Artur Rubinstein is selling a unique product for which it can theoretically charge all the traffic will bear. But monopoly power or not, all these personal services face cumulative higher costs and the necessity of charging endlessly higher prices.

Suppose, for example, that wages henceforth go up at only 4 percent a year. Rising productivity would enable manufacturers, utilities, and communication companies to absorb all or most of the increase, and the general price level would probably rise less than 2 percent a year. But those forlorn low-efficiency personal services will just have to charge more and more. In ten years their tariffs will be up 50 to 75 percent. The chateaubriand that is a bargain today at $7 will set the *feinschmecker* back $12 or $13; the $250 tailormade suit will fetch $450, the $8 symphony orchestra ticket $13 or more, and the $7 theatre ticket $12.

As it has in the past, this kind of escalation will be hard on, and baffling to, young executives and others who look forward to salary increases more than large enough to offset inflation. Suppose inflation settles down at the rate of 2 percent a year, and that a bright young executive gets raises averaging 4 percent a year after taxes. This is a high figure; according to a study by Professor Wilbur G. Lewellen for the National Bureau of Economic Research, recent executive pay gains have averaged only 3 percent after taxes. Such an incentive nevertheless drives the young executive to work hard and long and to use his brain power effectively. It also leads him to hope he can enjoy now and then a few more of the luxuries that other lucky people enjoy. To his dismay, he finds that their prices move up just about as fast as his ability to buy them. As a matter of fact, other demands on his salary, such as the expense of rearing children who consume a lot of low-efficiency services like schooling, may put these luxuries further than ever beyond his everyday reach.

The High Cost of Distribution

Some services are not sheltered from market discipline and the outstanding example is wholesale and retail trade which in 1968 provided no fewer than 16,700,000 jobs, or 35 percent of all service employment. Here competition has forced employers to find ways of improving their productivity, and they have come up with advances such as computerized warehousing and inventory controls, new accounting methods, and self-service. Wholesaling accordingly has achieved some impressive productivity increases and to a lesser extent so has retailing. A lot of the chain stores' success in reducing the cost of retailing, however, is the result not so much of real productivity increases as of making the manufacturer and the consumer perform services formerly performed by the storekeeper. The manufacturer prepackages, prelabels, and presells. The consumer does his own selection, delivery, and financing (i.e., he often pays a charge for credit).

There is little likelihood that large capital investments can pay off in improving the productivity of retail trade. It seems forever burdened with the rising cost of personal contacts. Unless some wholly new impersonal technique of distribution is invented, the cost of distributing goods is bound to keep on mounting faster than the cost of making them. Mail order is no alternative, for it uses enormous amounts of labor and so costs even more than store sales. Appropriately enough, therefore, employment in trade is rising considerably faster than employment in manufacturing. In 1948 trade employed 12 million people, 70 percent as many as manufacturing; by 1980 it will employ nearly 21 million, nearly as many as manufacturing. And by 1990 trade may employ more than manufacturing. Distribution accordingly will become more expensive, and offset to some extent the high productivity growth and lower cost of making goods.

But Comes the Revolution

The only way to counter the tendency of services to gorge themselves on time is to see that productivity in both goods and services grows as fast as the optimistic projections assume it will grow. This is easier said than done. You cannot cheerlead people into improving their efficiency. You can cheerlead them into being saved, or into joining a revolution to end poverty, or to put the "people" in charge of things. But once you've got your revolution, as all Communist regimes have discovered, the only way you can begin to tackle poverty is to shun cheerleading and set up a profit-and-loss system that rewards producers who use resources efficiently and penalizes those who do not.

The problem in the U.S. is to provide the equivalent of market incentives where markets are weak or nonexistent. One of the biggest opportunities for improving output per man-hour lies in raising productivity in government. As every schoolboy should know, government cannot deliver goods and services efficiently. It habitually sets up self-perpetuating bureaucracies that put statistical output ahead of everything else. Since government has no incentive to use resources efficiently, it doesn't do so.

Happily, a growing number of experts are exploring means of raising government efficiency. Professor John W. Kendrick of George Washington University, a leading authority on productivity trends and a long-time advocate of measuring and increasing government productivity, believes that pressures for improving it can be built into public administration; and he estimates that its productivity can be improved significantly. Others would turn government bureaus over to private business, putting them up for competitive bidding. In his recent book, *The Age of Discontinuity,* which was required reading in high Administration circles, Peter Drucker argues for what he calls the reprivatization of government, or farming out its routine functions to outside organizations, such as foundations and corporations. Drucker would establish a government agency something like the Bureau of the Budget, and would charge it with setting objectives, choosing means, and redefining tasks.

Built-in Handicaps

Maintaining the high rate of productivity growth in TUC and the goods industries is of prime importance. Only if these industries can maintain or come reasonably close to maintaining their recent average of 4 percent will there be enough manpower to keep real disposable income rising and at the same time power all the programs for improving the quality of U.S. life. It is a common assumption that a sufficient addition of capital, new technology, and better educated employees will automatically raise the productivity of the goods industries. This assumption is valid in mining, TUC, construction, and some manufacturing. Modern and efficient power plants operated by a few highly skilled workers almost automatically raise utility output per man-hour. Bigger volume and enormous drafts of capital do the same for communications. Better equipment, combined with plenty of competition, keeps up efficiency in freight transportation.

But other auguries are not quite so bright. Farming automatically expanded its productivity as marginal farmers left the land, but by 1980 agricultural employment will be down to about three million, close to an irreducible minimum, and any improvement in productivity there will release very few people for other activities. Mining too is close to rock bottom; by 1980 the whole sector will employ only 580,000. Manufacturing, on the whole, still looks promising. The productivity of some manufacturing industries, such as petroleum refining, responds handsomely to capital and technology. And in most of them—machinery, food, apparel, chemicals, metals, motor vehicles, and so on—measured productivity is linked to performance in the market. Even when the market is very imperfect, the goal of profitability abets the efficiency with which they combine resources. But manufacturing statistics turn up one trend that could handicap productivity growth: the number of employees producing what are essentially services has been steadily increasing, and now probably accounts for more than a third of the manufacturing labor force.

The goods industries, on the face of it, will probably have a harder time maintaining their productivity growth than they have had. Thus the job of raising living standards and improving greatly the quality of American life will not be as easy as many think. Even if the performance of the services gets better, the prospects for reducing the hours on life's treadmills very much will keep receding into the future.

The Background to the Guaranteed-Income Concept

Robert Theobald

Social critics often claim that the present need for economic and social reform stems from past failures in economic and social policy. There is, of course, much merit in this contention. It is, however, far more realistic to perceive present problems as resulting not from failures but from the extraordinary success of Western societies in fulfilling their drive for ever-greater mastery over nature and, in particular, developing the productive potential that today makes it possible to provide every individual in the rich countries with a decent standard of living while requiring a decreasing amount of toil from the vast majority of the population.[1]

The economic history of the past two hundred years may perhaps most properly be couched—to paraphrase H. G. Wells—in terms of a race between increasing production based on ever more complex and sophisticated technology and man's cultural inventiveness in devising and gaining acceptance of new methods of distributing and using this increasing production. It is surprising, therefore, that the mainstream of economics has only recently become concerned with the problems of balancing the available production with the rights of individuals and institutions to obtain this production. Throughout the nineteenth century it was rather generally accepted by economists that production and purchasing power—effective supply and potential demand—would automatically stay in balance. This assumption, called Say's law after its originator, dominated economic analysis until the great slump of the 1930s.

Innovations in techniques of distributing rights to resources have not, therefore, been based until recent years upon theoretical analysis but rather on pragmatic adjustments to the need to be able to sell what could be produced

1. It is, of course, true that those in the developing countries cannot be provided with a decent standard of living today. But this does not mean, as many argue, that the rich countries should produce everything they can and deliver it to the poor. We should have learned by now that excessive aid can be just as dangerous as too little. We must accept the bitter fact that poverty in the poor countries—as opposed to the rich—cannot be abolished in the near future. We must also recognize that we still have no strategy for the elimination of poverty in the underdeveloped countries. For an examination of this subject see: Robert Theobald, "Needed: A New Development Strategy," *International Development Review,* March, 1964.

or to obtain the labor force required for the production of quality goods. The lack of a theoretical basis for changes in techniques of distributing income inevitably led to widespread controversy about the impact and implications of each new measure designed to raise purchasing power or attract workers. Thus Ford's five-dollar day, rapid growth in consumer credit and advertising, social security, and unemployment compensation were, in the past, just as controversial as the guaranteed income is today.

The motivation of Ford in introducing the five-dollar day early in the twentieth century and thus doubling the wages of his workers is still far from clear. Some interpreters argue that his main aim was to increase the number of people who could afford to buy the cars that he was turning out in ever greater numbers. Some have concluded that he was motivated by a desire to obtain a more highly skilled and stable labor force, and some believe he wished to increase the welfare of his workers. It would certainly be unprofitable to re-evaluate Ford's motives at this point in time. It would be equally unprofitable to examine in this essay the implications of the fact that the pattern of income distribution that has resulted from Ford's initiative cannot be reasonably explained in terms of existing economic theory—and indeed destroys its validity.[2] It is important to recognize here only that Ford did introduce a mechanism that made it possible for the wages and salaries of workers to rise in parallel with production. This mechanism has been the chief factor responsible for ensuring that American purchasing power has kept in reasonable balance with American productive power during the last fifty years—with, of course, the exception of the Great Depression.

Two major developments in methods of distributing and using production occurred in the twenties. First, widespread use of consumer credit developed —people were allowed to purchase *before* they had earned the required funds. Second, manufacturers and distributors widened the range and scope of selling activities designed to cultivate new tastes. Despite these efforts, however, potential supply was so far ahead of effective demand by 1929 that the economy collapsed.

It was the Great Depression, which followed this collapse, that led economists to become deeply concerned with the problem of maintaining purchasing power. The change in the thrust of economic analysis is generally and correctly attributed to John Maynard Keynes's book *The General Theory of Employment, Interest and Money*. Nevertheless, it must be noted that there is a good deal of evidence suggesting that the brute facts of the Depression forced

2. Economic theory claims that each factor of production—land, labor, and capital—will be paid in accordance with its marginal (additional) contribution to production. Throughout the twentieth century, most of the increase in production has resulted from increased sophistication of equipment (i.e. capital) rather than through the harder work or the greater knowledge of the average worker. Thus most of the increase in production should, on the basis of theory, have gone to capital. This does not mean that we should distribute rights to resources by widening the ownership of capital—the proposal made by Lewis Kelso and Mortimer Adler. Rather we need a revision of theory on the basis of the new realities of a cybernated era.

politicians to move in the direction of increasing purchasing power before a full economic justification for this step had been found—and indeed even while a large proportion of the economic profession was still opposing this step and calling for decreases in government expenditure. Thus social security and the make-work schemes of the thirties were conceived as a response to social unrest rather than justified on economic grounds as a means of ending the recession through increasing demand.

It is also important to recognize that present developments in economic theorizing, which are generally believed to be an extension of Keynesian analysis, do not adequately reflect the spirit of Keynesian thought—as opposed to his technical conclusions. Keynes's main contribution to theory came when he proved that it was possible for unemployment to persist over long periods because effective demand would not necessarily rise as fast as potential supply. Modern economic theorists grasped this insight and set to work to devise policies that would lead to a sufficiently rapid increase in effective demand to balance increases in potential supply and thus ensure minimum unemployment. However, this is not the *only* policy proposal that can be derived from an interpretation of Keynesian analysis: society could equally well decide that it no longer wished to channel the quasi-totality of its efforts toward the goal of full employment but rather desired to seek a new social order that would allow us to take full advantage of the potential of emerging abundance and our ability to eliminate toil.

Keynes himself quite clearly hoped for the second development, arguing that

> when the accumulation of wealth is no longer of high social importance, there will be great changes in the code of morals. We shall be able to rid ourselves of many of the pseudo-moral principles which have hag-ridden us for two hundred years, by which we have exalted some of the most distasteful of human qualities into the position of the highest values. We shall be able to afford to dare to assess the money-motive at its true value. . . . All kinds of social customs and economic practices affecting the distribution of wealth and its rewards and penalties which we now maintain at all costs, however distasteful and unjust they may be in themselves . . . we shall then be free, at last, to discard.[3]

It is quite clear, therefore, that although present policy is justified on the basis of Keynesian analysis, Keynes would, in present conditions, reject many of the policy prescriptions being advanced, for he would hold that they perpetuated the worst of the values of the industrial age.

What methods have economists proposed to ensure that potential supply and effective demand would stay in balance? The first step toward this goal, which was accomplished around the end of the Second World War in almost all Western countries, was the passage of legislation pledging the efforts of governments to ensure that supply and demand would remain in balance and

3. J. M. Keynes, *Essays in Persuasion* (New York: Harcourt, Brace and Company, 1932), pp. 369–70, "Economic Possibilities for Our Grandchildren."

thus provide jobs for all: in the United States this was accomplished by the Employment Act of 1946.

This commitment to a full employment policy through balancing supply and demand has deepened in all Western countries in the years since the Second World War. The United States has undoubtedly been the last country to understand the full implications of this policy approach, but the first five years of the sixties have marked its final acceptance. It is now generally believed, not only by economists but by the vast majority of businessmen, that it is the responsibility of the government to ensure that the economy remains in balance—that the government should aim to balance the economy rather than to balance the budget. As Meno Lovenstein points out in his essay, the government has now essentially taken a commitment to "guarantee the national income" by ensuring that rights to all available productive resources are distributed.

The difference between this approach to the government's responsibility and that current in the nineteenth century, when it was believed that government damaged the operation of the economy whenever it intervened, is so vast as to need no stressing. Unfortunately, economic theory has not yet re-examined all the implications of the shift in approach. For example, if the government is deeply involved in guaranteeing the national income of the whole country, and if, as is inevitable, its interventions affect the pattern of income distribution, what goals should it adopt? Another facet of this problem results from the fact that a large number of people are unable to earn their living because they are too old, too young, too mentally or physically ill. How should they be provided with incomes and what amount of resources should they receive? Economics has few, if any, answers to these and similar questions.

The problem of providing incomes to those who are too old, too young, or too sick to hold a job is already urgent and is certain to become more so in coming years because of the inevitable shifts in patterns of age distribution. This reality is already causing the emergence of a new consensus that cuts across party lines and interest groups. This consensus is based on a belief that the government has already taken an implied commitment to provide a minimum level of income to all individuals, but that the present mosaic of measures designed to ensure this result is both excessively complex and unduly costly. It is argued that it would therefore be desirable to introduce a single plan that would meet the implied commitment of government as simply and cheaply as possible through the introduction of a guaranteed income floor for all those who either cannot, or should not, earn their living through holding a job.

There should be no need to justify payments to the physically and mentally ill: they cannot work and society should surely provide for them. Some justification, however, is often felt to be required for more adequate payments to the old, for one of the most sanctified of our work myths is that older people both could and should have saved enough to provide for their old age. This is, of course, merely a cynical fiction. Those who are old today worked in an era when their income was necessarily far lower than is paid for jobs demanding a comparable level of skills and application today. They needed to spend a large proportion of their income just to cover their expenses, including the education of their children. They were therefore able to save very little, if anything, whether directly or through insurance schemes. Today's labor force, however, would not be enjoying its present level of income without their hard work and that of earlier generations who had even less to show for their toil. Any fair distribution of the nation's resources should ensure that old people be allowed to share in the wealth they created. Their labor was, in fact, wealth, and it was invested in the national economy at a time when its value was at a premium. Today this group should be collecting their "earned interest."

It will, perhaps, help to put this question in perspective if we recognize that most of those presently being paid social security benefits are receiving more than the actuarial value of their contributions: i.e., they did not pay in enough money to cover the benefits they are receiving. Continuing expansion of the social security system makes it almost certain that it will not become actuarily sound at any point in the future. Thus we have already accepted, on a practical basis, that the old are entitled to a more adequate income than would be theirs on an insurance basis. The next step is to bring the logic of this position into the open and see what more needs to be done.

The question of income distribution among the young poses equally serious problems, for we have not yet been willing to accept the fact that we have extended the principle of parental support of the young far beyond the breaking point. In an agricultural or even an early industrial society, a child was wealth. After a few years of care, the child added to the family income rather than subtracted from it. In addition, the younger generation was expected to support their parents as they grew old. There was, thus, a rough balance between the economic responsibility of the parents and that of the children.

Let us contrast this with the situation today. Because of the demands of the new world in which we live, a child should be educated at least until he is twenty-one and perhaps until he is twenty-five or thirty. Despite the growing number of loans, grants, and scholarships, it is still a fundamental assumption of our society that the primary economic responsibility for the education and support of the child lies with the parent. However, the parent receives little financial return, for by the time the child leaves the educational process he is generally married and feels little obligation for the economic support of his parents. Parents should no longer be expected to underwrite the lengthy educational process that the future society requires of today's and tomorrow's young people. We must recognize that the student is already "working" as relevantly as the man in the factory or the office.

While the idea that we must find new ways of providing income to those who cannot, or should not, hold a job has received increasing support in recent years, the wider concept that *everybody* should receive a guaranteed income as a matter of right is still highly controversial. The proposal for a universal guaranteed income can be justified on the ground that the evolution described in the essays by Robert Davis and Ben Seligman ensures that most types of structured[4] jobs will be taken over, within a relatively brief period, by advanced machinery. This will necessarily be true because, in addition to the often substantial direct economic savings from the use of automatic machinery, machinery also appears more attractive than men for a wide range of noneconomic reasons. Machine systems do not get tired, they can carry out a particular task with a continued precision that cannot be demanded or expected of a human work force; they are incapable of immorality, they do not lie, steal, cheat, or goof off; they do not claim that their rights as human beings are being violated by factory work practices; they are not class-conscious; above all, they are not vocal in their criticism of management and they do not go on strike.

In the relatively near future, therefore, those who need to expand their plant to meet created demand will prefer to buy machines rather than to hire men: the machines they buy will be produced predominantly by other machines. The new machines purchased will be so much more efficient than earlier machinery that large numbers of existing firms using older machinery and thus employing many men will be forced to close down: they will be too inefficient to compete.

The process can be summarized as follows: created demand will lead to purchases of highly efficient and productive machine systems that need few men to control them: i.e., to the installation of cybernation. Thus, in the relatively near future, a policy of forcing rapid growth in demand in order to increase employment opportunities will actually lead to the opposite result: it will raise unemployment rather than lower it.

The conclusion that massive unemployment is inevitable is still rejected by most economists and policy-makers, who argue that increases in demand brought about, if necessary, by federal intervention to balance the economy can *always* be large enough to ensure that all the available labor will be used. Unfortunately, however, there is no economic theory or contemporary evidence to support this conclusion. The neoclassical theorizing of the last part of the nineteenth century and the beginning of the twentieth assumed that men and machines would cooperate with each other; today, however, they are competitive. Keynes, who is presently used as the justification for the assertion that demand and supply can be kept in balance, and jobs provided for all, should not be used for this purpose because he excluded from his analysis those

4. A structured job is one in which the decision-making rules can be set out in advance. While computer theorists agree that the computer can, by definition, take over all structured jobs, they still disagree on the proportion of jobs on the factory floor and in the office which can eventually be structured. Everybody agrees, however, that the process of replacement of men by machines is only beginning.

very factors that now threaten massive unemployment. "We take as given the existing skill and quantity of available labor, the existing quality and quantity of available equipment, the existing technique. This does not mean that we assume these facts to be constant, but merely that in this place and context, we are not considering or taking into account the effects and consequences of changes in them."[5]

In effect, therefore, economists have no valid theoretical structure to support their contention that unemployment can be avoided by increases in demand. To the noneconomist, such a statement will necessarily be shocking, but it is unfortunately valid. Economists, like many social scientists, have generally been far more concerned about theoretical rigor within a given pattern of assumptions than about the validity of the assumptions themselves; the development of theory has proceeded despite the ever decreasing relevance of the assumptions on which it is based. Economic predictions about unemployment rates will not be valid until the analysis from which they are drawn is based on a new and more relevant set of assumptions.

As minimum unemployment cannot be achieved in coming years, fundamental change in the socioeconomic system will be absolutely essential. As we have already seen, our present system is postulated on the belief that every individual who desires a job will be able to find one and that the jobs thus obtained will pay well enough to enable the individual to live with dignity. I am convinced that if we desire to maintain freedom, a guaranteed income will necessarily have to be introduced. In addition, during the period of transition from a scarcity to an abundance socioeconomy, we will have to consider the whole problem of income maintenance for those whose income level is above the minimum income floor in order to allow them to update their education and to minimize hardship when individuals lose their jobs because of further increases in technological sophistication. Although neither this essay nor this book can deal with the issue of income maintenance, it is necessary to stress that the need for an income maintenance program is just as great as the need for a guaranteed income floor.[6]

The economic controversy is not, however, the most important one. Just as Keynes foresaw that the issue of scarcity was not the long-run problem of mankind, he warned us against placing too much emphasis on strictly economic analysis: "Do not let us overestimate the importance of the economic problem, or sacrifice to its supposed necessities other matters of greater and more permanent significance."[7] The real question raised by the coming of cybernation is not whether we *can* provide jobs for everybody, but whether

5. J. M. Keynes, *The General Theory of Employment, Interest and Money* (New York: Harcourt, Brace and Company, 1936), p. 243.
6. Proposals for both a guaranteed income and income maintenance are set out in Part II and the Appendix of my book *Free Men and Free Markets* (Garden City, N.Y.: Doubleday & Company, Inc., 1965). This material is reprinted in the Appendix of this book, pp. 231–37.
7. Keynes, *Essays in Persuasion*, p. 373.

we *should* provide jobs for everybody: the question we need to examine is whether our present policy of providing income rights on the basis of job-holding is the best way to ensure that the urgent work of society will be accomplished.

Most economists, as well as government, management, and union leaders claim that the type of work that now needs to be done and will need to be done in the future can, and should, be turned into jobs for which a wage or salary can then be paid. This is the assumption that is explicitly challenged by those who support the guaranteed income. Job-holding within the increasingly bureaucratic structures whose growth can be expected, given the continuation of the present socioeconomic system, would certainly not be conducive to the self-development of the individual. In addition, and even more importantly, the lack of flexibility inherent in bureaucratic structures makes them unsuitable forms of organization for acting upon, or even perceiving, developments that would benefit the socioeconomic system.

The essayists in the third part of this volume argue, in effect, that many individuals are perfectly capable of perceiving what needs to be done to develop themselves and their society and that these individuals would act upon this perception if they had the funds that would free them from the necessity of holding a job. A parallel is often made with the ownership of capital: it is claimed that the possession of capital has not led to a general decline in individual and social responsibility and that there is therefore no reason why a guaranteed income should lead to a decline in individual and social responsibility. Comparisons with the dole are rejected; it is suggested that the dole results in degradation partly because it is seen by its recipients as "charity" rather than as a right, and partly because the techniques of distribution used in many areas of the country inevitably sap self-respect and initiative.

For society at large, and especially for those creative individuals now shackled by the absence of a guaranteed source of income, the situation would seem to be analogous to that which obtained at the time of the introduction of limited liability in the nineteenth century. Limited liability was introduced to encourage risk-taking by those investing in companies. The concept of a joint venture was replaced by the concept that a stockholder's liability for company debts no longer put a lien on his total wealth but only on the amount he invested in the company. Limited liability was a precondition for the taking of risks: it did not ensure innovation or risk-taking, but it did make them possible, thus allowing the economy and society to benefit from the self-interested acts of individuals.

A guaranteed income provides the individual with the ability to do what he personally feels to be important. This will allow risk-taking and innovation in areas where the existing and emerging needs of society are not being met by an otherwise efficiently functioning free-enterprise system. The guaranteed income is not mediated through the offices of any other individual or organization within the market system and therefore does not bring with it built-in pressures for the recipient to continue doing what is already being done through the market system.

The guaranteed income therefore involves a major shift in rights and obligations. Today we demand of an individual that he find a job, but we then provide him with the right to "pursue happiness." Tomorrow we will provide him with the right to receive enough resources to live with dignity, and we will demand of him that he develop himself and his society.

The guaranteed-income proposal is based on the fundamental American belief in the right and the ability of the individual to decide what he wishes and ought to do. This is surely the basic meaning of the phrase "private enterprise": that the individual should have the right to obtain enough resources to do what he believes to be important. In the past, the individual could go into business for himself and thus obtain resources. Today all the evidence shows that neither the self-employed businessman nor the small company can compete with the large corporation. The ideal of private enterprise can, therefore, be preserved only if the guaranteed income is introduced.

The guaranteed income will, in fact, lead to the revival of "private enterprise." Once the guaranteed income is available, we can anticipate the organization of what I have called "consentives": productive groups formed by individuals who will come together on a voluntary basis simply because they wish to do so. The goods produced by these consentives will not compete with mass-produced goods available from cybernated firms. The consentive will normally produce the "custom-designed" goods that have been vanishing within the present economy. The consentive would sell in competition with firms paying wages, but its prices would normally be lower because it would need to cover only the costs of materials and other required supplies. Wages and salaries would not need to be met out of income, as the consentive members would be receiving a guaranteed income. The consentive would be market-oriented but not market-supported.

We can anticipate that small market-supported firms will be enabled to survive by transforming themselves into market-oriented consentives. The opposite process will occur as consentives that make significant profits automatically turn into market-supported firms.[8] Thus the guaranteed income would help to bring about a reversal of the present trend toward similarity in type of goods and services, inflexibility in methods of production, and uniformity in productive organization.

At the present time we are committed as a society to the idea that we can and should provide jobs for all. This goal is no longer valid, and we should therefore provide everybody with an absolute right to a guaranteed income. This will, of course, mean that there will be far more unemployment in the

8. For a description of these processes see *Free Men and Free Markets,* Chapter 9.

future than there is today. We will, however, come to perceive unemployment as favorable rather than unfavorable. The individual and the society fear unemployment today for two reasons: first, because it usually involves the receipt of an inadequate income; second, because it threatens cessation of all activity that seems meaningful and indeed encourages antisocial activities. Once we have provided adequate incomes to all and have introduced the new policies required to develop each individual's potential, unemployment—which will then be redefined as the condition of *not* holding a job—will be seen to be highly desirable, for it will provide the individual with freedom to develop himself and his society.

The Psychological Aspects of the Guaranteed Income

Erich Fromm

This paper focuses exclusively on the *psychological* aspects of the guaranteed income, its value, its risks, and the human problems it raises.

The most important reason for the acceptance of the concept is that it might drastically enhance the freedom of the individual.[1] Until now in human history, man has been limited in his freedom to act by two factors: the use of force on the part of the rulers (essentially their capacity to kill the dissenters); and, more importantly, the threat of starvation against all who were unwilling to accept the conditions of work and social existence that were imposed on them.

Whoever was not willing to accept these conditions, even if there was no other force used against him, was confronted with the threat of starvation. The principle prevailing throughout most of human history in the past and present (in capitalism as well as in the Soviet Union) is: "He who does not work shall not eat." This threat forced man not only to *act* in accordance with what was demanded of him, but also to *think* and to *feel* in such a way that he would not even be tempted to act differently.

The fact that past history is based on the principle of the threat of starvation has, in the last analysis, its source in the fact that, with the exception of certain primitive societies, man has lived on the level of scarcity, both economically and psychologically. There were never sufficient material goods to satisfy the needs of all; usually a small group of "directors" took for themselves all that their hearts desired, and the many who could not sit at the table were told that it was God's or Nature's law that this should be so. But

---

1. Cf. my discussion of a "universal subsistence guarantee" in *The Sane Society* (New York: Holt, Rinehart & Winston, Inc., 1955), pp. 335 ff.

it must be noted that the main factor in this is not the greed of the "directors," but the low level of material productivity.

A guaranteed income, which becomes possible in the era of economic abundance, could for the first time free man from the threat of starvation, and thus make him truly free and independent from any economic threat. Nobody would have to accept conditions of work merely because he otherwise would be afraid of starving; a talented or ambitious man or woman could learn new skills to prepare himself or herself for a different kind of occupation. A woman could leave her husband, an adolescent his family. People would learn to be no longer afraid, if they did not have to fear hunger. (This holds true, of course, only if there is also no political threat that inhibits man's free thought, speech, and action.)

Guaranteed income would not only establish freedom as a reality rather than a slogan, it would also establish a principle deeply rooted in Western religious and humanist tradition: man has the right to live, regardless! This right to live, to have food, shelter, medical care, education, etc., is an intrinsic human right that cannot be restricted by any condition, not even the one that he must be socially "useful."

The shift from a psychology of scarcity to that of abundance is one of the most important steps in human development. A psychology of scarcity produces anxiety, envy, egotism (to be seen most drastically in peasant cultures all over the world). A psychology of abundance produces initiative, faith in life, solidarity. The fact is that most men are still geared psychologically to the economic facts of scarcity, when the industrial world is in the process of entering a new era of economic abundance. But because of this psychological "lag" many people cannot even understand new ideas as presented in the concept of a guaranteed income, because traditional ideas are usually determined by feelings that originated in previous forms of social existence.

A further effect of a guaranteed income, coupled with greatly diminished working hours for all, would be that the spiritual and religious problems of human existence would become real and imperative. Until now man has been occupied with work (or has been too tired after work) to be too seriously concerned with such problems as "What is the meaning of life?" "What do I believe in?" "What are my values?" "Who am I?" etc. If he ceases to be mainly occupied by work, he will either be free to confront these problems seriously, or he will become half mad from direct or compensated boredom.

From all this it would follow that economic abundance, liberation from fear of starvation, would mark the transition from a prehuman to a truly human society.

Balancing this picture, it is necessary to raise some objections against, or questions about, the concept of a guaranteed income. The most obvious question is whether a guaranteed income would not reduce the incentive for work.

Aside from the fact that there is already no work for an ever increasing sector of the population, and hence that the question of incentive for these people is irrelevant, the objection is nevertheless a serious one. I believe, however, that it can be demonstrated that material incentive is by no means the only incentive for work and effort. First of all there are other incentives: pride, social recognition, pleasure in work itself, etc. Examples of this fact are not lacking. The most obvious one to quote is the work of scientists, artists, etc., whose outstanding achievements were not motivated by the incentive of monetary profit, but by a mixture of various factors: most of all, interest in the work they were doing; also pride in their achievements, or the wish for fame. But obvious as this example may seem, it is not entirely convincing, because it can be said that these outstanding people could make extraordinary efforts precisely because they were extraordinarily gifted, and hence they are no example for the reactions of the average person. This objection does not seem to be valid, however, if we consider the incentives for the activities of people who do not share the outstanding qualities of the great creative persons. What efforts are made in the field of all sports, of many kinds of hobbies, where there are no material incentives of any kind! To what extent interest in the work process itself can be an incentive for working was clearly demonstrated for the first time by Professor Mayo in his classic study at the Chicago Hawthorne Works of the Western Electric Company.[2] The very fact that unskilled women workers were drawn into the experiment of work productivity of which they were the subjects, the fact that they became interested and active participants in the experiment, resulted in increased productivity, and even their physical health improved.

The problem becomes even clearer when we consider older forms of societies. The efficiency and incorruptibility of the traditional Prussian civil service were famous, in spite of the fact that monetary rewards were very low; in this case such concepts as honor, loyalty, duty, were the determining motivations for efficient work. Still another factor appears when we consider preindustrial societies (like the medieval European society, or half-feudal societies in the beginning of this century in Latin America). In these societies the carpenter, for instance, wanted to earn enough to satisfy the needs of his traditional standard of living, and would refuse to work more in order to earn more than he needed.

Secondly, it is a fact that man, by nature, is not lazy, but on the contrary, suffers from the results of inactivity. People might prefer not to work for one or two months, but the vast majority would beg to work, even if they were not paid for it. The fields of child development and mental illness offer abundant data in this connection; what is needed is a systematic investigation in which the available data are organized and analyzed from the standpoint of "laziness as disease," and more data are collected in new and pertinent investigations.

However, if money is not to be the main incentive, then work in its technical or social aspects would have to be sufficiently attractive and interest-

2. Cf. Elton Mayo, *The Human Problem of an Industrial Civilization*, 2d ed. (New York, The Macmillan Company, 1946).

ing to outweigh the unpleasure of inactivity. Modern, alienated man is deeply bored (usually unconsciously) and hence has a yearning for laziness, rather than for activity. This yearning itself is, however, a symptom of our "pathology of normalcy." Presumably misuse of the guaranteed income would disappear after a short time, just as people would not overeat on sweets after a few weeks, assuming they would not have to pay for them.

Another objection is the following: Will the disappearance of the fear of starvation really make man so much freer, considering that those who earn a comfortable living are probably just as afraid to lose a job that gives them, let us say, $15,000 a year, as are those who might go hungry if they were to lose their jobs? If this objection is valid, then the guaranteed income would increase the freedom of the large majority, but not that of the middle and upper classes.

In order to understand this objection fully we have to consider the spirit of contemporary industrial society. Man has transformed himself into a *homo consumens*. He is voracious, passive, and tries to compensate for his inner emptiness by continuous and ever increasing consumption (there are many clinical examples for this mechanism in cases of overeating, overbuying, over-drinking, as a reaction to depression and anxiety); he consumes cigarettes, liquor, sex, movies, travel, as well as education, books, lectures, and art. He *appears* to be active, "thrilled," yet deep down he is anxious, lonely, depressed, and bored (boredom can be defined as that type of chronic depression that can successfully be compensated by consumption). Twentieth-century industrial-ism has created this new psychological type, *homo consumens,* primarily for economic reasons, i.e., the need for mass consumption, which is stimulated and manipulated by advertising. But the character type, once created, also influ-ences the economy and makes the principles of ever-increasing satisfaction appear rational and realistic.[3]

Contemporary man has an unlimited hunger for more and more con-sumption. From this follow several consequences: if there is no limit to the greed for consumption, and since in the foreseeable future no economy can produce enough for unlimited consumption for everybody, there can never be true "abundance" (psychologically speaking) as long as the character structure of *homo consumens* remains dominant. For the greedy person there is always scarcity, since he never has enough, regardless of how much he has. Further-more he feels covetous and competitive with regard to everybody else; hence he is basically isolated and frightened. He cannot really enjoy art or other cultural stimulations, since he remains basically greedy. This means that those

3. The problem is all the more complicated by the fact that at least 20 per cent of the American population live on a level of scarcity, that some parts of Europe, especially the Socialist countries, have not yet attained a satisfactory standard of living, and that the majority of mankind, which dwells in Latin America, Asia, and Africa, is still living at hardly above starvation level. Any argument for less consumption meets with the argument that in most of the world *more* consump-tion is needed. This is perfectly true, but the danger exists that even in the countries that are now poor, the ideal of maximal consumption will guide their effort, form their spirit, and hence will continue to be effective even when the level of optimal (not maximal) consumption has been reached.

who lived on the guaranteed-income level would feel frustrated and worthless, and those who earned more would remain prisoners of circumstances, because they would be frightened and lose the possibility for maximum consumption. For these reasons I believe that guaranteed income without a change from the principle of maximal consumption would only take care of certain problems (economical and social) but would not have the radical effect it should.

What, then, must be done to implement the guaranteed income? Generally speaking, we must change our system from one of maximal to one of optimal consumption. This would mean:

A vast change in industry from the production of commodities for individual consumption to the production of commodities for public use: schools, theaters, libraries, parks, hospitals, public transportation, housing; in other words an emphasis on the production of those things that are the basis for the unfolding of the individual's inner productiveness and activity. It can be shown that the voraciousness of *homo consumens* refers mainly to the individual consumption of things he "eats" (incorporates), while the use of free public services, enabling the individual to enjoy life, do not evoke greed and voraciousness. Such a change from maximal to optimal consumption would require drastic changes in production patterns, and also a drastic reduction of the appetite-whetting, brainwashing techniques of advertising, etc.[4] It would also have to be combined with a drastic cultural change: a renaissance of the humanistic values of life, productivity, individualism, etc., as against the materialism of the "organization man" and manipulated ant heaps.

These considerations lead to other problems that need to be studied: Are there objectively valid criteria to distinguish between rational and irrational, between good and bad needs, or is any subjectively felt need of the same value? (Good is defined here as needs that enhance human aliveness, awakeness, productivity, sensitivity; bad, as those needs that weaken or paralyze these human potentials.) It must be remembered that in the case of drug addiction, overeating, alcoholism, we all make such a distinction. The study of these problems would lead to the following practical considerations: what are the minimum legitimate needs of an individual? (For instance: one room per person, so much clothing, so many calories, so many culturally valuable commodities such as a radio, books, etc.) In a relatively abundant society such as that of the United States today, it should be easy to figure out what the cost for a *decent* subsistence minimum is, and also what the limits for maximal consumption should be. Progressive taxation on consumption beyond a certain threshold could be considered. It seems important to me that slum conditions should be avoided. All this would mean the combination of the principles of a guaranteed income with the transformation of our society from maximal to

---

4. The need of restricting advertising and, even more, of changing production in the direction of greater production of public services are, in my opinion, hardly thinkable without a great deal of state intervention.

optimal individual consumption, and a drastic shift from production for individual needs to production for public needs.

I believe it is important to add to the idea of a guaranteed income another one, which ought to be studied: the concept of *free* consumption of certain commodities. One example would be that of bread, then milk, and vegetables. Let us assume, for a moment, that everyone could go into any bakery and take as much bread as he liked (the state would pay the bakery for all bread produced). As already mentioned, the greedy would at first take more than they could use, but after a short time this "greed-consumption" would even itself out and people would take only what they really needed. Such free consumption would, in my opinion, create a new dimension in human life (unless we look at it as the repetition on a much higher level of the consumption pattern in certain primitive societies). Man would feel freed from the principle "he who does not work shall not eat." Even this beginning of free consumption might constitute a very novel experience of freedom. It is obvious even to the non-economist that the provision of free bread for all could be easily paid for by the state, which would cover this disbursement by a corresponding tax. However, we can go a step further. Assuming that not only all minimal needs for food were obtained free—bread, milk, vegetables, fruit—but the minimal needs for clothing (by some system everybody could obtain, without paying, say one suit, three shirts, six pairs of socks, etc., per year); that transportation was free, requiring, of course, vastly improved systems of public transportation, while private cars would become more expensive. Eventually one could imagine that housing could be solved in the same way, by big housing projects with sleeping halls for the young, one small room for older, or married couples, to be used without cost by anybody who chose. This leads me to the suggestion that another way of solving the guaranteed-income problem would be by free minimal consumption of all necessities, instead of through cash payments. The production of these minimum necessities, together with highly improved public services, would keep production going, just as guaranteed-income payments would.

It may be objected that this method is more radical, and hence less acceptable, than the one proposed by the other authors. This is probably true; but it must not be forgotten that, on the one hand, this method of free minimal services could theoretically be arranged within the present system while on the other hand, the idea of a guaranteed income will not be acceptable to many, not because it is not feasible, but because of the psychological resistance against the abolishment of the principle "He who does not work shall not eat."

One other philosophical, political, and psychological problem has to be studied: that of freedom. The Western concept of freedom was to a large extent based on the freedom to own property, and to exploit it, as long as other legitimate interests were not threatened. This principle has actually been punctured in many ways in Western industrial societies by taxation, which is a form of expropriation, and by state intervention in agriculture, trade, and industry. At the same time, private property in the means of production is becoming increasingly replaced by the semipublic property typical of giant corporations. While the guaranteed-income concept would mean some additional state regulations, it must be remembered that today the concept of freedom for the average individual lies not so much in the freedom to own and exploit property (capital) as in the freedom to consume whatever he likes. Many people today consider it as an interference with their freedom if unlimited consumption is restricted, although only those on top are really free to choose what they want. The competition between different brands of the same commodities and different kinds of commodities creates the illusion of personal freedom, when in reality the individual wants what he is conditioned to want.[5] A new approach to the problem of freedom is necessary; only with the transformation of *homo consumens* into a productive, active person will man experience freedom in true independence and not in unlimited choice of commodities.

The full effect of the principle of the guaranteed income is to be expected only in conjunction with: (1) a change in habits of consumption, the transformation of *homo consumens* into the productive, active man (in Spinoza's sense); (2) the creation of a new spiritual attitude, that of humanism (in theistic or nontheistic forms); and (3) a renaissance of truly democratic methods (for instance, a new Lower House by the integration and summation of decisions arrived at by hundreds of thousands of face-to-face groups, active participation of all members working in any kind of enterprise, in management, etc.).[6] The danger that a state that nourishes all could become a mother goddess with dictatorial qualities can be overcome only by a simultaneous, drastic increase in democratic procedure in all spheres of social activities. (The fact is that even today the state is extremely powerful, without giving these benefits.)

In sum, together with economic research in the field of the guaranteed income other research must be undertaken: psychological, philosophical, religious, educational. The great step of a guaranteed income will, in my opinion, succeed only if it is accompanied by changes in other spheres. It must not be forgotten that the guaranteed income can succeed only if we stop spending 10 per cent of our total resources on economically useless and dangerous armaments; if we can halt the spread of senseless violence by systematic help to the underdeveloped countries, and if we find methods to arrest the population explosion. Without such changes, no plan for the future will succeed, because there will be no future.

---

5. Here too, the totalitarian bureaucratization of consumption in the Soviet-bloc countries has made a bad case for any regulation of consumption.
6. Cf. Fromm, *loc. cit.*

Economic Aspects of the Evolving Society

H. F. W. Perk

Conventional economic thought provides no help in trying to cope with the economics of abundance. A return to fundamentals is called for; one result is to shift our emphasis from "economy" to "ecology".

The relevance of Fuller's work is discussed, with particular emphasis on his concept of wealth. Evidence is given to support the conclusion that a global society of cumulative and accelerating abundance is attainable within one generation.

The United States could end poverty and achieve abundance for all in a much shorter period: five to ten years. Some suggestions are offered for realizing this goal.

On the one hand, a computing machine can be instructed to erase its memory whenever a fresh start seems needed. On the other hand, human beings are notoriously resistant to any comparable process—"brainwashing" and amnesia notwithstanding. Mostly, this is all to the good; and amnesia is considered a pathological situation. The negative aspects of the persistance of memory, however, concern me at the moment. For unless we can find ways to rid ourselves of all the cherished misinformation each of us has collected through the years, we shall be ill-prepared to make the fresh start the times require. Thus my first task must be to deal with some of the undesirable consequences of the persistance of human memory.

As far as my present topic is concerned, probably our greatest obstacle is the widespread acceptance of conventional economic theory—the belief that this body of knowledge provides us with ideas and information useful in our present circumstances. In fact, this hasn't been true for, at least, the last fifty years. Despite the devastating assaults of Veblen, Arnold, Galbraith, Schoeffler, Bazelon, and others, economists continue to play with their impervious system of irrelevancies, in full expectation that the rest of us are too awed to complain. Said Schoeffler,[1] summing up a penetrating critique of economic thought:

> The record of performance of economics in the field of prediction and policy-making has been and is so very poor because the concepts employed by economists are, with few exceptions, utterly unsuited to the requirements of their task ...

Here Schoeffler recounts evidence for this assertion, and concludes

1. Sidney Schoeffler, *The Failures of Economics,* Harvard University Press, Cambridge (1955) pp 40-41.

Unavoidably, therefore, predictions about economic reality which are produced with the aid of these techniques are quite undependable, and professional economics has been and continues to be a relatively ineffectual debating society.

If the "debating society" cannot offer us any reliable help, where shall we turn? I suggest we re-set to zero and proceed from fundamentals.

Perspective

Bertrand Russell once remarked that everything of concern to man could be subsumed by the three questions. Where did we come from? Where are we going? What do we do meanwhile? Consider the first: Where did we come from?

The physical universe may be viewed as a cosmic reservoir of energy in a continual state of flux. We perceive this energy primarily in the form of mass —in other words, our inventory of chemical elements. All but about one-tenth of one percent of this mass is hydrogen or helium. Thus, as far as the physical universe is concerned, man is simply a recurring pattern of relationships among some of the rarer chemical elements, but principally hydrogen, oxygen, and carbon. This pattern is replicated by a special code of the DNA molecule —a code which the biologists are about to decipher. There is no doubt that they will soon set about "improving" nature's patterns. This leads us to the next question. Where are we going?

The harnessing of atomic energy about twenty years ago, should have been ample evidence that mankind has arrived at a qualitatively different historical condition. At that moment, it became only a matter of time before man would have essentially unlimited energy at his disposal, to use for whatever purposes he preferred. Typically, man's first use of this newly-won capability was to destroy his fellow-men. The demonstration effect was apparently awesome enough to discourage any repeat performances—at least so far. Indeed it is even possible to hope that a basic lesson was learned: namely, man must exercise self-control, or he may well eliminate any possibility of the continuing recurrence of that rare pattern of elements we call *homo sapiens*. I trust the biologists, in particular, have learned this lesson well—*before* they succeed in re-writing the DNA code.

Mind has triumphed over matter: Man can evolve as he chooses. With essentially unlimited energy at our disposal, and the ability to transform this energy into any patterns, we are left with the basic ethical problem. What do we *choose* to do? And that is the point of Russell's third question, what do we do meanwhile? Mankind's entire future—and the remainder of this paper —thus must necessarily revolve around this question.

Frame of Reference

Moving now from the cosmic to the global, and from all time to our own times, I shall consider man's situation in this century. As Buckminster Fuller has shown us, less than one percent of mankind had achieved a high industrial standard of living at the beginning of the century. [2] By the end of World War I—as a by-product of investment in weaponry programs—this proportion had risen to six percent. Despite the great depression, the application of the "fall-out" from World War I weaponry increased the portion of mankind who lived in comparative affluence to twenty percent by the beginning of World War II.

The second world war—and the subsequent "cold war" era—has caused the trend to continue, and the proportion has now changed to approximately forty-four percent. Extrapolating the trend to the end of the century, we discover that one-hundred percent of mankind will then be served by a very high industrial standard of living—and this will have occurred simply as a by-product of high-priority investments in weaponry. I am making the enormous assumptions that investments will continue to take their present form —as a massive governmental underwriting of the economy—and that no actual use is made of the catastrophic destructive capability that is also generated.

But why take this chance? Why couldn't we accelerate the trend to affluence by a more appropriate selection of priorities and allocations of resources? Since we so obviously *could* do so, the answer obviously is that we *should* do so. It is a matter of choice; and the appropriate selection, as Ashby points out, is a hallmark of intelligence. According to Fuller, it is a choice between "killingry" and "livingry," between investing in "negative tools" for the limitation of human freedom, or "positive tools" for the expansion of the domain of human freedom.

As I read the signs, things are looking up. President Johnson, [3] noting that 1965 has been designated "International Cöoperation Year" by the United Nations, proposed that 1965 should also become the "Year of Science." Said the President:

> I propose to dedicate this year to finding new techniques for making man's knowledge serve man's welfare. Let it be a turning point in the struggle—not of man against man, but of man against nature. In the midst of tension, let us begin to chart a course toward the possibilities of conquest which bypass the politics of the cold war.

Is the politician catching up with the poet-philosopher-scientist?

The Second Emancipation Proclamation

Imagine the following situation. In response to the combined pressures of rising unemployment, greatly intensified civil rights activity, and imminent disarmament, President Johnson issues what comes to be called the "Second Emancipation Proclamation." This document declares that no society can be

2. Buckminster Fuller, *Ideas and Integrities,* Prentice-Hall, Englewood Cliffs, N.J. (1963) Chapter 15, p 192.
3. President Lyndon B. Johnson, Quoted in the *New York Times* (Thursday, June 11, 1964) pp 1 and 18.

a great society unless it is a free society; and that no man is truly free unless he is secure from want. Under the conditions which prevail in modern industrial society, this requires that every citizen—as a matter of right—be guaranteed sufficient purchasing power to meet his needs. And so on.

We have heard Alice Mary Hilton speak about Living Certificates. And I trust we are all familiar with the report of the Ad Hoc Committee on the Triple Revolution. This is meant to be a step beyond that. Consider, for example, some of the implications of issuing to everyone a universally acceptable credit card. Each individual could order any of the goods and services he desires. That would certainly put an end to poverty. In effect, everyone would become independently wealthy—like Barry Goldwater.

Such a credit card would support mass purchasing power—which would support mass consumption—which would support mass production—which would support the credit of the individual holding the card. We would have created a regenerative closed-loop system, embedded within the free energy of the cosmic reservoir, and continually improved by the leavening of applied intellect.

The credit card would eliminate the need for money[4]—since all transactions would be made by means of the card. A by-product of the transactions —kept track of by computers, naturally—would be data on the kind of goods and services that are in demand; and where the demand exists, appropriate instructions could be sent to the cybernated production and distribution facilities. Automatically, of course.

With the need for money the "paper economy" and its excrescences would disappear. By means of direct distribution, the universal credit card would eliminate the concept of buying and selling together with all the related practices of banking, accounting, insurance, advertising, and so forth. And with the industrial production-distribution system fully cybernated, probably fewer than twenty percent of the population would be required to keep the system functioning for one-hundred percent of the population.

Since "money is the root of all evil," as the old saying goes, there should be a substantial reduction in evil with the elimination of this root. With everyone independently wealthy the motive for much, if not all, crime is removed; and, obviously, without any poverty we would require none of the myriad services which presently cater to it. Thus much of the machinery of law, courts, police—indeed, of government generally—could be dispensed with. This should bring the residue of jobs down to perhaps ten percent or fewer of the present figure.

With ninety percent of the adult population free to engage in "producing the goods of civilization"—art, music, science, literature—and released from

---

4. Since money is a means of exchange it would simply be superseded by another means of exchange —one that is as suited (or better) to the age of abundance as money probably was to the age of scarcity. Ed.

laboring for realizing their full human potential, we would surely witness a flowering of culture without parallel in all history. And undoubtedly a large proportion of our most talented citizens would devote themselves to assisting the rest of the world to achieve the industrial base necessary for enjoying similar affluence and freedom. I do not fear the leisure society. I am impatient for its arrival.

What is Holding Us Back?

Certainly no technological cause restrains us. We confidently plan and implement all kinds of tasks that are far more complicated than raising the purchasing power of those who are at present without the advantages of a modern society. I should guess that we have not found a suitable ideological justification for it. Perhaps that staunch New Englander, Edward Bellamy, might help us out[5]

> How happened it . . . that you workers were able to produce more than so many savages would have done? Was it not wholly on account of the heritage of the past knowledge and achievements of the race, the machinery of society, thousands of years in contriving, found by you ready-made to your hand? How did you come to be possessors of this knowledge and this machinery, which represent nine parts to one contributed by yourself in the value of your product? You inherited it, did you not? And were not these others, these unfortunate and crippled brothers whom you cast out, joint inheritors, co-heirs with you? What did you do with their share? Did you not rob them when you put them off with crusts, who were entitled to sit with the heirs, and did you not add insult to robbery when you called the crusts charity?

Bellamy believed that all men born are heirs equally to the world's pre-existing and increasing wealth, and that a social system which deprives anyone of a fair share of the common wealth is both immoral and unjust. I would add that in an era of cybernation, it is also absurd and unworkable.

Prescription for a Cure

I have tried to show that conventional economic thought is worse than useless in facing the challenges and opportunities of the rapidly advancing age of abundance. It serves about as well as expecting a tribe of hunters and food gatherers to advise an agrarian people on sowing and reaping. To understand the problems inherent in the transformation from an industrial society organized on the basis of scarcity to a cybercultural society organized on the basis of abundance, we must look elsewhere for help.

I suggest that an unbiased investigation of our true circumstances would show that we have effectively tapped the inexhaustible, dynamic, cosmic reservoir of energy; that man can, by virtue of his triumphant intellect, discover the

5. Edward Bellamy, *Looking Backward,* New American Library, New York (1963) p 100.

means to do what he chooses to do; that we already have—inadvertently and unwittingly—achieved success for almost half of mankind; and that by the application of our basic wealth—cosmic energy plus human intellect and the tools created thereby—we could soon lead all mankind to the age of abundance.

A truly "free society" is quite conceivable—indeed, it is just around the corner. Plants do not buy their carbon dioxide from animals, nor do animals buy their oxygen from plants. Each contributes to, and draws from, his common environment of abundance, without payment, by virtue of being alive. We are closer than we realize to the age of cyberculture—when all men will contribute to, and draw from, a common environment of abundance, without payment, by virtue of being alive.

Our task is to help hasten the arrival and the realization of a cybercultural society.

Buckminster Fuller[6] has been a pioneer in anticipating and preparing for this eventuality, as he wrote in his proposal for a "World Design Science Decade, 1965–1975."

> We could progressively scrap the whole paraphernalia of civilization's present mechanical equipment, both production and end product, replacing these with improved mechanics, while netting not only a higher standard of living performance for ourselves, but enough raw materials and improved know-how to expand the physical apparatus of our standard of living to serve not just our U.S.A. seven percent but half the human family. The advantages gained by the latter would allow them in turn to develop new sources, sum totally providing the new advances of science and industry to serve all the peoples of the world. What to do about war surpluses here and abroad is clearly indicated in these truths. Scrap them one hundred percent and reprocess the chemical elements into higher and more appropriate use forms, thus generating common world wealth. Spend ingenuity on brand-new higher performance products, rather than on make-shift applications of obsolete gadgets.

That is a prescription calculated to cure what ails us. Let us hope our government can be persuaded to underwrite this program for "livingry" as an alternative to continued investment in overkill.

6. Buckminster Fuller, "World Design Decade (1965–1975)," Phase I (1964), Document 2. *The Design Initiative, World Resources Inventory,* Southern Illinois University, Carbondale (1963) p 136.

Moonlighting—An Economic Phenomenon

Harvey R. Hamel

The primary motivation appears to be financial pressure, particularly among young fathers with low earnings.

Moonlighting habits of the American worker have not increased or even changed much in recent years. The most recent survey of dual jobholding shows that 3.6 million workers, just under 5 percent of all employed persons, held two jobs or more in May 1966. This proportion was somewhat smaller than those revealed by the 1964 and 1965 surveys.

The typical multiple jobholder is a comparatively young married man with children who feels a financial squeeze. He has a full-time primary job and moonlights about 13 hours a week at a different line of work. Teachers, policemen, firemen, postal workers, and farmers are most likely to moonlight. Many of them work for themselves on their extra jobs (operating farms or small businesses) while many others are sales or service workers.

One of the major subjects explored in this article is the relationship between moonlighting and weekly earnings, data on which is available for the first time. There is also an analysis of the association between moonlighting and hours of work, an indication of some of the possible reasons for moonlighting, and a discussion of the industries and occupations of moonlighters.[1]

1. Data in the current report are based primarily on information from supplementary questions to the May 1966 monthly survey of the labor force, conducted for the Bureau of Labor Statistics by the Bureau of the Census through its Current Population Survey. The data relate to the week of May 8 through 14.

This is the seventh in a series of reports on this subject. The most recent was published in the *Monthly Labor Review,* February 1966, pp. 147-154, and reprinted with additional tabular data and explanatory notes as Special Labor Force Report No. 63, which also includes a complete listing of earlier reports and their coverage.

For purposes of this survey, multiple jobholders are defined as those employed persons who, during the survey, (1) had jobs as wage or salary workers with two employers or more; (2) were self-employed and also held a wage or salary job; or (3) worked as an unpaid family worker, but also had a secondary wage or salary job. The primary job is the one at which the greatest number of hours were worked. Also included as multiple jobholders are persons who had two jobs during the survey week only because they were changing from one job to another. This group was measured in the December 1960 survey and was found to be very small—only 2 percent of all multiple jobholders.

Persons employed only in private households (as a maid, laundress, gardener, babysitter, etc.) who worked for two employers or more during the survey week were not counted as multiple jobholders. Working for several employers was considered an inherent characteristic of private household work rather than an indication of multiple jobholding. Also excluded were self-employed persons with additional farms or business, and persons with second jobs as unpaid family workers.

A Quest for Higher Earnings

Why do over 3-1/2 million persons hold two jobs or more? The primary reason seems to be economic. Many moonlighters need, or believe they need, additional income. For some, a second job is a necessity. A second job enables others to live at a higher standard.

For still others, a second job may be the means by which they are able to maintain a standard of living that would otherwise be lost because of, for example, sudden large expenses, loss of wife's income, or a decline in earnings on the primary job.

Because financial reasons are a prime factor motivating moonlighters, the Bureau of Labor Statistics collected data on the usual weekly wage and salary earnings of dual jobholders on their primary job and of single jobholders. These data show that generally the level of a worker's earnings determines his propensity to moonlight. Multiple jobholding rates for men 25 to 54 years old are highest at the lowest earnings level—under $60 a week. As the level of earnings rises, the incidence of dual jobholding declines (see chart 1). The lowest rates were found among workers with the highest weekly earnings—$200 or more.

The close association between multiple jobholding and earnings is most evident from the data for married men 25 to 54 years old, the group for whom family financial responsibilities are usually the greatest. Among these men, the moonlighting rate for those earning less than $60 a week was 12.5 percent, more than twice as high as the 5.3 percent for men earning $200 or more a week.

Data available for the first time show that among men who are heads of households, there is a close relationship between the multiple jobholding rates, the number of young children, and usual weekly earnings. The moonlighting rate tends to increase with the number of children under age 18. The rate for men with at least five children was nearly twice that for men with no young children, as shown in the following tabulation:

Multiple jobholding rates for men who were heads of households, May 1966
Children under age 18

Total	7.9
None	5.4
1 child	8.3
2 children	9.1
3 or 4 children	9.8
5 children or more	10.3

Within each of these groupings, multiple jobholding rates tended to decrease as earnings increased. For example, among men who were household heads with three or four children, the rate was 16 percent for those who earned under $60 weekly, about double that for those with earnings of $200 or more.

Financial pressure, however, is not the only reason why workers moonlight. There are several other considerations. Some workers with a regular wage or salary job want to continue or try their hand at working for themselves on a part-time basis while still maintaining their basic source of income. One-third of the multiple jobholders are self-employed on their second job. They moonlight at their own business or devote a few hours to providing some professional service in their spare time without committing large resources or all their time to the venture. Moreover, the fact that half of this self-employed group operates a farm as their second job suggests that some of these dual jobholders have chosen not to abandon the farm way of life even though economic reasons force them to work at a full-time wage or salary job. Others may have moved to the country and taken advantage of the opportunity to do a little farming on the side.

Chart 1.

Multiple Jobholding Rates for Men 25 to 54 Years Old, May 1966

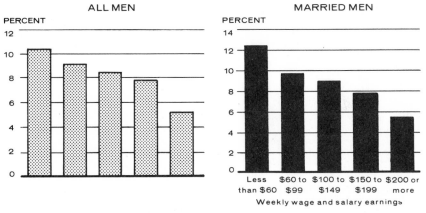

Some persons moonlight because they are interested in another line of work. They experiment with a second job, but still maintain their primary job until they determine whether they like the work on their new job and decide whether it is feasible to make a change to this new line of work. Still others moonlight because there is a shortage of their particular skill (for example, teachers and skilled craftsmen) and they find it very easy to make extra money.

The basic characteristics of moonlighters have remained about the same in the course of several BLS surveys. The majority are men. Their multiple jobholding rate is about three times that for women workers. (See table 1.) A smaller proportion of Negro than white workers were multiple jobholders.[2]

The incidence of holding two jobs or more was highest among men 25 to 44 years old. This age group accounted for 43 percent of all employed men, but over half of all men holding more than one job. Moonlighting was least likely among the very young (14 to 19 years old), most of whom are attending school, and among workers 65 years old and over. Married men were twice as likely to be moonlighters as single men.

2. Data for nonwhites will be reported as data for Negroes, who constitute about 92 percent of all nonwhites in the United States.

In sum, the data suggest that the typical moonlighter is a highly motivated and energetic young married man with a growing family, who works at two jobs or more primarily to provide additional income for his family but also for a variety of other reasons: to try his hand at working for himself; to keep busy; to obtain satisfaction; to experiment with another line of work; or to supply his skills that are in demand in his community. The moonlighter aspires to a better living and is willing to work hard to obtain his goal.

Work-Hours on Both Jobs

Although the rate of multiple jobholding has remained substantially the same in recent years, the question still arises as to whether a shortened workweek would lead to higher moonlighting rates among workers who are affected by the cutback in hours. There is no question that when hours are shortened the opportunity to hold an extra job increases. However, an individual's decision on how to use his free time—to moonlight or do something else—involves many factors other than the number of hours worked.

Table 1.
Employed Persons with Two Jobs or More, by Sex, 1956-1966

Month and year	Persons with two jobs or more			
	Number (thou-sands)	Multiple jobholding rate [1]		
		Both sexes	Men	Women
May 1966	3,636	4.9	6.4	2.2
May 1965	3,756	5.2	6.7	2.3
May 1964	3,726	5.2	6.9	2.1
May 1963	3,921	5.7	7.4	2.4
May 1962	3,342	4.9	6.4	2.0
December 1960 [2]	3,012	4.6	5.9	2.0
December 1959	2,966	4.5	5.8	2.0
July 1958	3,099	4.8	6.0	2.2
July 1957	3,570	5.3	6.6	2.5
July 1956	3,653	5.5	6.9	2.5

[1] Multiple jobholders as percent of all employed persons.
[2] Data for Alaska and Hawaii included beginning 1960.

One way of examining the relationship between moonlighting and the length of the workweek is to compare the dual jobholding rates of men working shorter hours with those on a longer workweek. The data show that in nonfarm industries persons who worked 35 to 40 hours on their main job were no more likely to be multiple jobholders than those who had worked 41 to 48 hours:

Hours worked on primary job	Multiple jobholding rates for men, May 1966		
	All indus-tries	Agricul-ture	Nonfarm industries
Total	6.5	8.7	6.3
1 to 21 hours	7.3	9.0	7.0
22 to 34 hours	10.3	14.1	9.6
35 to 40 hours	6.8	9.7	6.7
41 to 48 hours	6.7	14.6	6.4
49 hours or more	4.5	5.8	4.3

This suggests that reducing the workweek by only a few hours would not in and of itself substantially affect the incidence of multiple jobholding provided there was no cutback in earnings. No significant inverse relationship exists between moonlighting and the length of the workweek. This finding accords with the conclusions of a recent study of rubber workers in Akron, Ohio.[3] It seems reasonable, therefore, to assume that among full-time workers, factors other than the length of the workweek determine whether a man looks for a second job.

Men working part time (22 to 34 hours) were more likely to be moonlighters than men with a full-time job (but since most men work full time, the majority of multiple jobholders are full-time workers). The rate was lowest for men working over 48 hours a week on their main job. Dual jobholding rates for men who worked less than 22 hours weekly were relatively low, reflecting the fact that men working so few hours a week are mainly students or older men unlikely to be interested in a second job.

Typically, multiple jobholders worked full time on their principal job and part time on their extra job; about one-fourth worked part time on both jobs; and 8 percent worked full time on both. On the average, they worked a total of 52 hours, only 13 of which were on their second job. The 39 hours on the primary job paralleled the 39 hours that single jobholders worked on their only job. Of all multiple jobholders, those who were farmers or factory workers on their primary jobs worked the longest total workweeks—59 and 57 hours, respectively. Men worked much longer hours than women on their extra jobs, 14 compared with 9 hours. Men who had additional wage or salary jobs worked longer at these jobs than those who were self-employed on their extra jobs, 15 hours and 12 hours, respectively.

Moonlight Industries

One of the most significant aspects of moonlighting is the high incidence of self-employment. About 1.5 million or more than 2 out of 5 multiple jobholders operated their own farms or businesses or were self-employed professionals on the first or second job (chart 2). About half of them were farmers, typically holding down a regular blue-collar job and running their farms in their spare time (table 2). Workers who operated farms as their normal line of work were nearly twice as likely to have a second job as the average worker. About 25 percent of the 200,000 moonlighting farmers had second jobs as a hired hand on someone else's farm; 40 percent worked on construction or transportation jobs or in factories.

3. John Dieter found no statistically significant difference in multiple jobholding rates for Akron workers on a 36-hour workweek and those on a 40-hour workweek. He concluded that the high incidence of moonlighting in Akron for many years may reflect an established custom of these workers, and that other factors (primary job income, number of children in the family, and employment of the spouse) offered better explanations of moonlighting. See "Moonlighting and the Short Workweek," *The Southwestern Social Science Quarterly,* December 1966, pp. 309-315.

Chart 2.
Class of Workers of Primary and Secondary Jobs for
Multiple Jobholders, May 1966

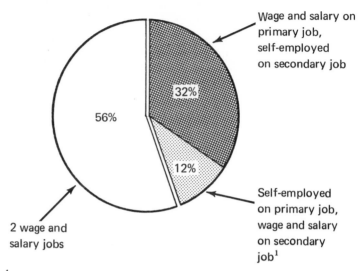

Wage and salary on
primary job,
self-employed
on secondary job

32%

56%

12%

2 wage and
salary jobs

Self-employed
on primary job,
wage and salary
on secondary
job[1]

[1] Includes a small proportion of multiple jobholders who were
unpaid family workers on their primary jobs.

On the other hand, the multiple jobholding rate for nonfarm self-employed workers was low. This reflected both their relatively high earnings and the fact that businessmen and self-employed professional people often do not have the time for a second job. The majority of the dual jobholders had two wage or salary jobs. Of salaried employees, public administration workers were more likely to moonlight than workers in any other major nonfarm industry. The dual jobholding rate is particularly high for postal workers (1 out of 10), a proportion which has remained consistently high over the years (table 3). Other nonfarm wage or salary workers with higher than average multiple jobholding rates included those working in educational services, entertainment and recreation, transportation, construction and forestry, fisheries, and mining.

One-third of all the secondary jobs were in either farm or nonfarm self-employment. Another 43 percent of the moonlighters had paid jobs in the trade or service industries, which can use many part-time workers. Usually, moonlighters did not work in the same industry on their second job as they did on their primary job. Except for service and trade workers, only a small proportion had two jobs in the same industry.

Table 2.

Type of Industry and Class of Worker of Primary and Secondary Jobs,
for Persons With Two Jobs or More, May 1966

[Numbers in thousands]

Type of industry and class of of worker of primary job	Total employed	Persons with two jobs or more		Type of industry and class of worker of secondary job					
				Agriculture			Nonagricultural industries		
		Number	Percent of total employed	Total	Wage and salary workers	Self-employed workers	Total	Wage and salary workers	Self-employed workers
Total - - - - - - - - - - - - - - - - -	73,764	3,636	4.9	721	139	582	2,915	2,335	580
Agriculture - - - - - - - - - - - - - - - -	4,292	335	7.8	120	83	37	215	212	3
Wage and salary workers - - -	1,326	388	6.6	56	19	37	32	29	3
Self-employed workers - - - -	2,253	200	8.9	49	49	(1)	151	151	(1)
Unpaid family workers - - - -	713	47	6.6	15	15	(†)	32	32	(2)
Nonagricultural industries - - - - - -	69,472	3,301	4.8	601	56	545	2,700	2,123	577
Wage and salary workers - - -	62,529	3,110	5.0	599	54	545	2,511	1,934	577
Self-employed workers - - - -	6,371	177	2.8	2	2	(1)	175	175	(1)
Unpaid family workers - - -	571	14	2.5	- - -	- - -	(2)	14	14	(2)

[1] Self-employed persons with a secondary business or farm, but no wage or salary job, were not counted as multiple jobholders.

[2] Persons whose primary job was as an unpaid family worker were counted as multiple jobholders only if they also held a wage or salary job.

NOTE: Because of rounding, sums of individual items may not equal totals.

There was a sharp difference in the kinds of second jobs held by white and Negro dual jobholders. About one-third of the white moonlighters were self-employed on the second job, and one-fourth worked in service industries. Among Negroes, however, fewer than 20 percent were self-employed and nearly half worked in service industries.

Occupations of Moonlighters

Multiple jobholding rates vary with the worker's main occupation. As in prior surveys, moonlighting rates in May 1966 were highest among men who were teachers—1 out of 5 had a second job (table 4). Some elementary and high school teachers may moonlight because they have an opportunity to take evening jobs at school in some professional activity, but other evidence suggests that the most likely explanation is their comparatively low earnings of teachers.[4] The dual jobholding rate for other male professional and technical workers is high, but less than half that of teachers.

A very high proportion of men employed in protective services (policemen, firemen, and guards) had an extra job in May 1966—1 out of every 6. Their flexible work schedules make moonlighting possible and their relatively low earnings often make it necessary. Other service workers (including barbers, cosmetologists, janitors, attendants, and other workers) also had higher than average moonlighting rates. Men who were managers, officials and pro-

4. Harold W. Guthrie suggests that the teaching profession is an economically deprived one and men teachers, particularly those who are married with a nonworking wife, must moonlight to maintain a standard of living commensurate with their professional status. See "Who Moonlights and Why?" *Illinois Business Review*, March 1965, p. 8.

prietors—an occupation group which typically works long hours and whose earnings are generally above average—were least likely to be multiple jobholders. Nonfarm laborers and retail sales workers were also unlikely to be multiple jobholders. Moonlighting rates were generally higher for white than Negro men, particularly among blue-collar and service workers.

Table 3.
Industry Group and Class of Worker of Persons With One Job
and With Two Jobs or More May 1966

Industry group and class of worker	Percent distribution			Multiple jobholding rate[1]
	Persons with one job	Persons with two jobs or more		
		Primary job	Secondary job	
All industries -	100.0	100.0	100.0	4.6
Agriculture -	5.6	9.2	19.8	7.8
Wage and salary workers -	1.8	2.4	3.8	6.6
Self-employed workers -	2.9	5.5	16.0	8.9
Unpaid family workers -	.9	1.3	(2)	6.6
Nonagricultural industries -	94.4	90.8	80.2	4.8
Wage and salary workers -	84.7	85.5	64.2	5.0
Forestry, fisheries, and mining - - - - - - - - - - - - - -	.8	1.0	.4	6.0
Construction -	5.2	6.5	4.2	6.1
Manufacturing -	27.0	23.8	6.2	4.4
Durable goods -	15.7	15.4	3.0	4.9
Nondurable goods -	11.3	8.4	3.2	3.7
Transportation and public utilities - - - - - - - - - - -	6.0	7.3	5.3	5.9
Wholesale and retail trade - - - - - - - - - - - - - - - - -	15.5	11.9	16.8	3.8
Wholesale -	3.1	2.8	1.2	4.5
Retail -	12.4	9.1	15.6	3.7
Eating and drinking places - - - - - - - - - - - - -	2.6	1.4	3.9	2.8
Other retail trade - - - - - - - - - - - - - - - - - - -	9.8	7.7	11.8	3.9
Service and finance -	25.3	25.4	26.6	4.9
Finance, insurance, and real estate - - - - - - - - -	4.0	3.9	4.2	4.8
Business and repair services - - - - - - - - - - - - - -	2.1	2.4	2.8	5.6
Private households -	3.6	.7	3.2	1.0
Personal services, except private households - -	2.2	1.7	2.2	3.9
Entertainment and recreation - - - - - - - - - - - -	.9	1.1	3.3	6.2
Educational services - - - - - - - - - - - - - - - - - - -	6.3	9.6	4.8	7.3
Professional services, except education - - - - - -	6.1	5.9	6.2	4.7
Public administration -	4.9	9.5	4.7	9.2
Postal services -	.8	1.7	.9	10.1
Other public administration - - - - - - - - - - - - -	4.1	7.9	3.8	9.0
Self-employed workers -	8.8	4.9	16.0	2.8
Unpaid family workers -	.8	.4	(2)	2.5

[1] Persons with two jobs or more as percent of all employed persons in industry of primary job.

[2] Persons whose only extra job has an unpaid family worker were not counted as dual jobholders.

A large proportion of the moonlighters (42 percent) earned their supplementary income as professional and technical workers or managers, or by operating their own farm or nonfarm businesses. Much smaller proportions of the moonlighters were craftsmen or operatives on their second than on their first job. One of the principal differences in the types of jobs held by white compared with Negro moonlighters is that a much larger proportion of Negroes work in lower paying service occupations, including private household service, while a much smaller proportion of Negro moonlighters hold white-collar jobs on either their main or their extra jobs.

The majority of second jobs were in occupations different from the moonlighter's main line of work, but usually within the same major occupation group as their first job. Half the professional and technical workers had a second job in the same occupation group, and half the farm laborers were farm workers on their second job. About one-third of the clerical and the service workers, and one-fourth of the managers and the craftsmen, had second jobs in the same broad occupation groups. On the other hand, the manual skills of farmers and blue-collar workers made a common moonlighting combination. Half the self-employed farmers had a second job in a blue-collar occupation and about one-fourth of the craftsmen, operatives, and laborers ran their own farm as a sideline.

Table 4.

Occupational Distribution of Persons With Two Jobs or More, and Rate of Multiple Jobholding, by Occupation and Sex, May 1966

Occupation group	Persons with two jobs or more—			
	Percent distribution		Multiple jobholding rate[1]	
	Primary job	Secondary job	Men	Women
All occupations- -	100.01	100.0	6.4	2.2
Professional, technical, and kindred workers - - - - - - - - - - - -	17.8	15.1	8.9	3.5
Medical and other health workers- - - - - - - - - - - - - - - -	1.8	1.6	8.3	2.1
Teachers, except college -	5.2	1.8	19.7	3.8
Other professional, technical, and kindred workers - - -	10.8	11.6	7.4	4.1
Farmers and farm managers -	5.5	16.1	9.5	2.2
Managers, officials, and proprietors, except farm - - - - - - - - -	7.8	10.6	4.2	2.1
Clerical and kindred workers -	10.4	7.4	6.5	2.1
Sales workers -	5.2	8.2	5.4	1.7
Retail trade -	2.1	4.9	4.4	1.3
Other sales workers -	3.1	3.3	6.1	3.8
Craftsmen, foremen, and kindred workers - - - - - - - - - - - - -	15.8	9.8	6.0	4.7
Operatives and kindred workers -	17.0	11.4	6.0	.9
Private household workers -	.7	2.2	(2)	1.1
Service workers, except private household - - - - - - - - - - - - -	11.7	11.4	9.6	2.7
Protective service workers -	3.8	1.3	16.8	(2)
Waiters, cooks, and bartenders - - - - - - - - - - - - - - - - - -	2.3	3.7	6.4	3.3
Other service workers -	5.7	6.3	7.5	2.4
Farm laborers and foremen -	3.2	3.0	6.7	6.2
Laborers, except farm and mine - - - - - - - - - - - - - - - - - - -	4.9	4.7	4.8	3.1

[1] Persons with two jobs or more as percent of all employed persons in occupation of primary job.

[2] Percent not shown where base is less than 100,000.

The Split-Level Challenge

Whitney M. Young, Jr.

There are two infallible ways to surprise the average American when discussing race. One is to describe in any depth the Negro middle class. Seldom publicized, it has burgeoned to include doctors, lawyers, engineers, designers, editors, government officials, scientists, teachers, and businessmen—representatives in all the professions that demand skill and sophistication. Well-read, well-traveled, well-dressed, its members are among the 20 percent of American blacks whose family income exceeds $10,000; the 40 percent who own their own homes; the 70 percent whose marriages are stable. In every way these Negroes are the equals—and often the superiors—of white people in those categories. Yet these are the potential neighbors whom white suburbanites "fear"; the scientists and professionals whom used-car salesmen feel will "lower the neighborhood's standards."

A second unfailing method of surprise is to portray the dimensions of deprivation in America's central cities—the rats, the odors, the hate, the despair. Walk a main street in a Negro ghetto. Here you will pass numerous marginal "mom-and-pop" stores, an occasional white-owned department store selling shoddier goods at higher prices than the big downtown establishments, boarded-up vacant shops with TO LET signs moldering from months of exposure, and knots of jobless men on stoops and street corners. Here and there is a skills training center or a new venture formed by ambitious young entrepreneurs, but by and large the scene is atypical of a booming economy. Most white Americans would associate such scenes only with the Great Depression, or played-out mining or mill towns.

The truest comparison might well be with another country—an economically underdeveloped nation. For the discrimination that has kept Negroes from the jobs, education, and skills needed in our highly advanced technological society has also created an economically deprived "nation" in the heart of our urban centers of economic power. It is against this backdrop that the place of black capitalism in ghetto development must be discussed.

The economic gap that separates white and black Americans is, despite all the efforts of recent years, *growing.* Last year, for example, amid much fanfare, the government announced that Negro family income, as a proportion of white income, rose sharply. Negro family income was now 59 percent that of white family income, said the federal press release—a 4 percent jump in only two years. But in 1952, Negro family income was 57 percent that for white families—so the proportionate gain was a mere two percentage points in a decade and a half.

And the *dollar gap* between the two groups actually grew. In 1950, median white family income was $3,445; median black family income, $1,869—a dollar gap of $1,576. By 1967, after widely heralded social reforms, white income had soared to $8,318. Negro income was $4,939. And the dollar gap was now $3,379. The seriousness of this situation is sharpened when we realize that black people, to a greater degree than whites, are urbanized. Better than half of all Negroes live in central cities, compared to only a fourth of whites. City life is more expensive, and the dollar gap means a dangerously lower standard of living.

The federal government, taking into consideration living costs and standards, estimates that an urban family of four must have an income of $9,243 to maintain a "moderate standard of living." Nothing fancy, no big cars or private schools—a *moderate* standard of living. That is nearly double the median black family income. In the black ghetto, income is even below the official "lower living standard," which, for a family of four, is $5,994.

Last year, we congratulated ourselves on an overall unemployment rate of 3.6 percent, although that figure, the lowest in fifteen years, was still triple that for countries such as Japan, Germany, and Sweden. But Negro unemployment was around 7 percent. And that does not include people who worked part-time when they wanted to work full-time, nor those who, in bitterness and despair, simply gave up and dropped out of the job market. Special surveys

Negro Vocational Representation
Percentage of all employees in field

Occupation	1940	1956	1966
Professional and technical	3.7	3.7	5.9
Managers, officials, proprietors	1.7	2.2	2.8
Clerical		3.8	6.3
Sales	1.2	1.8	3.1
Craftsmen and foremen	2.7	4.2	6.3
Semiskilled	5.8	11.3	12.9
Farmers and farm managers	15.2	8.5	6.1
Farm laborers and foremen	17.5	22.9	20.2

Source: U.S. Census

of ghetto areas found unemployment rates at 9 percent, and underemployment ranging up to half of all workers. Among black teen-agers—and the black population, with a median age of 21.7 years, is younger than the white— unemployment reached the disastrous rate of more than one out of every four!

We must also keep in mind that these are monthly averages, and behind the numbers are people—human beings, with families to support and dreams to fulfill. Typically, the actual *number* of individuals unemployed at one time during the year is triple the monthly average, so a 7 percent unemployment average means roughly one out of five workers experienced some unemployment during the year.

These figures are grim, but they don't begin to show the true extent of the racial gap: that Negroes experience poverty at a rate triple that for whites; that black high school graduates make *less* than whites with a grade school education; that infant mortality among blacks is double that for whites; that life expectancy is shorter; that housing conditions are far inferior; that in almost every area of life in these prosperous United States, the gap between white and black in the decencies of life is vast—and growing.

But what of the future? Surely all the training programs, the increased emphasis on education, the crumbling job barriers that are allowing blacks into corporate ranks and into unions heretofore closed to them, the growing emphasis on black-owned businesses—all these recent developments will culminate in the very near future and the gap will close and, on an economic level at least, whites and blacks will be equal.

No.

Given present trends and the absence of a clear national commitment to equality, the future looks as bleak as the past.

The main reasons are that black people are concentrated in central cities, and discrimination (largely in the form of zoning laws that exclude lower-income families) keeps them out of the suburbs and disproportionately concentrated in jobs most likely to be automated out of existence in the coming decade. The Commerce Department has estimated that twenty-five of the largest metropolitan areas will lose three million jobs by 1975. More than half of all new employment in the past decade took place outside the central cities; in the larger metropolitan areas, four out of five new jobs were created in the suburbs. A Rand Corporation study projects a 20 percent shift in employment from city to suburb by 1975, with 56 percent of metropolitan-area jobs in the suburbs.

The composition of the job force also signals trouble. One economist has written: "If non-whites continue to hold the same proportion of jobs in each occupation, . . . the non-white unemployment rate in 1975 will be more than five times that for the labor force as a whole." Harvard economist Otto Eckstein, meanwhile, estimates that although Negroes will be 12 percent of the 1985 work force, they will hold a smaller percentage of jobs in the professional, skilled, clerical, or sales fields—the very fields where employment is expected to grow. At the same time, he calculates, blacks will constitute a fourth of all laborers and farm workers and two of every five private household workers.

If full economic equality were achieved, there should be 1,330,000 Negro managers and proprietors; Eckstein estimates there will be only 420,000. For Negro salespeople, instead of an economic equality ratio of 830,000, he foresees fewer than half that—410,000. Instead of 2,160,000 black clerical workers, he projects only 1,510,000. Thus, while the economy expands and more jobs are created, Negroes will remain underemployed. Clearly, a massive national commitment to close the gap is mandatory. No nation can flaunt its affluence before a permanently deprived underclass and expect to survive.

Joseph Kennedy, father of the late President John F. Kennedy, said during the Depression that he would gladly part with half his fortune if he could be assured of keeping the other half in safety. Fortunately, sacrifices on such a scale are not necessary—yet. What is needed is a reallocation of resources, and a sharing of the power and the fruits of this society with those citizens now excluded from them.

Back in 1961, President Kennedy made a simple decision: the United States would land a man on the moon by the end of the decade. Plans were drawn, billions of dollars were allocated, personnel were trained—an unprecedented national effort was mounted. And now an American flag stands on the barren, crater-pocked moonscape, and the world hails the first men to walk on the surface of the moon. The same commitment is needed so that men can walk in dignity here on the barren streetscapes of the urban ghetto—a reorder-

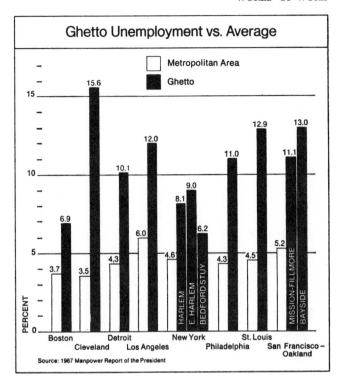

Ghetto Unemployment vs. Average

☐ Metropolitan Area
■ Ghetto

Source: 1967 Manpower Report of the President

ing of national priorities backed by dollars, jobs, schools, and housing, to bring about true equality for all American citizens.

I served on the Commission on Technology, Automation, and Economic Progress, and our report, *Technology and the American Economy,* identified more than five million new jobs that could be created in such key public service areas as health, education, safety, home care, and sanitation. These aren't make-work; they are meaningful jobs in crucial fields now starved for workers. Experiences in New Careers programs, training programs, and with the so-called hard-core unemployed indicate that the valuable human resources now wasted by a profligate society can be turned to productive uses.

It is necessary to take a split-level approach—integration and simultaneous economic development of the ghetto. Although the job-creating potential of the government and of private industry is paramount, it would be tragic to neglect the importance of building sound economic structures in the ghetto. Black people must themselves have access to capital and to industries that not only create jobs but also give the black community the power—the economic muscle—to compete on equal terms with other groups.

What black citizens—North and South—share to a degree not experienced by white Americans is an all-pervasive sense of powerlessness. A few years ago, when the Chicago Urban League surveyed the racial composition of that city's decision-makers, it concluded that "it is safe to estimate that Negroes held less than 1 percent of the effective power in the Chicago metropolitan area." Although 28 percent of the city's population is black, only two of the 156 top posts in the city's administration were held by blacks; a mere 227 of the 9,909 policy-making positions in the private sector were held by Negroes; and no blacks were in policy-making posts in the major non-financial corporations that dominate the city's economy.

Wide-scale support for black businesses—either of entrepreneurs or, preferably, of cooperative and community-run corporations—will help create an economically secure managerial class such as exists within the white community. Important as jobs in the larger society are—and creation of such jobs must be the main thrust of economic efforts toward equality—there is a pride and dignity in ownership that must be satisfied within the black community, as it is within the white.

If anything represents the black ghetto economically, it is the paucity of black-owned businesses and the marginal existence of those that do exist. A recent study by the Small Business Administration found that the "typical" minority-owned business is a one-man personal service or retail shop in the central city that has $20,000 or less in annual sales. About a third were family-run stores, and only one of ten black-run businesses employed more than ten workers.

In New York, one out of forty whites owns his own business; only one out of a thousand Negroes does. Newark, with a black majority, has only 10 percent of its businesses owned by blacks. Once again the situation is worsening. Between 1950 and 1960, while the total number of businesses was increasing, Negro business ownership declined by 20 percent. Forty years ago there were forty-nine Negro-owned banks in thirty-eight cities; now there are twenty in nineteen cities.

There are many ways in which this gap might be closed. Credit and insurance remain among the biggest problems for the black businessman. Banking and insurance "pools" can solve this by spreading the risk thinly among participating companies. Federal assistance to ghetto businesses should be greatly increased, and training and management-consulting services provided. Government at all levels can stimulate the development of strong, job-producing industries through loans and contracts that guarantee to buy products and services.

In *Beyond Racism* I proposed development of community corporations —Neighborhood Development Corporations—funded by a National Economic Development Bank financed by government-backed bonds bought, in part, by federally insured financial institutions. It is important that these new ghetto industries be just that—industries that create jobs and power and not just marginal hand-to-mouth retail operations.

Private business has an important role. It has the expertise, the resources, and the power to help build a healthy ghetto economy. I would like to see a self-imposed tithe of 2 percent of net corporate profits *invested* in ghetto enterprises to be operated and eventually owned by blacks. This would create a development fund of more than $2-billion a year in new factories, marketing facilities, professional services, or guaranteed purchases of goods on a subcontracting basis. It is done every day in other fields; subsidiaries are created and then spun off. Why not in the ghetto?

Appropriate tax and regulatory legislation could also free more of the $1-*trillion* in lendable assets of banks, savings and loan associations, and insurance companies. If only a fraction of these funds were made available for ghetto-based economic development corporations, the poverty-stricken ghetto could become a thing of the past.

I hope that the cynicism that pervades so much of modern life has not yet grown to exclude the demands of common morality. The black man was brought to these shores in chains, enslaved for 250 years, kept in peonage for most of the past hundred years, and now suffers disproportionately in all areas of life. Common decency demands that this situation be changed. The growing anger of the black masses, especially among the younger people who see

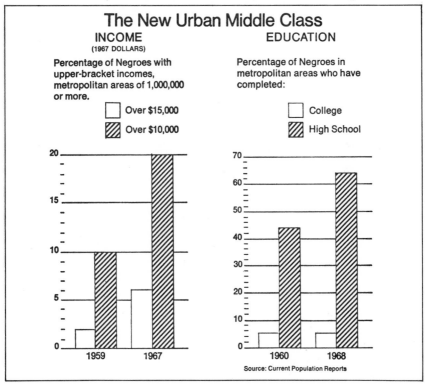

The New Urban Middle Class

INCOME
(1967 DOLLARS)

Percentage of Negroes with upper-bracket incomes, metropolitan areas of 1,000,000 or more.

☐ Over $15,000

▨ Over $10,000

EDUCATION

Percentage of Negroes in metropolitan areas who have completed:

☐ College

▨ High School

Source: Current Population Reports

through the hypocrisy of this society, will not lessen unless such a national commitment is swiftly implemented. Countering such protest with increased repression can only fan the flames of anger and eventually lead to the kind of police-state that will enslave all Americans.

Those in this morally underdeveloped nation who do not readily respond to the demands of conscience may be more responsive to self-interest. The black minority, disproportionately disadvantaged as it is, constitutes a huge market that cannot be ignored. Negroes earn about $30-billion, form up to 50 percent of the total consumer market for certain goods, and will be in the majority in about a dozen major cities well before 1980. Economic equality would mean enlarging this market by about $25-billion, and it would also mean the creation of a stable urban middle class.

Our economy can ignore this huge market and the implications of equality only at its peril. The question really becomes: Does a society dominated by white-run institutions have any kind of future in a world that is three-fourths non-white? And can a business-oriented society survive when its major centers of power and commerce are populated by poverty-stricken, angry black majorities?

Illiteracy: The Key to Poverty

Bernard Asbell

There is a man in Chicago who dares to think he has discovered the biggest cause of American poverty and how his city can begin to get rid of it. In fact, how any city can begin to get rid of it. His method—and this is not a flippancy—is as simple as teaching the A B Cs.

Raymond H. Hilliard, Director of Public Aid for Cook County, Illinois, recently was surprised to learn, after a long career of studying poverty, that most people who are extremely poor have in common a single, secret, crippling trait: they are virtually illiterate. He soon made other surprising discoveries. Impoverished illiterates and near-illiterates, no matter what their age, can be taught to read and write at insignificant cost ($5.50 a month), often in only a few months, and many can be made not only employable but employed. Perhaps most important of all, their education often leads to lifting the school interest and grades of their children.

"This," says Hilliard, "is the most hopeful thing I've ever had hold of."

To appreciate the seeming hopelessness against which Hilliard pits his hope, one must be willing to know how many Americans are poor and how poor they are. Not the underpaid, $1.25-an-hour minimum-wage poor, but the empty-pocket poor. If you take all the tens of millions of miserably rewarded sharecroppers and scavengers, mopslingers and laundry sorters, and even the jobseekers and their wives and kids who must make do on skimpy unemployment compensation checks, and *count them among the rich,* you still have seven and a half million souls who are *poor.* More than the combined populations of Los Angeles, Chicago, Pittsburgh, and Boston. Those are the Americans who live by the grace of public welfare. Their parents were poor and, unless something extraordinary is done rapidly, most of their kids will be poor. For the modern variety of "hard core" poverty has something in common with the elegance and security of established wealth. It is inherited.

One man who rose in a few months from hopelessness to hope is Sam Frost, a fifty-four-year old Negro ex-laborer with mighty shoulders and a stony, solemn, awesomely proud jaw. Frost (I have changed his name, but not his story) has fourteen children, the oldest still in school. In 1959, after a thirty-four-year history of steady employment, he was out of a job and could not for the life of him find a new one. Uneasiness turned to fear, then to horrible feelings of uselessness. Backaches and abdominal pains stabbed at him. These are the common occupational diseases of the unemployed; they seem to vanish only under the miracle drug of opportunity. Soon Frost reached the end of a downhill path: he and his large family landed on relief.

Two years ago Frost, like thousands of Chicagoans on relief, was given a test in the "three Rs." Like many others, he failed to show the formal learning of even a fifth-grade child. Soon he was "requested" to go to school two nights a week and do lots of homework in between (anyone refusing would forfeit his relief checks, but almost everyone went gladly). In a year and a half, Frost, an eager student, progressed from a near-illiterate to a possessor of an eighth-grade certificate—and more. His teacher, who volunteered to coach the student after classes, feels that Frost is almost ready to take an achievement exam for a full-fledged high-school diploma.

That still is not the most remarkable of Frost's accomplishments. Shell Oil Company admitted him to training as a gas-station attendant and taught him how to fill out a shift foreman's report, an intricate procedure for balancing all merchandise sold against money taken in. Frost mastered it better than some experienced foremen. Soon several station owners jointly hired Frost to circulate from station to station, combining the figures from shifts into daily round-the-clock reports, and finally into monthly reports. Also, he coaches station managers in better methods of record keeping. After more than five decades of ignorance, all this happened to Sam Frost in less than two years.

Now Milwaukee, Baltimore, Newark, and other cities are sending relief recipients to school. Like Chicago, they are teaching illiterates to read and write, others to qualify for grammar-school, even high-school, diplomas. These cities don't expect to find a Sam Frost behind every relief check. But they are convinced that only the thin walls of elementary schooling separate many good men from productive use of their native intelligence.

But is that intelligence really there to be developed and tapped? Anyone able to read these words will find it hard—almost impossible—to imagine the native cleverness one must have to make his way through a wordy world when he doesn't know how to read or write. In the way a blind man "sees" with his ears and fingers, or a deaf man "hears" by staring at lips, the illiterate must "read" his way around, not knowing one printed symbol from another.

Andrew Timmons had developed that special kind of cleverness. Standing near the entrance of the public housing project in Chicago where he lives with his wife and seven children, I pointed to a small wooden sign stuck into the dirt. It said, HELP US KEEP OUR LAWN BEAUTIFUL. I asked, "What does that sign say?" He replied confidently, "It says, 'Stay off the grass.' " I asked how he knew that. He said, "I just know that's what those signs always say."

At thirty-seven, Timmons (that is not his real name) distinguishes one street from another by its houses, not its street signs. When he once had a job downtown as a car washer—the only job he has found in the past nine years —he was able to find his way home on the Cottage Grove bus because he knows it is number 4. "I can read numbers," he assured me. When his wife sends him out for a can of tomato soup (which has only words on the label), he never brings home vegetable soup by mistake. He shops in the kind of grocery store where you ask for things, not where the customer selects from the shelf. Food is more costly there, but what can he do? Also, he has learned the ceremonial lies of the illiterate. Every culture has its ceremonial lies. Like the educated suburbanite who serves the best Scotch so his guests won't know he's broke, Timmons sometimes tucks a newspaper under his arm so his neighbors won't suspect he can't read. When people who work at desks give him papers to fill out (sometimes job applications), he says, "I just got my hands dirty. Could you put this in your typewriter and I'll tell you the answers?"

Timmons cannot, however, decipher a warning that says "Poison," a movie marquee, a big newspaper headline—or the tiny letters on newspapers' back pages that say "Help Wanted." He would be unable to compete for a job in a big, wordy city even if he knew where to find one. Timmons, his wife, and his seven children are on relief, all of them supported by his fellow citizens who *do* know how to read and write.

Why didn't Andrew Timmons—and the rest of the ignorant poor—learn when they had the chance? The fact is that while other children were going to school, Timmons, a native citizen of the land of free education and equal opportunity, was not given the chance to go.

"Where I was raised," he told me (he had grown up in Jasper County, Mississippi), "hardly none of the kids ever went to school a day. Nobody from the school made you go. My grandfather—he raised me 'cause my mother died when I was seven—figured going to school wouldn't help me pick cotton any better, so why go? I hardly ever thought a thing about it till I was about fourteen and saw some kids in a store looking at magazines and things and I wished I could do some of that. But it was too late."

Some did have the exceptional strength to defy such disadvantages, but even then defeat was almost inevitable. One of Timmon's housing-project neighbors—also on relief—is a spunky woman in her fifties. I'll call her Maybelle Masters. Growing up in Shelburne, Mississippi, she became determined not to mature into an ignorant, helpless adult. At fifteen, she enrolled in the first grade.

"Walking four miles in the mud was the only way to get to school," she told me. "Lots of kids didn't go because they didn't have the right kind of high-top boots. On rainy days, the school would be closed because the rain would come down through the roof."

At the age of twenty—a fifth-grader—Maybelle got married and quit school. According to the U. S. Census, she is literate because she reported attending school for five years, the minimum standard for "functional literacy." Yet she cannot read and write.

Why not? Didn't she try hard enough?

"When they ask me how long I went," says Mrs. Masters, "I say five years, but the truth is I didn't go even eighteen months. School was only open from January to April. Sometimes the cotton wasn't all picked in January, so you couldn't start school till the work was done. In April the planting started. You stopped going to school when the work started in the fields. So maybe I went two months a year, maybe three."

Five "years" of schooling left her virtually as uneducated as Andrew Timmons who had had none.

Every large city is loaded down with Andrew Timmonses, their burdened, bewildered women, and their ragged, benighted children, the inheritors of what Hilliard calls "infectious ignorance." Chicago is typical with 270,000 on relief. Some are southern Appalachian mountaineers (sometimes said to be the only white Anglo-Saxon Protestants who, as a class, are victims of discrimination and deprivation). Some are Puerto Ricans, Mexicans, American Indians. But overwhelmingly they are Negroes who come from farms of the Deep South or whose parents did.

Much as America enjoys regarding itself as a nation of universal education, we have known the image is not entirely genuine. The 1960 Census tells us that eight million adults over twenty-five—one out of every dozen—attended school less than five years, thus are defined as "functionally illiterate." That figure strikes most people as dismayingly large. But the statistic is far smaller than the truth. Selective Service officials, for example, reject 22 per cent of draft registrants for failing a simple mental test; in Southern states the percentage of failures varies from 35 up to 56. They aren't deliberate flunkers; the test is mined with devices for trapping malingerers. Testing officials say that the young men fail mainly because they can't read the questions.

In Oklahoma City, a meat packer and two unions agreed jointly to retrain 170 workers displaced by new machines. The publicized calamity of automation in the lives of packinghouse workers in Oklahoma City, Omaha, and Chicago was one of the first national alarms that deepened our latest dread of machines. The dread grew deeper when word got about that the retraining programs were not very successful. But hardly anybody publicized why. The Oklahoma City retrainers found 110—or 65 per cent—of the workers too uneducated "to show promise of benefiting from training." More bluntly, the workers couldn't pass qualifying tests for the courses because they couldn't read and do simple figuring. In Michigan, where Don Jones's life had been changed by learning a skill, a group of 761 unemployed were tested for retraining; 515—or 68 per cent—failed. In Chicago, 4,500 on relief were tested; 1,900 were unable to read the questions well enough to pass.

Those figures reveal the startling difference between a man like Don Jones and the "chronically unemployed." Jones was equipped with a basic education to make himself ready for the new kinds of jobs created by automation and the prosperity it brings. Illiterates are not. A large Chicago restaurant chain needed good help so badly it had to turn to Europe to recruit, yet in America 287 unemployed had to be tested to find 20 literate enough to train as cooks —reading recipes and the scribbled orders from waitresses. Hospitals, desperate for nurses, either registered or practical, have brought students from the Philippines, yet 500 unemployed American women had to be tested to find 30 literate enough for a class in practical nursing. In the recent past, a man became a janitor if he could qualify for nothing else. But today a janitor is no longer a mindless floorsweeper. He must operate cleaning machines of complexity, study a manual for making repairs. There are even training classes for janitors; in Washington, D. C., many applicants were turned down because, unable to read labels, they couldn't distinguish a box of detergent from rat poison.

Are new machines to blame? Are we to label these people "victims of automation"? Or shall we at last attribute their ignorance to the primitive world of preautomation, which has condemned so many people for so long to such drudgery as chopping cotton, and educating them for no more? Is automation victimizing them, or rescuing them? The new, humanizing demands of automation are changing illiteracy from an unfortunate—and often ignored —statistic to a serious national threat.

Hilliard, the Chicago welfare chief, started becoming aware of the secret, growing threat of illiteracy in January 1959. He was disturbed by a statistical chart on his desk and he called in his energetic, intensely inquiring research director, Deton J. Brooks, Jr. A recession had just ended. Employment was picking up. According to all past experience, the relief rolls, which had always reacted sensitively to ups and downs of business, should be declining. But they kept rising. Why?

Hilliard and Brooks decided on an intensive study of a large, crowded, poverty-stricken neighborhood called Woodlawn where 25 per cent of all households were on relief. One of their findings seemed to tower in significance above all others. According to the Census, 6.6 per cent of the relief recipients had five years of schooling or less. But standard tests in the three Rs revealed that measurements of literacy based on "years" in school are widely misleading. A startling 51 per cent of the able-bodied adults were found unable to read and write at a fifth-grade level. The remainder who tested higher were so little above functional illiteracy that the difference hardly mattered. Nearly all were too uneducated to get the simplest jobs in the modern labor market.

"Here then" said Hilliard, reporting his findings to a convention of welfare officials, "is the major cause of today's poverty. Here is the reason for the high cost of relief. Punishing these people for their poverty won't help, badgering them with investigations, violating their small rights of privacy, condemning them for alleged immorality, putting them in jail, calling them loafers and idlers and cheats and frauds, which few of them are, will avail nothing. . . . These only divert attention from real solutions."

As a real solution, Hilliard set about to teach fifty thousand relief recipients to read and write better. By December 1963 the eight thousand most urgently needing education were going to school, but money was lacking for the rest. The Chicago Board of Education began footing the entire cost, providing classrooms and paying regular school teachers $4.50 an hour for the two evenings each week that classes are held. To pay these costs, the Board of Education siphoned money that was earmarked for Americanizing the foreign-born. Later, Congress authorized the use of federal funds for teaching literacy to the unemployed.

The biggest initial problem, according to Hilliard, was arranging for baby sitters to free mothers for classes. If women's organizations were prepared to offer such help, he said, they would make a unique contribution to helping families rid themselves of poverty. But lacking such help, welfare caseworkers undertook the huge job of arranging mutual baby sitting among the student-mothers. Last summer, when classes were changed to daytime to save custodial costs in school buildings, the baby-sitting problem became so difficult many classes had to be cancelled.

Considering how many classes for teaching literacy to adults have been springing up around America, one would expect the techniques to be down to a science. In Yakima, Washington, the LARK (Literacy for Adults and Related Knowledge) Foundation has organized classes as far east as Michigan. In St. Louis the Adult Education Council has aggressively sought to educate illiterates. Indiana Central College started a course for teaching teachers of illiterates. Daily TV programs, one series produced in Philadelphia, another in Memphis, have been loaned to other cities in the hope that illiterates will tune in and learn to read. But everywhere, teachers are groping for effective teaching methods.

The absence of recognized teaching materials came as a shock to Robert

L. Dixon, a junior-high-school teacher supervising some Chicago welfare classes. He recalls the first orientation meeting of several hundred teachers.

"Before you ask what textbooks you are to use," the speaker said, "let me tell you that we have none. We know that the Little Red Hen won't do for adults, but we don't know what *will* do. Your students are not like immigrants who want to learn to speak English. For most of your students English is the only language they know. This challenge is new. We will have to find our way by experimenting."

Dixon faced his first class, composed entirely of Negroes like himself, with uneasiness. His chief tools were a piece of chalk and a blackboard. His twenty students sat in the unfamiliar pose of poising pencils over notebooks. There were two women for each man. Young people outnumbered the elderly.

"First I wondered how much they knew about the world," he told me. "Next, I wondered how much they knew that I don't know. I had to keep reminding myself that they had rich experiences I know nothing about. One had been a paint mixer, one a drill-press operator, a few housewives and mothers, each with full lives, some longer than mine. I had to remind myself not to lump them together with a simple label like 'illiterate.'

"Then I began to wonder how much they thought I knew, what they imagined book learning really is. This made me wonder—and this was the most troubling of all—how much they expected of me."

His students came with uneasiness, too.

"I wondered," said Eddie O'Brien, a father of twelve, "how much education a man needed so he could get himself a job, and how long it would take. I was forty-two already and didn't have much time."

O'Brien took it as no joke when one night Mr. Dixon distributed play money and set up a shelf full of commodities for "sale." As pupils "bought" things, the teacher spun a line of talk, meanwhile shortchanging each of his customers. They were first embarrassed, then stunned as Mr. Dixon revealed the utterly simple ways in which slick salesmen fleece the uneducated.

These ways already had cost O'Brien his last job. For seven years he had worked for a baking company, stacking crackers as they came down automatic conveyors and sweeping crumbs from the floor. He took home $84 a week.

"I wanted to be like the rest of the people," O'Brien told me. "I bought a used car, some furniture we needed, a TV set, clothes for the kids. Those salesmen, they always kept telling me it wasn't hard, just a dollar down, a dollar a week. Before I knew it, I was caught in the trick bag."

The "trick bag" had many hidden pockets. If educated people are often swindled for overlooking the fine print, how easy to skin someone unable to read even the big print. One day a lawyer informed O'Brien he could be saved from his excess number of creditors only by claiming personal bankruptcy. The lawyer would gladly arrange this at a cost of $300. Payments would be easy: $100 down, $40 a month. When the fourth payment for the lawyer fell due, O'Brien was unable to pay it. Thus he forfeited $220 he had already paid, and the lawyer abandoned the case. O'Brien was undefended when his creditors descended on the baking company to claim slices of his wages. He was fired for "excessive wage assignments," a phrase well known in slum districts. Soon his family was on relief.

O'Brien's teacher, Mr. Dixon, includes in his instruction the essentials of how to be a good employee, such as the idea of "job loyalty." But the idea is hard to get across. When Eddie O'Brien needed his job most—to get out of the debt he was tricked into—he was fired for getting into debt. Nobody stood by him. When Andrew Timmons worked at his only job in nine years—at the car washer's—his employer once hailed him from the steamy wash tunnel and sent him into the bitter cold of December to fetch coffee for the front office. Timmons caught a chill, developed pneumonia, and spent Christmas near death in a hospital. When he recovered, he found that another man had been given his job and the boss was "too busy" to talk things over. That's when Timmons began heading for the relief rolls.

"How does a man learn loyalty," Dixon asks, almost in futility, "when no one has ever been loyal to him? He doesn't see it and he can't even read about it. We expect him to know the importance of regular attendance at work. Yet all his life he's been told to come and told to go, on a day's notice or an hour's, always at the employer's convenience. He only knows what he has seen. And he hasn't seen much to support our lectures on the rewards of being a good employee."

Still the students come eager to learn. After the first embarrassment at being exposed to their friends as uneducated, many are seized with a desire to know how to spell their children's names. They recite the names to the teacher and become absorbed in the magical process of copying down the letters— sometimes the first meaningful syllables they have ever written in their lives. One woman, after only three months of tutelage, brought her teacher an elaborate chocolate cake. She carried it proudly and announced, "I got a book all about cooking and *read* how to make it."

Eddie O'Brien describes his sense of achievement differently: "I feel like a caged bird that all at once got out." His escape has indeed been dramatic, for he escaped into the comparatively exhilarating world of self-dependence and self-respect. O'Brien began averaging $450 a month driving a taxi. He is one of the most successful of almost five hundred drivers lifted from the literacy classes and relief rolls, and being trained for jobs by the Yellow Cab Company. Once they had demonstrated they could read street signs, they were taught Chicago's house-numbering system, the location of the city's seventy-eight most important buildings, and how to fill out trip reports. Also they were given special training in meeting the public—and how to buy sensibly on the installment plan.

Yellow Cab expects to train a thousand relief recipients, possibly more. Despite publicized unemployment, the company has constant trouble finding suitable men. After surveying its own employees, the company drew up a description of the man who seemed most likely to succeed at the wheel. He was middle-aged, had children, little formal education, was probably a Negro, and his last employment was as a laborer or packinghouse worker. These characteristics, the company learned, almost exactly described the able-bodied male on relief. They also described the man least likely to be aware of a shifting national pattern of jobs from the factory (such as packinghouses) to services (such as cab driving). He is the man most fearful of looking for a job different from the one he had before. Of the first 406 to be educated and lifted from the public's relief roll to Yellow Cab's payroll, 333 made a go of driving cabs. Of the remainder, many left for other, better jobs; a few entered military service. A handful were fired for careless driving.

Soon after the success was apparent, the Chicago Urban League brought the Yellow Cab story to the Shell Oil Company. Major oil companies have been troubled by a shortage of high-grade men as gas station attendants. Sales are lost by employees who don't seem to care, and the man who meets the customer can make or break the reputation of his company.

With trepidation, Shell undertook to train a group of men from the literacy classes. Strange things happened. In tests for spelling, for example, men continued to have trouble with words like "which," but racked up high scores in technical words like "detergency" and "differential." Company officials accepted this as a surprising sign that the men were burning late lamps to insure succeeding on their new jobs. It helped destroy company fears that "reliefers" were natural loafers. Still, the company was skeptical. Yet two months after the men went to work for neighborhood dealers, the company found that seven out of thirty-five had already been promoted to shift foreman, taking charge of men who had been on the job a year or more. Station owners reported that the new men were among their best employees. The training program is now permanent.

While the star trainee at Shell, Sam Frost, who became the roving bookkeeper, was working furiously to learn, an odd thing happened at home. His eighteen-year old son, a recent high-school dropout, began talking about going back to school; his grammar-school children headed for their homework whenever Daddy did.

This discovery was made in many families, some of which never had occasion to own a book, seldom a newspaper.

"Children do what their grown folks do," Mrs. Dorothy Slade told me, slightly awed by the behavior of her eight-year-old son. "Since I been going to school, it seems he just behaves better. I used to play solitaire and he'd want to take the cards to bed with him. I stopped playing because I don't think children should learn about cards. Now I pick up a book, and darn if he don't pick up one of his schoolbooks. Now look and see what's happening to his marks."

Mrs. Slade showed me report cards for two years. The previous year the child had earned an F (fair) in reading and G (good) in arithmetic. This year's grades were E (excellent) in both subjects.

The experience of Mrs. Slade and her son—duplicated in many households I looked into—helps explain Raymond Hilliard's enthusiasm for educating mothers, even though many might not become self-supporting at jobs. These mothers may slow the spread of "infectious ignorance." Mrs. Slade's son has no father to model himself after; he was born out of wedlock. The Woodlawn study revealed that 84 per cent of able-bodied adults on relief are women, most of them abandoned early in marriage. Hilliard's research director, Deton Brooks, emphasizes:

"In a matriarchal structure, the women are transmitting the culture. If the woman is illiterate, she transmits the values, the images, of an illiterate's world. She can't do otherwise. This is dangerous, extremely dangerous, for the future of these children and the society that may soon have to support them as illiterate adults."

Extending this reasoning, Hilliard is convinced that literacy classes strike at the roots of broken homes. Hilliard concludes:

"You can see a straight line operating from illiteracy to illegitimacy. The American culture teaches all men—including Negro men segregated from the main culture—that a father's job is to be a provider. The man who had been abandoned by his father and in turn abandons his children is convinced he can never succeed as a good American father. Where's his chance to provide? From the day he takes his vows he knows he can't fulfill his function, that his manhood has been taken away, that his marriage is doomed, and the insecurity of his woman has begun.

"Instead of heaping more contempt on this man, let's look for ways that will let him stay at home. Let's give him at least the meagerest education to help him find a decent job at a decent wage, and give him some assurance that he won't be the first one fired because of his color. Then you'll start to see a downslide in the illegitimacy rate. That's how you can extend the possible results—maybe very early results—of something as simple as teaching the A B Cs."

The unhappy, unproductive survival of an illiterate primitive with a large family costs an American city $300, $400, sometimes $500 a month. Perhaps it is sinful to take measure of this human misery in dollars, but that is the original sin in which the misery was born. For the uselessness of these people did not begin when the machines came. It was ordained when their native states—Mississippi, Kentucky, Alabama, Arkansas, and all the rest—saw no purpose in spending even $200 *a year* to school their cotton pickers and squirrel hunters.

Now that this sin against humanity has at least received overnight recognition, Americans would like to make overnight repentance, quickly, painlessly. But there is no way. In denying education, we have taught lessons of aimlessness, futility, inferiority and self-hatred. We have taught these lessons well; they will not quickly or easily be unlearned.

Occasionally a Sam Frost will make the effort against ignorance look like victory. The success of the Yellow Cab and Shell Oil experiments give us the flushed hope that perhaps there is some kind of overnight repentance.

But then one visits the literacy classes. The truth he must face is that great numbers of lost souls are hopeless in trying to cram the alphabet into their aging heads. They go to classes because they are told to go. They sit with their pencils poised because that's what others do. Perhaps they even engage in some fumbling inner struggle with themselves to try to study. But it is too late. In darkness they have lived; in darkness they will remain for what time they may have remaining.

What is to be done?

"You can't just ignore these people," goes a familiar argument. "Your long range optimism is all very well, but these people are out of work *now*. Your new machines may create new jobs, all right, but not for the unskilled and illiterate. What do you propose to do for *them—now?*"

What must be done is obvious. They must continue to be kept, as they are today, and have been for years, by the society that misused them. But the question is for how many generations must the chain of inherited dependence go on? When "liberals" reply with the facile demand that we "must take care of these people," are they not condemning them, in the name of compassion, to a continuation of the same dark, outcast lives to which "conservatives" often condemn them with contempt?

The chronically poor cannot be ignored, of course not. But their condition of ignorance cannot be accepted as static and unchangeable, either. Something indeed must be done for them now.

Of all the "somethings" that have been proposed—a moratorium on installing new machines; expansion of government make-work projects designed more to employ than to build; and the latest one, rearranging our national ethics to divorce the idea of income from the idea of work—only one plan addresses itself to what the trouble really is. If the trouble is ignorance, the solution must be education. All the rest is flailing of the arms that brings relief to nothing except the troubled social conscience of the arm flailer.

The education of illiterates and near-illiterates is not only mandatory but comparatively cheap. The cost of teaching one jobless man in Chicago for five years is smaller than his relief check for a single month. The schooling may provide his shortest, most practical route back to useful work. But the important thing is that the education is useful and a bargain *even if he doesn't learn.* For his simple acts of going to school, carrying home the first book he ever laid his hands on, clearing the kitchen table and licking his pencil to do homework, are, when exposed to the eyes of his slum-imprisoned children, the essential first acts in breaking the chain of dependence. These simple gestures are what may first reveal to a child that school-learning has some real connection with the real life of some real adult he knows.

Compelling as the need may be to help a man too ignorant to perform a job, the need is far more compelling to save his children from such crippling ignorance while they still may be saved.

Employment and Unemployment Developments in 1969

Paul O. Flaim and Paul M. Schwab

Job growth lost steam after strong surge in first quarter, and the unemployment rate inched upward from a 15-year low

Employment rose substantially in 1969, with about 2 million additional jobs being created. The most vigorous job growth, however—as well as the lowest rate of unemployment—was recorded in the early months of the year. In the ensuing months, the demand for labor slackened significantly under the impact of the Government's anti-inflationary measures, and the jobless rate moved to somewhat higher levels.

The slower pace of employment growth that prevailed after the first-quarter surge halted a sustained decline in the incidence of unemployment. After reaching a post-Korean war low of 3.3 percent as 1969 began, the jobless rate returned gradually to the 3.5 to 4.0-percent range of the previous 2 years. As 1969 drew to a close, however, the rate dipped again slightly below the 3.5 percent mark.

For the year as a whole the unemployment rate averaged 3.5 percent, slightly lower than the annual rate for 1968, which was the lowest since 1953. The number of unemployed persons remained at the 2.8 million level of 1968, despite a huge increase in the labor force.

Employment Growth

The year opened with an exceptionally strong demand for labor prevailing in nearly all sectors of the economy. This surge in the demand for workers, which had begun in the closing months of 1968, led to a 1.8-million increase in employment (on a seasonally adjusted basis) between September 1968 and March 1969. (See chart 1.)

This pace of employment growth could obviously not be sustained, because additional workers would be increasingly difficult to locate under the relatively tight labor market conditions that prevailed in the first half of 1969, and because the stepped-up Government efforts to restrain the economy were bound to have some dampening effect on the demand for labor. As the year progressed, the demand for labor did taper off considerably, as reflected both by the gradual rise in unemployment and by the much smaller employment increases which took place after the first quarter of the year.

The annual employment gain, however, was still very impressive. At 2.0 million it exceeded the average annual increases posted during the 1961–68 period of sustained economic expansion, and raised total employment to 77.9 million.

Who Got the New Jobs

Because of the very tight labor market situation that prevailed in early 1969, the additional workers required by the economy during the year had to be drawn almost entirely from outside the labor force. This was unlike the situation in previous years when a portion of the workers added to the employment rolls came from the ranks of the unemployed, which were being gradually whittled down from the highs of the early 1960's. With unemployment reaching a 16-year low in early 1969, it became increasingly difficult to find qualified workers—especially adult male workers—among the jobless.

Drawing workers from outside the labor force means essentially the hiring of large numbers of women and teenagers, which is exactly what took place in 1969. Employment of adult women increased by more than 1 million and teenage employment rose by 330,000. By contrast, adult male employment rose by only 550,000 over the year, although men make up about three-fifths of the civilian labor force. (See table 1.)

Although adult women have been making rapid advances in the job market for many years, the gain in employment achieved in 1969 was the biggest since World War II. Women 20 years old and over now hold slightly more than one-third of the Nation's jobs. This is a considerable advance compared with the situation in the immediate post-World War II period, when women accounted for only one-fourth of employment.

Of the 1.1 million additional jobs secured by women in 1969, about one-third were obtained by 20- to 24-year-olds, for whom the increase in population and labor force participation has been particularly rapid in recent years. Women 25 to 54 years old accounted for about one-half (560,000) of the year's gain in female employment. But even women 55 years and over posted a very sizable increase in employment in 1969 (200,000).

Teenagers (16- to 19-year-olds) accounted for 330,000, or about one-sixth of the employment increase. Annual job gains by this group have varied widely during the decade; increasing dramatically between 1963 and 1966, but falling off sharply in the next 2 years, reflecting primarily a leveling off in the growth of the teenage population, as well as stepped-up draft calls.

Employment of adult males increased by about 550,000 in 1969, which was about 100,000 less than the job gain posted by this group in each of the previous 2 years, but in line with the group's average annual gain since 1961. Of the 550,000 adult men added to the employment rolls, about two-fifths were men 20 to 24 years old. Product of the baby boom of the late 1940's, these young men are now coming into the labor force in increased numbers, and the influx will gain momentum if the present reduction of draft calls continues for a protracted period.

Full-time and Part-time Workers

About one-third of the employment increase in 1969 was accounted for by part-time workers. These are persons who customarily work less than 35 hours a week, either for personal reasons or because of the nature of the job. The number of such workers, which has been increasing at a much faster rate than total employment in recent years, passed the 11-million mark in 1969 and now accounts for about 15 percent of all employed persons. (See table 2.)

It should be emphasized that the rapid increase in part-time employment does not necessarily denote a scarcity of full-time employment opportunities.

Chart 1.
Employment and Unemployment, 1968-1969, Seasonally Adjusted

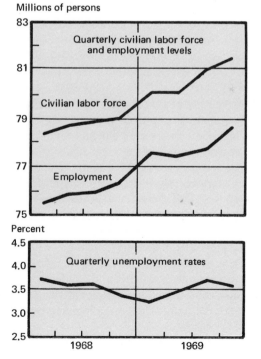

In some fields, the increased use of part-time workers is dictated by changing business patterns. A typical example is the greater reliance on part-time help made necessary by the suburbanization of the retail trade industry and the "open-every-evening" policy of most suburban stores. In other cases, employers must turn to part-time help simply because they cannot find workers who are available on a full-time basis, particularly during periods of peak demand. The great majority of the 11 million persons who usually worked only part time in 1969 were either not available for or did not want full-time work. Only about 1 million of them said that they preferred full-time but had found only part-time work, or had seen their workweek reduced below 35 hours because of a shortage of work.

Of the 10 million workers on voluntary part time, slightly over one-half (or 5.5 million) were adult women. The greater availability of part-time jobs has no doubt contributed significantly to the sharp increase in labor force participation by women in recent years. Adult men accounted for about 2 million of the persons usually working part time in 1969, with slightly more than a quarter 20 to 24 years old and thus likely to be in college. A roughly similar proportion were 65 years of age and over and apparently semiretired. The balance of the part-time workers (nearly 3 million) were teenagers, for whom such employment has almost doubled since 1963.

Occupational Developments

Blue-collar employment, which had grown only moderately in the previous 2 years, posted an impressive gain in 1969 despite the leveling off of industrial production that took place in the second half of the year. The number of workers in blue-collar occupations rose to 28.2 million, an increase of 700,000 over 1968.

An interesting development among these workers in 1969 was the sharp increase in the employment of operatives and laborers. These unskilled and semiskilled workers are relatively easier to find and train than skilled workers, and thus their employment is much more responsive to cyclical movements in the economy. With the vigorous tempo of economic activity and the tight labor market of early 1969, many employers had no practical alternative but to hire and train unskilled and semiskilled workers in order to meet production goals.

The increased demand for blue-collar workers in 1969 was clearly reflected in the unemployment rate for this group. Primarily on the strength of particularly low unemployment in the early months, the annual rate of unemployment for blue-collar workers edged down to a record low of 3.9 percent in 1969. However, with the pace of industrial production slackening in the second half of the year, the demand for blue-collar workers tapered off considerably. By the end of the year, their jobless rate had returned to the 4.0 to 4.5-percent range of the previous 3 years.

Even with the relatively strong showing of blue-collar employment, the proportion of workers engaged in white-collar work posted another increase in 1969. Over the year, white-collar employment advanced by 1.3 million to 36.8 million, resulting in nearly one-half of the Nation's jobs being white-collar.

Table 1.

Civilian Labor Force, Employment, and Unemployment, by Age, Sex, and Color, 1968 and 1969

[Numbers in thousands]

Age-sex-color group	Civilian labor force			Employment			Unemployment		
	1969	1968	Percent change, 1968-69	1969	1968	Percent change, 1968-69	1969	1968	Percent change, 1968-69
All Races									
Total, 16 years and over	80,733	78,737	2.5	77,902	75,920	2.6	2,831	2,817	.5
Men, 20 years and over	46,351	45,852	1.1	45,388	44,859	1.2	963	993	−3.0
Men, 20-24 years	5,282	5,070	4.2	5,012	4,812	4.2	270	258	4.7
Men, 25 years and over	41,068	40,782	.7	40,376	40,047	.8	692	735	−5.9
Women, 20 years and over	27,413	26,266	4.4	26,397	25,281	4.4	1,015	985	3.0
Women, 20-24 years	4,597	4,235	8.5	4,307	3,950	9.0	290	285	1.8
Women, 25 years and over	22,815	22,031	3.6	22,090	21,331	3.6	726	700	3.7
Both sexes, 16-19 years	6,970	6,618	5.3	6,117	5,780	5.8	853	839	1.7
White									
Total, 16 years and over	71,779	69,977	2.6	69,518	67,751	2.6	2,261	2,226	1.6
Men, 20 years and over	41,772	41,318	1.1	40,978	40,503	1.2	794	814	−2.5
Women, 20 years and over	23,839	22,821	4.5	23,032	22,052	4.4	806	768	4.9
Both sexes, 16-19 years	6,168	5,839	5.6	5,508	5,195	6.0	660	644	2.5
Negro and Other Races									
Total, 16 years and over	8,954	8,760	2.2	8,384	8,169	2.6	570	590	−3.4
Men, 20 years and over	4,579	4,535	1.0	4,410	4,356	1.2	168	179	−6.2
Women, 20 years and over	3,574	3,446	3.7	3,365	3,229	4.2	209	217	−3.7
Both sexes, 16-19 years	801	779	2.8	609	585	4.1	193	195	−1.0

Table 2.

Full-Time or Part-Time Status of Employment by Age and Sex, 1968 and 1969

[Numbers in thousands]

Full-time or part-time status	Total, 16 years and over			Male, 20 years and over			Female, 20 years and over			Both sexes, 16-19 years old		
	1969	1968	Percent change	1969	1968	Percent change	1969	1968	Percent change	1969	1968	Percent change
All employed persons	77,902	75,920	2.6	45,388	44,859	1.2	26,397	25,281	4.4	6,117	5,780	5.8
Persons usually working full time	66,596	65,277	2.0	43,100	42,721	.9	20,454	19,601	4.4	3,042	2,956	2.9
Working on full-time schedules	65,503	64,225	2.0	42,530	42,164	.9	20,053	19,219	4.3	2,921	2,842	2.8
Temporarily working 1-34 hours	1,093	1,052	3.9	570	557	2.3	401	382	5.0	121	114	6.1
Persons usually working part-time	11,306	10,644	6.2	2,288	2,139	7.0	5,944	5,681	4.6	3,074	2,823	8.9
Voluntarily working 1-34 hours	10,343	9,726	6.3	2,002	1,863	7.5	5,524	5,268	4.9	2,817	2,595	8.6
Involuntarily working 1-34 hours	963	918	4.9	286	276	3.6	420	413	1.7	257	228	12.7

Consistent with employment trends characterizing the entire post-World War II period, the latest annual increase in white-collar employment occurred almost exclusively among workers in the professional, technical, and clerical fields. Persons in managerial positions registered a relatively small increase in employment, while the number of sales workers, which has shown little growth in recent years, remained practically unchanged.[1]

The jobless rate for white-collar workers was 2.1 percent in 1969, practically the same as the record-low rate of 2.0 percent posted by this group in 1966 and 1968.

Employment of service workers rose by 150,000 in 1969, with all the growth taking place among those engaged in other than private household work (for example, restaurant work, protective services, etc.). Those engaged in private household work continued to decline (for the fifth consecutive year), falling by 110,000 to 1.6 million. In large part, this trend reflects the emergence of many new employment opportunities for these workers. This has particularly been the case for Negroes, whose occupational upgrading has been relatively rapid in recent years.

Occupational Advances of Negroes

Negroes and members of other minority races made significant progress on the occupational ladder in 1969. Although at year's end they still held a disproportionately large share of the Nation's least desirable jobs, their most rapid employment gains for the year were achieved in the higher-skill, higher-status occupations.

As table 3 shows, overall Negro employment increased by about 3 percent in 1969 (to 8.4 million), but the number of Negroes employed in white-collar work rose by about 10 percent. Moreover, this rise saw significant numbers of blacks secure jobs in the professional and managerial fields as well as in clerical and sales occupations.

Encouraging upward progress was also made by Negroes in blue-collar occupations. Nearly all the additional jobs they secured in the blue-collar sector were in the craftsmen and foremen group or as operatives. The number of Negroes employed as nonfarm laborers was practically unchanged over the year.

The exodus of Negroes from low-skill, low-status occupations is reflected in a decline in their employment as service workers (particularly as private household employees) and as farmworkers. With more attractive jobs opening up in other occupations, the number of Negroes employed in private households or on farms declined by about a tenth during the year.[2]

---

1. It should be noted that these data refer only to workers for whom sales work is the primary employment. Those multiple jobholders who are "moonlighting" as part-time sales workers but whose primary job is in another field are not counted as sales workers from an occupational standpoint.
2. For a longer-term look at the occupational advances of Negroes, see Claire C. Hodge, "The Negro Job Situation: Has It Improved?" *Monthly Labor Review,* January 1969, pp. 20-28.

Industry Developments

The 1968–69 advance in total employment was concentrated entirely in the nonagricultural sector of the economy. Employment in agriculture continued its long-term decline, falling by about 200,000 to 3.6 million. With the exception of 1968, when farm employment remained virtually unchanged, annual declines in agriculture have exceeded 100,000 in each of the past 10 years. In the past two decades, the number of farm jobs has been cut in half and agricultural employment has now dropped to less than 5 percent of employment. The main factors contributing to this fairly steady decline have been the continuing mechanization of farming processes and the availability of more attractive jobs in the nonfarm sector. The exodus from agricultural jobs in recent years has been particularly rapid for Negroes. In 1960 they held about 900,000 farm jobs, now only 400,000.

Total nonagricultural employment—including self-employed and unpaid family workers, and wage and salary workers—increased by about 2.2 million in 1969, reaching a record of 74.3 million. Despite a noticeably slower rate of growth in the second half, the year's gain in total nonfarm employment exceeded the increases of the previous 2 years.

Payroll employment in the nonagricultural sector advanced by 2.3 million in 1969, passing the 70-million mark for the first time. (Payroll employment excludes private household, self-employed, and unpaid family workers, but counts workers more than once if they hold more than one job.)[3] The 1969 increase in the number of payroll workers was about one-fifth greater than the gains the previous 2 years.

The growth in payroll employment was particularly rapid (seasonally adjusted) in the closing months of 1968 and the early months of 1969. During the September 1968–March 1969 period, monthly gains in payroll employment averaged about 250,000. These advances, moreover, were broadly based, being spread across most major industries. (See table 4.)

Beginning in the second quarter of 1969, however, the pace of employment growth slackened considerably in most major industries. For the remainder of the year, it showed only moderate gains, coinciding with other signs of a general deceleration in the nation's economy.

As has been the trend for several years, the bulk of new job opportunities in 1969 were provided by the service-producing industries—trade; services; government; transportation and public utilities; and finance, insurance, and real estate. It is these industries which provide most of the new job opportunities for women and teenagers. Even in these industries, however, employment growth had begun to slow down as 1970 approached.

3. See Gloria P. Green, "Comparing Employment Estimates From Household and Payroll Surveys," *Monthly Labor Review,* December 1969, pp. 9-20.

Substantial job growth was also exhibited by the goods-producing industries in 1969, particularly in the first half of the year. During the second half, however, employment growth in this sector—which includes manufacturing, construction, and mining—slackened considerably. The employment developments for each major industry are discussed briefly below.

Table 3.

Occupational Distribution of Employment by Color, 1968 and 1969 Annual Averages

Color and occupation	1969		1968		Percent change, 1968–69
	Level (in thousands)	Percent distribution	Level (in thousands)	Percent distribution	
WHITE					
All employed persons	69,518	100.0	67,751	100.0	2.6
White-collar workers	34,647	49.8	33,560	49.5	3.2
Professional and technical workers	10,074	14.5	9,685	14.3	4.0
Managers, officials, and proprietors	7,733	11.1	7,551	11.1	2.4
Clerical workers	12,314	17.7	11,836	17.5	4.0
Sales workers	4,527	6.5	4,489	6.6	.8
Blue-collar workers	24,647	35.5	24,063	35.5	2.4
Craftsmen and foremen	9,484	13.6	9,359	13.8	1.3
Operatives	12,368	17.8	12,023	17.7	2.9
Nonfarm laborers	2,795	4.0	2,681	4.0	4.3
Service workers	7,289	10.5	7,066	10.4	3.2
Farmworkers	2,935	4.2	3,062	4.5	−4.2
NEGRO AND OTHER RACES					
All employed persons	8,384	100.0	8,170	100.0	2.6
White-collar workers	2,197	26.2	1,991	24.4	10.4
Professional and technical workers	695	8.3	641	7.8	8.4
Managers, officials, and proprietors	254	3.0	225	2.8	12.9
Clerical workers	1,083	12.9	967	11.8	12.0
Sales workers	166	2.0	158	1.9	5.1
Blue-collar workers	3,591	42.8	3,462	42.4	3.7
Craftsmen and foremen	709	8.5	656	8.0	8.1
Operatives	2,004	23.9	1,932	23.6	3.7
Nonfarm laborers	877	10.5	874	10.7	.3
Service workers	2,239	26.7	2,315	28.3	−3.3
Farmworkers	356	4.2	402	4.9	−11.5

Manufacturing

Employment in the manufacturing industries continued to be a key indicator of the general pace of our economy. Although manufacturing employment rose by 350,000 in 1969, surpassing the 20-million mark for the first time, virtually all of the year's advance took place during the first half. Stepped-up Government efforts to halt the mounting pace of inflation weakened the demand for factory labor in the ensuing months, as evidenced by the trend of unemployment in the manufacturing industries. The jobless rate in this sector, which had dropped from 3.3 percent in 1968 to a 16-year low of 3.1 percent in the first half of 1969, moved to 3.5 percent in the last 6 months of the year.

The number of production workers employed in manufacturing rose by 230,000 to 14.7 million in 1969, approaching once again the 15-million record posted during World War II. Despite the latest gain, however, the ratio of production workers to all manufacturing employees slipped another notch to 73.2 percent.

Three-fourths of the gain in manufacturing employment in 1969 was concentrated in the durable goods industries. This skewed distribution of factory employment growth, unlike the 1967 and 1968 experiences, resembled the pattern of 1965–66, when the hard goods industries set the fastest pace of economic activity.

In 1969, 10 of the 11 durable goods industries registered employment pickups, whereas only 6 of the 10 soft goods industries recorded advances. Especially large employment gains were shown by the electrical equipment (60,000) and fabricated metal (60,000) industries. Machinery and primary metals also registered considerable increases over the year (50,000 and 40,000, respectively). Together, these four industries accounted for slightly more than half of the 1969 increase in manufacturing employment. Employment strength among these industries resulted largely from the combination of a continuing boom in capital equipment and firm demand for domestically produced steel. The only hard goods industry to register an employment decline over the year was ordnance (a drop of 10,000). This development was not unexpected in view of the cutbacks in defense spending implemented during the year.

Table 4.

Employees on Nonagricultural Payrolls by Industry, 1968 and 1969 (Seasonally Adjusted)

[Numbers in thousands]

Industry	Annual averages		Quarterly averages							
	1969 [1]	1968	1969				1968			
			4 [1]	3	2	1	4	3	2	1
Total	70,139	67,860	70,648	70,379	70,034	69,465	68,655	68,076	67,611	67,057
Mining	628	610	633	630	623	627	606	620	615	598
Construction	3,410	3,267	3,441	3,421	3,412	3,359	3,316	3,275	3,268	3,203
Manufacturing	20,121	19,768	20,054	20,232	20,142	20,061	19,898	19,808	19,743	19,625
Durable goods	11,881	11,624	11,807	11,986	11,891	11,846	11,698	11,649	11,605	11,544
Nondurable goods	8,240	8,144	8,247	8,246	8,251	8,214	8,201	8,159	8,138	8,081
Transportation and public utilities	4,449	4,313	4,487	4,482	4,450	4,375	4,351	4,325	4,283	4,292
Wholesale and retail trade	14,644	14,081	14,806	14,696	14,602	14,463	14,276	14,148	14,019	13,871
Wholesale trade	3,768	3,618	3,820	3,779	3,756	3,714	3,669	3,634	3,603	3,564
Retail trade	10,876	10,464	10,985	10,918	10,846	10,749	10,607	10,514	10,417	10,308
Finance, insurance, and real estate	3,558	3,383	3,607	3,578	3,543	3,502	3,450	3,396	3,357	3,326
Services	11,102	10,592	11,266	11,112	11,058	10,967	10,782	10,614	10,517	10,451
Government	12,227	11,846	12,354	12,226	12,203	12,112	11,977	11,889	11,808	11,691
Federal	2,756	2,737	2,721	2,759	2,767	2,762	2,714	2,748	2,740	2,722
State and local	9,471	9,109	9,633	9,467	9,436	9,350	9,263	9,141	9,068	8,969

[1] The 1969 annual averages and the data for the 4th quarter of the year are preliminary.

In contrast to the relatively strong performance among durable goods industries in 1969, employment among nondurable industries advanced only moderately (95,000). The bulk of this increase was accounted for by gains of 20,000 to 25,000 each in the printing, paper, chemical, and rubber industries. Smaller job increases were registered in the food processing and apparel industries.

Construction and Mining

Activity in the construction industry was exceptionally strong in early 1969, with large employment gains and a drop in the jobless rate to a 16-year low. As the year progressed, however, the building industry showed signs of increasing weakness, with housing activity, in particular, softening under the impact of tightening credit and high interest rates. Construction employment, consequently, showed no growth during the second half of 1969. Nevertheless, the 1968–69 employment gain in this industry remained impressive, with 140,000 new workers added to payrolls.

After 11 years of continuous declines in employment, the number of mining employees rose by 20,000 in 1969 to 630,000 workers. Largely responsible for this relatively strong showing were job pickups in the oil and gas extraction segment of the industry, and reduced strike activity in coal and metal mining.

Trade

Throughout most of 1969, employment gains in trade were off substantially from the sharp increases recorded during the September 1968–March 1969 period. Nonetheless, employment in this industry, which is a large user of part-time help, registered a substantial annual increase of 560,000 workers.

A significant portion of the 1968–69 gain in trade employment was accounted for by the wholesale sector, the number of wholesale trade employees increasing by 150,000. However, virtually all of this advance occurred during the first half of the year.

Services

Employment in services increased at a particularly rapid rate during the first quarter. Between March and August, however, employment remained relatively unchanged. One possible explanation for the absence of the usual March–August gains in services might have been the difficulty of obtaining seasonal workers in low-paying industries. Despite the mid-year lull, employment in services registered a year-to-year gain exceeding 500,000 workers. About two-fifths of this increase occurred in private medical and other health services, a field where employment has doubled over the past decade.

Table 5.
Unemployment Rates by Industry, 1968 and 1969 (Seasonally Adjusted)

Industry	Annual averages		Quarterly averages							
			1969				1969			
	1969	1968	4	3	2	1	4	3	2	1
Private wage and salary workers[1]	3.5	3.6	3.7	3.7	3.5	3.3	3.4	3.6	3.6	3.7
Construction	6.0	6.9	6.2	6.9	5.6	5.7	6.0	6.5	6.7	7.8
Manufacturing	3.3	3.3	3.7	3.3	3.2	3.1	3.1	3.3	3.2	3.4
Durable goods	3.0	3.0	3.6	2.9	3.1	2.6	3.0	3.0	2.9	3.1
Nondurable goods	3.7	3.7	3.9	3.8	3.4	3.7	3.4	3.7	3.8	3.9
Transportation and public utilities	2.2	2.0	2.5	2.0	2.3	2.0	2.0	2.4	1.7	1.9
Wholesale and retail trade	4.1	4.0	4.0	4.4	4.1	3.8	4.0	4.0	4.0	4.1
Finance and service industries	3.2	3.4	3.0	3.5	3.3	3.0	3.2	3.5	3.5	3.3
Government wage and salary workers	1.9	1.8	2.2	1.9	1.7	1.7	1.8	1.9	1.8	1.9
Agricultural wage and salary workers	6.0	6.3	6.0	7.9	5.4	5.2	5.3	7.9	6.6	5.4

[1] Includes mining, not shown separately.

Government

Employment in the government sector rose by 380,000 in 1969 to 12.2 million workers. This was well below employment pickups recorded in this sector in recent years. State and local government employment continued to expand, but at a reduced rate as the year progressed. Federal government employment, meanwhile, was virtually unchanged over the year, due largely to a stringent budget and severe staffing limitations.

Transportation and Finance

Elsewhere in the service-producing sector, transportation and public utilities showed impressive job strength over the year, with an employment rise of 140,000. Employment in finance, insurance, and real estate, where growth has been accelerating since 1965, increased by 180,000, to 3.6 million workers.

Hours of Work and Earnings

Despite the unusually high rate of economic activity which prevailed early in the year, the average weekly hours of American rank and file workers declined another notch in 1969. Hourly earnings continued to increase during the year, but the gains were canceled out by the steady rise in prices.

Workweek

For all production and other nonsupervisory workers on private payrolls, the average workweek in 1969 slipped 0.1 hour to 37.7 hours, representing the fourth consecutive year in which the workweek has declined. Among major industries, however, the year-to-year picture was mixed, with workweek declines in the trade and manufacturing industries offsetting increases in mining and construction.

In manufacturing, the average workweek inched slightly downwards to 40.6 hours. This was only the second year since 1960 in which the weekly hours of factory production workers have dropped. The only other decline took place in 1967, when a period of mild economic readjustment occurred after the rapid expansion in the preceding 2 years. Overtime for production workers in manufacturing averaged 3.6 hours a week in both 1968 and 1969. In the closing months of 1969, however, overtime hours were running somewhat below this average.

The workweek for employees of the trade industry continued its long-term decline in 1969, dropping another 0.4 hour to 35.6 hours. The shortening workweek in this industry, however, is clearly not a reflection of declining business. It is instead a reflection of the increased use of part-time help made necessary by the great expansion of retail stores in the growing suburbs of our cities. Since suburban stores must maintain late closing hours in order to serve their clientele, they can be staffed efficiently only through the hiring of many part-time workers.

The average workweek increased for workers in mining and construction by 0.4 and 0.6 hours, respectively. Even in these two industries, however, weekly hours tended to be higher during the first part of the year.

Chart 2.

Average Weekly Hours of Production Workers on Private
Nonagricultural Payrolls, 1968-1969, Monthly Averages

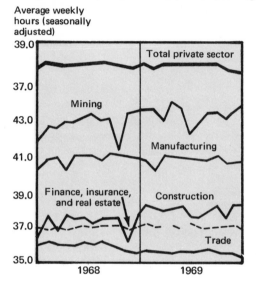

Earnings

Under the tight labor market conditions which prevailed during most of 1969, particularly in the first half, workers exerted increased pressure on employers to obtain higher wages and employers, in turn, often had to offer higher wages in order to maintain or increase their work force. This being the setting, the wage level in 1969 increased at an exceptionally fast pace set during the previous year, rising to $3.04 per hour— 6.7 percent above 1968.

On a year-to-year basis, the average gross weekly earnings for the Nation's rank and file workers rose about $6.90 (or 6.4 percent) to $114.60. Since the average workweek dipped slightly over the year, the increase in weekly earnings was attributable entirely to the higher hourly wage level.

Because of rapid increases in the price level, however, average gross earnings for all rank and file workers rose by only 1 percent in terms of constant (1957–59) dollars. Although the rate of consumer price increases appeared to taper off toward the close of 1969, the total increase represented the fastest annual rise in consumer prices since 1951, virtually erasing all wage gains achieved by workers.

Take-home pay (gross weekly earnings less Federal income and social security taxes) for the average worker with three dependents rose nearly 5 percent in 1969. Because of continued price pressures, however, real take-home pay remained virtually unchanged for this hypothetical breadwinner. His purchasing power, in fact, has not increased since 1965, despite the steady rise of his wage rates.

Among major industries, construction registered the sharpest rise in gross weekly earnings (10.1 percent), reflecting both a higher wage rate and a longer workweek. A similar combination of factors brought higher weekly earnings in mining and in finance, insurance and real estate. Larger average weekly earnings for manufacturing and trade nonsupervisory workers, meanwhile, stemmed entirely from increases in hourly earnings.

Unemployment Trends

As a result of the surge in demand for labor which began in the fall of 1968, the Nation entered 1969 with the unemployment rate (seasonally adjusted) at a 15-year low of 3.3 percent. At this level, the jobless rate was not only much lower than it had been during the early 1960's, when it hovered around 6 percent; it was substantially below the 3.5 to 4.0 percent range within which it had fluctuated during most of 1967 and 1968, years which have generally been viewed as representing relatively full employment.

Unfortunately, perhaps inevitably, the unusually high rate of economic activity, which sparked the strong demand for labor, also added fuel to the fires of inflation. As the unemployment rate was dropping to a record low for the decade, the rate of price increases began to rise, forcing Government to take fiscal and monetary measures that ultimately moderated the demand for labor and returned the unemployment rate to the 3.5 to 4.0 percent range.

While on an annual basis unemployment in 1969 showed little change from 1968, the quarterly averages show clearly how the jobless rate dropped as the economy surged ahead and how it returned to previous levels as the Government anti-inflationary restraints began to slow the pace of economic growth. As chart 1 shows, the seasonally adjusted unemployment rate had dropped from 3.7 percent in the first quarter of 1968 to a post-Korean War low of 3.3 percent in the first quarter of 1969. It then reversed the trend, rising gradually during the next two quarters and averaging about 3.7 percent for the second half of the year, despite a small decline in joblessness among marginal workers toward the close of the year.

The general upward turn in unemployment did not spread immediately to all industries. Within manufacturing, for example, this was the case only for the durable goods industries. For workers in nondurable goods production, the unemployment rate continued to decline until mid-year, before the trend reversed. By the fourth quarter, however, all major industries had somewhat higher unemployment than at the beginning of the year.

Although unemployment moved to generally higher levels in the second half of 1969, the jobless rate had still not exceeded the 4.0-percent level, once regarded as an interim index of full employment. While the number of unemployed also rose significantly, after having dipped to 2.7 million (on a seasonally adjusted basis) early in the year, the average unemployment level for 1969 (2.8 million) was virtually unchanged from the 1968 average.

Under present economic conditions, a further slowing of the rate of economic growth should not lead to as sharp increases in unemployment as those experienced during previous slowdowns. First, a much larger proportion of total employment is now in white-collar and service occupations, fields that are not very sensitive to changes in the general economy. Another factor that should militate against, or at least defer, any sharp increase in unemployment is the still relatively high levels of overtime work which prevail in many industries. Gradual elimination of overtime work in industries having to adjust production to lower levels of consumer demand should act as a buffer against layoffs of workers.

Another element, however, will add some uncertainty to the manpower and unemployment situation in the coming months. A stepped-up disengagement of American troops from Viet Nam and their subsequent demobilization might substantially swell the ranks of the jobseekers. How promptly these men could be absorbed by the job market would depend largely on the general health of the economy and on the impact of specific programs designed to assist their readjustment to civilian employment.

Jobless Trends for Major Groups

Paralleling the Nation's overall unemployment rate, the rates for most major groups in the labor force also moved from relatively low levels at the beginning of 1969 to generally higher levels by the end of the year. On an annual basis, however, even these rates showed little change from 1968. (See table 6.)

Adult Men

Unemployment rates for adult men, who make up the main body of full-time workers, continued at relatively low levels in 1969. The incidence of joblessness was particularly low among men 25 years of age and over. Although the unemployment rate for this group of experienced workers edged up slightly in the second half, as an annual average it remained below 2 percent for the second consecutive year. The strong demand for experienced workers was also reflected in the low jobless rate for married men, the most important group of breadwinners.

For men 20 to 24 years old, the unemployment rate fluctuated widely during 1969. The annual jobless rate for these young men, who are now entering the labor force in swelling numbers, remained at the 5.1 percent level of 1968, which was up from 4.6 percent in 1967. The extent of unemployment among them in the near future will depend not only on the availability of new jobs, but also on the rate at which they will be absorbed into and discharged from the Armed Forces.

Table 6.

Unemployment Rates for Major Labor Force Groups, 1967-1969

Group	1969	1968	1967
Total, all civilian workers	3.5	3.6	3.8
Men, 20 years and over	2.1	2.2	2.3
Men, 20 to 24 years	5.1	5.1	4.7
Men, 25 years and over	1.7	1.8	2.0
Married men	1.5	1.6	1.8
Women, 20 years and over	3.7	3.8	4.2
Women, 20 to 24 years	6.3	6.7	7.0
Women, 25 years and over	3.2	3.2	3.7
Teenagers, age 16–19 (both sexes)	12.2	12.7	12.9
White, total	3.1	3.2	3.4
Negro and other races, total	6.4	6.7	7.4

Adult Women

The unemployment rate for women inched up slightly during the year, after attaining a relatively low level in the first quarter. The annual rate however, was practically unchanged from 1968. The only significant 1969 improvement was registered by women 20 to 24 years of age—a group that has experienced the sharpest rise in labor force participation. Their jobless rate declined from 6.7 percent in 1968 to 6.3 in 1969. The rate for women age 25 years and over, on the other hand, remained at the 3.2 level of 1968.

Teenagers

Youths 16 to 19 years old continued to experience severe difficulties in securing employment in 1969, and their jobless rate remained substantially above 10 percent for the 16th consecutive year. While the rates for most adult worker groups attained very low levels in the first part of the year, the rate for teenagers did not decline much, hovering stubbornly around the 12-percent mark all year. The annual teenage rate was only slightly lower than in 1968 and thus not far below the levels of the early 1960's. Within the teenage group, the unemployment rate continued to be somewhat higher for girls than for boys. It also remained much higher for Negro than for white youngsters. Although many of the unemployed teenagers want only part-time work, the difficulties which these young persons encounter in finding a job remains one of the most vexing unresloved manpower problems, which assumes greater urgency due to the restiveness and alienation exhibited by members of this group in recent years.

Negroes

Relative to their white counterparts, Negroes and members of other minority races continued to experience serious problems in securing and holding a job. Although the Negro unemployment rate dropped to the lowest quarterly level for this decade in early 1969, it nevertheless continues to be about twice as high as the white rate. Averaged over the whole year, the Negro rate was 6.4 percent compared with 3.1 percent for the whites.

Several factors account for the disproportionately high incidence of unemployment among Negroes: they are handicapped in the job search by their lower median level of education and skills; their labor force includes a comparatively larger proportion of women and teenagers, two groups that are generally more vulnerable to unemployment than adult men; they are undoubtedly still the victims of some discriminatory practices.

Teenage Negro girls find it particularly hard to obtain a job. The unemployment rate for this group fluctuated around the 30-percent mark during 1969—nearly 3 times as high as the jobless rate for white girls. For Negro boys, the jobless rate hovered around the 20-percent mark—about double the white rate. The Negro jobless rates for adult males and adult females, 3.7 and 5.8 percent respectively, averaged somewhat less than double the rates for their white counterparts.

Occupational Groups

The unemployment rate for white-collar workers, who continued to expand their share of total employment in 1969, averaged 2.1 percent, slightly above the 2.0-percent level of 1968. The rate for blue-collar workers, on the other hand, declined slightly over the year, edging down from 4.1 to 3.9 percent. Within the blue-collar group, nonfarm laborers—the most unskilled group—again bore the highest unemployment in 1969. However, the latest annual rate for this group (6.7 percent) was somewhat lower than their jobless rate for 1968 (7.2 percent). (See table 7.)

For service workers, the jobless rate averaged 4.2 percent in 1969 compared with 4.4 percent in 1968. Within this group, however, workers engaged in private household work—an occupation declining rapidly in popularity—enjoyed the lowest unemployment rate on record in 1969. Farm workers, whose number is also declining steadily, had a jobless rate of 1.9 percent.

Characteristics of the Unemployed

The stereotype that most of the unemployed are men of prime working age who have lost their jobs does not represent the present unemployment situation. The composition of unemployment has changed substantially since the early 1960's, with primary male breadwinners now making up a substantially smaller proportion of total unemployment, and only two-fifths of the persons currently unemployed attributing their situation to job-loss.

Table 7.

Unemployment Rates by Occupational Group, 1967, 1968, and 1969

Occupational group	1969	1968	1967
Total	3.5	3.6	3.8
White-collar workers	2.1	2.0	2.2
Professional and technical workers	1.3	1.2	1.3
Managerial, officials, and proprietors	.9	1.0	.9
Clerical	3.0	3.0	3.1
Sales	2.9	2.8	3.2
Blue-collar workers	3.9	4.1	4.4
Craftsmen and foremen	2.2	2.4	2.5
Operatives	4.4	4.5	5.0
Nonfarm laborers	6.7	7.2	7.6
Service workers	4.2	4.4	4.5
Private household	3.6	3.9	4.1
All other	4.3	4.6	4.6
Farmworkers	1.9	2.1	2.3

Age-Sex-Color Distribution

Of the 2.8 million persons who were unemployed in 1969, nearly 1 million were adult men, another million were adult women, and 850,000 were teenagers. Of the unemployed adult men, one-half were 25 to 54 years old and thus likely to be their families' main breadwinners. About 570,000, or 20 percent of total unemployment, consisted of Negroes and members of other minority races. The following tabulation shows the percent distribution of the civilian labor force and unemployment in 1969:

Group	Civilian labor force	Unemployment
Total, all age groups	100.0	100.0
Adult men	57.4	34.0
Adult women	34.0	35.9
Teenagers	8.6	30.1
Total, all race groups	100.0	100.0
White	88.8	79.9
Negro and other races	11.1	20.1

The proportion of unemployment accounted by each group bore little relation to the group's share of the labor force. Because of the very high incidence of joblessness among teenagers and Negroes, these two groups accounted for disproportionately large shares of total unemployment.

Reasons for Unemployment

Data on the causes of unemployment indicate that job loss has accounted for only about two-fifths of recent unemployment. Most of the unemployed are persons who have either left their last job voluntarily to search for another one, or are entering or reentering the labor force, as shown in the following tabulation:

	1969	1968
Number unemployed (in thousands)	2,831	2,817
Percent	100.0	100.0
Lost last job	35.9	38.0
Left last job	15.4	15.3
Reentering labor force	34.1	32.3
Looking for first job	14.6	14.4

Only among adult men was job-loss the main reason for unemployment. Adult women cited reentering the labor force as the most common cause for unemployment. Looking for the first job is, understandably, the most common reason for teenage unemployment.[4]

Seeking Full-Time or Part-Time Work

About 700,000 (one-fourth) of the unemployed in 1969 sought only part-time work. These included 100,000 (about one-eighth) of the adult male unemployed, 200,000 (one-fifth) of the female unemployed, and 400,000 (nearly one-half) of the teenage unemployed. Most of the teenagers and many of the young adults seeking a part-time job are students. Most of the women seeking part-time work were housewives who wanted to boost their families' incomes while still maintaining their primary role as homemakers.

Household Status

Less than one-fourth of the unemployed were male heads of household. This is in sharp contrast to the situation in the early 1960's, when male heads of household accounted for well over one-third of the unemployed. Wives or other relatives of the household head constituted two-thirds of the unemployed, while female heads of household made up about 8 percent.

Occupational Distribution

Unemployment continued to be most prevalent among low-skill workers in 1969, with the unemployment rates for the individual occupations groups showing little change from 1968.

Although white-collar workers now hold practically one-half of the Nation's jobs, their unemployment rate continued relatively low (only 2.1 percent). Thus they accounted for only one-third of all the experienced unemployed. Blue-collar workers, being much more vulnerable to joblessness, accounted for about one-half of experienced unemployment. Within the blue-collar group, nonfarm laborers—the least skilled group—accounted for a particularly large proportion of the jobless.

4. For a detailed discussion of the reasons for unemployment, see Kathryn D. Hoyle, "Job Losers, Leavers, and Entrants—A Report On the Unemployed," *Monthly Labor Review,* April 1969, pp. 24-29.

Service workers had a jobless rate of 4.2 percent in 1969, accounting for nearly one-fifth of total unemployment, and farm workers had a jobless rate of only 1.9 percent, accounting for only 2 percent of the Nation's unemployed.

Duration of Unemployment

Most of the 2.8 million persons who were unemployed on average during 1969 were generally able to secure a job after searching for only a relatively short period. Only about one-third remained unemployed for more than 5 weeks and only one-eighth were still without a job after 15 weeks of search:

Duration of unemployment	Number (in thousands)		Percent distribution	
	1969	1968	1969	1968
Total unemployed -----------	2,831	2,817	100.0	100.0
Less than 5 weeks- --------------	1,659	1,594	57.5	56.6
5–14 weeks ------------------	827	810	29.2	28.8
15 weeks and over -------------	377	412	13.2	14.6
15–26 weeks --------------	242	256	8.5	9.1
27 weeks and over- -----------	133	156	4.7	5.5

Long-duration unemployment, which has been declining as a proportion of total joblessness for several years, was particularly low in early 1969. For the entire year, the number of persons remaining jobless for 15 weeks or more reached the lowest mark since the Korean War.

Geography of Unemployment

Newly available data show clearly that the burden of unemployment and underemployment was distributed very unevenly, not only among the various groups that make up the labor force, but also among geographic areas.[5] For example, the jobless rate is much higher in the West than in other areas of the country. It is also generally much higher for residents of central cities than for persons residing in suburbs.

Regional Pattern

Data for 1969 indicate that the West in general and the Pacific area in particular continue to carry a substantially higher unemployment burden than other regions of the country. This situation is probably attributable in large part to the continuous migration to the West of jobseekers from other areas of the country and to the initial delay they encounter in locating a job. The

5. See, for example, Howard V. Stambler, "New Directions in Area Labor Force Statistics," *Monthly Labor Review*, August 1969, pp. 3-9; Paul M. Schwab, "Unemployment by Region and in 10 Largest States;" *Monthly Labor Review*, January 1970, pp. 3-13; Paul O. Flaim, "Unemployment in 20 Large Urban Areas," *Employment and Earnings*, March 1969, pp. 5-18; Paul M. Ryscavage, "Employment developments in urban poverty neighborhoods," *Monthly Labor Review*, June 1969, pp. 51-56, Harvey J. Hilaski and Hazel M. Willacy, "Employment patterns by place of residence," *Monthly Labor Review*, October 1969, pp. 18-25.

following tabulation shows the percent of the civilian labor force unemployed or working part time for economic reasons, by region and color:

	United States	North-east	North Central	South	West
Percent unemployed:					
Total	3.5	3.2	2.9	3.6	4.9
Negro and other races	6.4	5.5	6.8	6.4	6.8
Percent limited to parttime work for economic reasons:					
Total	2.5	1.6	2.1	3.6	2.9
Negro and other races	5.1	2.6	2.9	7.5	3.1

Another interesting finding from the regional employment data concerns the high number of Negro workers in the South who are involuntarily limited to part-time work. While unemployment among southern Negroes does not exceed national averages, the percentage of Negro workers performing part-time work was twice as high in the South as in the other regions. The principal reason for this situation is that Negroes in the South are still heavily concentrated in the lowest skill occupations—such as household work or farm labor —where work is often not available on a full-time basis.

Metropolitan Areas

A special series of labor force data for the Nation's 20 largest metropolitan areas have shown clearly that unemployment is generally much higher in the central cities than in the surrounding suburban areas and that the rates also vary significantly from city to city. In 1969 the unemployment rate for the central cities of these 20 areas was 3.9 percent, while the rate for their suburban areas was only 3.0 percent. This compares with 1968 jobless rates of 4.1 and 2.9 percent, respectively. The suburban areas are mainly white, while some central cities are becoming predominantly Negro. Over one-third of total Negro unemployment in the Nation is concentrated in the central cities of these 20 areas.

The slight improvement in the unemployment situation for central city residents appears to reflect in part a decline in joblessness among persons residing in the poorest urban neighborhoods. Based on data for the 100 largest metropolitan areas, the jobless rate in the poorest one-fifth of the urban neighborhoods edged down from 6.0 percent in 1968 to 5.5 percent in 1969.

The metropolitan areas with the highest unemployment in recent years have been Los Angeles-Long Beach and San Francisco-Oakland in the West and Pittsburgh in the East. The jobless rate has been running well above 4 percent in each of these three areas. In a few other metropolitan areas, on the other hand, unemployment has been exceptionally low in recent years. In Boston, Dallas, Minneapolis-St. Paul, and Washington, D.C., for example, the jobless rate has averaged only around 2.5 percent.

Slums

In order to pinpoint the employment problems of persons residing in the poorest urban areas, the U.S. Department of Labor initiated a special study in July 1968 (known as the Urban Employment Survey) conducted in Concentrated Employment Program (CEP) areas of six large cities—Atlanta, Chicago, Detroit, Houston, Los Angeles, and Detroit.[6] Initial findings from this study indicate that the residents of these areas, who are mainly Negroes or Spanish-Americans, not only have generally high unemployment rates, but also are concentrated in the less desirable occupations that often provide only intermittent work and thus a low level of annual income. The study also revealed, however, that the situation varies widely from city to city. The areas with the highest unemployment, as was the case in Detroit, were not necessarily those with the lowest incomes. The lowest level of weekly earnings and family income was found among the residents of the Atlanta and Houston target areas.

Other Areas

While concern is justly focused on the troubled urban scene, unemployment problems are still in evidence even outside of urban areas.

The unemployment rate for workers residing in small towns runs somewhat above the national average; it was 3.9 percent in 1968 and 3.8 percent in 1969. The rate for workers residing on farms, on the other hand, was only 1.6 percent both in 1968 and 1969.

Workers living on farms, however, are much more likely to be employed only part time or as unpaid family workers. Although their low unemployment situation may not indicate economic problems, the continuous exodus of farm residents to the cities is a clear indication of the lack of reasonably attractive employment opportunities in these areas.

Other Employment Problems

Being without a job is not the only problem confronting a worker. He may, for example, want full-time work but be confined involuntarily to a part-time job where his earnings may not be commensurate with his capacity.

Involuntary Part-Time Work

In 1969, about 1 million workers on average wanted full-time work but were able to locate only a part-time job. Also, about 1.1 million workers

6. See Norman Root, "Urban Employment Surveys: Pinpointing the Problem," *Monthly Labor Review,* June 1968, pp. 65-66. The initial findings from this survey were summarized in BLS Report No. 370, October 1969. Individual reports for each of the six cities where the survey is being conducted may be obtained from the regional offices of the Bureau of Labor Statistics.

supposedly employed full-time were limited to less than 35 hours of work per week because of economic factors (shortages of material, reduced orders, etc.). The number of workers confined to part-time employment was particularly low during the first half of 1969, but in the slower second half of the year the number increased significantly. On an annual basis, their average number was about the same as in 1968.

Discouraged Workers

In addition to workers who are unemployed or underemployed, there is another group of persons who have long worried manpower experts: "discouraged workers" who want jobs but who feel that any search for work on their part would be futile. Since these persons do not take overt steps to look for work, they are not included in the unemployment count, being viewed as "out of the labor force."

Through special questions added to the Current Population Survey questionnaire in 1967, it is now possible to identify and count such persons on a regular basis. They averaged about 700,000 in both 1967 and 1968, but their number dropped to about 600,000 in 1969, reducing the proportion of "discouraged workers" relative to unemployed workers from 1 to 4 to 1 to 5.[7]

Discouragement over job prospects is a serious problem only among the very young and the old. Persons of prime working age—particularly men—have included very few discouraged workers in recent years. Out of approximately 210,000 adult men in 1968 and 180,000 in 1969 who wanted work but felt that they could not find a job, only about one-third were between 20 and 59 years of age. It can thus be said that discouragement over job prospects has kept relatively few persons of prime working age from the labor force in recent years.

7. Detailed data on persons not seeking work because of discouragement over job prospects—as well as other specific reasons—are now being published quarterly in *Employment and Earnings*, with the first series of tables having appeared in the December 1969 issue. For a discussion of these new data, see also Paul O. Flaim, "Persons not in the labor force: who they are and why they don't work," *Monthly Labor Review*, July 1969, pp. 3-14.

PART III

An Era of Change

INTRODUCTION

We can read in the Bible that "In the beginning, God created the heaven and the earth," and that it was good. No matter how it is thought that the earth was formed or evolved, there was a world with uncharted oceans, untilled soil, untouched natural resources and unharnessed energy.

In recent years many of the secrets of the earth have been disclosed. By the discovery of these secrets, man has created a civilization that sustains over three billion people. This tremendous feat has been accomplished not by adapting to the environment but rather by adapting the environment to fit our needs. The change came about, basically, during the last 150 years, and is so complete that man now appears to be the master rather than the slave.

Nature exists and thrives on a delicate thread of balance. If any link of that environmental balance is broken, a chain of events occurs in reaction. It appears that nature's balance has been disturbed to the point that the damage may be irreversible; or at least to the point where the quality of life is threatened and/or diminished.

This section deals with an era of change. Change caused by many factors such as (1) increased population, (2) increased efficiency in the production of goods and services, (3) the gap between the haves and have-nots, and (4) changing attitudes. The issues of the world today are numerous and complex. However, by familiarization with some of the issues, the reader will gain insight and awareness toward others.

The Complications of Change

Donald P. Lauda

> In an important sense this world of ours is a
> new world, in which the unity of knowledge, the
> nature of human communities, the order of so-
> ciety, the order of ideas, the very notions of
> society and culture have changed and will not
> return to what they have been in the past. What
> is new is new not because it has never been there
> before, but because it has changed in quality.
> One thing that is new is the prevalence of new-
> ness, the changing scale and scope of change
> itself, so that the world alters as we walk in it,
> so that the years of man's life measure not some
> small growth or rearrangement or moderation
> or what he learned in childhood, but a great
> upheaval ... To assail the changes that have
> unmoored us from the past is futile, and in a
> deep sense, I think it is wicked. We need to
> recognize the change and learn what resources
> we have.[1]

Recent technological advances have inspired curiosity not only in the realm of research and scholarly study but in the layman's world as well. The study of man's surroundings is not new, of course, but until today it has been primarily the concern of the research scientist and the educational elite. Now the advance of modern technology has become a matter of concern to the lay person, as well as the scholar. Mass media has brought an awareness to many people; it has introduced new concepts which create anxiety in many, rejection of ideas by some, but hopefully in most cases acceptance of those tangibles and intangibles which better mankind. The 20th century is the age of change, the age of electronics, computers and automation. Yet, not all can understand these concepts and this lack of knowledge can create disharmony in society.

This preoccupation with technological achievement is also at the apex of contemporary thought due to its economic interpretations. With the United States gross national product (GNP) hovering at one trillion dollars, man plays an ever widening role in modern technology. Whether he agrees with the philosophies of modern technology or not, he is literally "locked" in a world which is based on machines and scientific truths, which not only provide him with an income, but mold his family unit and direct his life. The question that man must now answer is not how good progress should be, but how much

1. Robert Oppenheimer, "Prospects in the Arts and Sciences," *Perspective USA*. Spring, 1955, 11: pp. 10-11.

259

progress is desirable. No event, social, economic, or spiritual, is so broad in scope as technological change. To most people the technology represents recent events, when actually technological change has been present on earth since the beginning of time. Change in the 20th century is more rapid but it is by no means a new concept. Modern means of reporting data, access to materials never before known to man, coupled with more scientific data on a readily available basis has created change at an ever-increasing rate.

Every element in man's culture is subject to change and in some cases change without notice. Acceptance of change will vary from immediate acquiescence to total rejection. It is this gap that creates many of the ills in modern society. Yet, it is this gap that may save mankind from self-destruction. If one studies the population pattern in the United States and other countries he finds that graphically the chart looks like a pyramid. The base of the pyramid represents the new born and the peak those people who are aged. The vast majority of the population is under the age of twenty-four in the United States. In the undeveloped countries of the world at least 40 percent of the population is under the age of fifteen.

It is the base of this pyramid that represents individuals who have not had time to acquiesce to technology. These persons were born in the 1950's and 1960's. If we accept the year of 1950 as when "radical technological change" began it is evident that these people were born into change. Contrarily, their parents accepted the changes slowly, tolerated the first, second, third and even the fourth order effects, and can view these consequences without questioning our priorities.

On the other hand, the youth who have not witnessed this growth have the capability to take a fresh look at society and can see society's ills. It is this reaction that has prompted many of the campus upsets in the past few years. If it were not for those forward-objective looking individuals en masse it is doubtful that our society would have the meager social legislation it has to control racial prejudice, pollution, and scholastic irrelevancy.

Nation after nation is caught in the modernization process of the world, each having special identifying characteristics based upon their geographical location, natural and human resources, economic stability, and date of entrance into the "era of modernization." Few areas of the earth have escaped invasion by man's desire to control the forces of nature, and to control human and material resources for the betterment of life on earth. It cannot be denied that some countries have progressed further than others but careful examination of this progress reveals that it is based upon certain universal factors; mainly economic, natural, political and most of all human. No society has or can every hope to progress in the ranks of modernization without these factors prevalent.

Change is inevitable; history has proven this over and over again. Ideally, with controlled change comes progress. Resistance to advancement stifles mankind, yet many people in the world follow the philosophy of the middle-aged farmer who said, "I don't mind progress but I sure hate change." Why does man resist the forces that are designed to make his life easier and more profitable? Is it because he does not understand change or could it be that he is bound by traditional or religious tenets?

Change has always been a part of human life. The only difference today is that it comes faster; carrying with it a backwash of shifting values, morals and religious beliefs. The student of modern technology cannot restrict himself to the study of the tangible items that affect life. He must also resolve in his mind how to handle the social consequences of these technological developments occurring throughout the world. For example, the United States was comfortable with its control of atomic power until recent decades. Now many other nations have the same power. This results in a revision in values concerning its use. Medical innovations have made organ transplant a reality but with such developments come moral, spiritual and economic judgments that must be made by man. As sophisticated as the computer has become the final diagnosis of society's ills is left up to man alone.

In societies that experience little technological change, values and human judgments become static, hence little social adjustment is necessary. However, with 20th century communications, capacity for travel and the desire for economic development, few if any areas of the world experience the tranquillity of such life.

Structure of Culture

Though every culture has fundamental elements, many sociologists divide the content into two basic components, the material and the non-material culture. On the material side of the ledger we find the tangible items that are part of our culture such as the machines, tools, and countless other items man uses either for self-perpetuation or pleasure. The non-material segment of our culture is more difficult to isolate. No inventory could completely isolate the ideas and norms of a given society. Many are subliminal, others are mere legend and still others are accepted only by a minute segment of the society's population. For the most part, however, man has accrued a set of heterogeneous concepts that he follows; certain ideas which control his behavior. Any discrepancy or diversity in the actions of an individual or group within this society is disruptive. If this difference takes the form of resistance to change then technological and social progress is hampered.

Thus far this discussion has dealt with the degree of man's ability to adjust to change with indications that in some cases he cannot readily adapt to it. It cannot be denied that industry and modern technology play an important role in forming the social problems of our time. The labor movement, for example, with its strikes, walkouts and occasional violence are a part of the process we call industrialization. Our nation's economy is determined by modern industry and the resulting social problems are so complex that they completely confuse the worker in these industries. Paramount among the complexities caused by automation are statistics about the unemployment rate, lack of vocational training and re-training for thousands, problems of working mothers, among others. These and countless other problems are the by-products of technological progress. The understanding of modern society requires first a comprehension of industrial institutions, their history, contemporary status, future prospects and the inevitable social consequences that will result as modernization and industrialization continue to unfold before us.

Culture may be analyzed in another perspective than the material and the non-material. Ralph Linton[2] in *The Study of Man* suggested that culture be studied in terms of three categories—universals, specialties and alternatives. With slight modifications his analysis will be followed here to help understand how technology affects a culture and vice versa.

An analysis of any population will reveal that there are those elements which are universally distributed and accepted by the people. For example, in a given society the populace will wear similar clothing, utilize the same type of transportation, eat the same types of food and abide by the norms determined by their culture. These concepts are accepted by most members of the society and are called universals. In some cases adherence to these universals is assured through laws, whether written or unwritten. Anyone who studies man and his relationship with culture must be aware that universals will vary from society to society; therefore, he should not let his own ethnocentrism color his judgments of other societies.

In these same societies there are elements which influence only a certain, though relatively large, portion of the population. This is usually true in regard to the adult population and their occupational choice or avocations. The United States Department of Labor's *Dictionary of Occupational Titles* (DOT) has over 35,000 different job classifications listed and defined. This data reveals that people are employed in different areas, the most common of which are the professional group and the technical group. Yet these two groups can be broken down into any number of different but related occupations. These various tasks performed by certain segments of the population are called the specialties. As a consequence of these specific groupings a variety of outlooks and social structures result yet at the same time most people have at least a general understanding of the role of these individual specialties. An assembler in a manufacturing plant may not be able to transplant a human heart but he does understand the physician's role in society, at least superficially. He also knows to a great extent what behavior is expected of the physician.

The third segment of culture, according to Linton, is distributed through only a small part of the entire society. In fact, it may not even be present in

---

2. Ralph Linton, *The Study of Man* (New York: D. Appleton-Century Co., Inc., 1936), pp. 271-287.

many groups, or if it is in evidence it may not be perceptible to the greater majority of that group. This fraction of a given society's population is known as the alternatives, that is those elements which depart from the commonly accepted norms and practices of the culture. Included among the ranks of the alternatives we would find the extremists and reactionaries. An example of an alternative might be a new fad, which might not be accepted by many people.

In many cases the alternatives eventually become universals, giving balance and harmony to culture. However, if a society has only a few alternatives it will have a relatively bland atmosphere with little dissension or trouble. As alternatives increase the culture becomes disrupted and must eventually make a choice whether to allow the universals or specialties to either absorb the alternatives or reject them. As man continually imposes one culture or sub-culture upon another he can expect disharmony. Could it be that modern technology and its many ramifications is responsible for the creation of new alternatives or at least the transmission of alternatives from culture to culture throughout the world through the marvels of mass media? One might also ask whether modern man has too heavily tipped the scales with alternatives which have created many of the social problems in the 20th century. When a rapid influx of alternatives begins to destroy established norms the inevitable outcome will be dissension among the people, especially those who are older and set in their ways of thinking and acting. At the same time, those who accept the new alternatives often experience frustration, insecurity and guilt feelings about departing from established norms.

Since the beginning of time, tangibles, once discovered, are presented to society for approval. Obviously many of these are never accepted by the population due to ignorance, fear, or other reasons. Before the items which apparently please society or which society is forced to tolerate, are accepted there is generally a period of time during which they are, as it were, on trial. Man being man is reluctant to accept materials without first analyzing them, experimenting with them, and above all, evaluating the new product or discovery in terms of its ability to increase his comfort. This interim period between the appearance of a material on the market and its acceptance by the society is known as cultural lag.

This period of acceptance is part of the process of change for it cannot be emphasized too strongly that change does not come readily. Man has subconscious armour built of ideas he uses to lengthen this period of transition. A flat refusal to accept change can be a tragedy in those cases where such a refusal endangers lives. Likewise, substitutes for a new innovation also may endanger man and his surroundings. The reader should not get the impression that change for the sake of change is desirable since it cannot be denied that this period of time also has its utility in so far as it allows man to refine, delete and analyze further his new development. It is, however, man's refusal to accept materials introduced to him even after they are proven trustworthy and worthwhile that concerns the student of modern technology.

The question that must be answered is how much change can a society adapt to and how fast must it come before the disharmony becomes so great that it destroys that society. It seems logical that to understand change one must understand what forces are at work resisting it. Table 1, below, lists a few of these forces. Although the list of factors which resist change is by no means complete, it does represent some of the causative factors which prevent man's ready adoption of technological events.

Table 1.
Forces that Work to Resist Change in Any Society

IDEAS	NORMS
Religious doctrine	Common law
Superstition	Statutes
Stereotypes	Mores
Myths	Customs
Misconceptions	Folkways
Ignorance	Group pressure
Values	
Fear	

Effects of Change

Assuming that change takes place normally and overcomes cultural lag the total process is still by no means complete. In fact, this cycle of technological growth is never complete while man is available to revise, re-place or reject according to his wishes. Some person or group must produce every invention known to man although it is predicted that soon the computer will serve this function. It is ironic that although we give credit to one person for an invention it probably would have been invented even if that person had not existed. If this were not so the United States Patent Office would be relieved of one of its most important functions; to determine who developed an idea or material first.

Change is a concept that is broad in scope, in many cases too broad for comprehension. Still it is a definite part of our culture and must be understood if we expect harmony to prevail among the people. Alternatives, if understood, do not degenerate a society but rather give it that vibrant quality that signifies progress. Could it be that a lack of understanding causes people to rebel against newly set norms? Are people going through life today actually saying, "Please don't bend, fold, staple or mutilate me"?

Solutions to the Acceptance of Technology

The questions man must ask himself are, Will the gap between technology and social acceptance continue to widen? Will the chasm lead to total social disharmony? Will destructive non-conformity be the basis for the present generation?

Today we cannot read a newspaper or periodical without being confronted with a statistic about modern technology. Yet, we rely on our line with the past and our accumulation of knowledge from a preceding era. Alvin Toffler[3] has stated, "... and simple observation of one's own friends and associates will confirm it ... that even the most educated people today operate on the assumption that society is relatively static. At best they attempt to plan by making simple straight-line projections of present trends. The result is unreadiness to meet the future when it arrives. In short, 'future shock'."

It would be wrong to leave the impression that all people resist change and resist it to the same degree. For the most part, people accept change: if this were not true the world would not have gone through the dramatic changes it has, especially since the Industrial Revolution. It has become common to say, "People resist change," but it is easier to support the idea that "People accept change." It would be ridiculous to state that every generation acts and lives like the one before it. However, the student of human nature should constantly keep in mind that change always remains and no man lives in a single generation. We might rather say that we all pass through many generations, technologically speaking, in our lifetime.

The key to the acceptance of change is the understanding of that portion which affects us directly. Since we cannot control change we seek to control the rate at which it affects our lives and our basic institutions. Man can accept change as long as it does not threaten his basic securities, he can welcome it if he understands its nature, but he cannot accept it if it is forced upon him by society.

Never before has society made as many attempts to control the forces that create change as it has in the 20th century. It has been only in the past few decades that we have established basic institutions for controlling change as well as promoting it. Yet, still other parameters may have to be placed on the changing world since our traditional institutions cannot withstand the influx of the innovation anticipated in the future. Whether these controlling forces can be imposed by the government, religious or military groups is a question that will have to be answered by the people. It is probable that entirely new institutions will have to be established; institutions that can withstand the pressures of concepts which today seem inconceivable.

The role of education in this era is the key to our adaptation to change. Educators have the sole responsibility of altering not only teaching methods but our concept of what is relevant and what is irrelevant. Although history reveals many examples of stagnation or deterioration in cultures it can safely be stated that cultural change progresses steadily and persistently. Stagnation and deterioration are only further evidence of reluctance to change and it is

3. Alvin Toffler, "Future as a Way of Life," *Horizon,* Summer, 1965, pp. 108-115.

the task of man in the 20th and 21st century to guide progress in the direction that benefits all men rather than allowing it to grow to uncontrollable proportions, that could eventually destroy man.

Jacques Ellul[4] put it so succinctly when he said, "But what good is it to pose questions of motives? of Why? All that must be the work of some miserable intellectual who balks at technological progress. The attitude of the scientists, at any rate, is clear. Technique exists because it is technique. The golden age will be because it will be. Any other answer is superfluous."

Questions for Study and Discussion

1. Can any meaningful generalizations be made about the impact of technology on the quality of the life of the individual citizen, his family and the community?

2. What adjustments, if any, are required of man so that he can benefit from the potentialities of the advances in science and technology?

3. It has been stated that technology does not change society, rather it destroys it. Is this true?

4. Define cultural lag giving a concrete example, as well as the means to avoid such a lag or at least to help man to adjust to the cause of the lag.

5. Why doesn't modern technology solve the problems of our slums, unemployment and other social conflicts of the 20th century?

6. How can the rate of technological achievement be measured?

7. Is it realistic to consider only the economic basis of change without concern for the social consequences?

8. What do rapid advances in modern technology imply for education at all levels?

9. What adjustments are called for as we attempt to adjust to the changes made in the past twenty years?

Source References

Allen, Francis R., Hart, Hornell, Miller, Delbert C., Ogburn, William F. and Nimkoff, Mayer F., *Technology and Social Change,* New York: Appleton, Century-Crofts, Inc., 1957.

Bois, J. Samuel, *Explorations in Awareness,* New York: Harper & Row Publishers, 1957.

Chase, Stuart, *The Proper Study of Mankind,* New York: Harper and Brothers, 1956.

Ellul, Jacques, "Future As a Way of Life," *Horizon,* Summer, 1965.

Fabun, Don, *The Dynamics of Change,* Englewood Cliffs: Prentice-Hall, Inc., 1967.

4. Jacques Ellul, *The Technological Society* (New York: Alfred A. Knopf, 1964), p. 436.

Hoffer, Eric, *The Ordeal of Change,* New York: Harper & Row, 1963.

Linton, Ralph, *The Study of Man,* New York: D. Appleton Century Company, Inc., 1936.

Mumford, Lewis, *Technics and Civilization,* New York: Harcourt, Brace & World, Inc., 1963.

Oppenheimer, Robert, "Prospects in the Arts and Sciences," *Perspectives USA,* Spring, 1955.

Toffler, Alvin, "Future As a Way of Life," *Horizon,* Summer, 1965.

Ways, Max, "The Era of Radical Change," *Fortune,* May, 1964.

The World Alters as We Walk in It

Don Fabun

One foot inextricably trapped in the clockwork mechanism of 19th Century science, and the other planted fearfully in the newly radiant soil of the 20th Century, Henry Adams, then 60 years old—and a student of change for forty of those—stood in the Gallery of Machines at the Great Paris Exposition of 1900 and saw—more clearly than most men of his time, or of now—one of the great fracture points in human history.

It was to be another sixty years and two World Wars later before the dimensions of the change he saw had become the common currency of popular journalism and awareness of change an accepted tool for survival.

For most of his life, Adams had been trying to make some sense out of history. It had not been an easy quest, for history appears to be what we want to make of it, and Adams found he could not make much of it.

In his *Education,* Adams said, "Satisfied that the sequence of men led to nothing and that the sequence of society could lead no further, while the mere sequence of time was artificial and the sequence of thought was chaos, he turned at last to the sequence of force, and thus it happened, after ten years pursuit, he found himself lying in the Gallery of Machines . . . his historical neck broken by the irruption of forces totally new."

What Adams had seen was that change is observed motion, and motion is the product of applied force. Man had begun by acquiring fire—and that was a force—and then sometime later the use of wind and water, and then he put water and fire together and turned the steam engine loose on the world. The period of time between each new acquisition and application of power was successively shorter. It was only a little after steam that he began applying electrical force and even more shortly after that discovered and put to work nuclear force.

By 1900, the power of the electrical age was best symbolized in the dynamo, a force that Adams found parallel to the force of the Christian Cross in the affairs of men. And, in the discoveries of Roentgen and Curie, in the hidden rays of the suprasensual, the dimensions of a vast new world powered in its course by the dance of electrons, were—at the turn of the century— already visible to those few pairs of eyes curious enough to see them.

It was possible to plot the progress of these new forces in graphs, and when this was done, and the different graphs compared, it was seen that—in almost any application of force you wanted to measure—there was a constant acceleration; the changes became larger and they occurred more frequently as we moved forward in time.

Several years ago, *Scientific American* plotted Adams' "Law of Accelera- tion." Graphs were made up of such processes as the discovery of natural forces, and the time lag between each successive discovery; tables were made that plotted the isolation of natural elements, the accumulation of human experience, the speed that transportation has achieved from the pace of a man walking to space satellites, and the number of electronic circuits that could be put into a cubic foot of space. In every case, the rising curves on the graphs showed almost identical shapes, starting their rise slowly, then sharper and sharper until, in our times, nearly every trend line of force is embarked on a vertical course.

Some of the idea of the dimensions of change in our times, and the acceleration of it, can be found in the fact that, "Half of all the energy consumed by man in the past two thousand years has been consumed within the last one hundred." Kenneth Boulding, the economist and writer, finds that, "For many statistical series of quantities of metal or other materials extracted, the dividing line is about 1910. That is, man took about as much out of mines before 1910 as he did after 1910."

The picture of our world that emerges is as if all the rockets at Cape Kennedy were to go off at once, in some grand Fourth of July, and their skyward-soaring trails were the trend lines of our exploding technology.

Writing on "The Era of Radical Change," in *Fortune* magazine, Max Ways has said, "Within a decade or two it will be generally understood that the main challenge to U.S. society will turn not around the production of goods but around the difficulties and opportunities involved in a world of accelerating change and ever-widening choices. Change has always been a part of the human condition. What is different now is the pace of change, and the prospect that it will come faster and faster, affecting every part of life, including personal values, morality, and religions, which seem most remote from technology.

"So swift is the acceleration that trying to 'make sense' of change will come to be our basic industry. Aesthetic and ethical values will be evolving along with the choices to which they will be applied. The question about progress will be 'how good?' rather than 'how much?' "

He goes on to point out that, "The break between the period of rapid change and that of radical change is not sharp; 1950 is an arbitrary starting date. More aspects of life change faster until it is no longer appropriate to think of society as mainly fixed, or changing slowly, while the tide flows around it. So many patterns of life are being modified that it is no longer useful to organize discussion or debate mainly around the relation of the new to the old.

"The movement is so swift, so wide and the prospect of acceleration so great that an imaginative leap into the future cannot find a point of rest, a still picture of social order."

We are told that 25 percent of all the people who ever lived are alive today; that 90 percent of all the scientists who ever lived are living now; the amount of technical information available doubles every ten years; throughout the world, about 100,000 journals are published in more than 60 languages, and the number doubles every 15 years.

We are told these things, but we do not always act as if we believed them. "The fact is," says Alvin Toffler in *Horizon* in the summer of 1965, "—and simple observation of one's own friends and associates will confirm it—that even the most educated people today operate on the assumption that society is relatively static. At best they attempt to plan by making simple straight-line projections of present trends. The result is unreadiness to meet the future when it arrives. In short, 'future shock'."

"Society has many built-in time spanners that help link the present generation with the past. Our sense of the past is developed by contact with the older generation, by our knowledge of history, by the accumulated heritage of art, music, literature and science passed down to us through the years. It is enhanced by immediate contact with the objects that surround us, each of which has a point of origin with the past, each of which provides us with a trace of identification with the past.

"No such time spanners enhance our sense of the future. We have no objects, no friends, no relatives, no works of art, no music or literature that originate in the future. We have, as it were, no heritage of the future."

And so not having one, and needing it, we will have to develop one. This can be done, perhaps, by examining the forces of change around us and by trying to understand how they originated, where they are likely to be going, and how we can to some extent, by guiding them, cushion ourselves against "future shock."

We might begin by seeing ourselves in a somewhat different relationship to time than we are accustomed to. We can agree that there is not much we can do to affect the past, and that the present is so fleeting, as we experience it, that it is transformed into the past as we touch it. It is only the future that is amenable to our plans and actions. Knowing this, we can draw a broad general outline of the kind of future world we feel we would be most happy in. And because we have now arrived at a stage in our development, or shortly will arrive there, where our most pressing problems are not technological, but political and social—we can achieve the world that we want by working together to get it.

The forces of change, which we will shortly begin to discuss, are amenable to our guidance. It we seem to be hurried into the future by a runaway engine, it may be that the main reason it is running away is that we have not bothered yet to learn how it works, nor to steer it in the direction we want it to go.

Era of Radical Change

Max Ways

Within a decade or two it will be generally understood that the main challenge to U.S. society will turn not around the production of goods but around the difficulties and opportunities involved in a world of accelerating change and ever widening choices. Change has always been part of the human condition. What is different now is the pace of change, and the prospect that it will come faster and faster, affecting every part of life, including personal values, morality, and religion, which seem most remote from technology.

The condition is man-made; and everybody has some share of responsibility for it, not the least, the U.S. industrialist. For many changes come about through the business system, which has an active role to play between the discoverers on one hand and the consumers on the other. Within corporations hundreds of techniques, arising from scores of separate scientific and technological disciplines, are drawn together through complex management structures. Here all kinds of values arising from the individual initiative and responsibilities of corporate managers and specialists are somehow integrated. A larger and more intricate mediation of values and purposes occurs in "the market," meaning thousands of interconnected markets, where the public exercises ever increasing power through billions of daily decisions. The resultant of all these corporate and consumer decisions alters the very conditions of life.

So swift is the acceleration that trying to "make sense" of change will come to be our basic industry. Aesthetic and ethical values will be evolving along with the choices to which they will be applied. The question about progress will be how good rather than how much. Already this shift away from purely materialist and quantitative criteria is well advanced. Change is called "excessive" when it appears to outrun ethical or aesthetic patterns. In the conflicts that arise on this point there are dangers not only for the business system but also for the democratic constitutional state and for the hope that the spirit of individual man can enlarge its freedom.

Not long ago Alvin Pitcher, of the University of Chicago's divinity-school faculty, writing with anguished eloquence in the *Harvard Business Review,* asked, "How much flux can man stand?" Not this much, he said, calling for a slowdown of automation and other socially disruptive and "needless" changes. Pitcher's anxiety was broader than the usual fear of such economic consequences as unemployment. He associated the present unprecedented mobility of American society with juvenile delinquency, with the dissolution of communities, with a barrenness in individual life. Man, who needs a measure of order and stability, is being dehumanized, he said, by excessive change.

Most American political and social issues today arise, like Pitcher's protest, out of concern over the pace and quality of change. In many cases, the protest is accompanied by proposals that government restore order by taking some additional degree of control. Historically, it would be ironic if the long struggle of the individual vs. the state should issue as a program for protecting the individual by having government take charge of change. Practically, it is quite hopeless to expect a central government to perform well a task requiring a high degree of flexibility, decentralization, and willingness to accept risk. But the argument against statism will not prevail as long as we think that responsibility for coping with change must be assigned either to the government or else to the naked, isolated individual.

"The Middle Tier"

This way of framing the choice is set up by a Jeffersonian tradition (endorsed by Lincoln, Richard Nixon, and Lyndon Johnson, among others), which holds that the government should do only those things that the people cannot do for themselves. But in the American society of 1964, what can the people as individuals do for themselves? Each man can grasp only a few of the disciplines in which knowledge is divided. No individual, by himself, can sustain his present level of living. Most obviously of all, no individual can cope with radical change. If it's a choice between the isolated individual and government—then government had better do it.

But, of course, that isn't the real choice. In Jefferson's day organizations other than government were thin on the ground, small and simple. In the last hundred and fifty years they have proliferated in numbers, grown huge in size, and, most important, have so evolved as to widen the scope of individuals working within them and of individuals dealing with them from outside. Organizations making up this "third area" or "middle tier" include, in addition to business corporations, local government services, voluntary organizations, labor unions, philanthropic foundations, and universities. These last generate in their research centers most of the scientific discovery that is later transferred into technological change. They also educate nearly all the managers of the sister organizations and the intellectuals who try to express the patterns of value emerging in the society—or to improve those patterns by criticism.

No judgment can be made that even the amazing organizational fertility of recent American society is or will be adequate to cope with radical change. The point is simply that if this "middle tier" of institutions is bypassed or fails, society will fail. Shifting the venue of the struggle by putting the responsibility on government does not solve the problem, because socialism, too, would need criteria on which to decide how much change was "excessive" or "needless" and what kind of change was "disruptive." The police power works well only where norms are settled.

Unless it is stopped by the deliberate exercise of power, the pace of change will continue to accelerate. *Fortune* is proud to publish in this issue, as it was to publish a similar list in 1955, some projections of development expected by a practical visionary, R.C.A.'s Chairman David Sarnoff (see page 281). Most of General Sarnoff predictions relate to developing technological application of scientific discoveries already made.

A study of General Sarnoff's list of what technology will be able to do will raise a number of non-technological questions of whether society *should* do them and, if so, how. There is, for instance, the possibility of improving the human race by deliberately altering genes. Aesthetic as well as ethical questions are involved in *that.* Will we wish to accept this opportunity at all and what should the rules be if we do?

No widely held system of ethics could possibly contain in definite, ready-to-use form the norms by which to evaluate many of the possibilities ahead. This is all the more true since the uses to which new developments will be put cannot always be foreseen, nor can their indirect effects on society. Ideas of "the good life"—in both its ethical and aesthetic meanings—will not be static. There will have to be a lot of tentative judgments, a process of probing and testing, a continuous comparison of purposes, consequences, and priorities.

Four Conditions of Change

Most people would agree with Alvin Pitcher that American society should be doing a better job of coping with change. Perhaps the performance would be better if Americans did not underestimate how different the present condition is from anything in the past. To stress the difference, four categories are set up—(1) gradual change, (2) revolution and disruption, (3) rapid change, (4) radical change.

Gradual change. Immense alterations in the human condition have occurred in the past without much occurring in the life of any one man. A language, for instance, can evolve from the most limited and primitive form to one rich and complex without any generation's being aware of the process. When stable societies are conscious of change, it is almost always of a particular change, directly affecting one part of life and requiring mere adjustments in some other parts. If a new trade route is opened or a new god appears in the pantheon, the social task is accommodation. Order is restored as nearly as possible to the previously existing equilibrium. Even those groups in a stable society that deliberately try to bring about a change have in mind from the start their terminal point, their new equilibrium.

Revolution and major disruption. These occur in all periods of history and the changes involved can be both rapid and general. But from the beginning a revolutionist thinks he knows the order he wants to establish. The wheels will turn and come to a point of rest. The bourgeoisie and its values will replace the aristocracy and its values; or the workers will replace the bourgeoisie. Thrust will be followed by consolidation. Such a disruption as foreign conquest tries to superimpose the order of the conqueror on the conquered. Subsequent politics are efforts to restore an equilibrium.

Rapid change. The period 1800–1950 in some Western countries saw for the first time an open vista of rapid and general change. No terminal point was in view once "the method of invention" had been invented. Debate over politics and social policy tended to polarize, with those who clung to the old challenging those attracted by the new. From 1850 until 1914, there was a growing tendency to see "progress" as a kind of inevitable and beneficent tide, lifting up all the familiar patterns of society as if they were a colony of houseboats, leaving them unchanged in themselves and in relation to one another. Dissenters saw the tide as inexorably receding, and the familiar institutions as headed for the mud. The period's most successful teacher of revolutionists, Karl Marx, also believed in inevitability. His new order would not be like any particular previous order but was to conform with patterns he thought he saw in the general laws of history; the victory of the proletariat was preordained by historical truth.

Radical change. The present period has been called "post-modern" and "post-industrial"—and even "post-Christian" and "post-Freudian." (It is unlikely to wear for long the belittling prefix "post.") The break between the period of rapid change and that of radical change is not sharp; 1950 is an arbitrary starting date. More aspects of life change faster until it is no longer appropriate to think of society as mainly fixed, or changing slowly, while a tide flows around it. So many patterns of life are being modified that it is no longer useful to organize discussion or debate mainly around the relation of the new to the old. So many old landmarks have been set in motion that they have become misleading as guides. Newness has become an even more treacherous beacon. In the late-nineteenth century or early-twentieth century, "to be up to date" was a boast. In 1964 the very phrase sounds dated, for everyone knows that to be up to date means to be on the verge of becoming out of date.

Nor can men in 1964 plant their feet firmly in a foreseen future. Inevitability disappears. While mere revolution (we have got to get used to thinking of revolution as "mere") can believe itself capable of forecasting a new order, radical change cannot do this. The movement is so swift, so wide, and the prospect of acceleration so great that an imaginative leap into the future cannot find a point of rest, a still picture of social order.

Robert, Charles, and Gabriel

Robert Oppenheimer has expressed the break with former eras quite succinctly. "This world of ours," he said, "is a new world, in which the unity of knowledge, the nature of human communities, the order of society, the order of ideas, the very notions of society and culture have changed, and will not return to what they have been in the past. What is new is new not because it has never been there before, but because it has changed in quality. One thing that is new is the prevalence of newness, the changing scale and scope of change itself, so that the world alters as we walk in it, so that the years of man's life measure not some small growth or rearrangement or moderation of what he learned in childhood, but a great upheaval."

Charles de Gaulle in Mexico put it more succinctly: "It so happens that the world is undergoing a transformation to which no change that has yet occurred can be compared, either in scope or in rapidity."

Historian Marshall M. Fishwick recalls that the Angel Gabriel in *Green Pastures* put it still more succinctly, "Everything nailed down is coming loose."

Yet there is a danger in that word "everything." If it means the particular works of man as we see them around us, it is true: most of these (e.g., old buildings, old machines) will quickly disappear; some (e.g., methods of work and education) will be modified; and the greatest (e.g., high achievements of art and intellect) will remain, unchanged in themselves, though perceived somewhat differently by future decades. But if "everything" is taken to include, say, the continuity of man's quest for order and right and truth and harmony, then it is not true. There are, indeed, two different ways in which we can blind ourselves to the meaning of the radical change around us: either we comfort ourselves with some such piece of obsolescent wisdom as "the more it changes, the more it stays the same," or else—and this is worse—we assume that the flux is total, that past, present, and future have nothing to do with one another, that no patterns are discernible, that no purpose if feasible because the "winds of change" are beyond control and, anyway, human values now have no endurance, no footing. Both these escapes from thinking tend to deny or diminish human responsibility, the first by assuming that the order "built into" society is beyond man's ability to destroy or improve, the second by confusing radical change with absolute randomness. To speak of change, however radical, presupposes some constants, some continuities, some patterns in that which is changing. The greater the change, the harder the reach needed to establish the patterns.

In looking for patterns, we do not start from scratch. Neither men nor society determine objective truth, and a change in society will not change truth. But the perception of truth by men—what they perceive and how they perceive it—is surely affected by changes in society. Many truths of history, science, philosophy, and religion have been perceived as transcending change. In a time of great social flux the truths perceived as transcendent may be fewer —and therefore more precious. They may be perceived as more abstract—and therefore more precious. They may be perceived as more abstract—and therefore harder to apply to actual life. A tremendous effort will be required to build the intermediate links between the novel patterns of a changing society and abstract, enduring truth. The "middle-tier" organizations of American society are trying to forge such links out of the new values and purposes that emerge in the course of practical operations.

The Multiversity

California, the fastest-changing state in the fastest-changing nation, is a good place to view this process. What would be the best focus for a history of California in the last twenty years? Climate? Citrus? Hollywood? Politics? The aerospace industry? The right focus is higher education, around which much of California's recent industrial and population growth has been organized. The University of California is a paradigm of how a society deals with radical change through "middle-tier" organizations. Its president, Clark Kerr, readily admits that the university is not organized around any central concept, derived from an ordered system of ideas. Whereas Cardinal Newman could write of *The Idea of a University* and Abraham Flexner could speak of "the idea of a modern university," Clark Kerr's recent book has to start at a different level; it is called *The Uses of the University.* Any claim to essential *present* unity is abandoned; Kerr calls it the multiversity—and not merely because it has several campuses. But the quest for unity is not abandoned. Kerr and his fellow administrators deal sensitively with all kinds of "markets"—the world academic community, faculty members with a thousand specialized interests, the public, the undergraduates, the business community, the dispensers of federal grants. Yet the University of California is not a mere supine field in which these various conflicting "forces" are in play. The university has a character. It has evolving purposes. It assumes responsibility for actively mediating and modifying the influences upon it.

If this institution has no fixed star to steer by, it at least has very efficient antennae with which to feel its way. In its unfolding policy there is a sense of emergence, of organizational evolution, of process rather than finished product. Its huge adult-education program gives a clue to its place in a changing society: one-third of all the lawyers in the state are now taking its courses; so are one-sixth of the doctors. Learning is not directed toward a terminal point. California's future is regarded as an open vista. Its quality will be determined not by set plan and not by blind chance, but largely by the quality of the men and women who enter the great stream of the state's system of higher education.

The Knowledge Industry

The University of California's annual operating expenditure of over half a billion is a tiny part of what Princeton's Professor Fritz Machlup calls "the knowledge industry." Using an overly generous definition that includes as "knowledge" everything from the printing on a pillbox to the Beverly Hillbillies, Machlup calculates that "the knowledge industry" accounts for 29 percent of the gross national product (as adjusted by him). Even with the fluff excluded, a very large part of all the work done in the U.S. can be realistically included in knowledge.

A minor fraction of "the knowledge industry" is the cost of original research, the production of new knowledge. Most of the industry is the processing and distribution of knowledge—all the teaching, the textbooks, the journalism, the advertising, and other forms of communication. The tremendous growth of these activities raises the question of why we need so much more organized communication than our ancestors had.

The answer lies in radical change as a condition of life. Most members of a stable society absorbed almost unconsciously all they needed to know about the life around them. A small group of educated men consciously shared a fixed body of organized knowledge. In short, only a small part of a stable society's total effort needed to be devoted to the "connective tissue" binding one man to another or connecting one aspect of life to another. With us it is otherwise. All parts of society are in motion and most men are in motion—changing jobs, changing residence, changing acquaintances, buying things they never bought before, confronting problems and seizing opportunities their fathers never heard of. In the radically changing society the connective tissue, the organized effort required to stay in touch, is enormous and every year will require a larger proportion of the total work. At bottom, this is an effort to consume change intelligently, to establish some pattern of order in the midst of flux.

This need is the real basis for doubting that our grandchildren will have nothing to do when they grow up except chase one another around swimming pools. The evidence that there will be more work and harder work ahead already piles up around us. Much quoted is a recent prediction that within a generation only 2 percent of Americans, as farmers and factory production workers, will be able to produce all the food and manufactured goods the whole population will want. This is supposed to be an astounding projection, one which clinches the argument that the future will have lots of leisure—or mass unemployment. The projection is not astounding at all when it is recalled that last year in the U.S. only 9 percent of the population, as farmers and factory production workers, produced all the food and manufactured goods. In the last twenty-three years the number of farmers and farmworkers has decreased by more than 45 percent while the number of teachers has almost doubled. The contrast between those two trends is a key to the nature of life in a radically changing society.

Indeed, it would be quite impossible for a society to cope with radical change if it had not already reduced to a minor fraction of the labor force the number of people required to produce *things*. For the condition of radical change is not one that can be handled by a tiny elite group of scientists, politicians, and economic administrators. From this viewpoint, the computer revolution is not a threat to employment but an opportunity for society to find among the displaced white-collar workers the people it needs for higher functions involved in the absorption of change. American society is behaving as if it senses this future need for what might be called "massive leadership," a labor force that would have more captains, majors, and colonels than privates. (Machines are the privates, computers the sergeants.)

Innovation and Risk

The rapid evolution of the American corporation in its external and internal aspects shows both the growing importance of connective tissue and the trend toward massive leadership. Peter F. Drucker, who for twenty years has been perhaps the shrewdest prophet foretelling the changing structure and functions of corporations, made "innovation" a key word in his *Landmarks of Tomorrow* (1957). To innovate is to plan, which can be done either by governments or by corporations. But planned innovation, however good, cannot eliminate risk. When innovation is channeled through autonomous, competing corporations, risk is encouraged and the social cost of unsuccessful innovation can be limited. Society can afford to have a corporation fail, but society cannot afford to have central government fail. Government economic planners, proceeding by law or fiat, have no flexible mechanism comparable to a market in which they can assess the probabilities of any given risk and measure its results.

Every year corporations have become more conscious of innovation as their central activity. Whole industries—chemicals is the best-known example —now derive half their revenues from products not in existence a generation ago. No product is safe in an innovating world. No corporation, however big or diversified, is safe. The very fact that most goods and services sold today are *not* necessities of life gives consumers a new leverage. Accordingly, corporations will all live dangerously and try desperately to communicate—to listen for potentialities in the market, to present themselves as worthy mediators of change.

The shift in emphasis from finished product to continuing process is apparent in the transformation that has come over that ancient and honorable function of business—salesmanship. *Fortune* (see "The Salesman Isn't Dead —He's Different," November, 1962) described how the salesman no longer sees his work as one separate sale after another. His is a continuous service connecting the emerging potentialities of his customer's needs with the emerging potentialities of his own business.

The Creation of Wants

The rise of "marketing" is an example of evolving connective tissue, a process in which competing corporations by a delicate (and sometimes indelicate) process of probing play their part in society's choice of what it wants. The function is especially important in the introduction of new goods and services, an activity condemned in some quarters as "creating wants." This moral condemnation is derived from the ethics of the stable society—but not necessarily from the basic principles of Christianity. An innovating society has to form new wants, new purposes, as surely as it has to make new things. Created needs appeared even in the stable societies. Calories, sex, and a dry place to sleep—these are the primal wants; the rest is all contrivance. Who, prior to its invention, was demanding the printing press or the mass literacy that (very slowly) followed it? What barefoot multitude raised barricades crying, "Give us shoes"? What throne was toppled by an unwashed populace clamoring for soap?

Present-day critics of created wants retrospectively approve the fact that literacy, shoes, and soap finally came into the possession of most people. Progress, it would seem, is all right if it is detached from deliberate thought and organized purpose. This view rather harshly implies, of course, that it was better for whole generations to have died unlettered, unshod, and unclean rather than have men exercise the wit and will that would have hastened the creation of new markets.

The Evolution of Market Norms

Obviously, commercial risks are not the only ones involved in the business of creating new wants. Hideousness abounds in American cities, and some of it is newly minted, innovating hideousness proceeding from the organizations of the middle tier. The juvenile mass market in printed pornography has vastly expanded in recent years. In both these cases, commercial success encourages innovations that are held aesthetically or ethically wrong. Law and other instruments of community decision—outside of markets—must be evolved rapidly enough to set bounds upon innovation. But most of the hope that changes will be ethically and aesthetically better lies within the market itself. Both consumers and corporations share in the responsibility for developing improved norms. In the static society the applicable norms were considered as fixed (or gradually changing); whoever produced a new thing had only to obey the thou-shalt-not represented by the norms. Now the norms themselves have to be strengthened, refined, or reinterpreted as innovation proceeds.

Conventions inherited from previous eras inhibit corporate managers from telling a stockholders' meeting that they raised the charwomen's pay because they thought it "right" or built the factory that way because they thought it "beautiful." Around the vast activity of commercial styling, packaging, and designing can be heard a lot of cheaply cynical talk and a lot of sloppily sentimental talk. The first speaks of these aesthetic activities as gimmicks to trap the suckers; the second regards them as worthy evasions of the discipline of profits. But, in fact, the degree of actual integration of profits, ethics, and aesthetics may be considerably greater than the talk indicates. Managers often proceed as if the customers had a subliminal perception that a company with a rationally humane labor policy and a clearly designed letterhead would be more likely than not to deliver a reliable piece of farm machinery. The managers are right to make this assumption—right profitwise and otherwise.

This is not to say that the ethical and aesthetic situation in the "middle tier" organizations is "good" or that it will necessarily improve. The assertion is only that business organizations are deeply involved in values higher than materialist values, that they increasingly know this, and that they are not to be theoretically excluded from the possibility of performing vital social functions in the strengthening of these values.

The importance of this apparently modest proposition is suggested by a contrast with another society. Writing in the March *Fortune*, Henry Anatole Grunwald, a recent traveler to the Soviet Union, noted: "While in Russia I realized that in the West not only ads but also goods and services themselves try to please me, appeal to me, hence *talk* to me. The meal in the U.S. airplane, the new Detroit car design, the comfortable mass-produced shoe, the air-conditioned New York movie house—they all communicate with me. Perhaps they do so only to get my money away from me; but they still communicate. In Russia, things don't talk to people." The difference can hardly be found in a genetically transmitted superiority of American taste over Russian. The difference is in the way production and distribution are organized. Decision makers of the Soviet production system are isolated from the punishment of a market that allows a broad range of individual choice. Things don't talk to people because people can't effectively talk back. An economic system rigidly and centrally organized around the *necessities* of life can be expected to display in design and style the mute arrogance of take-it-because-you-must. Increasingly, the U.S. economy will be organized around options—take-it-if-you-like-it-better-than-something-else.

Do Corporations Crush?

The internal aspect of corporations is evolving as significantly as their relation with the rest of society. In the first half of the twentieth century the typical corporation was a rather simple pyramid with a base of workers who performed repetitive tasks, similar one to another; semiskilled was the word for them. At the top were a few managers; power was the word for them. Within the managerial group and between it and the workers the flow of authority was all one way—from top to bottom. A very plausible case could be made that this structure tended to crush the initiative and individual spirit of those who worked within it.

This picture is not true of many corporations today, and will be true of fewer tomorrow. The armies of undifferentiated workers have been replaced by better-trained men with more carefully defined responsibilities; specialization is the word for them. The managerial group has expanded hugely, and divides into specialists and generalists. The former have authority based on knowledge. They include not only scientists and technologists, but increasing numbers of experts stemming out of the social arts and sciences—communicators, psychologists, lawyers. The responsibility of the generalists, who appear at many levels, is to integrate and transcend the specialists, in a continuous process of changing operations and evolving purposes. Power there is, but it is a highly "constitutionalized" form of power, appearing in hundreds of focuses and running sideways and up as well as down. The logic of the new structure of corporations requires initiative at every focus of responsibility.

The men to match the logic are emerging. For several months *Fortune* has been examining "the young executive," successful managers moving toward the top rungs. Advance reports of the forthcoming series of articles indicate that the young executives are notably uncrushed by corporate life. They are not conformists and they are not cogs. They are conscious of their need to cooperate with one another, but also conscious that organization enlarges their individual range of action. The more alert among corporations are striving consciously to develop the internal atmosphere in which initiative flourishes.

Is this form of working life—in which millions will be engaged within a few decades—"dehumanizing"? A fellow editor remembers a scene of splendid isolation. One of his first jobs required him to enter the bottom of a silo where, as the corn poured down, he walked round and round, hour after hour, treading it down. Progress for him occurred as the corn rose toward the top of the silo where the air was better; then he started at the bottom of another silo. The task was not spiritually barren. He recited Shakespeare's sonnets to himself, an exercise of great aesthetic merit and one that conferred lasting illumination upon his literary style.

His present job requires exasperating cooperation as he debates with fellow editors. It requires discipline of thought and hard-won knowledge. His social function might be described as relating current trends to a body of intermediate patterns of order called political economy. He is part of the "knowledge industry." On the whole, he thinks his job has been "humanized" since the bygone days at the bottom of the silo. Some machine, no doubt, has taken over the old job of treading down the corn, releasing a man to lend a mind—not a foot—to the task of achieving an intelligible cohesion in a world that technology is changing.

The Uses of Oceans

Five years ago Edwin H. Land of Polaroid said: "Industry should address itself now to the production of a worthwhile, highly rewarding, highly creative, inspiring daily job for every one of a hundred million Americans." Of course, "industry" is not a collective and did not consciously formulate that as a national program. But in its decentralized, competitive way it has moved in the direction Land indicated.

The reader of General Sarnoff's technological predictions should let his imagination run on the non-technological tasks required by each advance. When Sarnoff speaks of "floating sea farms," one should not envision only the ship manned by seaborne fishboys (each with a Ph.D.) herding their charges with electronic lariats while the tape on deck plays, "Git along, little porgies, git along." The readers should also think of the hordes of lawyers, back in Bermuda and La Jolla and Pango Pango, who will be sweating and striving —and doubtless conniving—to work out the private-property law for the development of the oceans. What a gorgeously Gothic edifice of complex harmony that might become! No peasant-like "absolute" ownership in fee simple, carving the sea into vertical lots; instead, a subtle separation by *uses* so that, in the same cubic mile of ocean, one company would be running halibut, another would be artificially encouraging the growth of marketable plankton, a third would be mining the ocean floor while the ships of all nations freely traversed the surface.

The Uses of History

This article has been focused on the mediation of emerging values performed by the middle-tier organizations in an era of radical change. It needs to be said again that, ultimately, these values find their footing in transcendent truths drawn from history, philosophy, science, and religion. Each of these sources of order now has a somewhat different—but not less important— bearing on society.

History, for instance, is much less useful than it was as a source of directly applicable precedent. "The prevalence of newness" introduces too many variables. Yet we see men turning more and more to history—and they are right. The present "lessons" are to be found at a higher level of abstraction. Americans, for instance, can learn who they are and that their destiny has to do with building an order of freedom in a world of change. Their Constitution is sufficiently abstract to have withstood more social change than any other form of government in history, a fact of some encouraging significance in dealing with the future. History, one notices, keeps being rewritten—and this also is right. Each changing generation will have different needs—created needs—from the record of the past. E. H. Carr says history is "a dialogue between the events of the past and progressively emerging future ends. The historian's interpretation of the past, his selection of the significant and the relevant, evolves with the progressive emergence of new goals."

In the area of organized knowledge the notorious damage to unity—not mainly attributable to technological change—remains formally unrepaired. The disarray looks as bad as seventy years ago when Woodrow Wilson called attention to the disordered "constitution of learning," with its "separate baronies." But there may be more actual integration of knowledge than appears. C. P. Snow has recently noted that in the U.S. even "the two cultures" of science and the humanities are drawing closer. Meanwhile, separate scientific disciplines increasingly talk seriously to one another, their integration mediated by seminars, conferences, federal projects, interdepartmental university structures—and by business corporations. Scientists learn the bureaucratic arts and become formidable "operators" in Washington and New York. What draws the disciplines together is the growing sense that individual achievement can be enlarged by organized purpose. The effective condition of knowledge today is not unity; neither—and this is certain—is it chaos.

The absence of an accepted integrating philosophy is felt particularly keenly in the field of law—especially the law of nations—where the social invention of effective institutions has lagged far behind the pace of change. Chaos could come in that door—but it hasn't yet.

Many individuals in recent years have reacted to extreme social flux by turning to the order, unity, and transcendence represented by religious truth. In any age, awareness of the world's transiency is a fundamental of religious motivation and belief. But men of the traditional religions should be aware of the growing attraction of religious systems very different from their own.

The Sensorium Awaits

Scientific humanism has been a waxing way-of-belief and has been blamed for inculcating man-worship and, thus, for some of the twentieth century's more avoidable calamities. Innocent or guilty, it has taken a new lease on life. There is even an effort to project a future integration of science and religion in an extraordinary work, *The Integration of Human Knowledge,* by Oliver L. Reiser, professor of philosophy at that expanding powerhouse of learning, the University of Pittsburgh. The book is explicitly pantheist and rests heavily —very heavily—on mathematics and biology in modernizing the ideas of Pythagoras. Reiser predicts an evolution toward á synthesis of thought and value in a "giant organism of a world sensorium." This is an impressive, internally consistent work—in short, no joke. If the actual efforts to evolve intermediate patterns of order break down, a lot of people, avid for a restoration of unity, are going to head for Dr. Reiser's sensorium, rather than for the traditional religions.

Perhaps it might be prudent for men of the traditional religions to stop thinking of them as traditional. In the old-new polarity that developed in the period 1800–1950, many defenders of religion aligned themselves with "things as they had been." Today it is much more widely recognized that "eternal verity" does not mean the same as "old order." No preference for the stable society is symbolically expressed by the God of the Long March out of Egypt or by Christ, the Disturber, the Swordbringer.

Man is right to make what order he can in the midst of change, even though society may lack a generally accepted integrating system of ideas and values. It is possible that an integration of ideas, a philosophy, will evolve. It has done so before within the perimeter of Christian belief.

One can foresee, at least, many occasions for humility ahead. But not necessarily failure of society, and not necessarily success. The question, "How much flux can man stand?" assumes that the degree of change should be measured against some fixed point of human tolerance. But man and his society evolve. He can stand as much flux as his developing intellectual, moral, and organizational achievement can keep up with. So far—though the pace is furious—he has not lost the race with his own expanded capacity to innovate.

Must We Rewrite the Constitution to Control Technology?

Wilbur H. Ferry

I shall argue here the proposition that the regulation of technology is the most important intellectual and political task on the American agenda.

I do not say that technology *will* be regulated, only that it *should* be.

My thesis is unpopular. It rests on the growing evidence that technology is subtracting as much or more from the sum of human welfare as it is adding. We are substituting a technological environment for a natural environment. It is therefore desirable to ask whether we understand the conditions of the new as well as we do those of the old, and whether we are prepared to do what may be necessary to see that this new environment is made suitable to men.

Until now, industrial man has only marginally and with reluctance undertaken to direct his ingenuity to his own welfare. It is a possibility merely—not a probability—that he will become wise enough to commit himself fully to that goal. For today the infatuation with science and technology is bottomless.

Here is where all the trouble begins—in the American confidence that technology is ultimately the medicine for all ills. This infatuation may, indeed, be so profound as to undercut everything of an optimistic tone that follows. Technology is the American theology, promising salvation by material works.

I shall argue that technology is merely a collection of means, some of them praiseworthy, others contemptible and inhumane. There is a growing list of things we *can* do that we *must not* do. My view is that toxic and tonic potentialities are mingled in technology and that our most challenging task is to sort them out.

A few cautionary words are in order.

First, I am aware of the distinctions between science and technology but intend to disregard them because the boundary between science and technology is as dim and confused as that between China and India. Besides, it is impossible to speak of public regulation of technology while according the mother-lode, science, a privileged sanctuary. At the same time, it must be granted that the scientists have been more conscientious than the technologists in appraising their contributions and often warning the community of the consequences of scientific discovery.

Next, I shall use everyday examples. Some will therefore consider my examples superficial. But it appears to me to be better to illustrate the case by situations about which there is considerable general knowledge. I shall rely on well known contemporary instances of technological development chiefly to show the contrast between their popular aspects, including popular ideas about control, and those less well known side effects that in the long run threaten to cancel out promised benefits.

The first point to be made is that technology can no longer be taken for granted. It must be thought about, not merely produced, celebrated, and accepted in all its manifestations as an irrepressible and essentially benign human phenomenon. The treason of the clerks can be observed in many forms, but there is no area in which intellectuals have been more remiss than in their failure to comprehend technology and assign it its proper place in humane society. With many honorable exceptions—I give special recognition to Lewis Mumford, who for forty years has been warning against the castration of spirit by technique—the attitude of the physical scientists may be summarized in advice once proffered to me, "Quit worrying about the new scientific-technical world and get with it!" And the disposition of the social scientists, when they notice technology at all, is to suggest ways of adjusting human beings to its requirements. Kenneth Keniston says in *The Uncommitted:* "We have developed complex institutions to assure (technology's) persistence and acceleration (and we) seldom seek to limit its effects."

We are here near the core of the issue. Technology is not just another historical development, taking its place with political parties, religious establishments, mass communications, household economy, and other chapters of the human story. Unlike the growth of those institutions, its growth has been quick and recent, attaining in many cases exponential velocities. Federal expenditure for research and development in 1940 was $74,000,000—less than 1 percent of total government spending. In 1966 it was $16 billion—15 percent of federal spending. This is not history in the old sense, but instant history. Technology has a career of its own, so far not much subject to the political guidance and restraints imposed on other enormously powerful institutions.

This is why technology must be classed as a mystery and why the lack of interest of the intellectuals must be condemned. A mystery is something not understood. Intellectuals are in charge of demystification. Public veneration is the lot of most mysteries, and technology is no exception. We can scarcely blame statesmen for bumbling and fumbling with this phenomenon, for no one has properly explained it to them. We can scarcely rebuke the public for its uncritical adoration, for it knows only what it is told, and most of the information comes now from the high priests and acolytes of technology's temples. They are enraptured by the pursuit of what they most often call truth, but what in fact is often obscene curiosity, as when much of a nation's technological quest is for larger and more vicious ways of killing—the situation today.

There is an analogy between the rise of modern economics and that of the new technology that one would have thought intellectuals would be especially eager to examine. Technological development today is in the enshrined position in political-economic theory that was accorded to economic development in the nineteenth and early twentieth centuries. Unguided and self-directed technology is the free market all over again. The arguments justifying *laissez faire* were little different from those justifying unrestrained technology. The arguments in both cases are either highly suspect or invalid. The free market dwindles in real importance, though the myth remains durable enough. But we know now that the economic machine needs to be managed if it is not to falter and behave eccentrically and needlessly injure people. That we have not yet conquered the political art of economic management only shows how arduous and thought-demanding a process it is, and why we should get after the equivalent task in technology at once.

Quite a lot of imaginative writing has been done about the world to come, whether that world develops from the technological tendencies already evident or is reconstructed after a nuclear war. This future-casting used to be known as Utopian writing. Utopias today are out of fashion, at least among novelists and poets, who are always the best guides to the future. With only two exceptions, the novels I have read tell of countries that no one here would care to live in for five minutes.

The conditions imagined are everywhere the same. High technology rules. Efficiency is the universal watchword. Everything works. All decisions are made rationally, with the rationality of the machines. Humans, poor folk, are the objects of the exercise, never the subjects. They are watched and manipulated, directed, and fitted in. The stubborn few in whom ancient juices of feeling and justice flow are exiled to Mars or to the moon. Those who know *how* are the ones who run things; a dictator who knows *all* reigns over all; and this dictator is not infrequently a machine, or—more properly—a system of procedures. I need go no further, for almost everyone is familiar with Orwell's *1984*.

I proceed to examples of the benign and malignant capacities of technology. I am aware that many will find unacceptable my treatment of technology as a semi-autonomous force. These critics say that *tonic* and *toxic* are words to apply to human beings, to ignorant or wise statesmen, to thoughtless or conscientious engineers, to greedy or well-intentioned entrepreneurs. To holders of this viewpoint, there is no intrinsic flaw or benefit or value in technology itself. But I hope to demonstrate that technology has an ineluctable persistence of its own, beyond the reach of all familiar arguments based on the power structure.

My first example is privacy, today a goner, killed by technology. We are still in the early days of electronic eavesdropping, itself an offshoot of communications research, and at first celebrated as a shortcut to crime control. But now no office, schoolhouse, or bedroom is any longer safe from intrusions. A good many people, including Senators, casino operators, felons, and executives on holiday with their secretaries have been made conscious of possible bugs in their cocktail olives and automobiles as well as in their telephones. A good many others were aroused when it was disclosed some time ago that the FBI possesses the fingerprints of tens of millions of citizens. What are we to think of the proposal for a National Data Center, which will have the capacity and perhaps the responsibility to collect every last bit of information concerning every citizen? Not only tax records, but police records, school grades, property and bank accounts, medical history, credit ratings, even responses to the Kinsey sexual behavior questionnaire.

To its credit, Congress has already taken a cautious look. A subcommittee of the House Committee on Government Operations took several hundred pages of testimony in the summer of 1966 on "The Computer and Invasion of Privacy." Referring to the programs designed to "help America" under the efficient guidance of the Data Center, Subcommittee Chairman Cornelius Gallagher said: "Such programs should not be at the cost of individual privacy. What we are looking for is a sense of balance. We do not want to deprive ourselves of the rewards of science; we simply want to make sure that human dignity and civil liberties remain intact. . . . Thought should be given to these questions now, before we awaken some morning in the future and find that . . . liberty as we know it has vanished."

Chairman Gallagher then said he did not doubt that some way of reconciling the claims of efficiency and privacy would be found. To me, however, this is by no means a foregone conclusion. It ought to be against public policy to take any chance whatever with the little privacy remaining to Americans.

We have been reading a lot recently about the greatest intrusion on privacy yet dreamed up, in terms of numbers of people affected. I refer to the supersonic transport plane, a multibillion-dollar folly to which the nation is now apparently committed irrevocably. In a few years' time, the sonic boom of the SST will daily and nightly waken sleepers; worsen the condition of the sick; frighten tens of millions; induce neuroses; and cause property damage beyond estimate.

At least three European countries are considering putting the traveling thunderclap of the sonic boom on the forbidden list by passing legislation which would prevent SSTs from flying over their territories. The position of these countries on this issue is people first, machines second.

The idea has been wafted about by the Federal Aviation Authority of the United States—which has been more than ordinarily slippery on the issue of the SST—that we will spend the billions required for SST but forbid its use overland in this country. I don't believe it for a moment. Overland flight is where the big profits are to be made. If, as seems to be the case, SSTs will be built here, there can be no doubt that tens of millions of Americans will be subjected to sonic boombardment.

The doctrine of the United States is that whatever can be done must be done; otherwise, the United States will fall behind in the technological race. That is the thesis. Therefore, if the SST can be built, it must be built. This technological imperative is bolstered by dozens of irrelevant arguments in support of SST. It is said that other nations will gather the glory and profit and jobs resulting from SST manufacture. American manufacture of SST will help the balance of payments. These arguments are as popular as they are off the mark. Against them are many equally valid.

It has not occurred to many that the argument should be about the superiority of SST, all things considered, as a means of getting from here to there. It should be about the benefits to the thousands and the disbenefits to the millions. The pursuit of super-speed is being conducted by experts who might better be working to make present aviation super-safe. The socially necessary tasks to which these nimble minds might be turned are uncountable if we should take seriously the proposition that people must come first, machines second.

The deep irony is that we, the tax-payers, are paying for this unprecedented attack on ourselves. The unalterable fact is that the privacy and right to quiet of millions of Americans will shortly be sacrificed to an undertaking that thereby becomes fundamentally senseless. Their welfare goes down before the desires of a few hundred or thousand people who may ultimately be able to get from Los Angeles to New York in half the present time.

When SST proponents are asked to justify the assault on the bodies and minds of human beings, the customary answer is, "They'll get used to it." Some technologists, however, are more direct. Speaking of the sonic boom, Engineer Charles T. Leonard gives this prescription: "A greatly more tolerant populace than is presently assumed to be the case ... may well become mandatory if the SST is to realize its full potential."

It may turn out this way: We may be compelled to become tolerant of every and all techniques—but at what human expense we may not appreciate for generations. Silicosis among miners was not discovered until long after they had become used to dust-laden mineshafts. Neurophysiologists warn that the growing din of modern life is already making us deaf, and ravaging sensibilities and nervous systems. This is part of the price already being exacted by technology; and with SST we are choosing, as a nation, to raise the price enormously.

"Choosing" is perhaps the wrong word, although authority for the SST has been tentatively granted by Congress. And a silly Senate has recently authorized a further $143 million in development funds—at the exact time it was reducing expenditures on programs for people. But it is not a true choice, for reasons already given. Congress, lacking the understanding of the evils of technology because of the slackness of the intellectuals, merely has been swept along in the technological madness.

That public servants can act with good sense and foresight when informed about the impact of technology is illustrated by the City Council of Santa Barbara. Responding to incessant boombarding of that quiet city, the council recently passed an anti-boom ordinance.

In only one case, that of atomic energy, has this country had enough imagination about results to put a stiff bridle on technology. The Atomic Energy Commission came into being partly because of the lethal potentialities of the new force and partly because of a few leaders—mainly scientists—who were able to convince Congress that this cosmic threat should never be a military monopoly.

The ineffectuality of efforts toward smog control in the last twenty years is instructive. In the first few years, not enough was known to do anything about it. Air pollution was considered an unavoidable evil of modern life, as the ear pollution of the SST is now said to be by its proponents. For the last ten years the air pollution problem has been clearly identified, yet there is as much smog as ever, or more. Federal, state, county, and city governments all are working on the control of air pollution, so it is idle to say that public attention is lacking. We gain little yardage by declaiming against the automobile and petroleum interests, though assuredly their products are the main source of the garbage-laden air. Technology is the villain.

The fact that so much of the smog control effort is going into scrubbing the atmosphere obscures the real scope of the problem. For instance, Frank Stead, in *Cry California*, says that the way to deal with it is "to serve legal notice that after 1980 no gasoline-powered motor vehicles will be permitted to operate in California." So far, so good. The non-emission-producing automobile would be a clear gain for urban areas, and not only for California. At the moment, the automobile industry is making piteous sounds about giving up the gasoline engine, explaining week after week how costly and difficult it is going to be to produce a substitute.

Now, it is hard to think that a new kind of automobile is an insuperable technical challenge to a nation that can dock ships in space. Designing a fume-free car would seem a far more worthy objective for government research than placing a man on the moon or re-creating the deadly plague, another of our bloodiest technical preoccupations. Yet the absurdly small sums allocated for federal research in new motor car design show how serious we are about alternatives.

We must not, incidentally, be misled by the optimistic publicity now being emitted by auto and petroleum industry centers. The dean of smog-studiers, Professor A. J. Haagen-Smit of Caltech, says, "I have yet to see a smog control plan that gives me any confidence we will some day have reasonably clean air."

Mr. Stead says little about the larger question, that of the entire transportation technology. He has hold of a very sharp technical thorn, but it is only one of a large cluster. Suppose, for instance, a way were found to dissipate the atmospheric peculiarities that lead to air pollution. Replacing internal combustion by electricity may lower the incidence of emphysema and eye trouble. By itself, however, it will do nothing about the equally troubling questions of urban congestion and dedication of more and more land, rural and urban, to asphalt. Not a little of the furor in Watts arose from lack of inexpensive transportation to jobs and recreation. One will say, "What about rapid transit?" The answer is, yes, of course, but still that is not the resolution, as the situations of those cities with well-developed transit systems attest. Buses, subways, and commuter trains may only compound the misery, as any visitor to New York will be able to testify.

What is needed is a firm grasp on the technology itself, and an equally clear conviction of the primacy of men, women, and children in all the calculations. This is a resounding prescription, and I regret to admit that I am more clear about ultimate steps than I am about how to do what needs to be done in the near future.

I am convinced only that political institutions and theory developed in other times for other conditions offer little hope. We now have, by courtesy of the 89th U.S. Congress, a Department of Transportation whose task is, in the words of President Johnson, "to untangle, to coordinate, and to build the national transportation system that America is deserving of." Under what authority, and by what means?

The mind wanders to the lengths of asking what would happen if the new department might one day soon feel itself compelled to limit by fiat the manufacture of cars and trucks; to coerce car owners by tax or otherwise to use public transportation; to close state and city borders to visitors approaching by car; to tear up rather than to build freeways, garages, bridges, and tunnels.

I turn to my final example of technological invasion. American business executives a half dozen years ago wakened to the existence of a multibillion-dollar market—education. It was hard to ignore. Today's real growth industry is education. The $4 billion we spent on it at the end of World War II has grown to $50 billion plus—an annual rate of increase of more than 12 percent. New corporate marriages have been hastily arranged. Large hardware companies wed large software companies. The object is profits, not education, although the public relations experts have got together on a prothalamion designed to convey the notion that these new matrimonial arrangements aim basically at the welfare of the educational enterprise, from the grades to the graduate schools. As always, the central claim is efficiency. Mass education it is said, requires mass production methods. The result is already discernible, and may be called technication. The central image of technication is the student at the console of a computer.

Our educational purposes have never been very clear. Technication may compel removal of the ambiguities and establishment of straightforward aims. But who will undertake this task? How shall we assure that the result is the betterment of children and not the convenience of machines? Are we really all that crazy about efficiency, or what we are told is efficiency? Already, tests are being devised that can be applied and graded by machines, thereby getting the cart squarely in front of the horse. I am not pressing the panic button but the one next to it. I am not denying that certain advantages to education are offered by the new technology. I'm repeating that tonic and toxic technology are here mixed in unknown proportions.

The forces of technication are already infiltrating our grade schools, encountering little resistance. Once again we are in the area of narrow choices. How shall we distinguish between what helps and what hurts? I know that education has suffered from lack of research for years, and that much of what is projected may well modernize anachronistic practices. We have no standards as to what shall be admitted, what rejected. The temptations to rely unquestioningly on technology are very great. The possibilities that are said to be inherent in the new gadgetry are dazzling. We are told that the high costs of technication will bar widespread use for a long while. This is what was said of television in the early days.

The perils are manifest. One of them lies in adopting the totally wrong notion that an educational system can be thought of in terms like those of a factory for producing steel plate or buttons. Another peril is to that indefinable relation between teacher and taught: Dare we think of it as a mere holdover from another world, as subject to the junkpile as the horse-drawn fire engine has been? A third peril is that the ends of education, already a near-forgotten topic, will be gobbled up by the means.

Webster College President Jacqueline Grennan speaks for education, not technication, when she asks for the development "not of one voice of democracy but of the voices of democracy." The great need, she says, "is to enable an individual to find his own voice, to speak with it, to stand by it. . . . Learning is not essentially expository but essentially exploratory."

Technication means standardization. The history of factories shows the benefits and limits of standardization. Factories are fine for producing things, but their record with people is terrible. We cannot expect to hear the voices of democracy emerging from education factories; we can hear only the chorus. Technication, as Robert M. Hutchins observes, will "dehumanize a process the aim of which is humanization."

The effect on the taught is crucial. The rebellion at Berkeley centered on the indifference of multiversity's mechanism to the personal needs of the students. When the protesters pinned IBM cards to their jackets—an act duplicated on campuses throughout the land—they were declaring against impersonality and standardization; and it cannot be said too often that impersonality and standardization are the very hallmarks of technology.

I have offered not-very-penetrating illustrations of the way technology is raising conspicuous questions about the social and personal welfare of Americans. Behind all these matters, as I remarked at the outset, are dangerous convictions that science and technology provide the panacea for all ailments. It is curious that this conviction should be so widespread, for life today for most people appears to be more puzzling and unsatisfactory and beset with unresolved difficulties than ever before. For most people—but not, I suppose, for the scientists and technologists, the priesthood of the modern theology that is more and more ruling the land, and from whose ingenious devices and fateful decisions we must find a way to make effective appeal.

One must nevertheless be grateful to those few members of the sanhedrin who keep pointing out the dangers as the nation turns doubtful corners. Dr. Murray Gell-Mann of the California Institute of Technology says that "society must give new direction to technology, diverting it from applications that yield higher productive efficiency and into areas that yield greater human satisfaction. . . . Carl Kaysen of the Institute for Advanced Studies at Princeton emphasizes that government institutions are no longer equal to the job of guiding the uses of technology.

Scientists and technologists are the indubitable agents of a new order. I wish to include the social scientists, for whose contributions to the technological puzzle I could find no space in this paper. Whether the political and social purpose of the nation ought to be set by these agents is the question. The answer to the question is no. We need to assign to their proper place the services of scientists and technologists. The sovereignty of the people must be reestablished. Rules must be written and regulations imposed. The writing must be done by statesmen and philosophers consciously intent on the general welfare, with the engineers and researchers summoned from their caves to help in the doing when they are needed.

How specifically to cope? How to regulate? Answers are beginning to filter through. Not many years ago it was considered regressive and ludditish even to suggest the need for control of technology. Now a general agreement is emerging that something must be done. But on what scale, and by whom?

E. J. Mishan, the British economist, calls for "amenity rights" to be vested in every person. He says, "Men [should] be invested by law with property rights in privacy, quiet, and clear air—simple things, but for many indispensable to the enjoyment of life." The burden would be on those offending against these amenities to drop or mend their practices, or pay damages to victims. Mishan's argument is scholarly and attractive, though scarcely spacious enough for the problems of a federal industrial state of the size of the United States. It does not seem likely that we can maintain our amenities by threat of tort suits against the manifold and mysterious agents, public and private, that are the "enemy."

The most comprehensive and thoughtful approach to the problem of regulation is that of U.S. Congressman Emilio Q. Daddario, chairman of the House Subcommittee on Science, Research, and Development. Representative Daddario starts with the necessity for "technological assessment," which he characterizes as urgent. It will amount to a persisting study of cause-effect relationships, alternatives, remedies. Representative Daddario does not speak of tonic and toxic, but of desirable, undesirable, and uncertain effects.

The subcommittee's study is only beginning, but it is based on some of the convictions that animated the writing of this article. Thus, the introduction to the first Congressional volume on technological assessment speaks of the dawning awareness of "the difficulties and dangers which applied science may carry in its genes" and of "the search for effective means to counter them."

It would be unfair to summarize the scope and method of this promising document in a sentence or two. I must leave it to those interested to look further into a first-rate beginning. It is too early to guess whether Congressman Daddario's group will come out where I do on this matter, but it seems unlikely. The subcommittee will probably come out for certain statutory additions to the present political organization as the proper way to turn back or harness technique's invading forces. There is ample precedent.

We can regard the panoply of administrative agencies and the corpus of administrative law as early efforts in this direction. They have not been very effective in directing technical development to the common good, although I do not wish to minimize the accomplishments of these agencies in other ways. Perhaps they have so far prevented technology from getting wholly out of hand. But it is very clear from examples like the communications satellites that our statutory means for containing technology are insufficient.

America is not so much an affluent as a technical society; this is the essence of the dilemma. The basic way to get at it, in my judgment, would be through a revision of the Constitution of the United States. If technology is indeed the main conundrum of American life, as the achieving of a more perfect union was the principal conundrum 175 years ago, it follows that the role and control of technology would have to be the chief preoccupation of the new founding fathers.

Up to now the attitude has been to keep hands off technological development until its effects are plainly menacing. Public authority usually has stepped in only after damage almost beyond repair has been done: in the form of ruined lakes, gummed-up rivers, spoilt cities and countrysides, armless and legless babies, psychic and physical damage to human beings beyond estimate. The measures that seem to me urgently needed to deal with the swiftly expanding repertoire of toxic technology go much further than I believe would be regarded as Constitutional.

What is required is not merely extensive police power to inhibit the technically disastrous, but legislative and administrative authority to *direct* technology in positive ways: the power to encourage as well as forbid, to slow down as well as speed up, to plan and initiate as well as to oversee developments that are now mainly determined by private forces for private advantage.

Others argue that I go too far in calling for wholesale revision of our basic charter. They may be right. Some of these critics believe that Constitutional amendment will do, and that what is needed is, in effect, reconsideration of the Bill of Rights, to see that it is stretched to cover the novel situations produced by technology. This is a persuasive approach, and I would be content as a starter to see how far Constitutional amendments might take us in protecting privacy and individual rights against the intrusions of technique.

But I also think that such an effort would soon disclose that technology is too vast, too pervasive to be dealt with in this way. The question is not only that of American rights, but of international relations as well, as Comsat illustrates. Technology is already tilting the fundamental relationships of government, and we are only in the early stages. A new and heavy factor has entered the old system of checks and balances. Thus, my perception of the situation is that the Constitution has become outdated by technical advance and deals awkwardly and insufficiently with technology's results.

Other critics tell me that we are sliding into anarchy, and that we must suffer through a historical period in which we will just "get over" our technological preoccupations. But I do not face the prospect of anarchy very readily.

So that my suggestion of fundamental Constitutional revision is not dismissed as merely a wild gasp of exasperation, I draw attention to the institutions dominating today's American scene which were not even dimly foreseen by the Founding Fathers. I refer to immense corporations and trade unions; media of communication that span continent and globe; political parties; a central government of stupendous size and world-shattering capabilities; and a very un-Jeffersonian kind of man at the center of it all.

It seems to me, in face of these novelties, that it is not necessarily madness to have a close look at our basic instrument in order to determine its ability to cope with these utterly new conditions, and especially with the overbearing novelty of technique. Technology touches the person and the common life more intimately and often than does any government, federal or local; yet it is against the aggrandizement of government that we are constantly warned. Technology's scope and penetration places in the hands of its administrators gigantic capabilities for arbitrary power. It was this kind of power the Founding Fathers sought to diffuse and attenuate.

Constitutional direction of technology would mean planning on a scale and scope that is hard now to imagine. Planning means taking account, insofar as possible, of the possibilities of technique for welfare. It means working toward an integrated system, a brand-new idea in this nation.

I recognize all the dangers in these suggestions. But leaving technology to its own devices or to the selfish attentions of particular groups is a far more hazardous course. For it must not be forgotten that the enormous proliferation of technology is today being planned by private hands that lack the legitimacy to affect the commonwealth in such profound measure.

The wholesale banning of certain techniques becomes absolutely necessary when technical development can no longer help but only harm the human condition. Scientists Jerome Wiesner and Herbert York exemplify this dictum in its most excruciating aspect when they say:

> Both sides in the arms race are . . . confronted by the dilemma of steadily increasing military power and steadily decreasing national security. It is our considered professional judgment that *this dilemma has no technical solution.* . . . If the great powers continue to look for solutions in the area of science and technology only, the result will be to worsen the situation.

Though many lives are being wrecked, though the irrationality and human uselessness of much new technology is steadily becoming more evident, we are not yet over the edge. I close with Robert L. Heilbroner's estimate of the time available:

> . . . the coming generation will be the last generation to seize control over technology before technology has irreversibly seized control over it. A generation is not much time, but it is *some* time. . . .

Law and the Nation

Allan J. Topol

The dangerous illusion that science and technology provide a panacea for all ills has been denounced in these pages [SR, March 2] by W. H. Ferry. He has persuasively argued for increased government control over technological development. In his view, the present Constitution of the United States does not permit the necessary legislative and administrative action. Therefore, he argues, the Constitution should be rewritten.

Mr. Ferry's concept of "toxic technology" and his concern over its public veneration are challenging and convincing. However, I question his conclusion that Congress cannot adequately deal with these problems under the present Constitution.

While the institutions dominating today's American scene were not and could not have been foreseen by the drafters of the Constitution, those men were on the whole intensely pragmatic. They realized that the document they were writing had to be simple; only the most general provisions could be considered. No effort was made to deal specifically even with eighteenth century problems. Simplicity and generality were essential not only to make the Constitution useful to a posterity whose problems could not be foreseen; simplicity and generality were also necessary if the Constitution was to be accepted by the many conflicting interests in the country.

Consistent with these objectives, the Constitution was phrased to give Congress power "to regulate commerce with foreign nations, and among the several states. . . ." This very vague, broad, and brief statement, known as the "commerce clause," has been the basic source of congressional authority for controlling the considerable technological development of the last 175 years.

Constitutional interpretation is not determined by the elusive intent of its draftsmen. Rather, the Constitution means what the Supreme Court says it means, and the Court varies its interpretation with changing social, economic and technological conditions in the country.

One of the earliest major technological innovations in the history of the Republic was the invention of the steamboat. Prior to Robert Fulton's famous voyage in the *Clermont* in 1807, trade within the United States suffered from a lack of effective transportation. The population of the country was moving westward into the Ohio and Mississippi valleys. The concomitant economic growth demanded something superior to flatboats, barges, and the other vessels of the colonial period.

Fulton and another pioneer of steamboating, Robert Livingston, attempted to exploit Fulton's invention by obtaining from the New York State

Legislature the grant of an exclusive right to operate steamboats on New York's waters. Fulton and Livingston then granted Aaron Ogden the exclusive right to steamboat navigation across the Hudson River between the City of New York and the City of Elizabethtown and other cities in New Jersey. By that time, however, the significance of the steamboat had been recognized by the Congress, which had adopted a Federal Coasting Act to provide federal control over steamboats. Under this law, Thomas Gibbons had obtained a license to operate steamboats. He navigated his boats over the same route that Fulton and Livingston had sold Ogden exclusive rights to. Ogden sued Gibbons to drive Gibbons off the Hudson.

The stake in this action was the development of the free trade and commerce generated by the invention of the steamboat. If New York could restrict the use of this remarkable invention in the waters of New York to the grantee of a monopoly, then many of the benefits that could be derived from the steamboat would be lost. Cognizant of these factors, the Supreme Court of the United States in the 1824 case of *Gibbons v. Ogden,* held that the New York law violated the Federal Coasting Act and was therefore invalid. Gibbons was permitted to remain in business. Trade generated by the steamboat prospered at an astonishing rate.

In numerous other ways, Congress acted affirmatively to insure full utilization of new technology in water transportation. Navigability of rivers and harbors was improved, and locks, dams, and levies were constructed. Channels of rivers were closed, courses of rivers were diverted, locks and dams were removed. The congressional power to take each step was upheld by the Supreme Court under the commerce clause.

Technological developments relating to overland transportation likewise came under congressional control. Congress organized a corporation to construct a national road—the Cumberland Road—from the valley of the Potomac across the Alleghenies to the valley of the Ohio. This highway, with a crushed stone surface, facilitated movement of goods being manufactured by the burgeoning machines of the eastern seaboard. Bridges were likewise constructed by corporations authorized by Congress. The Supreme Court had little difficulty in concluding that the commerce clause was sufficient to permit the construction of both the roads and the bridges.

By 1830, more than one-fourth of the population lived west of the Appalachians, and American businessmen realized the need for still better overland transportation than the roads and bridges made possible. Technologists began to experiment with various types of rails and means of power to propel cars on the rails. Out of this experimentation the great "iron horse" emerged with new prospects for the country's economic expansion. By the middle of the nineteenth century the American population was spread from coast to coast. By 1869 the first transcontinental railroad was completed.

With the development of the railroad came a whole series of new problems and cries for governmental help in dealing with them. Farmers contended that the rates railroads charged were unfair, excessive, and discriminatory. Businessmen and shippers joined the farmers in bringing pressure on state legislatures for corrective measures, and the legislatures in several midwestern states responded with stringent controls. This action only aggravated the problem by producing different regulations in the several states and interfering with the free flow of trade.

The necessity for federal regulation was clear, and the adequacy of the 100-year-old Constitution again was tested. Using the commerce clause as a lever once more, Congress enacted the Interstate Commerce Act in 1887, prohibiting unjust and unreasonable rates and outlawing rebates and rate discrimination. Enforcement of the act was entrusted to a five-member Interstate Commerce Commission. The commission was authorized to conduct investigations, to issue cease and desist orders, and to seek court injunctions.

For a period of approximately twenty years after passage of the Interstate Commerce Act, the Supreme Court exhibited considerable hostility toward federal regulation of railroad rates. However, none of the decisions questioned the authority of Congress, under the commerce clause, to authorize the commission to control rates. Rather, the Court interpreted the intent of Congress to be against granting the commission this power. An administrative agency had never existed before, and the Court simply reflected widespread deference to *laissez-faire* economics.

The Court had misinterpreted the congressional intent, however, and in 1906 the Hepburn Act was passed to clarify the situation. By the time suits arising from the Hepburn Act came up for decision, the initial hostility toward the commission had waned. In the 1910 *Chicago Rock Island and Pacific Railway* case, an Interstate Commerce Commission order imposing maximum rates was sustained with hardly any discussion of Congress's authority to provide for such regulation.

As time passed, the commerce clause of the horse-and-buggy-day Constitution proved its elasticity repeatedly. For example, Congress enacted a Federal Safety Appliance Act, restricting railroads to the use of certain equipment (such as driving wheel brakes and automatic couplers) deemed less hazardous to employees and passengers than other types of equipment. In short, the Supreme Court upheld the right of Congress under the commerce clause to legislate for safety on the railroads. Subsequently, Congress also asserted control over the liability of railroads for the injuries of their employees, over the labor disputes of railroads and their employees, and over the loss or damage to property carried by railroads.

In the twentieth century, the development of the modern airline industry is still another example of a major revolution in transportation created by technology. With this new dimension in trade and travel, new problems were added, and there were new demands for governmental regulation. Airplanes could not be permitted to roam the sky at will. Passenger safety and service became matters of public concern. Congress recognized the national responsibility for regulating movement through the air, and applied the commerce clause of the Constitution to it. In the words of Justice Robert Jackson, planes "move only by federal permission, subject to federal inspection, in the hands of federally certified personnel and under an intricate system of federal commands."

Without becoming embroiled in the question of whether or not Congress *should exercise control* over the supersonic transport plane, one can say with absolute certainty that Congress *has the power to act*. The SST could, for example, be flatly prohibited by Congress from making flights over the United States because of its effect upon individuals on the ground.

Transportation is not the only area in which Congress has exercised control over the products and side effects of technological development. The dramatic revolution in communication also has occurred within the restraints of the commerce clause.

Once Samuel Morse had used his telegraph key to transmit from Baltimore to Washington the news of Polk's nomination for the Presidency of the United States, there could be no doubt that governmental action would be required for full utilization of this new technology. State governments soon granted monopolies (repeating the history of the steamboat) which, if allowed to stand, would have drastically limited the value of the telegraph. And the Supreme Court fixed the issue in 1877 in the *Pensacola Telegraph* case and held (as it did with steamboats) that state monopolies conflicted with the federal act. Rejecting notions that the Constitution would have to be rewritten in the light of a new technology, the Court explicitly declared that the powers granted the Congress must keep pace with the progress of the country and adapt themselves to new developments.

Similarly, the development of radio and television raised wholly novel problems in the control of technology. Some agency had to be empowered to assign frequencies to radio stations, for without such priorities each station would be free to interfere with every other station's broadcasts. State regulation was obviously inadequate. Radio waves do not stop at state lines. Congress responded initially with the Radio Act creating a Radio Commission and bestowing broad licensing and regulatory powers upon the commission. The Supreme Court concluded that there was not even any question as to whether this could be done under the commerce clause.

A particularly striking example of congressional encouragement of maximum utilization of technology in this area was the 1962 Act requiring all television sets sold in interstate commerce to have a capability of receiving broadcasts over UHF (ultra-high-frequency) channels. Technology had made available for TV channels 14 through 83, in addition to channels 2 through 13, and Congress insured the use of all of them.

Most dramatically, the Communications Satellite Act of 1962 authorized United States participation in a global communications network. The constitutionality of this legislation, affirmatively exploiting the development of technology to the limits of earth's atmosphere, has not been challenged.

Federal control over advances in communications resulting from technology has not been limited to radio and television regulation, or even earth satellites. Concern for the individual's right of privacy prompted the Federal Communications Commission in 1966 to adopt rules prohibiting the use of any radio device for eavesdropping unless all parties to a private conversation consent to the use of such devices.

In numerous other areas technological innovation created new products, new methods of operation, new relationships. More importantly, the new technology permitted certain efficiencies to be achieved by larger industrial units which smaller units were incapable of achieving. These led to the creation of industrial giants known as "trusts" and an accompanying outcry against them by farmers and small businessmen. Congress responded with the Sherman Act in 1890 broadly outlawing monopolies and agreements in restraint of trade or commerce among the several states across the economic spectrum.

As in the case of railroad regulation, the Supreme Court at first was reluctant to permit this novel and extensive federal regulation of business activities. In the 1895 *E. C. Knight* case, the Court held that the Sherman Act could not be used to suppress the "sugar trust" because at most manufacturing of—not commerce in—sugar had been monopolized, as the commerce clause required for congressional control. Five years later the Court recognized that its distinction between manufacturing and commerce was plainly contrary to economic reality and the Sherman Act was held applicable to an agreement between pipe manufacturers. And the remnants of the *E. C. Knight* decision were fully laid to rest in the 1905 *Swift* case, in which the Court concluded that Congress could outlaw an agreement between meat packing houses.

While some economists and antitrust lawyers have expressed considerable dubiety over both the wisdom and effectiveness of some Sherman Act applications, the congressional authority to effectively control business agreements in restraint of trade has not been subject to serious question for six decades.

As industrial operations matured in the twentieth century and labor organizations developed as a by-product of the new technology, the resiliency of the commerce clause was once again demonstrated. Congress affirmatively adopted collective bargaining as the mechanism for settling labor disputes, and more recently positive measures requiring union democracy have been enacted. Federal standards of maximum hours and minimum wages received the blessings of a Congress concerned with social problems unknown in the day of a simpler technology. And the Supreme Court approved.

The "farm problem" is another classic example of federal action under the commerce clause to deal with problems raised by a developing technology. Rapid advances in farm machinery, fertilizers, chemicals, and electrification made it possible for a farmer to produce more food on a given acreage. This increased productivity quickly led to an increased supply of farm products which in turn resulted in reduced income for farmers. No one would suggest that Congress has solved this technologically spawned social-economic problem. However, it is only the wisdom of the measures enacted and not the authority of Congress under the commerce clause to deal with this problem which is open to question.

Most recently, the commerce clause was employed to support the public accommodations sections of the Civil Rights Act of 1964, which prohibit discrimination by hotels, restaurants, and motion picture houses. In upholding the constitutionality of this legislation, the Supreme Court recognized that Congress was using the commerce power to legislate against a moral wrong.

In this and in similar legislation it is only necessary for there to be some relationship between the conduct being regulated and interstate commerce. While this may sound like a formidable obstacle, it is not so in fact. The Supreme Court has said that conduct local in nature can be controlled if it exerts an effect on interstate commerce regardless of whether the effect is direct or indirect. And recent decisions have made it clear that in the present day national economy there is practically no activity even remotely commercial which does not satisfy this test.

The problem of congressional control over pollutions of various kinds, by-products of modern technology, is illustrative of the adequacy of the present Constitution. Leaving aside the question of whether or not Congress should exercise control in these areas and the relative merits of alternative approaches to these problems, the fact that Congress can act is inescapably clear. Congress has already enacted legislation to control water pollution and air pollution; noise pollution could also be controlled under the commerce clause.

One may be justified in feeling a sense of frustration at the absence of effective control over technology in some areas of our present society. However, the Constitution provides no barrier to what is needed.

Drastic Change

Eric Hoffer

It is my impression that no one really likes the new. We are afraid of it. It is not only as Dostoyevsky put it that "taking a new step, uttering a new word is what people fear most." Even in slight things the experience of the new is rarely without some stirring of foreboding.

Back in 1936 I spent a good part of the year picking peas. I started out early in January in the Imperial Valley and drifted northward, picking peas as they ripened, until I picked the last peas of the season, in June, around Tracy. Then I shifted all the way to Lake County, where for the first time I was going to pick string beans. And I still remember how hesitant I was that first morning as I was about to address myself to the string bean vines. Would I be able to pick string beans? Even the change from peas to string beans had in it elements of fear.

In the case of drastic change the uneasiness is of course deeper and more lasting. We can never be really prepared for that which is wholly new. We have to adjust ourselves, and every radical adjustment is a crisis in self-esteem: we undergo a test, we have to prove ourselves. It needs inordinate self-confidence to face drastic change without inner trembling.

The simple fact that we can never be fit and ready for that which is wholly new has some peculiar results. It means that a population undergoing drastic change is a population of misfits, and misfits live and breathe in an atmosphere of passion. There is a close connection between lack of confidence and the passionate state of mind and, as we shall see, passionate intensity may serve as a substitute for confidence. The connection can be observed in all walks of life. A workingman sure of his skill goes leisurely about his job and accomplishes much though he works as if at play. On the other hand, the workingman new to his trade attacks his work as if he were saving the world, and he must do so if he is to get anything done at all. The same is true of the soldier. A well-trained soldier will fight well even when not stirred by strong feeling. His morale is good because his thorough training gives him a sense of confidence. But the untrained soldier will give a good account of himself only when animated by faith and enthusiasm. Cromwell used to say that common folk needed "the fear of God before them" to match the soldierly cavaliers. Faith, enthusiasm, and passionate intensity in general are substitutes for the self-confidence born of experience and the possession of skill. Where there is the necessary skill to move mountains there is no need for the faith that moves mountains.

As I said, a population subjected to drastic change is a population of misfits—unbalanced, explosive, and hungry for action. Action is the most obvious way by which to gain confidence and prove our worth, and it is also a reaction against loss of balance—a swinging and flailing of the arms to regain one's balance and keep afloat. Thus drastic change is one of the agencies which release man's energies, but certain conditions have to be present if the shock of change is to turn people into effective men of action: there must be an abundance of opportunities, and there must be a tradition of self-reliance. Given these conditions, a population subjected to drastic change will plunge into an orgy of action.

The millions of immigrants dumped on our shores after the Civil War underwent a tremendous change, and it was a highly irritating and painful experience. Not only were they transferred, almost overnight, to a wholly foreign world, but they were, for the most part, torn from the warm communal existence of a small town or village somewhere in Europe and exposed to the cold and dismal isolation of an individual existence. They were misfits in every sense of the word, and ideal material for a revolutionary explosion. But they had a vast continent at their disposal, and fabulous opportunities for self-advancement, and an environment which held self-reliance and individual enterprise in high esteem. And so these immigrants from stagnant small towns and villages in Europe plunged into a mad pursuit of action. They tamed and mastered a continent in an incredibly short time; and we are still in the backwash of that mad pursuit.

Things are different when people subjected to drastic change find only meager opportunities for action or when they cannot, or are not allowed to, attain self-confidence and self-esteem by individual pursuits. In this case, the hunger for confidence, for worth, and for balance directs itself toward the attainment of substitutes. The substitute for self-confidence is faith; the substitute for self-esteem is pride; and the substitute for individual balance is fusion with others into a compact group.

It needs no underlining that this reaching out for substitutes means trouble. In the chemistry of the soul, a substitute is almost always explosive if for no other reason than that we can never have enough of it. We can never have enough of that which we really do not want. What we want is justified self-confidence and self-esteem. If we cannot have the originals, we can never have enough of the substitutes. We can be satisfied with moderate confidence in ourselves and with a moderately good opinion of ourselves, but the faith we have in a holy cause has to be extravagant and uncompromising, and the pride we derive from an identification with a nation, race, leader, or party is extreme and overbearing. The fact that a substitute can never become an organic part of ourselves makes our holding on to it passionate and intolerant.

To sum up: When a population undergoing drastic change is without abundant opportunities for individual action and self-advancement, it develops a hunger for faith, pride, and unity. It becomes receptive to all manner of proselytizing, and is eager to throw itself into collective undertakings which aim at "showing the world." In other words, drastic change, under certain conditions, creates a proclivity for fanatical attitudes, united action, and spectacular manifestations of flouting and defiance; it creates an atmosphere of revolution. We are usually told that revolutions are set in motion to realize radical changes. Actually, it is drastic change which sets the stage for revolution. The revolutionary mood and temper are generated by the irritations, difficulties, hungers, and frustrations inherent in the realization of drastic change.

Where things have not changed at all, there is the least likelihood of revolution.

The Chaotic Society:
The Social Morphological Revolution

Philip M. Hauser

Presidential address, 63rd annual meetings of the American Sociological Association, Boston, Massachusetts, August 28, 1968.

Society as a whole has been viewed historically from many perspectives. It has been envisaged among other ways as "the great society," "the acquisitive society," and "the affluent society." Contemporary society, whether observed globally, nationally, or locally, is realistically characterized as "the chaotic society" and best understood as "the anachronistic society."

Contemporary society is realistically characterized as chaotic because of its manifest confusion and disorder—the essential elements of chaos. On the international scene, to draw upon a few examples, consider the situation in Vietnam, Czechoslovakia, the Middle East, and Nigeria. On the national level consider the United States, France, the United Kingdom, and almost any country in Latin America or Africa. On the local level, in the United States, consider New York, Chicago, Los Angeles, Detroit, Cleveland, Memphis, Miami, and over 100 other cities which have been wracked by violence.

Contemporary society can be best understood when it is viewed as an anachronistic society. To be sure society at any time, at least during the period of recorded history, has been an anachronistic society. For throughout the millennia of the historical era, society, at any instant in time, comprised layers of culture which, like geological strata, reflected the passage and deposits of time. Confusion and disorder, or chaos, may be viewed in large part as the resultant of the dissonance and discord among the various cultural strata, each of which tends to persist beyond the set of conditions, physical and social, which generated it.

In some ways chaos in contemporary society differs from that in earlier societies only in the degree. But there are a number of unique factors in contemporary chaos which make it more a difference in kind. First, contemporary society, as the most recent, contains the greatest number of cultural layers, and, therefore, the greatest potential for confusion and disorder. Second, contemporary society, by reason of the social morphological revolution, possesses cultural layers much more diverse than any predecessor society and, therefore, much greater dissonance. Third, contemporary society unlike any predecessor contains the means of its own destruction, the ultimate weapon, the explosive power of nuclear fission. Fourth, fortunately, contemporary society unlike any predecessor possesses the knowledge, embodied in the emerging social sciences including sociology, that affords some hope for the dissipation of confusion and the restoration of order before the advent of collective suicide and societal annihilation. It is a moot question, however, as to whether society yet possesses the will and the organization to utilize available knowledge to this end.

By reason of these considerations the theme of this annual meeting of the American Sociological Association is most appropriate—"On the Gap Between Sociology and Social Policy." For sociology, as well as the other social sciences, provides knowledge, even though limited, permitting an understanding of society, contemporary and historical; and, in consequence, offers some hope for rational action towards the resolution of the chaos which afflicts us.

It is my central thesis that contemporary society, the chaotic and anachronistic society, is experiencing unprecedented tensions and strains by reason of the social morphological revolution. The key to the understanding of contemporary society lies, therefore, in an understanding of the social morphological revolution. Moreover, it is a corollary thesis that comprehension of the social morphological revolution points to the directions social engineering must take for the reduction or elimination of the chaos that threatens the viability of contemporary society.

I am mindful of the fact that "the social morphological revolution" is not a familiar rubric to the sociological fraternity—nor to anyone else. It is a neologism, albeit with a legitimate and honorable ancestry, for which I must plead guilty. I offer two justifications for injecting this abominable rhetoric into the literature. First, I am convinced that it contains useful explanatory power that has not yet been fully exploited in macro-social considerations and empirical research or social engineering activities. Second, it is appropriate that the discipline of sociology possess a revolution of its own. After all, the agronomists have the "agricultural revolution," economists the "commercial" and "industrial" revolutions, natural scientists the "scientific revolution," engineers the "technological revolution," and demographers the "vital revolution." Each of these revolutions is obviously the invention of scholars seeking a short and snappy chapter for a book title to connote complex and highly significant patterns of events. Sociologists, even if they have not formally recognized it to date, have the "social morphological revolution" and perhaps it is in order formally to acknowledge and to baptize it.

What is this social morphological revolution and what are its antecedents?

To answer the second of these questions first, let me repeat that its ancestry is legitimate and honorable. Durkheim, encapsulating earlier literature, provided in a focused way insight into the implications of the most abstract way of viewing a society, namely, by size and density of its population. In his consideration of the structure of the social order Durkheim used the term "social morphology." Wirth in his classical article "Urbanism as a Way of Life," drawing on Aristotle, Durkheim, Tonnies, Sumner, Willcox, Park, Burgess and others, explicitly dealt with the impact of size, density and heterogeneity of population on human behavior and on the social order.

The social morphological revolution refers to the changes in the size, density and heterogeneity of population and to the impact of these changes on man and society. As far as I know, the term was first published in my Presidential Address to the American Statistical Association in 1963. It was used in conjunction with my explication of the "size-density model." This model provides a simplistic demonstration of the multiplier effect on potential human interaction of increased population density in a fixed land area and, therefore, can appropriately be described as an index of the size and density aspects of the social morphological revolution. . . . It should not be too surprising that white racism is now breeding or exacerbating black racism, and, therefore, intensified hostility and conflict. Furthermore, the paralysis of government in the United States, as described above, further compounds the crisis and offers little hope of any short-run resolution of tension and conflict. This nation, on its present course, may well be in for an indefinite period of guerilla warfare on the domestic as well as on the international front.

Other Examples of Cultural Lag. There are many other examples of cultural lag in American society ranging from the trivial to the significant. In the trivial category is the persistence of the string designed to keep collars closed against inclement weather before the advent of the pin and the button. This string has become the necktie—a relatively harmless, if not always asthetic vestige which has acquired a new function—decoration. But other vestiges are not as harmless. They include the constitutional right to bear arms —admittedly necessary in 18th century America but a dangerous anachronism in the last third of 20th century America. They include also the inalienable rights of labor to strike and of management to shut down and employ the lockout, often through trial by ordeal of the public. In twentieth century mass society, labor's right to strike and management's right to lockout may be described as the rights of labor and management to revert to the laws of the jungle—to resolve their conflicts of interest by means of brute force. The same can be said of the so-called right of the students to impose their views through the employment of force—or of any person or group who fails to resolve conflicts of interest in a mass society by an adjudicative or democratic procedure.

Cultural atavisms are replete, also, in the administration of criminal justice, for many of the governing codes and procedures are of pre-social morphological revolution origin and constitute a menace to a mass society.

Finally, and by no means to exhaust the universe of cultural lags, mention should be made of organized religion as a living museum of cultural atavisms adding to the confusion and disorder of contemporary life. Sunday morning Christians have learned to honor and revere the messenger, his mother and his colleagues; they have learned to observe the ritual and practices of their churches which have endured for two millenia; but they have not received, or certainly they have not heeded, the message. For the message of the Judeo-Christian tradition is found in the concept of the Fatherhood of God—which implies the brotherhood of man. And comparable things can be said of the adherents of the other religions—the Jews, the Muslims, the Hindus, the Buddhists, and so on.

Interestingly enough, the concept of the brotherhood of man, apart from its supernatural context, is an excellent example of an ancient ethical principle which has great applicability to contemporary as well as to previous societies. Although I have pointed to cultural survivals which create confusion and disorder this is not to be interpreted to mean that all that is the product of the past is incompatible with the present. In fact, it may be argued that the increased interdependence and vulnerability of the mass society place a greater premium on this moral principle than any earlier society ever did. This is an example of a principle of mass living that has not yet taken hold despite its longevity—a principle the adoption of which in deed, as well as in word, may be prerequisite to the continued existence of mankind.

Before departing from the subject of religion, I cannot, as a demographer, refrain from calling attention to the cultural dissonance represented by Pope Paul VI's recent encyclical "Of Human Life" which ignores the findings of empirical demography. This example of cultural lag closely parallels that afforded by the Roman Catholic Church during the reign of Pope Paul V, which, some three centuries ago, similarly ignored the findings of empirical astronomy—and produced the Galileo incident.

Among the most serious consequences of the failure of contemporary American society to keep pace with the social morphological revolution is evident in the deficiencies of the process of socialization. Bronfenbrenner illuminates this problem in his comparative study of education in the United States and the Soviet Union. In the USSR the child is so inbred with a sense of belonging and obligation to the society of which he is an infinitesimal part that he tends to lack initiative and creativity. In the United States, in contrast, the child is so little the recipient of a sense of membership in, and responsibility to, the social order that, although he develops great initiative and creativity, his attitude is essentially one of concern with how he gets his and unconcern with others. We have yet to achieve the golden mean of these extremes to produce human products consonant with the requirement of the mass society —human beings with a balance of initiative, creativity and social responsibility.

On the international front, there is similar evidence of cultural lag. Most grave in its consequences, obviously, is the failure to achieve the resolution of national conflicts of interest by means other than physical force. Vietnam, the Middle East, and Nigeria are but a few timely reminders of this fact. The social morphological revolution has generated a highly interdependent, vulnerable and shrunken world increasing the probability and intensifying the nature of conflicts of interest. But the traditional means of resolving international tensions and hostilities, namely, war, in a society which possesses the hydrogen bomb, carries with it the threat of the ultimate disaster—even the extinction of mankind. Nevertheless, contemporary diplomatic policies and contemporary military postures, are more the product of societies of the past than of the present.

To be sure, some progress has been made in the evolving of machinery for the peaceful resolution of international disputes as exemplified by the League of Nations, the World Court, the United Nations and the Specialized Agencies. But it is not yet certain that the United Nations will not follow the League of Nations into oblivion as is actually desired by some of our most anachronistic organizations—such as the Daughters of the American Revolution and the John Birch Society. If the plague of deleterious cultural survivals which afflict contemporary society cannot be effectively dealt with, it may well be that nuclear holocaust will yet be the means to undo both the process and the products of the social morphological revolution.

Finally, on the international front, at least mention must be made of the cleavages between the have and have-not nations, between the socialist and communist nations, and between the factions within these blocs. The great disparities in levels of living among the nations of the world and the great international ideological differences, in part products of the differential impact of the social morphological revolution, constitute the most serious threats to peace and are harbingers of potential disaster. It remains to be seen whether contemporary society can muster the will to utilize available knowledge in a manner to override ideological, structural and procedural atavisms to cope with these problems. In this year, officially proclaimed by the United Nations as Human Rights Year, it is a sad commentary on the role of this nation that the Congress has reduced foreign aid appropriations to an all-time low. And it is an even sadder commentary on the state of international affairs that the world spends well over 100 billion dollars annually for the military while the developing nations, after a disastrous "Development Decade," still starve for capital and other resources to achieve their economic development goals. . . .

A hopeful indication that the social morphological revolution is producing mechanisms for the resolution of the social problems it has precipitated lies in the bill now before Congress calling for the establishment of a parallel Council of Social Advisors, and an annual Social Report to the nation. Furthermore, the Department of Health, Education and Welfare has through its Advisory Panel on Social Indicators been engaged, on instruction from the President of the United States, in the preparation of a prototype Social Report.

The unprecedented period of high level economic activity uninterrupted by depression or recession that this nation has recently experienced is certainly related to the existence and activities of the Council of Economic Advisors. We have recently experienced a costly inflation and we are now threatened by a possible recession mainly because the Congress, a repository of cultural lag, has not heeded, or tardily heeded, the recommendations of the Administration based on the findings of the Council of Economic Advisors.

It is my judgment that had this nation possessed a Council of Social Advisors since 1947, along with the Council of Economic Advisors, and had the recommendations of such a Council been heeded by the Administration and the Congress, the "urban crisis" which sorely affects us would not have reached its present acute stage.

It is the role of the social sciences including sociology to generate the knowledge on the basis of which social policy and social action may be directed to the solution of our problems. The primary function of the social scientist is research, the production of knowledge. It is not the function of the social scientist, *qua* scientist, to be a social engineer. To be sure, many of us social scientists have in the early stage of the development of the social sciences been called upon to perform both roles. But there can be no question about the fact that the two roles are distinct ones; and that each, in the long run, will be better performed as separate and specialized activities.

More specifically, it is the role of the social scientist, including the sociologist, to develop and produce the "social indicators" which will permit effective social accounting. Fortunately, the social morphological revolution has generated much in the way of social statistics and other types of knowledge which are already quite impressive even if still deficient and in relatively early stages of evolution.

Social accounting will become possible only after consensus is achieved on social goals. The development of social goals is not a scientific function. It is not a social engineering function. It is a function that must be performed by society as a whole, acting through its political and other leaders. In a democratic society it presumably reflects the desires of the majority of the people.

Although a majority of the people must fix the goals of a society, the social scientist and the social engineer are in a strategic position to participate in goal formation. They must work closely with political and other leaders to help develop a broad spectrum of choices bolstered, as far as possible, by knowledge about the requirements and consequences of specific goals. I have elsewhere proposed one set of social goals for consideration—published in a recent report of the Joint Economic Committee of the Congress.

Man is the only significant culture-building animal on the face of the earth. He not only adapts to environment he creates environment to which to adapt. Man has created a world in which mankind itself is the crucial environment—a mankind characterized by large numbers, high densities and great heterogeneity. He is still learning how to live in this new world he has created. The product of the chief components of the social morphological revolution —the population explosion, the population implosion and population diversification—together with rapid technological and social change—is contemporary society, a chaotic society, an anachronistic society. It is a society characterized by dissonant cultural strata—by confusion and disorder. It is also a society which for the first time in human history possesses the capacity to destroy itself—globally as well as nation by nation.

In addition to the acceleration in the rate of technological and social change, and partly in response to it, society has acquired a greater capacity for social change. Virtually instantaneous world wide social interaction is possible with modern means of communication; and the mass media, bolstered by communication satellites and new educational hardware, create new opportunities for the modification and creation of attitudes and behaviorisms consistent with the realities of the contemporary world. But, although the capability for social change has undoubtedly increased, adequate and effective mechanisms for the control of social change, for accommodation and adaptation to the changing social milieu, as well as to the changing material world, has yet to be evolved. Planning, as a mechanism for rational decision-making, is still in its infancy and has yet to develop an integrated approach with apposite administrative, economic and social planning along with physical planning. Progress is being made in this respect, however. In this nation, for example, planning has become a respectable word now if modified by the term "city"; but when modified by such terms as "metropolitan," "regional," or "national," it is still considered a dangerous thought in some quarters. But planning, in ever broader contexts, will undoubtedly be a first step in the dissipation of confusion and the restoration of order.

That we live in a chaotic world should not be too surprising in view of the perspective provided by calendar considerations. Only 12 human generations have elapsed since the "modern era" began. Only 7 human generations have elapsed since this nation was founded. Only 6 generations have elapsed since mankind acquired the means to permit the proliferation of cities of a million or more inhabitants. Only 2 generations have elapsed since the onset of significant internal migratory flows of Afro-Americans. Less than 2 generations have elapsed since the United States became an urban nation. Less than one generation has elapsed since the advent of the explosive power of the atom. Little more than a decade has elapsed since the Supreme Court decision outlawing *de jure* segregation in America's schools—and a clearcut judicial decision on *de facto* segregation is yet to come.

Furthermore, only two human generations have elapsed since Durkheim and Weber and, to confine my attention to my own teachers and colleagues, less than one since Burgess, Ogburn, Redfield, and Wirth. The social sciences, in general, and sociology in particular, are still emergent sciences. It was only during the century, roughly from about 1750 to 1850, that the physical sciences achieved the respectability and acceptance that paved the way, through engineering, for the transformation of the physical and material world. It was only during the century, roughly from 1850 to 1950, that the bio-medical sciences achieved a similar status that paved the way, by means of bio-medical engineering, for the remarkable increase in longevity and health. It is to be hoped that the century from 1950 to 2050 will be the period during which the social sciences including sociology will achieve a level of respectability and acceptance that will pave the way for social engineering to eliminate the chaos that characterizes contemporary society. The question is whether mankind can muddle through without collective suicide before rational decision-making overtakes the confusion and disorder of our tottering transitional society.

The Growing Importance of Human Ecology

Lawrence E. Hinkle, Jr., M.D.

"Ecology" is an "in" word these days. You hear it used by people in government, business and academic life, by people who are engaged in activities as varied as the control of air pollution, the use of pesticides, the development of highways, the control of population growth, and the study of disease.

"Ecology" is the study of the "oikos": the "neighborhood," the "dwelling place," the "habitat." Specifically it is the study of the *interrelations* between organisms and their environment.

By its very nature all ecology is complex, but human ecology is especially complex, because of the complex nature of man. Not only must man adapt to the food he eats, the air he breathes, and the bacteria and pollens he encounters, but he must also adapt to his family, his community, his job, and to many other facets of his society, the people in it, and the rapidly changing technology he has created.

It has been this rapid development of technology that has led to the present growing concern about human ecology:

As we have created new pesticides and have applied them to our fields in order to obtain a greater yield of food crops, we have found that when we destroyed the pests, we destroyed other plants and animals that we wish to have. Furthermore, some of our chemicals have been eaten by fish or animals, which people in turn have consumed, and the chemicals have found their way into human systems.

As we have developed electric dishwashers so we could get the housewife out of the kitchen, we have found that we needed new detergents to make them work well. Unfortunately, some of these new detergents were immune to the bacterial action that destroys soap. They passed through our septic systems and our sewage disposal plants. Soon we had suds in our streams and foam in our drinking water.

As we have developed automobiles to get us around more quickly, incinerators to burn our trash, and power plants to supply us with our ever-growing needs for electricity, we have found that there is a haze in the air over our cities, and on still days our eyes smart because of the smog that we have created.

But these are only several of the simpler and more obvious effects of some of our interactions with our environment. In many ways, some subtle and some not, the whole pattern of human life is being changed by technology. If you compare the life of a farmer of five generations ago with that of an American working man of today, you can quickly see contrasts at almost every point.

Consider the farmer of the year 1800. His major threats to existence were infection, malnutrition and injury. To him the dangers he faced were immediate: an angry bull or a swarm of bees. Or they were unpredictable: pestilence, drought or storm. There were few options for him. He could meet his challenges by hard work, by prayer, or not at all.

Compare this man with an American workman of today. The major threats to this man's existence are no longer infection or malnutrition; he is much less likely to die or be disabled by the effects of an injury. No longer are he and his children prey to typhoid, dysentery, cholera, smallpox, diphtheria, pneumonia and tuberculosis. They do not starve or get rickets, scurvy, pellagra or protein malnutrition. They are not nearly so likely to die from the effects of compound fractures, osteomyelitis, or appendicitis. Freed of these causes of early mortality, this man and his family live much longer than their ancestors. They die at a later age of heart disease, cancer and stroke, of suicide and alcoholism, and with a variety of metabolic disorders and an increasing number of diseases which apparently are caused by the disturbances of his own defense mechanisms.

Diseases an Outgrowth of Environment

Physicians today—and especially industrial health specialists—are interested in human ecology because they believe that the diseases of modern man may be in part an outgrowth of his modern environment. Physicians suspect that the fatty atherosclerotic deposits in human arteries, which lay the ground for heart attacks and strokes, may be related to our abundant food supply, and possibly to our rich supply of protein and animal fat. They believe that our abundant food supply is also a factor in the prevalence of obesity, diabetes and high blood pressure.

Physicians do believe that the lack of physical activity among modern men contributes to the weakness of the muscles of their hearts, as well as to the muscles of their bodies, and that this prevents their hearts from developing that rich supply of nutrient vessels which would make them better able to survive the effects of the closing of their coronary arteries by atherosclerosis. The necessity for meeting deadlines and time schedules, and for stifling aggression, may have physiological consequences that make men more likely to have heart attacks. These features of modern life are an important reason why so many people have symptoms of anxiety, fatigue and insomnia. They may be one reason why the smoking of cigarettes and the consumption of alcohol is so high in our society, and why people are so reluctant to give up these habits. Smoke that men inhale may cause some forms of severe chronic lung disease as well as cancer. Other chemicals that men inhale, eat or drink may also cause serious illnesses.

The modern physician believes this, but he does not "know" it with the assurance that he "knows" that tubercule bacilli are involved in tuberculosis. The evidence relating to the causes of modern diseases is not as convincing or complete as the evidence relating to the causes of diseases of yesteryear. In the classical infectious diseases and nutritional disorders, the role of a single environmental agent was of such overwhelming importance that other factors could, in effect, be disregarded. It did not matter that pulmonary tuberculosis was a disease of the urban workers, or that pellagra was a disease of the southern sharecroppers; if you could control the infectious agent or supply the missing vitamins, the disease would disappear.

This does not seem to be true of coronary heart disease, for example. An abundant diet, a sedentary life, a large amount of smoking and the effects of meeting deadlines and struggling to get ahead have all been implicated as causes of heart attacks, but none of these seem to be "the cause" of the disease. There may be a common thread running through all of these causes, but such a thread has not yet been discovered. As of this moment it appears that a number of "causes" cooperate to produce the conditions under which coronary heart disease and many other modern diseases disappear.

The physician who is interested in the effects of human ecology upon disease likes to study people in their natural habitat as they go about their daily lives. He often needs large numbers of people. By studying different people involved in the same life pattern, or similar people engaged in different life patterns, he can gain some idea of the effects that a pattern of life has upon a man's health, and the method by which this effect is produced.

This is why we at the Division of Human Ecology at the Cornell University Medical College in New York City have asked a number of men in the Bell System to collaborate with us during the past five years in our studies of human ecology in relation to coronary heart disease. These men, selected from various Bell System companies, have undergone many different kinds of diagnostic tests, and have answered questions about their health, their activities, and their habits.

Several years ago, more than 100 men from the New York Telephone Company underwent a long series of psychological tests. Some of these men had had heart attacks; others had had a completely clean bill of health and showed no evidence of characteristics that are thought of as predisposing to coronary disease. In 1963 and 1964 more than 300 men from the New Jersey Bell Telephone Company went through an elaborate series of diagnostic tests, filled out a long questionnaire, and went through a number of interviews designed to find out about their past health, the health of their parents, brothers and sisters, their daily routines, and their habits of exercise, smoking, eating, drinking and taking medicine. These men also participated in a seven-hour test during which their hearts were monitored constantly. Now four years later, they are coming back to Cornell—New York Hospital for another series of tests.

Studies Yield Mass of Data

As you might expect, studies such as these yield a great mass of data. For each man in the New Jersey study, there are more than 80 punched cards of data to be analyzed by computers. In this study, every episode of disability caused by coronary heart disease among men in the telephone companies has been reported anonymously, along with information that would allow us to estimate how various factors of geography and work history have influenced the occurrence of heart attacks.

Some results of the studies to date can be summarized briefly. Although coronary disease is the most common cause of death among active male Bell System employees, the death rate from that cause here is slightly lower than that for men of comparable age in the nation as a whole. It is not peculiarly a disease of higher level managers and executives. The investigation so far shows no indication that such factors of work experience as promotions, transfers, or new job assignments, *taken by themselves,* have had any marked effect on the risk of death from coronary disease. The results of the studies are not yet final. Continued study will shed further light on the relation between life patterns and the incidence of coronary disease among men in the Bell System.

All of these studies have been concerned with the effects of the human environment on human health; but it would be wrong to give the impression that human ecology is concerned primarily with health. At many institutions throughout the country, scientists from many disciplines are studying many facets of human ecology: how different patterns of water utilization may affect the ecology of a river valley; how different patterns of industrial production affect the pollution of the atmosphere; how the heat produced by cities affects the weather around them; how population planning may affect the characteristics of national populations. There are studies of transportation patterns, housing designs, pesticides, and the effects of time change during air travel.

No single aspect of these studies is new. What is new is an awareness among modern scientists of the complex interrelations that exist between society, technology, people and the surrounding world.

Formerly those concerned with developing a new airplane considered only those features of the plane which would affect its flying qualities. Now, as the new supersonic transport is being designed, scientists are considering how its exhaust will contaminate the atmosphere; how its noise will disturb the people around the airport; what kind of new airports will be needed; how these airports may disturb population patterns, transportation, and the economy of the region in which they may be located. Such considerations of human ecology are becoming an ever greater part of the science of today.

Population and Panaceas:
A Technological Perspective

Paul R. Ehrlich and John P. Holdren

Today more than one billion human beings are either undernourished or malnourished, and the human population is growing at a rate of 2 percent per year. The existing and impending crises in human nutrition and living conditions are well-documented but not widely understood. In particular, there is a tendency among the public, nurtured on Sunday-supplement conceptions of technology, to believe that science has the situation well in hand—that farming the sea and the tropics, irrigating the deserts, and generating cheap nuclear power in abundance hold the key to swift and certain solution of the problem. To espouse this belief is to misjudge the present severity of the situation, the disparate time scales on which technological progress and population growth operate, and the vast complexity of the problems beyond mere food production posed by population pressures. Unfortunately, scientists and engineers have themselves often added to the confusion by failing to distinguish between that which is merely theoretically feasible, and that which is economically and logistically practical.

As we will show here, man's present technology is inadequate to the task of maintaining the world's burgeoning billions, even under the most optimistic assumptions. Furthermore, technology is likely to remain inadequate until such time as the population growth rate is drastically reduced. This is not to assert that present efforts to "revolutionize" tropical agriculture, increase yields of fisheries, desalt water for irrigation, exploit new power sources, and implement related projects are not worthwhile. They may be. They could also easily produce the ultimate disaster for mankind if they are not applied with careful attention to their effects on the ecological systems necessary for our survival (Woodwell, 1967; Cole, 1968). And even if such projects are initiated with unprecedented levels of staffing and expenditures, without population control they are doomed to fall far short. No effort to expand the carrying capacity of the Earth can keep pace with unbridled population growth.

To support these contentions, we summarize briefly the present lopsided balance sheet in the population/food accounting. We then examine the logistics, economics, and possible consequences of some technological schemes which have been proposed to help restore the balance, or, more ambitiously, to permit the maintenance of human populations much larger than today's. The most pertinent aspects of the balance are:

1. The world population reached 3.5 billion in mid-1968, with an annual increment of approximately 70 million people (itself increasing) and a doubling time on the order of 35 years (Population Reference Bureau, 1968).

2. Of this number of people, at least one-half billion are undernourished (deficient in calories or, more succinctly, slowly starving), and approximately an additional billion are malnourished (deficient in particular nutrients, mostly protein) (Borgstrom, 1965; Sukhatme, 1966). Estimates of the number actually perishing annually from starvation begin at 4 million and go up (Ehrlich, 1968) and depend in part on official definitions of starvation which conceal the true magnitude of hunger's contribution to the death rate (Lelyveld, 1968).

3. Merely to maintain present inadequate nutrition levels, the food requirements of Asia, Africa, and Latin America will, conservatively, increase by 26 percent in the 10-year period measured from 1965 to 1975 (Paddock and Paddock, 1967). World food production must double in the period 1965–2000 to stay even; it must triple if nutrition is to be brought up to minimum requirements.

Food Production

That there is insufficient additional, good quality agricultural land available in the world to meet these needs is so well documented (Borgstrom, 1965) that we will not belabor the point here. What hope there is must rest with increasing yields on land presently cultivated, bringing marginal land into production, more efficiently exploiting the sea, and bringing less conventional methods of food production to fruition. In all these areas, science and technology play a dominant role. While space does not permit even a cursory look at all the proposals on these topics which have been advanced in recent years, a few representative examples illustrate our points.

Conventional Agriculture. Probably the most widely recommended means of increasing agricultural yields is through the more intensive use of fertilizers. Their production is straightforward, and a good deal is known about their effective application, although, as with many technologies we consider here, the environmental consequences of heavy fertilizer use are ill understood and potentially dangerous[1] (Wadleigh, 1968). But even ignoring such problems, we find staggering difficulties barring the implementation of fertilizer technology on the scale required. In this regard the accomplishments of countries such as Japan and the Netherlands are often cited as offering hope to the underdeveloped world. Some perspective on this point is afforded by noting that if India were to apply fertilizer at the per capita level employed by the Netherlands, her fertilizer needs would be nearly half the present world output (United Nations, 1968).

On a more realistic plane, we note that although the goal for nitrogen fertilizer production in 1971 under India's fourth 5-year plan is 2.4 million metric tons (Anonymous, 1968a), Raymond Ewell (who has served as fertilizer production adviser to the Indian government for the past 12 years) suggests that less than 1.1 million metric tons is a more probable figure for that date. Ewell[2] cites poor plant maintenance, raw materials shortages, and power and transportation breakdowns as contributing to continued low production by existing Indian plants. Moreover, even when fertilizer is available, increases in productivity do not necessarily follow. In parts of the underdeveloped world lack of farm credit is limiting fertilizer distribution; elsewhere, internal transportation systems are inadequate to the task. Nor can the problem of educating farmers on the advantages and techniques of fertilizer use be ignored. A recent study (Parikh et al., 1968) of the Intensive Agriculture District Program in the Surat district of Gujarat, India (in which scientific fertilizer use was to have been a major ingredient) notes that "on the whole, the performance of adjoining districts which have similar climate but did not enjoy relative preference of input supply was as good as, if not better than, the programme district. . . . A particularly disheartening feature is that the farm production plans, as yet, do not carry any educative value and have largely failed to convince farmers to use improved practices in their proper combinations."

As a second example of a panacea in the realm of conventional agriculture, mention must be given to the development of new high-yield or high-protein strains of food crops. That such strains have the potential of making a major contribution to the food supply of the world is beyond doubt, but this potential is limited in contrast to the potential for population growth, and will

---

1. Barry Commoner, address to 135th Meeting of the AAAS, Dallas, Texas (28 December 1968).
2. Raymond Ewell, private communication (1 December 1968).

be realized too slowly to have anything but a small impact on the immediate crisis. There are major difficulties impeding the widespread use of new high-yield grain varieties. Typically, the new grains require high fertilizer inputs to realize their full potential, and thus are subject to all the difficulties mentioned above. Some other problems were identified in a recent address by Lester R. Brown, administrator of the International Agricultural Development Service: the limited amount of irrigated land suitable for the new varieties, the fact that a farmer's willingness to innovate fluctuates with the market prices (which may be driven down by high-yield drops), and the possibility of tieups at market facilities inadequate for handling increased yields.[3]

Perhaps even more important, the new grain varieties are being rushed into production without adequate field testing, so that we are unsure of how resistant they will be to the attacks of insects and plant diseases. William Paddock has presented a plant pathologist's view of the crash programs to shift to new varieties (Paddock, 1967). He describes India's dramatic program of planting improved Mexican wheat, and continues: "Such a rapid switch to a new variety is clearly understandable in a country that tottered on the brink of famine. Yet with such limited testing, one wonders what unknown pathogens await a climatic change which will give the environmental conditions needed for their growth." Introduction of the new varieties creates enlarged monocultures of plants with essentially unknown levels of resistance to disaster. Clearly, one of the prices that is paid for higher yield is a higher risk of widespread catastrophe. And the risks are far from local: since the new varieties require more "input" of pesticides (with all their deleterious ecological side effects), these crops may ultimately contribute to the defeat of other environment-related panaceas, such as extracting larger amounts of food from the sea.

A final problem must be mentioned in connection with these strains of food crops. In general, the hungriest people in the world are also those with the most conservative food habits. Even rather minor changes, such as that from a rice variety in which the cooked grains stick together to one in which the grains fall apart, may make new foods unacceptable. It seems to be an unhappy fact of human existence that people would rather starve than eat a nutritious substance which they do not recognize as food.[4]

Beyond the economic, ecological, and sociological problems already mentioned in connection with high-yield agriculture, there is the overall problem of time. We need time to breed the desired characteristics of yield and hardiness into a vast array of new strains (a tedious process indeed), time to convince farmers that it is necessary that they change their time-honored ways of cultivation, and time to convince hungry people to change the staples of

---

3. Lester R. Brown, address to the Second International Conference on the War on Hunger, Washington, D.C. (February 1968).
4. For a more detailed discussion of the psychological problems in persuading people to change their dietary habits, see McKenzie, 1968.

their diet. The Paddocks give 20 years as the "rule of thumb" for a new technique or plant variety to progress from conception to substantial impact on farming (Paddock and Paddock, 1967). They write: "It is true that a *massive* research attack on the problem could bring some striking results in less than 20 years. But I do not find such an attack remotely contemplated in the thinking of those officials capable of initiating it." Promising as high-yield agriculture may be, the funds, the personnel, the ecological expertise, and the necessary years are unfortunately not at our disposal. Fulfillment of the promise will come too late for many of the world's starving millions, if it comes at all.

Bringing More Land Under Cultivation. The most frequently mentioned means of bringing new land into agricultural production are farming the tropics and irrigating arid and semiarid regions. The former, although widely discussed in optimistic terms, has been tried for years with incredibly poor results, and even recent experiments have not been encouraging. One essential difficulty is the unsuitability of tropical soils for supporting typical foodstuffs instead of jungles (McNeil, 1964; Paddock and Paddock, 1964). Also, "the tropics" are a biologically more diverse area than the temperate zones, so that farming technology developed for one area will all too often prove useless in others. We shall see that irrigating the deserts, while more promising, has serious limitations in terms of scale, cost, and lead time.

The feasible approaches to irrigation of arid lands appear to be limited to large-scale water projects involving dams and transport in canals, and desalination of ocean and brackish water. Supplies of usable ground water are already badly depleted in most areas where they are accessible, and natural recharge is low enough in most arid regions that such supplies do not offer a long-term solution in any case. Some recent statistics will give perspective to the discussion of water projects and desalting which follows. In 1966, the United States was using about 300 billion gal of water per day, of which 135 billion gal were consumed by agriculture and 165 billion gal by municipal and industrial users (Sporn, 1966). The bulk of the agricultural water cost the farmer from 5 to 10 cents/1000 gal; the highest price paid for agricultural water was 15 cents/1000 gal. For small industrial and municipal supplies, prices as high as 50 to 70 cents/1000 gal were prevalent in the U.S. arid regions, and some communities in the Southwest were paying on the order of $1.00/1000 gal for "project" water. The extremely high cost of the latter stems largely from transportation costs, which have been estimated at 5 to 15 cents/1000 gal per 100 miles (International Atomic Energy Agency, 1964).

We now examine briefly the implications of such numbers in considering the irrigation of the deserts. The most ambitious water project yet conceived in this country is the North American Water and Power Alliance, which proposes to distribute water from the great rivers of Canada to thirsty locations all over the United States. Formidable political problems aside (some based on the certainty that in the face of expanding populations, demands for water will eventually arise at the source), this project would involve the expenditure of $100 billion in construction costs over a 20-year completion period. At the end of this time, the yield to the United States would be 69 million acre feet of water annually (Kelly, 1966), or 63 billion gal per day. If past experience with massive water projects is any guide, these figures are overoptimistic; but if we assume they are not, it is instructive to note that this monumental undertaking would provide for an increase of only 21 percent in the water consumption of the United States, during a period in which the population is expected to increase by between 25 and 43 percent (U.S. Dept. of Commerce, 1966). To assess the possible contribution to the *world* food situation, we assume that all this water could be devoted to agriculture, although extrapolation of present consumption patterns indicates that only about one-half would be. Then using the rather optimistic figure of 500 gal per day to grow the food to feed one person, we find that this project could feed 126 million additional people. Since this is less than 8 percent of the projected world population growth during the construction period (say 1970 to 1990), it should be clear that even the most massive water projects can make but a token contribution to the solution of the world food problem in the long term. And in the crucial short term—the years preceding 1980—*no* additional people will be fed by projects still on the drawing board today.

In summary, the cost is staggering, the scale insufficient, and the lead time too long. Nor need we resort to such speculation about the future for proof of the failure of technological "solutions" in the absence of population control. The highly touted and very expensive Aswan Dam project, now nearing completion, will ultimately supply food (at the present miserable diet level) for less than Egypt's population growth during the time of construction (Borgstrom, 1965; Cole, 1968). Furthermore, its effect on the fertility of the Nile Delta may be disastrous, and, as with all water projects of this nature, silting of the reservoir will destroy the gains in the long term (perhaps in 100 years).

Desalting for irrigation suffers somewhat similar limitations. The desalting plants operational in the world today produce water at individual rates of 7.5 million gal/day and less, at a cost of 75 cents/1000 gal and up, the cost increasing as the plant size decreases (Bender, 1969). The most optimistic firm proposal which anyone seems to have made for desalting with present or soon-to-be available technology is a 150 million gal per day nuclear-powered installation studied by the Bechtel Corp. for the Los Angeles Metropolitan Water District. Bechtel's early figures indicated that water from this complex would be available at the site for 27-28 cents/1000 gal (Galstann and Currier, 1967). However, skepticism regarding the economic assumptions leading to these figures (Milliman, 1966) has since proven justified—the project was shelved after spiralling construction cost estimates indicated an actual water cost of 40-50 cents/1000 gal. Use of even the original figures, however, bears out our contention that the *most* optimistic assumptions do not alter the verdict that technology is losing the food/population battle. For 28 cents

1000/gal is still approximately twice the cost which farmers have hitherto been willing or able to pay for irrigation water. If the Bechtel plant had been intended to supply agricultural needs, which it was not, one would have had to add to an already unacceptable price the very substantial cost of transporting the water inland.

Significantly, studies have shown that the economies of scale in the distillation process are essentially exhausted by a 150 million gal per day plant (International Atomic Energy Agency, 1964). Hence, merely increasing desalting capacity further will not substantially lower the cost of the water. On purely economic grounds, then, it is unlikely that desalting will play a major role in food production by conventional agriculture in the short term.[5] Technological "break-throughs" will presumably improve this outlook with the passage of time, but world population growth will not wait.

Desalting becomes more promising if the high cost of the water can be offset by increased agricultural yields per gallon and, perhaps, use of a single nuclear installation to provide power for both the desalting and profitable on-site industrial processes. This prospect has been investigated in a thorough and well-documented study headed by E. S. Mason (Oak Ridge National Laboratory, 1968). The result is a set of preliminary figures and recommendations regarding nuclear-powered "agro-industrial complexes" for arid and semiarid regions, in which desalted water and fertilizer would be produced for use on an adjacent, highly efficient farm. In underdeveloped countries incapable of using the full excess power output of the reactor, this energy would be consumed in on-site production of industrial materials for sale on the world market. Both near-term (10 years hence) and far-term (20 years hence) technologies are considered, as are various mixes of farm and industrial products. The representative near-term case for which a detailed cost breakdown is given involves a seaside facility with a desalting capacity of 1 billion gal/day, a farm size of 320,000 acres, and an industrial electric power consumption of 1585 Mw. The initial investment for this complex is estimated at $1.8 billion, and annual operating costs at $236 million. If both the food and the industrial materials produced were sold (as opposed to giving the food, at least, to those in need who could not pay),[6] the estimated profit for such a complex, before subtracting financing costs, would be 14.6 percent.

The authors of the study are commendably cautious in outlining the assumptions and uncertainties upon which these figures rest. The key assumption is that 200 gal/day of water will grow the 2500 calories required to feed one person. Water/calorie ratios of this order or, less have been achieved by

5. An identical conclusion was reached in a recent study (Clawson et al., 1969) in which the foregoing points and numerous other aspects of desalting were treated in far more detail than was possible here.

6. Confusing statements often are made about the possibility that food supply will outrun food demand in the future. In these statements, "demand" is used in the economic sense, and in this context many millions of starving people may generate no demand whatsoever. Indeed, one concern of those engaged in increasing food production is to find ways of increasing demand.

the top 20 percent of farmers specializing in such crops as wheat, potatoes, and tomatoes; but more water is required for needed protein-rich crops such as peanuts and soybeans. The authors identify the uncertainty that crops usually raised separately can be grown together in tight rotation on the same piece of land. Problems of water storage between periods of peak irrigation demand, optimal patterns of crop rotation, and seasonal acreage variations are also mentioned. These "ifs" and assumptions, and those associated with the other technologies involved, are unfortunately often omitted when the results of such painstaking studies are summarized for more popular consumption (Anonymous, 1968b, 1968c). The result is the perpetuation of the public's tendency to confuse feasible and available, to see panaceas where scientists in the field concerned see only potential, realizable with massive infusions of time and money.

It is instructive, nevertheless, to examine the impact on the world food problem which the Oak Ridge complexes might have if construction were to begin today, and if all the assumptions about technology 10 years hence were valid *now*. At the industrial-agricultural mix pertinent to the sample case described above, the food produced would be adequate for just under 3 million people. This means that 23 such plants per year, at a cost of $41 billion, would have to be put in operation merely to keep pace with world population growth, to say nothing of improving the substandard diets of between one and two billion members of the present population. (Fertilizer production beyond that required for the on-site farm is of course a contribution in the latter regard, but the substantial additional costs of transporting it to where it is needed must then be accounted for.) Since approximately 5 years from the start of construction would be required to put such a complex into operation, we should commence work on at least 125 units post-haste, and begin at least 25 per year thereafter. If the technology *were* available now, the investment in construction over the next 5 years, prior to operation of the first plants, would be $315 billion—about 20 times the total U.S. foreign aid expenditure during the past 5 years. By the time the technology *is* available the bill will be much higher, if famine has not "solved" the problem for us.

This example again illustrates that scale, time, and cost are all working against technology in the short term. And if population growth is not decelerated, the increasing severity of population-related crises will surely neutralize the technological improvements of the middle and long terms.

Other Food Panaceas. "Food from the sea" is the most prevalent "answer" to the world food shortage in the view of the general public. This is not surprising, since estimates of the theoretical fisheries productivity of the sea run up to some 50-100 times current yields (Schmitt, 1965; Christy and Scott, 1965). Many practical and economic difficulties, however, make it clear that such a figure will never be reached, and that it will not even be approached in the foreseeable future. In 1966, the annual fisheries harvest was some 57 million metric tons (United Nations, 1968). A careful analysis (Meseck, 1961) indicates that this might be increased to a world production of 70 million metric tons by 1980. If this gain were realized, it would represent (assuming no violent change in population growth patterns) a small per capita *loss* in fisheries yield.

Both the short- and long-term outlooks for taking food from the sea are clouded by the problems of overexploitation, pollution (which is generally ignored by those calculating potential yields), and economics. Solving these problems will require more than technological legerdemain; it will also require unprecedented changes in human behavior, especially in the area of international cooperation. The unlikelihood that such cooperation will come about is reflected in the recent news (Anonymous, 1968d) that Norway has dropped out of the whaling industry because overfishing has depleted the stock below the level at which it may economically be harvested. In that industry, international controls were tried—and failed. The sea is, unfortunately, a "commons" (Hardin, 1968), and the resultant management problems exacerbate the biological and technical problems of greatly increasing our "take." One suspects that the return per dollar poured into the sea will be much less than the corresponding return from the land for many years, and the return from the land has already been found wanting.

Synthetic foods, protein culture with petroleum, saline agriculture, and weather modification all may hold promise for the future, but all are at present expensive and available only on an extremely limited scale. The research to improve this situation will also be expensive, and, of course, time-consuming. In the absence of funding, it will not occur at all, a fact which occasionally eludes the public and the Congress.

Domestic and Industrial Water Supplies

The world has water problems, even exclusive of the situation in agriculture. Although total precipitation should in theory be adequate in quantity for several further doublings of population, serious shortages arising from problems of quality, irregularity, and distribution already plague much of the world. Underdeveloped countries will find the water needs of industrialization staggering. 240,000 gal of water are required to produce a ton of newsprint; 650,000 gal, to produce a ton of steel (International Atomic Energy Agency, 1964). Since maximum acceptable water costs for domestic and industrial use are higher than for agriculture, those who can afford it are or soon will be using desalination (40-100 + cents/1000 gal) and used-water renovation (54-57 cents/1000 gal [Ennis, 1967]). Those who cannot afford it are faced with allocating existing supplies between industry and agriculture, and as we have seen, they must choose the latter. In this circumstance, the standard of living remains pitifully low. Technology's only present answer is massive externally-financed complexes of the sort considered above, and we have already suggested there the improbability that we are prepared to pay the bill rung up by present population growth.

The widespread use of desalted water by those who *can* afford it brings up another problem only rarely mentioned to date, the disposal of the salts. The product of the distillation processes in present use is a hot brine with salt concentration several times that of seawater. Both the temperature and the salinity of this effluent will prove fatal to local marine life if it is simply exhausted to the ocean. The most optimistic statement we have seen on this problem is that "*smaller plants* (our emphasis) at seaside locations may return the concentrated brine to the ocean if proper attention is paid to the design of the outfall, and to the effect on the local marine ecology." (McIlhenny, 1966). The same writer identifies the major economic uncertainties connected with extracting the salts for sale (to do so is straightforward, but often not profitable). Nor can one simply evaporate the brine and leave the residue in a pile—the 150 million gal/day plant mentioned above would produce brine bearing 90 million lb. of salts daily (based on figures by Parker, 1966). This amount of salt would cover over 15 acres to a depth of one foot. Thus, every year a plant of the billion gallon per day, agro-industrial complex size would produce a pile of salt over 52 ft deep and covering a square mile. The high winds typical of coastal deserts would seriously aggravate the associated soil contamination problem.

Energy

Man's problems with energy supply are more subtle than those with food and water: we are not yet running out of energy, but we are being forced to use it faster than is probably healthy. The rapacious depletion of our fossil fuels is already forcing us to consider more extensive mining techniques to gain access to lower-grade deposits, such as the oil shales, and even the status of our high-grade uranium ore reserves is not clearcut (Anonymous, 1968e).

A widely held misconception in this connection is that nuclear power is "dirt cheap," and as such represents a panacea for developed and under-developed nations alike. To the contrary, the largest nuclear-generating stations now in operation are just competitive with or marginally superior to modern coal-fired plants of comparable size (where coal is not scarce); at best, both produce power for on the order of 4-5 mills (tenths of a cent) per kilowatt-hour. Smaller nuclear units remain less economical than their fossil-fueled counterparts. Underdeveloped countries can rarely use the power of the larger plants. Simply speaking, there are not enough industries, appliances, and light bulbs to absorb the output, and the cost of industrialization and modernization exceeds the cost of the power required to sustain it by orders of magnitude, regardless of the source of the power. (For example, one study noted that the capital requirement to consume the output of a 70,000 kilowatt plant—about $1.2 million worth of electricity per year at 40 percent utilization and 5 mills/kwh—is $111 million per year if the power is consumed by metals industries, $270 million per year for petroleum product industries [E. A. Mason, 1957].) Hence, at least at present, only those underdeveloped countries which are short of fossil fuels or inexpensive means to transport them are in particular need of nuclear power.

Prospects for major reductions in the cost of nuclear power in the future hinge on the long-awaited breeder reactor and the still further distant thermonuclear reactor. In neither case is the time scale or the ultimate cost of energy a matter of any certainty. The breeder reactor, which converts more nonfissile uranium (^{238}U) or thorium to fissionable material than it consumes as fuel for itself, effectively extends our nuclear fuel supply by a factor of approximately 400 (Cloud, 1968). It is not expected to become competitive economically with conventional reactors until the 1980's (Bump, 1967). Reductions in the unit energy cost beyond this date are not guaranteed, due both to the probable continued high capital cost of breeder reactors and to increasing costs for the ore which the breeders will convert to fuel. In the latter regard, we mention that although crushing granite for its few parts per million of uranium and thorium is possible in theory, the problems and cost of doing so are far from resolved.[7] It is too soon to predict the costs associated with a fusion reactor (few who work in the field will predict whether such a device will work at all within the next 15-20 years). One guess puts the unit energy cost at something over half that for a coal or fission power station of comparable size (Mills, 1967), but this is pure speculation. Quite possibly the major benefit of controlled fusion will again be to extend the energy supply rather than to cheapen it.

A second misconception about nuclear power is that it can reduce our dependence on fossil fuels to zero as soon as that becomes necessary or desirable. In fact, nuclear power plants contribute only to the electrical portion of the energy budget; and in 1960 in the United States, for example, electrical energy comprised only 19 percent of the total energy consumed (Sporn, 1963). The degree to which nuclear fuels can postpone the exhaustion of our coal and oil depends on the extent to which that 19 percent is enlarged. The task is far from a trivial one, and will involve transitions to electric or fuel-cell powered transportation, electric heating, and electrically powered industries. It will be extremely expensive.

Nuclear energy, then, is a panacea neither for us nor for the underdeveloped world. It relieves, but does not remove, the pressure on fossil fuel supplies; it provides reasonably-priced power where these fuels are not abundant; it has substantial (but expensive) potential in intelligent applications such as that suggested in the Oak Ridge study discussed above; and it shares the propensity of fast-growing technology to unpleasant side effects (Novick,

7. A general discussion of extracting metals from common rock is given by Cloud, 1968.

1969). We mention in the last connection that, while nuclear power stations do not produce conventional air pollutants, their radioactive waste problems may in the long run prove a poor trade. Although the AEC seems to have made a good case for solidification and storage in salt mines of the bulk of the radioactive fission products (Blanko et al., 1967), a number of radioactive isotopes are released to the air, and in some areas such isotopes have already turned up in potentially harmful concentrations (Curtis and Hogan, 1969). Projected order of magnitude increases in nuclear power generation will seriously aggravate this situation. Although it has frequently been stated that the eventual advent of fusion reactors will free us from such difficulties, at least one authority, F. L. Parker, takes a more cautious view. He contends that losses of radioactive tritium from fusion power plants may prove even more hazardous than the analogous problems of fission reactors (Parker, 1968).

A more easily evaluated problem is the tremendous quantity of waste heat generated at nuclear installations (to say nothing of the usable power output, which, as with power from whatever source, must also ultimately be dissipated as heat). Both have potentially disastrous effects on the local and world ecological and climatological balance. There is no simple solution to this problem, for, in general, "cooling" only moves heat; it does not *remove* it from the environment viewed as a whole. Moreover, the Second Law of Thermodynamics puts a ceiling on the efficiency with which we can do even this much, i.e., concentrate and transport heat. In effect, the Second Law condemns us to aggravate the total problem by generating still *more* heat in any machinery we devise for local cooling (consider, for example, refrigerators and air conditioners).

The only heat which actually leaves the whole system, the Earth, is that which can be radiated back into space. This amount steadily is being diminished as combustion of hydrocarbon fuels increases the atmospheric percentage of CO_2 which has strong absorption bands in the infrared spectrum of the outbound heat energy. (Hubbert, 1962, puts the increase in the CO_2 content of the atmosphere at 10 percent since 1900.) There is, of course, a competing effect in the Earth's energy balance, which is the increased reflectivity of the upper atmosphere to incoming sunlight due to other forms of air pollution. It has been estimated, ignoring both these effects, that man risks drastic (and perhaps catastrophic) climatological change if the amount of heat he dissipates in the environment on a global scale reaches 1 percent of the incident solar energy at the Earth's surface (Rose and Clark, 1961). At the present 5 percent rate of increase in world energy consumption,[8] this level will be reached in less than a century, and in the immediate future the direct contribution of man's power consumption will create serious local problems. If we may safely rule

8. The rate of growth of world energy consumption fluctuates strongly about some mean on a time scale of only a few years, and the figures are not known with great accuracy in any case. A discussion of predicting the mean and a defense of the figure of 5 percent are given in Gúeron et al., 1957.

out circumvention of the Second Law or the divorce of energy requirements from population size, this suggests that, whatever science and technology may accomplish, population growth must be stopped.

Transportation

We would be remiss in our offer of a technological perspective on population problems without some mention of the difficulties associated with transporting large quantities of food, material, or people across the face of the Earth. While our grain exports have not begun to satisfy the hunger of the underdeveloped world, they already have taxed our ability to transport food in bulk over large distances. The total amount of goods of *all* kinds loaded at U.S. ports for external trade was 158 million metric tons in 1965 (United Nations, 1968). This is coincidentally the approximate amount of grain which would have been required to make up the dietary shortages of the underdeveloped world in the same year (Sukhatme, 1966). Thus, if the United States *had* such an amount of grain to ship, it could be handled only by displacing the entirety of our export trade. In a similar vein, the gross weight of the fertilizer, in excess of present consumption, required in the underdeveloped world to feed the additional population there in 1980 will amount to approximately the same figure—150 million metric tons (Sukhatme, 1966). Assuming that a substantial fraction of this fertilizer, should it be available at all, will have to be shipped about, we had best start building freighters! These problems, and the even more discouraging one of internal transportation in the hungry countries, coupled with the complexities of international finance and marketing which have hobbled even present aid programs, complete a dismal picture of the prospects for "external" solutions to ballooning food requirements in much of the world.

Those who envision migration as a solution to problems of food, land, and water distribution not only ignore the fact that the world has no promising place to put more people, they simply have not looked at the numbers of the transportation game. Neglecting the fact that migration and relocation costs would probably amount to a minimum of several thousand dollars per person, we find, for example, that the entire long-range jet transport fleet of the United States (about 600 planes [Molloy, 1968] with an average capacity of 150), averaging two round trips per week, could transport only about 9 million people per year from India to the United States. This amounts to about 75 percent of that country's annual population *growth* (Population Reference Bureau, 1968). Ocean liners and transports, while larger, are less numerous and much slower, and over long distances could not do as well. Does anyone believe, then, that we are going to compensate for the world's population growth by sending the excess to the planets? If there were a place to go on Earth, financially and logistically we could not send our surplus there.

Conclusion

We have not attempted to be comprehensive in our treatment of population pressures and the prospects of coping with them technologically; rather, we hope simply to have given enough illustrations to make plausible our contention that technology, without population control, cannot meet the challenge. It may be argued that we have shown only that any one technological scheme taken individually is insufficient to the task at hand, whereas *all* such schemes applied in parallel might well be enough. We would reply that neither the commitment nor the resources to implement them all exists, and indeed that many may prove mutually exclusive (e.g., harvesting algae may diminish fish production).

Certainly, an optimum combination of efforts exists in theory, but we assert that no organized attempt to find it is being made, and that our examination of its probable eventual constituents permits little hope that even the optimum will suffice. Indeed, after a far more thorough survey of the prospects than we have attempted here, the President's Science Advisory Committee Panel on the world food supply concluded (PSAC, 1967): "The solution of the problem that will exist after about 1985 *demands* that programs of population control be initiated now." We most emphatically agree, noting that "now" was 2 years ago!

Of the problems arising out of population growth in the short, middle and long terms, we have emphasized the first group. For mankind must pass the first hurdles—food and water for the next 20 years—to be granted the privilege of confronting such dilemmas as the exhaustion of mineral resources and physical space later.[9] Furthermore, we have not conveyed the extent of our concern for the environmental deterioration which has accompanied the population explosion, and for the catastrophic ecological consequences which would attend many of the proposed technological "solutions" to the population/food crisis. Nor have we treated the point that "development" of the rest of the world to the standards of the West probably would be lethal ecologically (Ehrlich and Ehrlich, 1970). For even if such grim prospects are ignored, it is abundantly clear that in terms of cost, lead time, and implementation on the scale required, technology without population control will be too little and too late.

What hope there is lies not, of course, in abandoning attempts at technological solutions; on the contrary, they must be pursued at unprecedented levels, with unprecedented judgment, and above all with unprecedented attention to their ecological consequences. We need dramatic programs now to find

9. Since the first draft of this article was written, the authors have seen the manuscript of a timely and pertinent forthcoming book, *Resources and Man*, written under the auspices of the National Academy of Sciences and edited by Preston E. Cloud. The book reinforces many of our own conclusions in such areas as agriculture and fisheries and, in addition, treats both short- and long-term prospects in such areas as mineral resources and fossil fuels in great detail.

ways of ameliorating the food crisis—to buy time for humanity until the inevitable delay accompanying population control efforts has passed. But it cannot be emphasized enough that if the population control measures are *not* initiated immediately and effectively, all the technology man can bring to bear will not fend off the misery to come.[10] Therefore, confronted as we are with limited resources of time and money, we must consider carefully what fraction of our effort should be applied to the cure of the disease itself instead of to the temporary relief of the symptoms. We should ask, for example, how many vasectomies could be performed by a program funded with the 1.8 billion dollars required to build a single nuclear agro-industrial complex, and what the relative impact on the problem would be in both the short and long terms.

The decision for population control will be opposed by growth-minded economists and businessmen, by nationalistic statesmen, by zealous religious leaders, and by the myopic and well-fed of every description. It is therefore incumbent on all who sense the limitations of technology and the fragility of the environmental balance to make themselves heard above the hollow, optimistic chorus—to convince society and its leaders that there is no alternative but the cessation of our irresponsible, all-demanding, and all-consuming population growth.

Acknowledgments

We thank the following individuals for reading and commenting on the manuscript: J. H. Brownell (Stanford University); P. A. Cantor (Aerojet General Corp.); P. E. Cloud (University of California, Santa Barbara); D. J. Eckstrom (Stanford University); R. Ewell (State University of New York at Buffalo); J. L. Fisher (Resources for the Future, Inc.); J. A. Hendrickson, Jr. (Stanford University); J. H. Hessel (Stanford University); R. W. Holm (Stanford University); S. C. McIntosh, Jr., (Stanford University); K. E. F. Watt (University of California, Davis). This work was supported in part by a grant from the Ford Foundation.

References

Anonymous. 1968a. India aims to remedy fertilizer shortage. *Chem. Eng. News,* 46 (November 25): 29.

———. 1968b. Scientists Studying Nuclear-Powered Agro-Industrial Complexes to Give Food and Jobs to Millions. *New York Times,* March 10, p. 74.

———. 1968c. Food from the atom. *Technol. Rev.,* January, p. 55.

10. This conclusion has also been reached within the specific context of aid to underdeveloped countries in a Ph.D. thesis by Douglas Daetz: "Energy Utilization and Aid Effectiveness in Nonmechanized Agriculture: A Computer Simulation of a Socioeconomic System" (University of California, Berkeley, May, 1968).

————. 1968d. Norway—The end of the big blubber. *Time,* November 29, p. 98.

————. 1968e. Nuclear fuel cycle. *Nucl. News,* January, p. 30.

Bender, R. J. 1969. Why water desalting will expand. *Power,* 113 (August): 171.

Blanko, R. E., J. O. Blomeke, and J. T. Roberts. 1967. Solving the waste disposal problem. *Nucleonics,* 25: 58.

Borgstrom, Georg. 1965. *The Hungry Planet.* Collier-Macmillan, New York.

Bump, T. R. 1967. A third generation of breeder reactors. *Sci. Amer.,* May, p. 25.

Christy, F. C., Jr., and A. Scott. 1965. *The Commonwealth in Ocean Fisheries.* Johns Hopkins Press, Baltimore.

Clawson, M., H. L. Landsberg, and L. T. Alexander. 1969. Desalted seawater for agriculture: Is it economic? *Science,* 164: 1141.

Cloud, P. R. 1968. Realities of mineral distribution. *Texas Quart.,* Summer, p. 103.

Cole, LaMont C. 1968. Can the world be saved? *BioScience,* 18: 679.

Curtis, R., and E. Hogan. 1969. *Perils of the Peaceful Atom.* Doubleday, New York. p. 135, 150-152.

Ennis, C. E. 1967. Desalted water as a competitive commodity. *Chem. Eng. Progr.,* 63: (1): 64.

Ehrlich, P. R. 1968. *The Population Bomb.* Sierra Club/Ballantine, New York.

Ehrlich, P. R., and Anne H. Ehrlich. 1970. *Population, Resources, and Environment.* W. H. Freeman, San Francisco (In press).

Galstann, L. S., and E. L. Currier. 1967. The Metropolitan Water District desalting project. *Chem. Eng. Progr.,* 63,(1): 64.

Gúeron, J., J. A. Lane, I. R. Maxwell, and J. R. Menke. 1957. *The Economics of Nuclear Power. Progress in Nuclear Energy.* McGraw-Hill Book Co., New York. Series VIII. p. 23.

Hardin, G. 1968. The tragedy of the commons. *Science,* 162: 1243.

Hubbert, M. K. 1962. Energy resources, A report to the Committee on Natural Resources. National Research Council Report 1000-D, National Academy of Sciences.

International Atomic Energy Agency. 1964. Desalination of water using conventional and nuclear energy. Technical Report 24, Vienna.

Kelly, R. P. 1966. North American water and power alliance. In: *Water Production Using Nuclear Energy,* R. G. Post and R. L. Seale (eds.). University of Arizona Press, Tucson, p. 29.

Lelyveld, D. 1968. Can India survive Calcutta? *New York Times Magazine,* October 13, p. 58.

Mason, E. A. 1957. Economic growth and energy consumption. In: *The Economics of Nuclear Power. Progress in Nuclear Energy,* Series VIII, J. Gúeron et al. (eds.). McGraw-Hill Book Co., New York, p. 56.

McIlhenny, W. F. 1966. Problems and potentials of concentrated brines. In: *Water Production Using Nuclear Energy,* R. G. Post and R. L. Seale (eds.). University of Arizona Press, Tucson, p. 187.

McKenzie, John. 1968. Nutrition and the soft sell. *New Sci.,* 40: 423.

McNeil, Mary. 1964. Lateritic soils. *Sci. Amer.,* November, p. 99.

Meseck, G. 1961. Importance of fish production and utilization in the food economy. Paper R.11.3, presented at FAO Conference on Fish in Nutrition, Rome.

Milliman, J. W. 1966. Economics of water production using nuclear energy. In: *Water Production Using Nuclear Energy.* R. G. Post and R. L. Seale (eds.). University of Arizona Press, Tucson, p. 49.

Mills, R. G. 1967. Some engineering problems of thermonuclear fusion. *Nucl. Fusion.* 7: 223.

Molloy, J. F., Jr. 1968. The $12-billion financing problem of U.S. airlines. *Astronautics and Aeronautics,* October, p. 76.

Novick, S. 1969. *The Careless Atom.* Houghton Mifflin, Boston.

Oak Ridge National Laboratory. 1968. Nuclear energy centers, industrial and agro-industrial complexes, Summary Report. ORNL-4291, July.

Paddock, William. 1967. Phytopathology and a hungry world. *Ann. Rev. Phytopathol.,* 5: 375.

Paddock, William, and Paul Paddock. 1964. *Hungry Nations,* Little, Brown & Co., Boston.

————. 1967. *Famine 1975!* Little, Brown & Co., Boston.

Parikh, G., S. Saxena, and M. Maharaja. 1968. Agricultural extension and IADP, a study of Surat. *Econ. Polit. Weekly,* August 24, p. 1307.

Parker, F. L. 1968. Radioactive wastes from fusion reactors. *Science,* 159: 83.

Parker, H. M. 1966. Environmental factors relating to large water plants. In: *Water Production Using Nuclear Energy,* R. G. Post and R. L. Seale (eds.). University of Arizona Press, Tucson, p. 209.

Population Reference Bureau. 1968. Population Reference Bureau Data Sheet. Pop. Ref. Bureau, Washington, D.C.

PSAC. 1967. *The World Food Problem.* Report of the President's Science Advisory Committee. Vols. 1-3. U.S. Govt. Printing Office, Washington, D.C.

Rose, D. J., and M. Clark, Jr. 1061. *Plasma and Controlled Fusion,* M.I.T. Press, Cambridge, Mass., p. 3.

Schmitt, W. R. 1965. The planetary food potential. *Ann. N.Y. Acad. Sci.,* 118: 645.

Sporn, Philip. 1963. *Energy for Man,* Macmillan, New York.

————. 1966. *Fresh Water from Saline Waters.* Pergamon Press, New York.

Sukhatme, P. V. 1966. The world's food supplies. *Roy. Stat. Soc. J.,* 129A: 222.

United Nations. 1968. *United Nations Statistical Yearbook for 1967.* Statistical Office of the U.N., New York.

U.S. Dept. of Commerce. 1966. *Statistical Abstract of the U.S.* U.S. Govt. Printing Office, Washington D.C.

Wadleigh, C. H. 1968. Wastes in relation to agriculture and industry. USDA Miscellaneous Publication No. 1065. March.

Woodwell, George M. 1967. Toxic substances and ecological cycles. *Sci. Amer.,* March, p. 24.

The Control of Population

Stuart Chase

Profile of the Hungry World

Latin America, Asia, and Africa comprise the Hungry World, where, considering the area as a whole, population is now outrunning subsistence, and where the formula $L=O/P$ shows the standard of living going down. Dr. Roy E. Brown of the University of East Africa has drawn a profile of this vast area, setting forth its major characteristics. His vantage point for viewing it is of course unexcelled.

Its economy, he says in *Science,*[1] is based on subsistence agriculture; 90 percent of its people are rural.

The average per capita income is less than $200 a year.

Fewer than half the adults can read or write.

Technicians are short in all fields.

Birth rates are high, and despite a high rate of infant mortality, population for the whole area is growing at about 3 percent a year—a rate which doubles in 23 years.

Half the population is under 15 years old.

Unemployment in the villages is causing a mass movement to the cities, where unemployment is usually worse. (This corresponds to a similar trend in the U.S., as we have noted.)

Reforms are seriously handicapped by superstitions, taboos, fatalism, rigid diet customs—rice eaters, for instance, reject wheat.

The lower death rates achieved through preventive medicine add to the number of old, sick, and hungry people, and thus tend to make the population crisis more severe.

Finally, as Dr. Brown observes, the rich nations are giving the poor nations little effective help. The crisis has not yet been squarely faced by either group.

---

1. July 15, 1966.

This brings up an interesting comparison. Karl Marx predicted that, under capitalism, the rich would grow richer and the poor poorer. He was of course talking about economic classes—and he was wrong. The rich in open societies are not gaining too much after the income tax gets through with them, while the sometime "masses" have risen from the lower brackets to the middle class, and become the driving force of the affluent society. Even the truly poor today—in the United States this means about 20 percent of families—have enough to eat, in bulk if not in vitamins. When it comes to nations, however, the Marxian prediction holds. While affluent societies grow richer, at least in gross national product, the Hungry World grows hungrier, poorer, and more illiterate.

Following the formula $L = O/P$, the most vulnerable nations are India, Pakistan, China, Indonesia, Iran, Turkey, Egypt, Colombia, and Peru.[2] They may have to face famine conditions within five to ten years. India, indeed, is already facing them.

Somewhat less vulnerable, says Dr. Raymond Ewell, are Burma, Thailand, the Philippines, Mexico, Chile, and native countries below the Sahara in Africa. Their deep crisis will come later, but the equation is against them too.

Let us look at typical nations on three continents: India (Asia), Egypt (Africa), and Mexico (Latin America).

India

We begin with the most vulnerable nation of all. Mr. Asoka Mehta, chief economic planner for the Indian Government, agrees with Dr. Ewell. The Indian birth rate, he says, is now 42, the death rate 19, which means a national increase of 23, or 11 million more persons a year. In reporting these figures, Mr. Mehta throws up his hands. The food supply, he says, is seriously inadequate, great quantities of grain must be imported, unemployment is severe, and the nation is short 70 million dwellings. Shortages will grow worse, he says, despite our efforts to develop the economy. Before plans can be effective the birth rate must be cut in half—down to around 20—not far from the present rate in both the United States and Russia.

Meanwhile as I write, in the province of Kerala, mobs are burning and rioting because they have little or no rice to eat. As rice eaters, they are psychically allergic to wheat, and even more allergic to the corn (maize) and sorghum which the U.S. has been shipping to India.

Egypt

From Asia we turn to a vulnerable country in Africa. Egypt illustrates how, even when productivity is increased, the problem remains. In 1830 she had a population estimated at three million.[3] A dam was built in the Nile above

2. Paper for Population Crisis Committee, 1966.
3. Patrick Seale and Irene Beeson in *The New Republic,* May 7, 1966.

Cairo, which irrigated more land and, with improvements in agricultural techniques, raised the population to 10 million by the end of the nineteenth century.

The first Aswan Dam was constructed by the British in 1902. It brought in another million acres for irrigation, and helped to treble the population to today's 30 million. Now the great new Aswan Dam, the building of which will flood priceless sculpture of the Pharaohs, is nearing completion. Can its irrigation canals feed Egypt's 800,000 babies a year, or will it encourage even more babies? At prevailing growth rates, population will double to 60 million by 1980, and the great monuments will have been inundated to no avail.

President Nasser, well aware of what has been happening to his country, had inaugurated a substantial birth control program before the Israeli War of 1967. New clinics are appearing, gleaming white against the dark mud huts of the Nile villages. There are, however, serious cultural and psychological difficulties. For centuries, Egyptian babies have been welcomed as insurance for the future support of their parents. A plentiful brood, moreover, has been cherished as proof of the father's virility. War losses in 1967 put additional strains on Egypt's economy.

Mexico

In the early 1930's the population of this colorful country was estimated at 16 million. In 1966 it was 40 million. No fewer than 75 million people are predicted by 1980, if the current growth rate continues. Should it persist unabated, there would be 800 million Mexicans south of the Rio Grande a hundred years from now, exceeding the present population of China!

The birth rate in Mexico was around 50 per thousand in 1930, and is not much less today. But with the help of medical technology the death rate has dropped from 27 to 12. What is Mexico doing about her birth rate? Very little. During my visit there in 1966, I was told that the government was apathetic and the Church generally opposed to birth control, especially in the villages.

Mexico in the mid-1960's had perhaps the best economic record of any Latin-American nation, and was not much concerned with its demography. What really got attention, I found, was the rising gross national product. How long, however, can a prosperity in pesos cope with a 3.6 percent growth rate in people?

Birth Control

In the late 1960's India appears on the brink of disaster, Egypt is but a few years away, and Mexico only a few years beyond that—*if current rates continue.* The vulnerable nations face a reckoning in the next decade. Will it be orderly or chaotic? The only orderly and permanent solution lies in the planned control of population. Planned control is not a new idea. The Greeks and the Romans practiced it from time to time by exposing babies, especially girl babies and those with physical defects. Other societies have led the aged off to die, with due ceremony. Crude methods of abortion and contraception are probably as old as the race.

The exposure of children and old people is repugnant today in all cultures, while abortion is repugnant in many. Infanticide is a crime, but is not unknown. The modern crime of so-called "child battering," in which distracted parents beat their children, sometimes to death, is increasingly reported by American social workers. There are far kinder ways to control population. Applied science, which brought on the crisis, is now hard at work seeking to mitigate it by new and improved techniques—oral pills, the uterine loop, the temporary sterilization of men and women. A postintercourse pill is in the laboratory. All this came into the headlines in 1965, when an area hitherto dark was suddenly illumined, and the public debate over the issue has remained active and growing.

The goal of the birth control movement is a balanced society where man and his environment are in reasonable equilibrium, where children are wanted and cared for. As we observed in the last chapter, if a given society desires a 70-year life span, with modern medicine and a low death rate, *it must limit its birth rate.* Fortunately, intelligent people throughout the world, like the chief planner of India, are beginning to grasp this simple logic.

The late Margaret Sanger was a pioneer in birth control and had the courage to go to jail for it. She began her crusade with a clinic in Brooklyn in 1916, not so much as a solution to world problems, as the result of personal problems she had known in her own family. In a Senate hearing in 1932 she was denounced as a corrupter of public morals, whose program "will rob the nation of military power, even for defensive purposes."

To her, the reform would liberate women from what amounted to biological slavery. Her mail was full of heart-breaking letters from mothers reduced to despair by continuous pregnancies, sick and hungry children. She was told of many cases where a woman died young from bearing too many babies, leaving a motherless brood. To Mrs. Sanger and her colleagues, birth control seemed to have a vital function, too, in preventing war—by reducing the "cannon fodder" of the military establishment. She finally became the honorary head of a great world movement—"Planned Parenthood—World Population," with a long list of prominent and distinguished sponsors, including ex-Presidents, publishers, cabinet members, and many physicians.[4]

Dr. John Rock, a gynecologist and a Catholic, who was instrumental in developing a safe oral pill, appeared before a Senate committee in 1965 to say: "The growth of population on this planet presents a lethal threat to all that civilization has achieved." In 1967 he took the position that failing birth

4. Active promoters and sponsors in the U.S. now include ex-Presidents Eisenhower and Truman, President Johnson, General William H. Draper, Jr., Cass Canfield, Senators Ernest Gruening and Joseph Clark, Hugh Moore, Harry Emerson Fosdick, Dr. Mary S. Calderone, George Kennan, Marriner Eccles, Elmo Roper, Chester Bowles, Stewart Udall, John W. Gardner, and many more distinguished Americans.

control the United States had reached the limit of its ability to educate its children, while the underdeveloped nations, three-quarters of humanity, had already *exceeded* that limit. He was saying in effect that the family cannot adequately perpetuate the species in the face of the population explosion.

These are strong words, and they come from a Catholic doctor. For many years the Church has been the most formidable institution arrayed against contraceptives. Now the Church is ready to discuss the question, and an eminent Catholic layman of the Harvard Medical School is mixing pills. It is as if the Berlin Wall had suddenly been thrown down!

The United Nations, the World Bank, U.S. Government departments, the U.S. War on Poverty, Food-for-Freedom, and innumerable organizations here and abroad are now combining to carry on the work which Margaret Sanger so gallantly began. Programs, some of them ambitious, are under way in India, China, Egypt, Pakistan, South Korea, Taiwan, Kenya, Tunisia, and Turkey.

A Gallup poll in 1965 indicated that Americans are beginning to understand this crisis. Seventy percent of interviewed adults said that birth control information should be available to all married persons who want it; 50 percent went so far as to propose it for unmarried persons. A slim majority felt that population was a serious problem in the United States, and a larger majority felt it a serious world problem. Two-thirds believed that the federal government should aid states and cities with birth control programs. Educated respondents, ironically, were far ahead of the poorly educated in their replies, and of course it is the uneducated who need the information most.

Meanwhile Professor Robin Barlow of the University of Michigan says that programs for malaria control should be accompanied by programs for birth control. Otherwise, he says, the economic gains of improved health "can be turned into an economic loss." (This is exactly what is happening in Mauritius.)

An energetic campaign has undoubtedly at last begun, but the total effect to date is hardly more than a launching pad. Robert C. Cook estimates that births, the world around, must be reduced by at least 20 million a year within a decade. This is a formidable goal. More than a billion adults in the Hungry World must be reached and be shown the techniques that are available. The total outlay for research so far is small—only a tiny fraction of the cost to put a man on the moon.

Will More Food Save the Day?

The idea that all we need to do is to grow more food dies hard. More food postpones the crisis; *it does not solve it.* Even with the most advanced technical aid, the supply of food in the world cannot hope to keep up with population at present rates of growth; Malthus stands vindicated. In 1965, when national food stocks were near their peak, the United States had enough surplus wheat in storage to feed the world for just two weeks. More food can certainly be grown, it can be dredged from the sea; eventually it can even by synthesized.

These processes, however, will take time, and enormous amounts of capital, while the effect can only be temporary.

According to a U.S. Government task force, in the five years from 1960 to 1965 population in the Hungry World increased twice as fast as its output of food. Before World War II, low-energy areas exported large amounts of grain, net, to the high-energy nations. Now the net traffic is moving the other way. To make matters worse, prices have been rising for manufactured goods imported by the poorer countries, while prices of the raw materials which they grow and export have been declining.

Secretary of Agriculture Orville L. Freeman sums it up:

> There are three basic benchmarks to which the rate of increase in food production can be usefully related: *first,* the rate of increase needed to keep pace with population growth; *second,* the rate of increase needed to attain target rates of economic growth while maintaining stable prices; *third,* the rate needed to eliminate the serious malnutrition common to most of the developing countries. By all three criteria, the rate of food increase has been inadequate. . . . *We are losing the war on hunger.*

The Case of Japan

Before we despair, however, let us look at a great nation which has reduced its birth rate by half in the last 15 years, and pared its growth rate to that of Western Europe. Japan has demonstrated that population can be controlled, at least in a disciplined society.

Japan must support 100 million people in an area about the size of California, only one-sixth of which is arable. After her defeat in World War II, soldiers came home from all over the Pacific and the birth rate jumped. A serious food shortage loomed, and something had to be done. In 1948 the Diet adopted the "Eugenic Protective Law," with competent medical advice, and the announced purpose of promoting family welfare. The effect on the whole social structure was destined to be profound.

A desperate situation called for a desperate remedy. Abortion was made legal; one gynecologist could sanction the operation. The law applied to very poor families, to women who were ill, and to victims of rape. Legal abortions rose to a peak of more than a million in 1955 and then declined, as contraceptives began in part to replace them. One result was the virtual elimination of illegitimate births.

By the late 1950's hundreds of health centers were established throughout Japan to give contraceptive information and supplies, while 50,000 nurses were specially trained. Big Business cooperated effectively by setting up clinics along with company housing. Courses in birth control became as popular as courses in cooking and child care. Businessmen did not fail to notice, furthermore, that the factory accident rate declined. They account for it by the fact that workers slept better, with fewer crying babies in the small Japanese homes.

By 1964 the birth rate had fallen from 34 per thousand to 17—*cut in half in 15 years.* This compared with an average European birth rate of 18, an

American and a Russian rate of just over 20. The Japanese death rate also declined along with everybody else's, but even so the growth rate fell below one percent a year, about that of Western Europe. (Japan was included with Western Europe in Dr. Keyfitz's group 1, cited earlier.)

Japan is now dealing with a labor shortage, the reverse of unemployment. The number of young people coming on the job market is going down—while in the United States the number is going up. This has forced Japanese businessmen to pay higher wages. Older workers, meanwhile, are no longer "fired at 40"—or 50 or 60. The retirement age is up, too, along with an increase in training programs for both old and young.

In short, Japan has taught the world two important lessons: the benefits to the economy of a low growth rate in people, and, even more significant, the proof that a whole population can be deliberately planned and controlled. We must remember, however, that Japan is a special case—a high-energy society in low-energy Asia. To do what Japan has done requires a literate, disciplined people, without strong superstitions, or theological principles governing the biology of reproduction.

Our Gravest Problem

Many students consider the population crisis the trend most to be feared and outmaneuvered. It is not alone the problem of the Hungry World; we are all in this together, rich nations and poor alike. One can anticipate, however, a vigorous attempt, by people with little knowledge and imagination, to seal off the poor nations, halt foreign aid programs, and let two-thirds of mankind starve its way to some kind of primitive equilibrium. I doubt the success of any such movement, however vocal its advocates. It is too late; technology has bound us too firmly together. Telstar lights the television sets of all the continents, no nation is self-supporting, jet planes weave in and out of every capital, the United Nations is a growing symbol of a world united in material fact. One might even hazard the guess that the most urgent business of the United Nations for some years to come will not be so much the policing of cease-fire agreements as promoting smaller families.

The total world-wide growth rate is the result of the birth rate less the death rate, disregarding in- and out-migration. If a given community had 1,000 souls at the beginning of the year, and 50 were born and 40 died during the year, there would be 1,010 people at the end. This is an increase of 10 per thousand, or a growth rate of one percent.

This rate can be raised by improved medical care acting on the death rate and making people live longer. It can be reduced harmlessly by birth control, or violently by famine, plague, and war. A thermonuclear war, however, would probably poison the pool of human genes by massive fallout, with effects disastrous beyond calculation.

The long-term goal becomes increasingly clear: it is an average of not much over two children per family. (Ireland for the moment has a negative growth rate; so has East Germany, doubtless due to out-migration.) The result will be a *growth rate close to zero,* if a steady-state world is to be won.

This is probably the most important finding in my whole study; a growth rate close to zero.

Let Arnold Toynbee have the last word. He told the World Food Congress in Washington in June, 1963:

> We must aim at a figure that will allow a substantial part of our time and energy to be spent, not on keeping ourselves alive, but on making human life a more civilized affair than we have succeeded in making it so far.

Can We Survive the Madding Crowd?

Bernard Asbell

One of the engaging superstitions of our time, a three-part myth, is that overpopulation is just around the corner, that a shortage of food will do us in by the millions, and that only mass reduction of births (especially among the proliferating poor) will prevent disaster.

This is not an essay to encourage complacency about our rapidly growing population. The dangers facing us are real, perhaps more imminent than most of us think. But to deal with them properly, we need to catch up on some newly-emerging scientific research that contradicts widely-held assumptions.

First of all, while there must be some level of population that would constitute *over*population of our finite-size planet, none of us has any idea what that level is. What will threaten us in advance of overpopulation is *crowding,* a wholly different idea. We are starting to learn something about crowding, thanks to a handful of scholars who are pooling an unlikely mixture of insights ranging from anthropology to biochemistry. One thing they are finding is that under some circumstances millions, perhaps scores of millions may live in densely-packed harmony (say, in stacked dwellings of well-designed apartment houses in a Boston-to-Washington megalopolis). But under other circumstances the old saw, "two's company but three's a crowd," may be a sound scientific warning.

Second, food supply is, at best, only indirectly linked with our possible doom. If we were to die of crowding by the millions, that would happen long before the food supply ran out. We would die not of hunger, but of shock, lowered resistance, rampaging disease, nervous breakdowns and, possibly, mass mayhem and widespread murder. Of the last, we are already witnessing early warning signals in our cities—collections of humanity that are not over-populated but, in certain neighborhoods, dangerously crowded.

Finally, campaigns for birth control may do little to lessen the oppression of crowding, at least in the short run. In fact, there is evidence that a runaway birthrate among the poor is not so much a cause of crowding as a result of it.

Crowding is a specific happening, clinically observable and definable. In simplified terms, crowding occurs when organisms are brought together in such manner and numbers as to produce *physical* reactions of stress. Important among these reactions is stepped-up activity of the adrenal glands. When these reactions to stress are widespread and sustained, they are followed by physical weakening, sometimes rage and violence or extreme passivity, a rise in sexual aberrations, and a breakdown of orderly group behavior. What may follow is a tidal wave of deaths, ending when the population is no longer crowded.

Those things have happened time and again, in various combinations, to all kinds of animals, from lemmings in Scandinavia to deer trapped on an island in Maryland. Ethologists—those who study group behavior of animals —have been cautious about projecting their findings on man. But Dr. Edward T. Hall, prominent Northwestern University anthropologist, and some colleagues in psychiatry, are stitching together evidence that the animal in man may be governed by somewhat the same system of stress reactions. Hall discusses this in *The Hidden Dimension,* his definitive book on the subject of crowding.

"If man does pay attention to animal studies," says Hall, "he can detect the gradually emerging outlines of an endocrine servomechanism not unlike the thermostat in his house. The only difference is that instead of regulating heat the endocrine control system regulates the population."

John J. Christian, an ethologist also trained in medical pathology, was a pioneer in discovering that population buildup, leading to stress, brings on an endocrine reaction and, finally, population collapse—which he calls a "die-off."

What Killed the Sika Deer?

About a mile out in Chesapeake Bay lies a small patch of land, half a square mile, called James Island. It is uninhabited, or at least was until 1916 when someone adorned the island by releasing four or five Sika deer. The deer were fruitful and multiplied until, by 1955, they had procreated a herd of almost 300—about one per acre, extremely dense for deer.

In that year, Christian visited the island, bringing a hypothesis and a gun. He shot five animals and made detailed examinations of their adrenal glands, thyroid, heart, lungs, gonads and other tissues. Their organs appeared normal in every way except one. The adrenal glands were immensely oversized, bulging like overused muscles. When animals are under frequent or sustained stress, their adrenals—which are important to regulation of growth, reproduction and defenses against disease—become overactive and enlarged. If this abnormality was related to crowdedness—and if the herd population was still growing—clearly James Island was in for an interesting time. Christian waited and watched.

For the next two years, herd size stayed about the same. Then in the third year, 1958, more than half the herd inexplicably dropped dead. The island was strewn with 190 carcasses in two years, chiefly females and young, leaving 80 survivors.

What had killed so many? It was not malnutrition, for food was abundant. The coats of the dead deer shone healthily, their muscles were well developed, plenty of body fat. For that matter, if the epidemic of whatever-it-was was so severe, how come 80 survived it—and now appeared robust?

After the die-off, Christian revisited the island in 1960, shot a few more animals and examined them. For one thing, they were substantially—more than 30 percent—larger in body size than those shot at the climax of the crowding. But the more striking thing was that their adrenals were *half* the size of those examined earlier—back to normal. In young deer, they were one-fifth the size of their overstressed counterparts.

"Mortality evidently resulted," Christian later reported to a symposium on crowding, "from shock following severe metabolic disturbance, probably as a result of prolonged adrenocortical hyperactivity. . . . There was no evidence of infection, starvation, or other obvious cause to explain the mass mortality." Subsequently, he says, it was found that the hyperactivity had in all probability resulted in potassium deficiency.

A landmark study it was, finding out how those deer died. But it doesn't tell us how they *lived,* what their behavior was like just before the agonies of emotional and physical stress killed them. A search for clues brings us to another study, which by chance was taking place at the same time in the same state.

Why the Rats Went Berserk

In a stone barn at the outskirts of Rockville, Maryland, John B. Calhoun began breeding populations of Norway rats, a deliberate creation of crowding. In each of several rooms, Calhoun set up four pens, connecting them in a row by ramps arching over their separating walls. In the wild, these animals normally organize in sexually-balanced groups of about 12. The penned rats soon multiplied to an adult population of 80, almost twice the comfortable number of 12 to a pen.

Rats are busybodies, and so got in the habit—at least at first—of scurrying over the ramps from pen to pen. Also they were conditioned to eat in the presence of others, cheek to cheek. So the two central pens, where they were most likely to find companions, became popular "eating clubs." Calhoun at times observed as many as 60 eaters crowding into a single inner pen.

This crowding soon led to what Calhoun calls a "behavioral sink." (A striking term. Webster's *New World Dictionary* defines the noun *sink* as: 1. a cesspool or sewer; 2. any place or thing considered morally filthy or corrupted.)

A single dominant male took charge of each of the less-populated end pens, preventing the entrance of other males, but freely permitting the comings and goings of his females. *His* females. The lord rat of each end pen established a harem of a half-dozen or more females.

Because of these end-pen harems, females were distributed among the four pens fairly evenly. Males, however, were overwhelmingly crowded into the middle pens. And their natural manners, under the stress of crowding and shortage of ladies, gave way to havoc. The more dominant males took to violence. They would suddenly go berserk, attacking females, juveniles and passive males, by biting their tails, sometimes severing them entirely. The floor was almost always bloody from these carryings-on, which Calhoun had never before seen among the species. Then there emerged a group of males that made sexual advances on unreceptive females, often those not in heat, and later on other males, and finally juveniles. Their ability to perceive appropriate sex partners seemed to have vanished.

Two other types of male emerged, almost opposite in levels of activity. "The first," Calhoun reports, "were completely passive and moved through the community like somnambulists. They ignored all the other rats of both sexes, and all the other rats ignored them. ... To the casual observer the passive animals would have appeared to be the healthiest and most attractive members of the community. ... But their social disorientation was nearly complete."

Perhaps the strangest type was what Calhoun called the "probers," who moved in packs of three or four. They would confound a female by courting her as a group, harass lactating females and upset nests of pups.

End of Motherhood

Under these strains, motherhood in the crowded pens began going to pieces. Mothers grew sloppy about nestbuilding, often losing interest and leaving it incomplete. Litters got all mixed up, so no mother seemed to know whose babies were whose—nor seemed to care. Frequently, abandoned young were cannibalized by groups of male probers.

Of 558 born at the height of the sink, only one in four survived weaning. Miscarriages were common. Autopsies of females revealed tumors of the uterus, ovaries, fallopian tubes and mammary glands. And, hardly surprising, adrenals were conspicuously enlarged. As on James Island, the stress took its greatest death toll on the young and the female, contributing heavily to halting the population growth.

There is reason to believe that the behavioral sink could have been prevented without increasing available space—if Calhoun had divided the same space into a greater number of smaller pens and closed them off from one another. Thus a small group of rats, say, the instinctive group of about 12, although pressed for space, would have its own inviolate territory. An English ethologist, H. Shoemaker, tried this with canaries. First he placed a large number in a single large cage. A hierarchy developed in which the dominant birds interfered with the nesting of low-ranking families. Then he transferred them to small cages so that each adult male, including the low-ranking, was master over his family's territory. Brooding then proceeded more normally.

The canary experiment, simple as it is, may have vast significance in considering the development of—and prevention of—behavioral sinks in cities of human beings. Edward Hall, who as an anthropologist is chiefly concerned with human behavior, emphasizes the critical importance of architecture in avoiding the stress of urban crowding. He also emphasizes that architectural needs vary greatly from culture to culture; that one man's company may be another man's crowd. And the alarming thing is that, while some human populations are already showing clear signs of crowding stress, we are doing next to nothing about learning to design territories—proper homes and neighborhoods—to prevent behavioral sinks among the crowded poor.

One of the few studies linking human dwelling space with stress was made by a French couple, Paul and Marie Chombart de Lauwe, among working-class families of France. First they tried to correlate behavior with the number of residents per dwelling unit. This revealed little. Then they got the idea of considering the number of *square meters per person* in the home, regardless of the number who lived in it. They found that when each person had less than 8-to-10 square meters, instances of physical illness and criminal behavior were double those in less crowded homes. Thus human crowding was clearly linked with illness and violence.

Space and violence were linked in a different kind of study recently completed by a Columbia University psychiatrist, Augustus F. Kinzel. Small both in size and subject scope, the study was limited to 14 men in a federal prison. Eight had histories of violent behavior; six were considered nonviolent. Standing his subject in the center of a bare room, Dr. Kinzel would say, "I'm going to step toward you. Tell me to stop when you feel I'm too close." He would try this from several directions. It turned out that all the men seemed encircled by an invisible "buffer zone" which, when intruded upon, made them feel intensely uncomfortable. The violent men felt "crowded" at an average distance of three feet, the nonviolent at half that distance. Perhaps the most important finding of the study was that the violent men were more sensitive to approach from the rear. Their reactions were often clearly physical, particularly among the violent types. Some reported tingling or "goose pimples" across their shoulders and backs. Some literally stepped away with clenched fists as Kinzel entered the buffer zone, even though he was hardly within touching distance. They accused Kinzel of "rushing" them. One commented, "If I didn't know you, I might be ready for anything."

Just as the violent and nonviolent prisoners felt crowded at different distances, the reactions of French workers establish no rule to measure what constitutes crowding in other ethnic groups, even other classes. Crowding differs from people to people, class to class. Hall describes customs of various national groups—particularly the Japanese, Germans and Arabs—to show the great variations in their sense of proper space between persons.

Most Japanese, for example, are happiest when family and friends are huddled together in the center of a room, or all making body contact under a huge quilt before a fireplace. They feel it is congenial for whole families to sleep close together on the floor. Their dwelling spaces are small but as variegated in purpose as the many rooms of a large American house. The Japanese change the size, moods and uses of their rooms by rearranging screens and sliding their doors open and shut. Their concepts of "togetherness" and "aloneness" are so different from ours that the Japanese language contains no word that translates into our word "privacy." Yet a Japanese has strong feelings against two houses having a common wall. If his house is not separated from his neighbor's by a strip of land, no matter how narrow, he feels crowded. It is an important sign of his territorial integrity. When strangers are out of sight, Japanese are entirely untroubled by their sounds. In a Japanese inn, where a Westerner would toss and turn angrily at the sounds of a party in the adjoining room, a Japanese would sleep unmindful of it.

In contrast, Germans are especially sensitive to intrusion by sounds of strangers, one reason their hotels often have double doors and thick walls. A German has a strong sense of his own space—*lebensraum,* another word not readily translatable—and is disturbed if that space is not respected. In an office, he keeps his door closed. American visitors often misread this trait as something unfriendly. On the other hand, Germans regard the American habit of leaving doors open as unbusinesslike. Hall tells of an American camp for German prisoners-of-war in which men were bunked four to a small hut. The men went to great lengths to find materials for building partitions to separate themselves. German families, too, require clear definitions of territory. During the postwar housing shortage, American occupiers blithely ordered Berliners to share kitchens and baths, having no idea of the extreme stress—and violence —their order invited. New arrangements had to be made, Hall reports, "when the already overstressed Germans started killing each other over the shared facilities."

Arabs are happiest among crowds of people, a high noise level of conversation. They require great human involvement, closeness. Conversing, they look at each other piercingly, with much touching of hand to hand, hand to body. In his home, however, an Arab prefers spaciousness—large, high rooms with a commanding view—or he feels crowded. For all his love of involvement, the Arab needs privacy too. The way he gets it is by falling silent, retreating into himself. To talk to an Arab who appears converstionally withdrawn is to exercise bad manners in the extreme—an act of aggression certain to induce stress.

These examples provide the sketchiest of hints of how complex, delicate and explosive the matter of human crowding can be. Little is formally known about the elements of crowding that are at work in the impoverished ghettoes of American cities—except that they *are* at work, rapidly creating behavioral sinks.

"It is fairly obvious," says Hall, "that American Negroes and people of Spanish culture who are flocking to our cities are being very severely stressed. Not only are they in a setting that does not fit them, but they have passed the limits of their tolerance to stress. The United States is faced with the fact that two of its creative and sensitive peoples are in the process of being destroyed and like Samson could bring down the structure that houses us all."

Factories of Stress

When the stress of newly-urbanized Negroes is discussed, the solutions proposed are almost always limited to ending discrimination, improving education, providing jobs—and housing that is seldom described beyond being low-cost. Without these social improvements, clearly stress will not be eliminated. But Hall's thesis is that these alone cannot halt the growth of behavioral sinks. Space—and the architecture of that space—must be designed for the specific cultural needs of these urban newcomers. A great deal is now known about how *not* to design this space, but little about ways it should be designed.

For example, high-rise apartments, no matter how low-cost, for people recently of an agrarian tradition, are factories of stress. "It's no place to raise a family," a typical tenant complains. "A mother can't look out for her kids if they are 15 floors down on a playground. When I want to go up or down, I think twice because it may take me half an hour to get the elevator."

For a starting point in planning proper spaces for urban newcomers, Hall urges planners to consider that "Puerto Ricans and Negroes have a much higher involvement ratio than New Englanders and Americans of German or Scandinavian stock." As an example of architecture for "involvement," Hall recommends a look at the Spanish plaza and the Italian piazza, "whereas the strung-out Main Street so characteristic of the United States reflects . . . our lack of involvement in others."

One enterprising planner, Neal Mitchell, a professor of design at Harvard, has worked out a novel way of finding out what impoverished Negroes want in their housing. He consults with impoverished Negroes. Mitchell bought $80 worth of doll houses and furniture and invited poor people to arrange it according to their preferences. He found—contradicting the assumptions of almost all low-cost housing architects—that nobody wanted a dining room. They wanted a kitchen large enough to eat in. They also complained that public housing apartments they lived in were much too small. Yet, Mitchell reports, "every single person who played with our game wound up with a smaller square footage than the one they thought insufficient. It was just a question of design."

Next, Mitchell brought out blocks marked "house," "school," "church," "store" and so forth, and let people arrange their communities. Most people in the ghetto, he found, reject the suburban single-family house. One welfare mother told him, "That green front yard is useless. I want to sit out on my front steps and see all those neighbors. I want to be close enough to holler at them."

"I Think We are Going to Pull Through . . ."

Thus that welfare mother confirms Hall's suggestion of "architecture for involvement."

Mitchell is hopeful: "I think we are going to pull through because, you know, there is one thing about this country: it is flexible and it is willing to learn."

If Edward Hall and his "crowding" colleagues are more apprehensive, it is because they fear that our willingness to learn may lose in a race against the onrushing development of the urban sink.

The Specter of World Hunger

D. W. Brooks

Quoting from the first chapter of Genesis: "God created man in his own image, and he blessed them and said to them, be fruitful and multiply and *fill* the earth and subdue it." Whether we believe the Bible literally or not, this prophecy is coming true during the twentieth century. In fact, it is coming true during this generation, right before our eyes. It is freely predicted that the population will double between 1965 and 2000. If this happens, certainly the earth will then be full and, probably, overflowing.

The report of the President's Academy of Science on the world food problem, together with a number of other scientific studies which have been made of this problem in recent years, indicate that by 1975 we will begin to cross the line into mass hunger. By 1980 we will have crossed the line, and from then until 2000 we will gradually see mass hunger overcome most of the world, including Asia, Africa, South America and finally even Europe. Probably the only two areas of the world that will survive will be North America, and Australia and New Zealand.

We hear lots about the industrial revolution in this country, but it has been rather tame compared to the agricultural revolution, by far the most productive segment of our economy. For example, from 1956 to 1966 agriculture increased its per capita productivity 77%, or an average of almost 8% a year, whereas industry increased its productivity only 25% during the ten year period, or only approximately one-third as fast as agriculture. Under these conditions, we would think that the U. S. would be in position to feed a large part of the world. The facts are, that even with all our tremendous increase in productivity—with the population doubling in a 35-year period—the population growth will far out-distance anything we can do in food production in this country.

Even with the present research, food production, breeding, fertilization and insect-control programs in the world, it is hopelessly doomed unless we can solve this great mass need for food which is fast moving upon us.

One of my experiences with industry and labor came about in a conversation with Walter Reuther, when we were on the War Mobilization Board. He stated that early in his life he was undecided whether he wanted to become a chicken farmer or a labor leader. I mentioned to him that industry has largely maintained its prices, or increased them substantially, while in contrast, because of its tremendous productivity, agriculture had greatly lowered its prices in the previous 20 years. I used chicken as an illustration.

Fifteen years ago, broilers were bringing 35 cents per pound in parts of the South, and today they are bringing 12 cents per pound, or approximately one-third as much. If labor and industry had done as well, the $3500 automobile would today be selling for $1200. This shows the impact that the high productivity of the American farmer has had on the consumer. Except for this impact, our cost of living would be tremendously higher than it is at present. In fact, agriculture has largely furnished all or practically all of any stabilization we have had in the cost of living throughout the last twenty years. For example, consider chickens. We have been able to lower the amount of food that it takes to produce one pound of meat from four pounds of food to two pounds in 15 years. We have lowered the time it takes to produce a three pound broiler from 14 weeks to seven weeks. These are but a few of the stories that can be told about modern agriculture.

One of several research projects now going on concerns feeding cattle to see if we can have some of the same break-through in cattle production that we have had in poultry. Fish production is another possible high quality protein feed, and we are carrying on research there through breeding and nutrition.

Communist Agriculture

Moving on to world production, certainly the Communists have not had any answers concerning agricultural productivity. Except for food supplied by the West, a large part of the Communist world would have already been starved out. Certainly the Communist pattern of agricultural production is no answer to the food problem. In fact, it has deterred the production of food instead of speeding it up.

The commune system of agricultural production, whereby they try to make farmers work together in communes, has been disastrous to the Communist world. Certainly the Capitalist system of agricultural production has been much better. My studies of Russian agriculture indicate their production is at least fifty years behind ours, with little hope of catching up.

In China, except for some of the changes in their commune system whereby they allocated to the individual farmer approximately one acre of land to farm, no doubt they would have already had a complete internal collapse. In fact, the crops were terrible all during the Great Leap Forward period of 1959, 1960 and 1961. By the spring of 1962, their food supply was so low their people, who were being fed in the commune kitchens, had to reduce their calorie intake to 1200 to 1400 calories per day. The food situation was so critical that many of the people who were studying the internal Chinese situation closely advised me that, as agricultural advisor to the President, I should ask him in the spring of 1962 to place ships, loaded with wheat, up and down China's coast because China was going to collapse. However, because they gave a little freedom to the farmer, had better weather conditions, and purchased wheat from the outside, they at least temporarily averted this collapse. But certainly there is no evidence that they have permanently settled their problems.

Even in Cuba, when Castro took over, agricultural production went into a nose-dive. Except for help from Russia and resources from outside, Cuba would have collapsed had it been forced to depend on its own productivity.

Free-World Agriculture

When we consider the free world, we still have great problems. Even Japan, one of the great industrial marvels of our time, is relatively weak in agricultural production, and consequently, the largest purchaser of agricultural products from the U. S.

When we study India, we encounter a tremendous amount of present hunger. Except for the food which we have given them in the past several years, we would have already had mass hunger in India. The crop was better in 1968, and they are very much elated; frankly, rather foolishly so, because they feel that because they harvested one good crop they will continue to have a good crop from now on. But those of us who have studied India, and understand agricultural science, certainly do not agree with this viewpoint. We feel that unless an almost impossible effort is made, certainly India will soon again face mass hunger.

I could move then to many other countries around the world, including Africa and South America, where similar conditions prevail. Here, population is going up at a rate of 2-1/2 to 3% a year and agricultural productivity in many of these countries is not progressing any, even losing in some countries. At best, it is averaging only 1% in most years.

Under these conditions, I think it would be very easy to concede that we are in an impossible situation and that mass hunger will in fact begin to engulf the world, possibly beginning by 1975—certainly between 1975 and 1980 at the latest.

(Editor's Note: The author's conclusions seem to be born out by that well-known English novelist, C. P. Snow *(The Two Cultures—The Corridors of Power),* who, when he delivered the John F. Green Lecture at Westminster College, Fulton, Missouri, last November stated, "I have to say that I have been nearer to despair this year, 1968, than ever in my life. We may be moving —perhaps in ten years—into large-scale famine. Many millions of people are going to starve. We shall see them doing so upon our television sets.")

But frankly (although I may be unduly optimistic), I do not think it necessarily has to be this way. I think an all-out effort should be made to meet the problems and I fully understand many of the problems involved.

(Editor's Note: Another who has this viewpoint is Former U. S. Secretary of Defense, Robert McNamara, now head of the World Bank. Mr. McNamara recently stated in his first speech as head of that Bank that "There is every reason for hope. In the past few generations, the world has created a productive machine which could abolish poverty from the face of the earth. Who can fail to see the immense prospects that lie ahead for all mankind, if we have but the work and the will to use our capacity fully?")

For example, eight or ten of us are now trying to build a large fertilizer complex in India. We think it will increase India's productivity approximately 6%, or enough to take care of two years increase in population growth. But, we have met almost impossible handicaps. The government of India seemed not to understand what we were trying to do and created some real problems for us. We estimated the project would cost $120,000,000, of which $55,-000,000 must be in dollars, because we will need that much machinery and equipment from outside. Although we can probably generate a large part of the balance in rupees inside India, we must have some way to get the $55,-000,000 back to one of the large banks in this country which has agreed to lend the money.

When we first discussed with the Indian government a contract for building this plant, we received a very violent reaction about signing anything labeled "contract." After a great deal of further discussion, they said they would probably agree to it if we would label it "agreement" instead of contract. This illustrates the frustration you face in dealing with a situation of this kind.

We, of course, were hopeful that we could enable several million farmers in India to own this plant, which would make it possible for them to buy, pay for and eventually own farms in India. The nation desperately needs to have a prosperous middle class, in order to bring stability to the country, as well as high productivity.

Unfortunately, many of the countries of the world formerly under colonial rule somehow equated colonialism and capitalism as one and the same thing. Consequently, they have had a tendency to try to move all these governments to the socialistic system, whereby most of the new plants are built by the government. We think this method will be disastrous and will, in the years ahead, *guarantee* mass hunger, the same as has taken place in the Communist countries. Therefore, we feel these countries' citizens must break the grip of their governments, which are strangling many of the economies of the developing countries.

Those of us who are working on this problem believe that in the end this will probably be our greatest struggle, and unless the governmental grip is broken then we are doomed to failure. But, we believe that time and hunger will gradually force governments to alter their policies and make it possible for their farmers to produce. We know that we now have enough scientific knowledge to win this battle against starvation. Our real problem is to encourage governments to move along fast enough, and to train farmers quickly enough to do the job technically.

A Modern Agricultural Miracle

One of the reasons why I believe this possible is because I have seen it happen, in one of the most difficult to develop countries in the world, Formosa. A Chinese classmate of mine at the University of Georgia was in charge of all agricultural research on the mainland of China. He was able to get off the mainland with Chiang Kai-shek, bringing twelve of his Chinese assistants. They convinced Chiang Kai-shek that the reason they lost the mainland of China was because of the manner in which agriculture was handled. They then proposed a system whereby tenants would be permitted to purchase farms and pay for them over a ten-year period.

They trained these tenants in the same way that we train our farmers over here, through our agricultural agent system. As part of the plan, the landowners agreed to invest the funds which were paid to them by the tenants into industry, which was far more profitable than their former ownership of the farms. Consequently, we have witnessed a miracle of agricultural and industrial production moving *side by side* there. In a country that does not have rich soil, we now see a productivity that is moving ahead so rapidly that on a twelve months basis they have the highest yield of any place on earth. They are increasing their productivity much faster than the increase in population. No country of Asia had a more difficult problem.

In watching this miracle take place, I know that a similar approach in other countries could likewise succeed. The only question is whether the governments involved will permit and promote this kind of development. We must somehow encourage a break in the stranglehold that certain governments have on their farmers. This, along with continuous research, in my opinion, gives us hope that the job can be done.

A team from Cornell University working in the Philippines is doing the type of research work necessary to solve this problem. They were able to develop a rice that has a yield of three to four times the yield of rice produced heretofore. One of the reasons they succeeded was because on Formosa the research people to whom I referred had already produced a short-stemmed low rice that could take fertilizer. When it was crossed with various long-stemmed varieties and aided with fertilizer, the yield spurted. Likewise, American research has developed miracle wheat and miracle corn that can help feed the world.

At this point, I would like to comment on Indonesia, whose government under Sukarno was overthrown. Indonesia, suddenly overnight, moved from a Communist nation back to a capitalist nation. We have great responsibility to help this nation recover economically, and to help them with their food problems.

Most of my scientific friends, who have made the closest study of this problem, have said to me that I am foolish about all this, and far too optimistic. They say that it is much too late to save the world from mass hunger, that it is already here, that it cannot be prevented, and that those of us who are working on the problem should select the countries we are certain we can save, save those, and forget the rest of the world.

Right or wrong—although I recognize my friends might be more realistic than I—I am not in agreement with this philosophy. I think the job can be done. I do not think we should pitch in the sponge yet and say that mass hunger is going to engulf most of the world and *there is nothing* we can do about it.

Since research now indicates that mental capacity is permanently affected when protein is inadequate early in life, this is our immediate urgent problem: to get sufficient protein in some way to the youth of the world.

(Editor's Note: Many world food authorities feel that there is no way that sufficient protein can be supplied to the world's people. Regardless, in a national magazine article last November, Addeke Boerma, Director-General of the United Nations Food and Agricultural Organization, was reported to have said that even though protein supplies may falter, recent advances with high-

yield cereals and rice strains will keep the world from mass famine. Mr. Boerma was quoted as saying, "The three basic ingredients required are capital, technology and organization. They are becoming available in increasing measure, and I believe that the agriculture of the developing countries is now reaching the point of 'take-off.' The situation remains precarious, but there are no symptoms of a long death.")

The Universities, like Emory, through their tremendous medical and other scientific research, can be very effective in helping solve this great humanitarian problem. They can be effective in pointing out the needs, and helping solve the food problems, of the world by direct contact with its hungry people.

The Protein Path:
Hunger Begins with a Hungry Plant

Don Fabun

"Whatever is necessary," says W. H. Ferry in another context, "is possible." It will be noted that he did not say "probable," only possible. Here we will examine solutions that appear to be possible (because necessary) in the alleviation of the world food crisis. At the present time, their probability seems quite low.

Earlier we presented a list of sub-problems that, added together, constitute the world food crisis. They lead to some seemingly inescapable conclusions. One is that a piecemeal attack, though helpful, is not likely to solve the over-all problem—at least not within the next ten years. The sub-problems are too inter-related, too dependent on multiple and simultaneous solution for that.

Perhaps, then, we should examine other forms of solution that might, just possibly, "buy the time" necessary to find integrated solutions to the sub-problems.

Nearly all of the solutions proposed for the sub-problems are quantitative in nature—their aim is to produce more food for more people in more places. Yet it has been well said, "Hunger begins with a hungry plant." The hungry plant is deficient in proteins for human use. Could we suggest that a qualitative approach might help reduce the magnitude of the quantitative problem, and

EDITOR'S NOTE: The preceding article was especially adapted from a talk given by the author before a meeting of the Emory University Board of Trustees and Board of Visitors, on Oct. 16. 1968.

that perhaps we do not need such vast quantities of new food, so much as to increase the nutritional quality of foods that already exist?

There are a number of currently feasible ways to raise the protein content of already existing foodstuffs, or at least the consumption of them, without planting another acre, without substituting the tractor for the bullock, without changing age-old food habits in any substantial way.

In most of the diet-deficient areas of the world, there are sufficient local sources of protein-rich foods that are not being used because the people who live there do not know about proteins. This is pretty excusable, because many people in the "developed" countries don't know, either. There are examples of how trained nutritionists, moving into back-country villages, have significantly raised the protein intake of local citizens, just by using whatever was growing nearby. But we will start with fish, because of all the world's existing sources of protein, this is, in most areas, one of the most neglected. As Gilbert and Sullivan said (in "Patience") . . . "There's fish in the sea, no doubt of it, as good as ever came out of it."

The plain truth is that out of the 25,000 known species of fish, only a few dozen are used directly or indirectly as food for man. In the protein-deficient countries—Asia, Africa, and Latin America—the consumption of fish products (marine and fresh water) is only about seven pounds per person each year. And yet the fish are there, and it is conservatively estimated that the annual catch of salt-water fish could be raised to far more than the 55 million tons projected as the annual catch in 1969, without the danger of destroying the oceanic food chain.

This would involve a tremendous effort; new ways of raising and harvesting fish instead of hunting them; new types of ships, new port and processing and storage facilities. No such effort is now underway, but this is not to say that it could not be made.

Probably the most efficient use of a greatly increased fish harvest would be in the form of Fish Protein Concentrate, which may be defined as "any inexpensive, stable, wholesome product of high nutritive quality, hygienically produced from fish." The fish, in its entirety, is ground up, reduced to eliminate the water and fish oil, and the resulting product emerges as a powder-like concentrate that is about 80 per cent protein. One hundred metric tons of fish yields 15 metric tons of concentrate, equivalent to 12 tons of 100 per cent protein. The concentrate can be used as a food additive to traditional foods, although some problems of taste and consumer acceptance remain. About 25 grams of Fish Protein Concentrate, used as an additive to already available foods, provides sufficient protein intake per day for an adult.

Costs are high, but not prohibitive. A 90,000-ton capacity plant, capable of producing enough protein for one million people per year, would require a capital investment of about $2.5 million, plus two fishing vessels of 120 feet in length, costing about $3 million, plus dock, handling facilities and distribution. Some $7 million over-all would meet the protein supplement needs of a million persons per year, at a cost of less than two cents per person per day.

Another important source of both whole fish and Fish Protein Concentrate lies in the protein that can be yielded by inland fisheries—the growing of fish in ponds, rice paddies, lakes, reservoirs, rivers, and coastal estuaries. It should be pointed out that some species of fish, as protein producers, are as efficient as chickens, and considerably more so than pigs and other domesticated animals raised for food.

At the present time, inland fishery catches supply about 16 million tons —or 30 percent of the world's total fish yield. Unfertilized ponds yield from 50 to 500 pounds of fish per acre per year; fertilized ponds from 150 to 1,500 pounds; fed and fertilized ponds from 2,000 to 5,000 pounds per acre, when agricultural wastes are used. Where ponds are enriched with grains and seed meals, the yield is from 1,000 to 16,000 pounds per acre. These gains are important; the yield is not only high in protein content, but where ponds are fertilized, one pound of fertilizer yields from five to eight pounds of fish.

The point is, not only are there many already existing bodies of water not being used for fish raising and harvesting, but building new ponds is relatively inexpensive. If the average cost for pond construction is $500 per acre, then the cost of new ponds to meet recommended consumption of inland water fish as a protein supplement for the world's population by 1975 would average only about $80 million annually.

Another possibility is to raise fish along with rice, in those areas where rice is the main source of food. There are approximately 197 million acres planted in rice each year. About 30 per cent of these are covered with water long enough to raise fish along with the rice. In Indonesia alone, there are 150,000 acres of combined rice-fish cultures, producing an average of 300 pounds per acre—a total of 45 million pounds of fish annually. In addition, it is estimated that about 150,000 tons of wild fish, not deliberately raised, are produced in rice paddies each year. Intensive programs to increase the raising and harvesting of fish in rice cultures, where it is feasible, could do much to supplement protein intake in those large areas of the world where rice is the principle source of food.

To this could be added the deliberate use of salt water estuaries for the production of fish and shell-fish. Where these areas are used—as on the Asia mainland, and in Japan, Formosa and Indonesia—yields are from 400 to 2,000 pounds of fish product per acre each year. In many areas of the world, these rich potential sources of protein are not being exploited at the present time. Unfortunately, it doesn't appear that much is likely to be done about them, either, although the costs are minimal and the results significant.

Already growing, already being harvested each year, is a low-cost source of high quality protein that is equal to all the protein produced by all the livestock in the world. Most of it is fed to animals or used as fertilizer; large amounts are discarded as waste.

This rich, virtually untapped source is the protein that can be extracted from soy beans, cottonseed and peanuts. Had the 1965–66 protein content of these been processed into protein for human use, it would have been the equivalent of 33.3 million tons of protein, compared to a total of 22 million tons produced from meat, milk and eggs. Each additional one percent increase in production of the 1966 soy bean crop would have yielded enough to correct the lysine deficiency in one million metric tons of wheat.

Concentrate from soybeans ranges from 40 to 95 percent protein content, at a cost of from 14 to 40 cents per pound. This material is amazingly versatile; it can be used as an additive to cereal products, in high protein beverages, baked goods, desserts, and processed into a wide range of meat products that, when properly prepared, are virtually indistinguishable from the "real" thing. Moreover, these foods can be produced in factories at costs that range from one-half to one-fifth what it costs to produce similar animal products.

Since soy beans grow best in temperate areas, it would probably be most economical to produce them there and ship the concentrate to the diet-deficient tropical countries. However, in some places it could be produced economically on the spot. According to Agricultural Chemistry (April, 1967), the USDA had developed a simple hand process for making soybean flour in areas where there is neither skilled labor nor electrical power. Five men can produce 300 pounds of soya flour in an eight-hour day, enough to supply the daily protein requirements for 1,600 adults. The flour can be used in beverages, yeast bread, corn bread, noodles, unleavened bread, mush, gruels, soups and desserts.

While not as versatile as soy bean, cottonseed still represents a highly useful, virtually untapped source of protein. It has the great virtue that, unlike soy beans, cottonseed comes from a plant that is indigenous to protein-poor tropical areas of Asia, Africa and Latin America. Upon removal of oil and the seed coat from the cottonseed, a concentrate with 50 to 55 percent protein can be obtained. This can be further concentrated—to 70 percent protein—by extracting the carbohydrates. Either concentrate can be used directly for feeding humans, or indirectly as cattle feed.

The great advantage here is that the cotton is already grown and harvested for its fibers and oils so that the cottonseed is a by-product; it can be produced economically by techniques already developed, and in the areas where additional protein is most urgently needed.

Peanuts, which are called groundnuts in most countries, offer a rich and economical source of protein. They grow well in both temperate and tropical areas and are indigenous to the Caribbean Islands, South America, and Africa. Although peanuts can be, and are, eaten directly by man, about 65 percent of the world's production is crushed for oil, which furnishes about 20 percent of the world's trade in edible oils. After the oil is removed, there remains a 50 percent protein concentrate that could be used for human food. Further processing, to remove the carbohydrates, raises the protein content to 70 percent. Although peanut protein is deficient in three amino acids, it can be fortified with synthetic materials.

Again, here is a widely produced crop, already growing, the by-products of which are rich in protein; and the crop grows in areas that are most in need of protein enrichment. Concentration of the protein itself is a relatively simple task, requiring neither high cost equipment nor technical skills. Certainly its use would provide a rich source of badly-needed protein at much less cost than extensive irrigation or new-land farming projects. However, little is currently being done about it.

Leaf Protein

Another virtually unused source of protein is the leaves and grasses which are usually wasted, or at best fed to animals. Nearly all of the essential proteins consumed by man are derived from amino acids originally created in the leaves of green plants, and those that remain in the leafy parts are of the same composition as other plant proteins.

Leaf proteins can be introduced into man's diet either by eating them directly (like spinach, turnip greens, lettuce, etc.) or by processing the leaves into high protein concentrates. In the latter, the leaves are pulped and pressed, and the resulting juice, when processed to a dry state, contains about 65 percent protein. The fiber from which the juice was pressed can be used as animal food.

The pressed cake of leaf protein concentrate is green in color and tastes like spinach or tea. Further processing, if desired, can remove most of the color and flavor, as is done in Israel; or in the green state, it can be used as a high-protein filler in native dishes such as ravioli. If a number of kinds of leaves are pressed together, the leaf protein produced appears to be better than cereal proteins and as good as animal proteins other than eggs and milk.

Leaf protein can be produced by hand methods, or with low power input, at low cost, since the raw material is usually considered a waste product anyway. On a village basis, using an input of two tons of feed and an output of 60 kilograms of protein an hour, protein would cost six to seven cents per kilogram (with labor figured at 15 cents an hour—which would be high in many agrarian areas). So far leaf protein has been produced in Jamaica, India, Uganda, New Guinea and Israel, but no large scale research, development or application projects are underway at present, nor, apparently, contemplated.

Whether oilseeds are used, or leaves, or both—there exists at the present time both the raw materials and the techniques for the large-scale production of high quality protein from products largely wasted, using production facilities well within the means of world industry. Precious little is being done about it.

Synthetic Amino Acids

As detailed earlier, the cereal foods that constitute the largest part of the world's diet, particularly in the diet-deficient countries, are deficient in several "essential" amino acids. These nitrogen compounds are "essential" because the body cannot manufacture them itself. Since the grains are already being raised, processed, distributed and consumed, it would appear that one of the most promising approaches to the problem of world malnutrition would be to "fortify" grain cereals with the missing, or deficient, amino acids, produced synthetically.

Fortunately, such a technology already exists. The three essential amino acids that are deficient in most cereal grains are called lysine, threonine and trytophan. Lysine can be produced by mass-production factory methods for as little as $1 per pound. The others presently can be synthesized for $2.50 to $4.50 per pound, but in mass production probably could be produced for prices close to that of lysine. These synthetic materials, different combinations or patterns of nitrogen, hydrogen and carbon, can be produced as colorless, odorless, tasteless crystals.

Their introduction into the already existing food production system would probably be most effective at a central processing point, such as a flour or corn meal mill, where the process would be no different than the current fortification by vitamins and minerals. Thus the people who purchase the flour (or the commercial products made from it) would not in any way have to change their regular food consumption habits. The difference is that they would be eating foods nearly as rich in essential proteins as milk or meat. To extend this system to those areas where the cereals do not pass through central processing, but are prepared on the farm or in the village itself, presents a different sort of problem. However, since what is required is not more difficult than the addition of salt, except that the flavor is not changed, there should be no taste, religious, or cultural obstacles.

Here, in the production, distribution and introduction of synthetic essential amino acids lies an approach that appears to be well within the means of modern industrial processing. Projected figures show that such a program would be considerably less expensive than the use of Fish Protein Concentrate (although that approach should also be used in areas where fish products are plentiful). If all the wheat and flour shipped in the 1965 Food for Peace program had been fortified with lysine, the total cost for lysine fortification would have been about $32.5 million for lysine vs. about $84 million for Fish Protein Concentrate.

What is argued here is not that a crash program be directed toward the mass production of synthetic amino acids, but a three-pronged program that involves oilseed production, Fish Protein Concentrate and synthetic aminos. All are technologically possible, all involve mass production and distribution methods well within the state of the art in advanced industrial systems. The combination of these could go a long way toward ameliorating the crucial protein shortage in the diet-deficient areas of the world while longer-term solutions are being worked out. Except for what are virtually pilot-plant sized operations, no large scale programs for the production of synthetic amino acids are currently underway in the technologically advanced nations today.

Single Cell Protein

The idea that crude petroleum and paper can be turned into food for humans is one that most people reject out of hand; not because it isn't true, but because they equate petroleum with fuel for automobiles and newspapers as something to be read and not eaten. Someplace along the line our educational system has failed to make it clear that both petroleum and paper are derived from living products and have in them chemical compounds that can be turned into foodstuffs whose protein composition does not differ materially from milk, beefsteak, or other common foods.

Most of the protein compounds in the world have been produced by the activities of microorganisms. When they are grown on substrates of once-living compounds they produce very nearly the same kinds of protein materials as do plants and animals higher on the evolutionary scale. These microorganisms —principally yeasts and bacteria—can grow on such various substrates as sucrose, molasses, soybean oil, kerosene, petroleum distillate No. 2, waste paper, surplus and spoiled fruits and vegetables, bagasse (wasted agricultural stalks), waste products of the lumber industry, and methane gas. In some cases it is obvious that the use of these materials would not only make important contributions to the production of protein, but would help in solving some of the world's waste product disposal problems.

Other advantages of what is called "single cell protein" production are that the microorganisms grow very rapidly, doubling their weight every five hours or less, which is several thousand times faster than farm animals can synthesize protein. The microorganisms can be grown in tanks requiring no soil or sunlight or rainfall, and they belong to the plant kingdom, so that when harvested and introduced into traditional food systems, they do not run up against religious or traditional obstacles.

The techniques for single cell protein production already are well known. Louis Champagnat, who has developed a sizable pilot scale operation in Lavera, France, has described the basic process (in *Scientific American,* October, 1965) thus:

"The medium is similar to that for the growing of yeasts on sugar, except that oil is substituted for the sugar. Nitrogen is added to the medium in the form of ammonia salts; phosphorus and potassium are supplied in the form of general fertilizers; trace elements and growth vitamins are added. The product is more than 50 percent protein.

"When the yeasts grown on petroleum have been dried and purified, the concentrate is in the form of a powder or whitish flakes with no pronounced odor or taste . . . Like concentrated protein from meat, fish, yeast or soybeans, the protein from petroleum can be transformed into many different foods."

Dr. Champagnat has calculated that with an outlay of some 40 million tons of petroleum (a small fraction of the total crude oil produced annually in the world), 20 million tons of pure protein could be produced per year. Other advantages are that the organisms feed on the wax in the crude oil, which makes the oil suitable for diesel engines and domestic heating, that petroleum is available in most of the diet-deficient areas, and that the processing equipment can be added relatively easily and economically to the existing 700 refinery complexes throughout the world, thus reducing distribution costs.

The cost of protein thus produced is from 20 cents to 40 cents per pound, still more expensive than oilseed meal containing 40 to 50 percent protein. However, the fact that waste products are used (which are expensive to dispose of, and in any event contribute to the growing air and water pollution program) should be taken into account.

A number of major petroleum firms have undertaken feasibility studies of the single cell protein process, and at least one American firm has been awarded a research project (by the U.S. Public Health Service) to study the production of protein from newspaper and agricultural waste products.

It has been estimated that a research program launched now and costing $10 to $30 million over a period of two to four years might launch a single cell protein industry whose contributions could be significant by the 1975–1980 period. No study of such proportions has so far been undertaken.

Genetic Manipulation

The approaches we have been discussing all involve, at some point, the concentration of proteins in such a way that they can be added to the bulk protein deficient foods that make up most of the diet in the world today. Obviously, if the fortification step could be eliminated, raising the world protein availability could be greatly accelerated, and substantial cost savings might be achieved.

One way of doing this would be to increase the natural protein content of the world's principal cereal crops—wheat, corn, and rice. This can be done through deliberate "genetic manipulation," i.e., choosing and breeding strains of plants that have higher protein content.

This already has been done with corn, for instance, where a superior variety of corn (opaque-2) has been discovered and developed. It contains about 65 percent more lysine, more trytophan and a better amino acid balance than ordinary hybrid corn. Since corn is an important part of the human diet in much of Latin America, parts of Africa, southern and eastern Europe and parts of Asia, the importance of this discovery is of supreme importance. However, it takes considerable time to develop a sufficient amount of seed corn of a new variety to be able to make it available to·farmers on a global basis, and even longer to break down established patterns in order to get it introduced. Thus, while holding enormous promise, the genetic approach involves a period of five or ten years of concentrated effort; an effort not being made at the present time.

Another genetic approach is to create new species of plants that have desirable growing and protein characteristics. In June, 1967, the University of Manitoba announced successful production of the world's first man-made grain, called triticale—a cross between durum wheat and rye. It is a new grain, not a hybrid of an old one. It grows well in very dry country and has a protein quality four percent higher than wheat. Work began on it in 1954—it may be on the market as seed grain by 1970.

The search for improved nutritional characteristics in other cereals has either been initiated or planned, including programs in Mexico for wheat and maize, in the Philippines for rice, and in India for sorghum and millet.

Like everything else mentioned in this section, the technology is there; the will to implement the technology is sadly lacking. We have the way, but not the will. And that is the crux of the world food crisis.

Life, Liberty and the Pursuit of Privacy

Alan F. Westin

The phenomenon of privacy has been important to man ever since he moved into a cave. Even animals stake out claims to territorial privacy. Here, the author probes the primitive roots of privacy, and follows their evolutionary growth into our electronic age. Professor Westin is Director of the Center for Research and Education in American Liberties of Columbia University and Teachers College. His writings on privacy, a central concern of his in books and articles for more than a decade, have been quoted by the Supreme Court and used as the basis for much legislation. He is a member of the District of Columbia Bar, a member of the National Board of Directors of the American Civil Liberties Union and the National Committee of the Anti-Defamation League.

To its profound distress, the American public has recently learned of a revolution in the techniques by which public and private authorities can conduct scientific surveillance over the individual. In chilled fascination, the press, television programs and popular books have described new means of telephone tapping, electronic eavesdropping, hidden television-eye monitoring, "truth measurement" by polygraph devices, personality testing for personnel selection, and growing dossiers of personal data about millions of citizens. As the late 1960s arrived, it was clear that American society had developed a deep concern over the preservation of privacy under the new pressures from surveillance technology.

In my view, the modern claim to privacy derives first from man's animal origins and is shared, in quite real terms, by men and women living in primitive societies.

One basic finding of animal studies is that virtually all animals seek periods of individual seclusion or small-group intimacy. This is usually described as the tendency toward territoriality, in which an organism lays private claim to an area of land, water or air and defends it against intrusion by members of its own species. A meadow pipit chases fellow pipits away from a private space of 6 feet around him. Except during nesting time, there is only one robin on a bush or branch. Antelopes in African fields and dairy cattle in an American farmyard space themselves to establish individual territory. For species in which the female cannot raise the young unaided, nature has created the "pair bond," linking temporarily or permanently a male and a female who demand private territory for the unit during breeding time. Studies of territoriality have even shattered the romantic notions that when robins sing or monkeys shriek, it is solely for the "animal joy of life." Actually, it is often a defiant cry for privacy, given within the borders of the animal's private territory.

Ecological studies have demonstrated that animals also have minimum needs for private space without which the animal's survival will be jeopardized. Since overpopulation can impede the animal's ability to smell, court or be free from constant defense reactions, such a condition upsets the social organization of the animal group. The animals may then kill each other to reduce the crowding, or they may engage in mass suicidal reductions of the population, as lemmings do. Experiments with spacing rats in cages showed that even rats need time and space to be alone. When they were deliberately crowded in cages, patterns of courting, nest building, rearing the young, social hierarchies and territorial taboos were disrupted. Studies of crowding in many animals other than rats indicate that disruption of social relationships through overlapping personal distances aggravates all forms of pathology within a group and causes the same disease in animals that overcrowding does in man—high blood pressure, circulatory diseases and heart disease.

The Veil of the Tuareg

Anthropological studies have shown that the individual in virtually every society engages in a continuing personal process by which he seeks privacy at some times and disclosure or companionship at other times.

A sensitive discussion of this distance-setting process has been contributed by Robert F. Murphy of Columbia University. Murphy noted that the use of "reserve and restraint" to provide "an area of privacy" for the individual in his relations with others represents a "common, though not constant" factor in all social relationships. The reason for the universality of this process is that individuals have conflicting roles to play in any society; to play these different roles with different persons, the individual must present a different "self" at various times. Restricting information about himself and his emotions is a crucial way of protecting the individual in the stresses and strains of this social interaction. Murphy also notes that creating social distance is especially important in the individual's intimate relations, perhaps even more so than in his casual ones.

Murphy's work among the Tuareg tribes of North Africa, where men veil their faces and constantly adjust the veil to changing interpersonal relations, provides a visual example of the distance-setting process. Murphy concluded that the Tuareg veil is a symbolic realization of the need for privacy in every society.

Another element of privacy that seems universal is a tendency on the part of individuals to invade the privacy of others, and of society to engage in surveillance to guard against anti-social conduct. At the individual level, this is based upon the propensity for curiosity that lies in each individual, from the time that as a child he seeks to explore his environment to his later conduct as an adult in wanting to know more than he learns casually about what is "really" happening to others. Gossip, which is only a particular way of obtaining private information to satisfy curiosity, seems to be found in all societies.

Curiosity is only half of the privacy-invading phenomenon, the "individual" half. There is also the universal process of surveillance by authorities to enforce the rules and taboos of the society. Any social system that creates norms—as all human societies do—must have mechanisms for enforcing those norms. Since those who break the rules and taboos must be detected, every society has mechanisms of watching conduct, investigating transgressions and determining "guilt."

The importance of recognizing this "social" half of the universal privacy-invading process is similar to the recognition of the individual curiosity—it reminds us that every society which wants to protect its rules and taboos against deviant behavior must have enforcement machinery.

Privacy in Democracies

It is important to realize that different historical and political traditions among contemporary democratic nations have created different types of overall social balances of privacy. Britain has what might be called a "deferential democratic balance," based on England's situation as a small country with a relatively homogeneous population, strong family structure, surviving class system, positive public attitude toward government, and elite systems of education and government service. This combination has produced a democracy in which there is a great personal reserve between Englishmen, high personal privacy in home and private associations, and a faith in government that bestows major areas of privacy for government operations. There is also a tradition of tolerating nonconformism which treats much deviant political and social conduct as permissible private action.

West Germany today has what might be called an authoritarian democratic balance. The Bonn Republic defines privacy in a nation where the traditions of democratic self-government came late; authoritarian patterns are deeply rooted in German family structure and social life; both law and government are permeated by high public respect for officialdom and experts; and neither German law nor government showed high capacity, until the post-World War II period, to enforce a meaningful system of civil liberties restraints on government surveillance practices or harassment of dissent. The result is a democratic state in which privileged elements having the authority of family, wealth and official position often enjoy substantial privacy and government enjoys great rights of secrecy; but the privacy of the critic and the nonconformist is still not secure in West German life.

Where does the United States fall in this spectrum?

American individualism—with its stress on unique personality in religion, politics and law—provides a major force for privacy in the United States. This attitude is derived from such factors in American national experience as frontier life, freedom from the feudal heritage of fixed class lines, the Protestant religious base of the nation, its private property system, and the English legal heritage. Along with the individualist stress has gone a complementary trait of associational life—the formation of numerous voluntary groups to pursue private and public goals. An outcome partly of our heterogeneous immigrant base and partly of the American's search for group warmth in a highly mobile, flexible-status society, associations have long been a distinctive aspect of our culture, with well-established rights of privacy against government surveillance or compulsory public disclosure. A final value supporting privacy is the American principal of civil liberty, with its belief in limits on government and private power, freedom of expression and dissent, and institutionalized mechanisms, particularly the legal system and independent courts, for enforcing these rights.

Were these the dominant values of the American sociopolitical tradition, the privacy balance in the United States might be called wholly libertarian. But, from colonial days down to the present, foreign and native analysts have observed other powerful tendencies in American life that press against privacy and support restrictive rules of disclosure and surveillance. The classic American belief in egalitarianism and "frontier democracy" gives rise to several trends: a denial of various "status rights" to privacy that once were attached to European aristocratic classes and are now claimed by elite groups of culture, intellect and science; a propensity toward "leveling curiosity" in social and political life that supports inquisitive interpersonal relationships; and a demand for external conformity of a high order, in the name of a middle-class system in which the blessings of equality and opportunity carry with them a heavy burden of ideological and social conformity.

The United States is thus a democracy whose balance of privacy is continually threatened by egalitarian tendencies demanding greater disclosure and surveillance than a libertarian society should permit.

Privacy and the Senses

Privacy also differs from nation to nation in terms of the impact of culture on interpersonal relations. The most extensive recent work on this theme comes from the cultural anthropologist Edward Hall, who states that people in different cultures experience the world differently not only in terms of language, but also with their senses. They "inhabit different sensory worlds," affecting the way they relate to one another in space, in matters ranging from their concepts of architecture and furniture arrangement to their setting of social distance and interpersonal contact.

To compare these differences, Hall studied a number of contemporary cultures to see how their notions of sensory pleasure and displeasure affected their definitions of interpersonal space. First, he compared the dominant norms of American society, as set by the white middle and upper classes, with three European cultures with which the American middle and upper classes are most closely linked historically and culturally—Germany, England and France.

Germans, Hall found, demand individual and enclosed places to achieve a sense of privacy. This need is expressed in closed doors to business and government offices, fenced yards and separate closed rooms in the home, discomfort at having to share facilities with others, and strict "trespass" rules regulating the person-to-person distance on social, business and ceremonial occasions.

In contrast, Americans are happy with open doors in offices, do not require fencing or screening of their homes to feel comfortable, and are far more informal in their rules of approach, order and distance. An American does not feel that a person walking close to a group or a home has "intruded" on privacy; Germans, on the other hand, will feel this a trespass.

English norms of privacy, Hall found, lie between the American and the German. The English accomplish with reserve what Germans do with doors, walls and trespass rules. Because English children in the middle and upper classes do not usually have separate rooms but share in the nursery with brothers and sisters until they go away to boarding school and live in dormitories, the Englishman grows up with a concept of preserving his individual privacy within shared space rather than by solitary quarters. He learns to rely on reserve, on cues to others to leave him alone. This habit is illustrated in later life by the fact that many English political and business figures do not have private offices; members of Parliament, for example, do not occupy individual offices, and they often meet their constituents on the terrace or in the lobbies of the House of Commons. Englishmen speak more softly and direct the voice more carefully so that it can be heard only by the person being spoken to, and the eyes are focused directly during conversation. Where an American seeking privacy goes to a private room and shuts the door, an Englishman stops talking, and this signal for privacy is respected by family, friends and associates. By contrast, when an American stops talking, it is usually a sign that something is wrong among the persons present.

Hall found that the influence of Mediterranean culture set the French apart from the American, English and German patterns. Mediterranean peoples pack more closely together in public, enjoy physical contact in public places, and are more involved with each other in sensory terms than more northern peoples. On the other hand, while the American brings friends and acquaintances into his home readily, the French home is reserved for family privacy and is rarely opened to outsiders, even co-workers of long standing or social acquaintances.

Individual Privacy

Recognizing the differences that political and sensory cultures make in setting norms of privacy among modern societies, it is still possible to describe the general functions that privacy performs for individuals and groups in Western democratic nations. These functions can be grouped conveniently under four headings: personal autonomy, emotional release, self-evaluation, and limited and protected communication. Since every human being is a whole organism, these four functions constantly flow into one another, but their separation for analytical purposes helps to clarify the important choices about individual privacy that American law may have to make in the coming decade.

Personal Autonomy

Each person is aware of the gap between what he wants to be and what he actually is, between what the world sees of him and what he knows to be his much more complex reality. In addition, there are aspects of himself that the individual does not fully understand but is slowly exploring and shaping as he develops. Every individual lives behind a mask in this manner; indeed, the first etymological meaning of the word "person" was "mask," indicating both the conscious and expressive presentation of the self to a social audience. If this mask is torn off and the individual's real self bared to a world in which everyone else still wears his mask and believes in masked performances, the individual can be seared by the hot light of selective, forced exposure. The numerous instances of suicides and nervous breakdowns resulting from such exposures by government investigation, press stories, and even published research constantly remind a free society that only grave social need can ever justify destruction of the privacy which guards the individual's ultimate autonomy.

The autonomy that privacy protects is also vital to the development of individuality and consciousness of individual choice in life. Leontine Young has noted that "without privacy there is no individuality. There are only types. Who can know what he thinks and feels if he never has the opportunity to be alone with his thoughts and feelings?" This development of individuality is particularly important in democratic societies, since qualities of independent thought, diversity of views, and nonconformity are considered desirable traits for individuals. Such independence requires times for sheltered experimentation and testing of ideas, for preparation and practice in thought and conduct, without fear of ridicule or penalty, and for the opportunity to alter opinions before making them public. The individual's sense that it is he who decides when to "go public" is a crucial aspect of his feeling of autonomy.

Emotional Release

Life in society generates such tensions for the individual that both physical and psychological health demands periods of privacy for various types of emotional release. At one level, such relaxation is required from the pressure of playing social roles. Social scientists agree that each person constantly plays a series of varied and multiple roles, depending on his audience and behavior situation. On any given day, a man may move through the roles of stern father, loving husband, car-pool comedian, skilled lathe operator, union steward, water cooler flirt, and American Legion committee chairman. Like actors on the dramatic stage, individuals can sustain roles only for reasonable periods of time, and no individual can play indefinitely, without relief, the variety of roles that life demands. There have to be moments "off stage" when the individual can be "himself": tender, angry, irritable or dream-filled. Such moments may come in solitude; in the intimacy of family, peers or woman-to-woman and man-to-man relaxation; in the anonymity of park or street; or in a state of reserve while in a group. Privacy in this aspect gives individuals, from plant workers to presidents, a chance to lay their masks aside for a rest. To be always "on" would destroy the human organism.

Self-Evaluation

Every individual needs to integrate his experiences into a meaningful pattern and to exert his individuality on events. For such self-evaluation, privacy is essential.

At the intellectual level, individuals need to process the information that is constantly bombarding them, information that cannot be processed while they are still "on the go."

The evaluation function of privacy also has a major moral dimension. While people often consider the moral consequences of their acts during the course of daily affairs, it is primarily in periods of privacy that they take a moral inventory of on-going conduct and measure current performance against personal ideals. For many persons this process is a religious exercise. Even for an individual who is not a religious believer, privacy serves to bring the conscience into play, for, when alone, he must find a way to continue living with himself.

A final contribution of privacy to evaluation is its role in the proper timing of the decision to move from private reflection or intimate conversation to a more general publication of acts and thoughts. This is the process by which one tests his own evaluations against the responses of his peers. Given the delicacy of a person's relaxations with intimates and associates, deciding when and to what extent to disclose facts about himself—and to put others in the position of receiving such confidences—is a matter of enormous concern in personal interaction, almost as important as whether to disclose at all.

Mental Distance

Limited and Protected Communication

The greatest threat to civilized social life would be a situation in which each individual was utterly candid in his communications with others, saying exactly what he knew or felt at all times.

In real life, among mature persons, all communication is partial and limited, based on the complementary relation between reserve and discretion that has already been discussed. Limited communication is particularly vital in urban life, with its heightened stimulation, crowded environment, and continuous physical and psychological confrontations between individuals who do not know one another in the extended, softening fashion of small-town life. Reserved communication is the means of psychic self-preservation for men in the metropolis.

Privacy for limited and protected communication has two general aspects. First, it provides the individual with the opportunities he needs for sharing confidence, and intimacies with those he trusts—spouse, the family, personal friends and close associates at work. The individual discloses because he knows that breach of confidence violates social norms.

In its second general aspect, privacy through limited communication serves to set necessary boundaries of mental distance in interpersonal situations ranging from the most intimate to the most formal and public. In marriage, for example, husbands and wives need to retain islands of privacy in the midst of their intimacy if they are to preserve a saving respect and mystery in the relation. Successful marriages usually depend on the discovery of the ideal line between privacy and revelation and on the respect of both partners for that line. In work situations, mental distance is necessary so that the relations of superior and subordinate do not slip into an intimacy which would create a lack of respect and an impediment to directions and correction. Thus, physical arrangements shield superiors from constant observation by subordinates, and social etiquette forbids conversations or off-duty contacts that are "too close" for the work relationship.

The balance of privacy and disclosure will be powerfully influenced, of course, by both the society's cultural norms and the particular individual's status and life situation. In American society, for example, which prefers "activism" over contemplation, people tend to use their leisure time to "do things" rather than to rest, read and think in privacy. And, in any society, differences in occupation, socioeconomic level and religious commitment are broad conditioning factors in the way each person allots his time and tunes his emotional wavelength for privacy.

In general, however, all individuals are constantly engaged in an attempt to find sufficient privacy to serve their general social roles as well as their individual needs of the moment. Either too much or too little privacy can create imbalances which seriously jeopardize the individual's well-being.

Organizational Privacy

Having discussed privacy thus far in terms of individuals, we turn now to privacy and group life. The approach adopted here involves making two judgments about the issue of "organizational privacy."

First, the legal and social claims to privacy given to organizations by American society are more than a protection of the collective privacy rights of the members as individuals. Organizational privacy is needed if groups are to play the role of independent and responsible agents that is assigned to them in democratic societies. Among these are: the satisfaction of needs for affiliation in large-scale society; the expression of basic interest felt by subgroups in the community; the operation of civic enterprises by private rather than government management; criticism of government policies; and measurement of public sentiment on issues and policies between elections. Just as with individuals, and subject to the same process of social limitation, organizations need the right to decide when and to what extent their acts and decisions should be made public.

Second, the term organization will be used to include public as well as private bodies. All organizations—from law firms and fraternal groups to political parties, courts, juries, legislatures and executive agencies—are similar in that they have an organizational purpose, a separate entity, and internal rules and procedures. Government agencies have the same basic need to be free from constant and immediate public exposure as do corporations, unions, universities, religious bodies and civic groups. Each government agency must also resist intrusion into its privacy by other government agencies under our separation-of-powers, checks-and-balances system of government. Though the traditional democratic belief in an open governmental process should operate to weight the balance between privacy and disclosure in favor of earlier disclosure and greater visibility for certain aspects of government's decisional process, it should not be seen as denying the claim to privacy.

The most helpful way to analyze the functions privacy performs for organizations in a democratic society is to apply the same categories used for individuals.

The lack of privacy for certain core secrets can threaten the independence or autonomous life of an organization much as it does that of an individual. The diplomatic, military, economic and scientific secrets of government agencies are protected by law because public disclosure of such information under conditions of international conflict could threaten national security and survival. Business groups often have trade secrets—special processes or formulas —on which their commercial success rests. The law will usually protect these secrets against disclosure to competitors by former employees or through business espionage, and against demands for access by labor unions or legislative committees. Wilbert Moore and Melvin Tumin have noted that privacy for confidential business decisions is an absolute requirement of a competitive economy.

Where to set the boundaries of organizational privacy remains a continuing topic of debate. "Full exposure" has been advocated by commentators who believe that the identities of those who attempt to influence public policy, such as organized lobbyists, ought to be known to the public in a free society. Claims to membership privacy have at times been rejected by the courts, the decision turning on the illegitimacy of those organizations' goals and methods of operation. Yet this requirement of visibility has been rejected in other areas where it too narrowly constricts organizational autonomy, as in the labor union, civil rights and political-lobbying areas. These rulings indicate that society must constantly set a balance between those ultimate secrets it feels may legitimately be kept private and those it does not. An enterprising press or social-science study may increase public wisdom by the penetration of the "inside affairs" of groups, but too much exposure can create distrust and hamper group activity. Permitting too much classification of information by government agencies can jeopardize democratic control over public policy; but too little may endanger national security.

Idealized Portraits

Just as individuals need privacy to obtain release from playing social roles and to engage in permissible deviations from social norms, so organizations need internal privacy to conduct their affairs without having to keep up a "public face." This involves, in particular, the gap between public myth and organizational reality.

For the same basic reasons that standards of moral expectation are set for individuals, society builds images of how universities, churches, labor unions, corporations and government agencies ought to operate. These idealized portraits are usually based on notions of rational decision-making, fair-minded discussion, direct representation of membership viewpoints by the leadership, dedication to public over personal interest, and orderly control of the problems assigned to the organization's care. In fact, much of the behavior of both private and public organizations involves irrational decision-making procedures, harsh and/or comic discussion of "outside" people and causes, personal motivations for decisions, and highly disorderly procedures to cope with problems seen by the organizations as intractable or insoluble. Despite press and social-science exposures of the true workings of organizations, society at large persists in believing that these are departures from a norm and that properly led and dedicated organizations will adhere to the ideal procedures.

Given this penchant of society for idealized models and the far different realities of organizational life, privacy is necessary so that organizations may do the divergent part of their work out of public view. The adage that one should not visit the kitchen of a restaurant if one wants to enjoy the food is applied daily in the grant of privacy to organizations for their staging processes. Privacy affords the relaxation which enables those who are part of a common venture, public or private, to communicate freely with one another and to accomplish their tasks with a minimum of social dissembling for "outside" purposes. Without such privacy the operations of law firms, businesses, hospitals, welfare agencies, civic groups, and a host of other organizations would be seriously impaired.

Of course, society decides that certain phases of activity by some organizations are so charged with public interest that they must be carried out in the open. This is illustrated by rules requiring public agencies or private organizations to conduct certain proceedings in public (such as regulatory-agency hearings or union elections), to publish certain facts about their internal procedures (such as corporate accounting reports and other public-record requirements for private groups), and to open their premises to representatives of the public for periodic inspections of procedures (such as visiting committees of universities and government inspectors checking safety practices or the existence of discrimination in personnel policies).

Secrecy in Philadelphia

Just as individuals need privacy to evaluate what is happening to them and to decide how to respond, so organizations need privacy to plan their courses of action.

Planning by organizations involves both periods of reflection for considering long-range implications of organizational policies and the frank process of internal debate needed to reach day-to-day decisions. In both situations privacy is essential if the individuals involved are to be able to contemplate and to express their views with primary loyalty to the organization.

It is useful to recall that the Constitution of the United States was itself written in a closed meeting in Philadelphia; press and outsiders were excluded, and the participants were sworn to secrecy. Historians are ageeed that if the convention's work had been made public contemporaneously, it is unlikely that the compromises forged in private sessions could have been achieved, or even that their state governments would have allowed the delegates to write a new constitution. Once the constitution had been drafted, of course, it was made public and its merits were freely debated as part of the ratification process.

The privacy involved in the writing of the American Constitution suggests the importance of confidentiality of organizational decisions until agreement has been reached, and confidentiality for a reasonable time thereafter of the way in which they were reached. Today this issue is most often discussed in terms of the federal executive branch and the question of legislative power to compel disclosure of policy positions taken by executive officials.

In one of the recent public debates over the propriety of publishing former presidential aides' accounts of recent intra-executive positions, Adolf A. Berle, Jr., has written:

"A President must talk to his staff. He can get the best from them—and they can best function—only when exchange is wholly candid. In the reviewer's experience, great decision-making usually boils down to a tired chief of state on one side of the desk and a trusted friend or aide on the other. If at that point the chief of state must consider not only the decision involved but also the possible effect of revelation of himself, his emotions and his thinking—concerning men, political effects of possible measures, his personal hopes and fears—frankness will necessarily be inhibited."

The other aspect for organizational decision-making is the issue of timing —when and how to release the decision—which corresponds to the individual's determination whether and when to communicate about himself to others. Groups obviously have a harder time keeping decisions secret. The large number of persons involved increases the possibility of leaks, and the press, competitors and opponents often seek energetically to discover the decision before the organization is ready to release it. Since most organizational decisions will become known eventually, privacy is a temporary claim —a claim of foundations, university administrations, political parties and government agencies to retain the power of deciding for themselves when to break the seal of privacy and "go public."

While the timing problem is not unique to government (advance news of a corporate decision is worth a great deal in the stock market and may harm the company's plans), its scope is greatest in governmental life. A major need is to prevent outsiders from taking unfair advantage of a government decision revealed through secret surveillance, careless leaks, or deliberate disclosure by a corrupted employee.

The basic point is obvious: privacy in governmental decision-making is a functional necessity for the formulation of responsible policy, especially in a democratic system concerned with finding formulas for reconciling differences and adjusting majority-minority interests. Nevertheless, drawing the line between what is proper privacy and what becomes dangerous "government secrecy" is a difficult task. Critics have complained that the public often has a right to know what policies are being considered and, after a decision is taken, to know who influenced the result and what considerations moved the governmental leaders.

The People Watchers

Surveillance is obviously a fundamental means of social control. Parents watch their children, teachers watch students, supervisors watch employees, religious leaders watch the acts of their congregants, policemen watch the streets and other public places, and government agencies watch the citizen's performance of various legal obligations and prohibitions. Records are kept by authorities to organize the task of indirect surveillance and to identify trends that may call for direct surveillance. Without such surveillance, society could not enforce its norms or protect its citizens, and an era of ever-increasing speed of communication, mobility of persons, and coordination of conspiracies requires that the means of protecting society keep pace with the technology of crime. Yet one of the central elements of the history of liberty in Western societies has been the struggle to instill limits on the power of economic, political and religious authorities to place individuals and private groups under surveillance against their will. The whole network of American constitutional rights—especially those of free speech, press, assembly and religion; forbidding the quartering of troops in private homes; securing "persons, houses, papers and effects" from unreasonable search and seizure; and assuring the privilege against self-incrimination—was established to curtail the ancient surveillance claims of governmental authorities. Similar rules have evolved by statute, common law and judicial decision to limit the surveillance powers of corporations and other private agencies.

Why Worry?

Though this general principle of civil liberty is clear, many governmental and private authorities seem puzzled by the protest against current or proposed uses of new surveillance techniques. Why should persons who have not committed criminal acts worry whether their conversations might be accidentally overheard by police officers eavesdropping on public telephone booths or at public places used by suspected criminals? Why should truthful persons resist verifying their testimony through polygraph examination? Shouldn't anyone who appreciates the need for effective personnel placement accept personality testing? And aren't fears about subliminal suggestion or increased data collection simply nervous response to the new and the unknown? In all these instances, authorities point to the fact that, beyond the benefits of the surveillance for the organization or the community, the individual himself can now prove his innocence, virtue or talents by "science" and avoid the unjust assumptions frequently produced by "fallible" conventional methods.

The answer, of course, lies in the impact of surveillance on human behavior. This impact can best be understood by distinguishing three main types of modern surveillance.

First is surveillance by observation. Writings by leading social scientists have made it clear that observation by listening or watching which is known to the subject necessarily exercises a restrictive influence over him. In fact, in most situations this is exactly why the observational surveillance is set up—to enforce the rules. When a person knows his conduct is visible, he must either bring his actions within the accepted social norms in the particular situation involved or decide to violate those norms and accept the risk of reprisal. Sociological writing has stressed that there are degrees of observation in various group types (work forces, government agencies, and the like) which prevent the group's members from performing effectively.

Though the destructive effect of near total observation and compulsory public confessions is associated in the public mind with totalitarian systems, as depicted with chilling effect in Huxley's *Brave New World* and Orwell's *1984,* the histories of utopian community experiments also document the disintegrative effect of complete observation over individual and group life. Robert Owen's famous community of New Lanark contained what Owen called a "silent monitor" system to watch the conduct of workers. This feature of life in several utopian communist communities of England and America in the 19th century led Charles Nordhoff to observe in 1875 that the absence of the "precious" thing called "solitude" was one of the key factors in the failure of these experiments.

Why is the prospect of total physical surveillance so psychologically shattering to the individual? If a factory is wired complete with listening and watching devices, workers know that every station cannot be monitored all the time. Yet no individual has any way of knowing when he is under observation and when not. The particular dehumanizing feature of this situation is not the fact that the surveillance is done by machine techniques rather than by direct human observation, but that the person-to-person factor in observation—with its softening and "game" aspects—is gone.

On the other hand surveillance may be such a vital means of providing physical security, as in our public places, the properly controlled use of new watching and listening devices may be desirable. This point was made by Margaret Mead in a recent essay. After noting that city life offers "extraordinary possibilities" for anonymity from neighbors, relatives and community controls, Dr. Mead stated that the desire for "personal privacy" is being confused with a notion of "privacy from the law." This confusion is based on the erroneous assumption that there is no obligation to create institutions of social protection in cities to replace the public safety provided by social surveillance of known persons in small communities. New listening, watching and recording devices in apartment buildings, police street monitoring, and similar situations are to be welcomed. "The devices we have rejected because they can be (and have been) used to invade individual privacy can also be used to ensure the public safety, without which privacy itself becomes a nightmare isolation."

Surveillance by Extraction

A second main type of surveillance is extraction—entry into a person's psychological privacy by requiring him to reveal by speech or act those parts of his memory and personality he regards as private. The issue of personality testing for personnel selection by industry and government provides a useful subject for studying the social effect of extraction.

The basic objection on private grounds to the typical personality test used in personnel selection today—with its questions on such topics as sex and political values—is that many individuals do not want to be sorted and judged according to standards that rest on the unexplained evaluations of professional psychologists in the employ of "institutional" clients. Liberals fear that a government or industrial psychologist will enforce conformist or elitist norms. Conservatives fear that school or government testing might not only "reward" liberal ideology and penalize conservative ideas but also "implant" ideas through the testing process itself. Negroes are concerned that psychologists might enforce standards of personality that penalize minority groups and that the personality test might enable the "white power structure" to accomplish covertly discrimination it can no longer carry out openly. In all these situations the assertion of privacy serves to say to those in power: "If you make evaluative decisions openly, questioning me directly and justifying your decisions openly, I can fight out publicly your right to judge me in a certain way, and American society will decide our conflicting claims. But if you invoke 'science' and 'expertise' and evaluate me through personality tests, the issue becomes masked and the public cannot judge the validity and morality of these evaluative decisions. Thus, where such basic issues as political ideology, religion and race are at stake, the selection process must be objective and public, and I assert my right of privacy to close my emotions, beliefs and attitudes to the process of job evaluation in a free society."

Finally, from the literature of psychology and psychiatry, as well as from personal experience, critics of personality testing know that many individuals go through life with personal problems and conflicts that they keep under control. These "managed" conflicts may involve sex, struggles over self-image, careers, and similar matters. Most of these people can grow old without having these conflicts become serious enough to impair their capacities at work, in the family or as citizens. If these capacities are impaired, of course, the individual needs help; he may seek it himself, or it may be offered to him when his difficulties become observable. The problem presented by the spread of personality testing is that it may, by the pressures of testing and of rejection in selection, bring to the surface personality conflicts that might otherwise never have become critical in the individual's life, and may thus precipitate emotional crises. It can be argued that it is healthy to bring such problems to the surface and to lead the disturbed individual to professional help. Perhaps we are moving toward an age of preventive mental health by personality testing, when individuals will get their emotional "check-up" just as they have their bodies, eyes and teeth checked. Before we accept this trend in American life, however, we had better be more certain than we are now that we can cure the wounds opened by such a process, or that awareness is a good thing even though a cure is not always possible.

A third type of surveillance, which has not yet been studied by social scientists because of its recent development, is what I would call reproducibility of communication. Through the new recording and camera devices, it is now simple to obtain permanent pictorial and sound recordings of subjects without their knowledge. This may be done by the person with whom the subject is talking or acting, or a secret recording may be made by a third party. The special character of this surveillance is that it gives the person who conducted the surveillance the power to reproduce, at will, the subject's speech or acts. When a person writes a letter or files a report, he knows that he is communicating a record and that there is a risk of circulation; thus he exercises care and usually tries to say what he really means. But in speech that is overheard and recorded, all the offhand comments, sarcastic remarks, indiscretions, partial observations, agreements with statements to draw out a partner in conversation or to avoid argument, and many similar aspects of informal private intercourse are capable of being "turned on" by another for his own purpose.

Something Is Taken Away

The right of individuals and organizations to decide when, to whom, and in what way they will "go public" has been taken away from them. It is almost as if we were witnessing an achievement through technology of a risk to modern man comparable to that primitive men felt when they had their photographs taken by visiting anthropologists: a part of them had been taken and might be used to harm them in the future.

American society now seems ready to face the impact of science on privacy. Failure to do so would be to leave the foundations of our free society in peril.

Technology and Theology

Myron B. Bloy, Jr.

Technology, considered both as the cumulative weight of an increasing proliferation of radical innovations, in what might be called the "economics" of man's existence, and also as an "objective spirit" or certain life-style, is the major force which is shaping the emerging culture. What are the most potent aspects of its power for cultural change? On the positive side, I would argue that technology, far from enslaving man as some writers (especially Jacques Ellul) aver, releases us into new dimensions of freedom from ancient restraints. Furthermore, far from depersonalizing man as others have argued, technology commits us to a much more profound awareness of other persons than has heretofore been generally possible.

For the two-thirds of the world where grinding poverty, incessant work, debilitating disease, life-long hunger, and early death are still the essential conditions of existence, technology is the primary instrument of freedom. I would not mention this obvious fact if so many Western critics didn't cavalierly denigrate it; our cultural critiques should never become so effete that this fact is overlooked. For us, however, technology is opening up subtler forms of freedom. Consider first three examples of how technological innovations affect our traditional value system. The so-called "prudential ethic" which has used the triple threat of infection, conception, and detection to enforce extramarital chastity has been all but knocked out by three new technologies— penicillin, Enovid, and the automobile and motel. Man's traditional habit of finding his self-identity through his work in the productive enterprise is being challenged by the astounding productive capacities of cybernation—the meshing of automation and cybernetic devices into a single productive process. And the family, heretofore the almost exclusive value-forming power for children, has been seriously eroded by rapid, easily accessible transportation and television which brings the whole raw world into the family living-room.

Furthermore, the spirit of technology, given philosophical form as pragmatism, has undermined every metaphysically-fixed value system; as William James has said of the pragmatist:

> He turns away from abstraction and insufficiency, from verbal solutions, from bad a priori reasons, from fixed principles, closed systems, and pretended absolutes and origins. He turns towards concreteness and adequacy, towards facts, towards action, and towards power. That means the empiricist temper regnant and the rationalist temper sincerely given up. It means the open air and the possibilities of nature, as against dogma, artificiality, and the pretense of finality in truth.[1]

1. William James, *Pragmatism: A New Name for Some Old Ways of Thinking.* (New York: Longmans, Green, 1907), p. 51.

Those of us with a vested interest in the traditional culture are bound to be dismayed by these developments, but, in fact, each one of the changes I have described does enlarge man's freedom over heretofore implacably contingent factors of his existence. The prudential sex ethic, based on nothing but fear, made moral morons of us all; now we can make decisions about sexual behavior on the basis of a positive understanding of sexuality. Similarly, by breaking the stranglehold that the necessity of productivity has always had on man's self-identity, cybernation is giving us the opportunity to evolve richer, more satisfying models of self-identity for ourselves. And the family, which has traditionally exacted an often tyrannous value-conformity in exchange for the security it provides the child, is now in a position to become a supportive setting in which the young have the freedom to explore value systems other than those of their parents. Finally, there is no gainsaying the sense of exhilaration and release in James' description of man freed from the ideological straightjackets of the traditional culture.

We are not only freed by technology from many of the material, social, psychological, and spiritual restraints which so pinched and enslaved most men in the past, we are also made much more aware of the presence and plight of our fellowmen than heretofore. Consider, for example, James Reston's description of how television and the airplane counted in the Selma freedom movement: "We are told by our philosophers and sociologists that our machines are enslaving and debasing us, but in this historic battle over voting rights these machines are proving powerful instruments for equality and justice." Television forced on our awareness the plight of fellow human beings, in fact made them our neighbors, and mass transportation allowed us to go to their assistance. Because we are so inescapably aware of the sufferings of so many more neighbors, we often draw the false conclusion that suffering itself abounds as never before and technology is often (ironically enough) blamed; actually our burgeoning awareness of every man as neighbor is possibly the first step in the solution of suffering which before now was hardly known outside its immediate context.

But modern electronic developments may be fostering the "age of man's encounter with man" in more subtle and compelling ways than those just described. For example, Marshall McLuhan has argued that students are increasingly restive under the traditional episodic, piece-meal curriculum, with the "linear" approach to learning, designed to add up after four years spent abstracted from society to something meaningful, because television has formed their perceptive modes from an early age to in-depth expectations, to deep involvement in human interactions. Literate man tends to be so horrified by the shoddy content of most television programming that he fails to see how television, whether its content is "good" or "bad," defines for those who live with it from an early age the scale of "reality" for all their future encounters with the world. Man as a discrete individual, participating in the world from a rational, detached point of view, is giving way to "re-tribalized" man, who lives almost completely *with* others, in almost total involvement, and for whom the point of view has little meaning. This new man fills out the shape of his life only insofar as he is involved dynamically with others.

Now, freedom and awareness are precisely necessary conditions for the growth of man towards his moral maturity. The freer we are the more responsibility we can take for our behavior, and the more aware we are of the presence and plight of the other person the more opportunity we have to exercise that responsibility. When man is enslaved by superstitions and brutalities of nature and lives in ignorance of the real plight of his fellow men, the possibilities of growth towards moral maturity are severely limited. But if the freedom and awareness which technology has fostered are necessary conditions for growth towards a culture of moral maturity, they are certainly not sufficient causes for such growth. Freedom *from* restraints of one kind or another only achieves its inherent meaning when it becomes freedom *for* fulfilling in action a normative commitment to the neighbor; and an awareness of the presence and plight of our fellow men only becomes creative when spurred by that normative commitment, it occasions more sensitive decisions in their behalf. Without some strong guiding norm freedom collapses into chaos and awareness into the anxiety of "information overload." The ironic dilemma we face is that the same spirit of technology which has increased our freedom and awareness has also decreased our ability to make normative commitments and thus destroys for many of us our ability to exercise that freedom and awareness.

Daniel Bell uses the phrase "eclipse of distance" to describe the dissolution of normative commitments; he says,

> The underlying social reality, the stylistic unity of the culture of the past hundred years lies, I would argue, in a structural form of expression that I have called the "eclipse of distance," of psychic, social and esthetic distance. Modern culture began as an effort to annihilate the contemplative mode of experience by emphasizing *immediacy, impact, simultaneity,* and *sensation.* It is today at the point of breaking up all fixed points of reference in formal genres.[2]

This is not the place to enter into an analysis of how man experiences normless existence, but the following two images are suggestive. The prophet Amos, after cataloguing many of the terrible things—such as darkness at noon, sackcloth and baldness, the mourning for an only son—which will befall the people because they have departed from the Lord's purposes for them, sums up their doom in this haunting description of normless existence:

> "Behold, the days are coming," says the Lord God,
>   "when I will send a famine on the land;
> not a famine of bread, nor a thirst for water,
>   but of hearing the words of the Lord,

---

2. Daniel Bell, "The Disjunction of Culture and Social Structure: Some Notes on the Nature of Social Reality," *Daedelus,* 94:220 (Winter 1965). Collected in *The Revolutionary Imperative,* Austin, ed., and in *Science and Culture,* Holton, ed. (Houghton Mifflin, 1965).

They shall wander from sea to sea,
  and from north to east;
they shall run to and fro, to seek the word of the Lord,
  but they shall not find it.
In that day the fair virgins and the young men shall
  faint for thirst." (Amos 8:11–13)

And Arthur Miller, seeming almost to paraphrase Amos' words of doom for our ears, summarizes his *Death of a Salesman* in "the image of private man in a world full of strangers, a world that is not home nor even an open battleground but only galaxies of high promise over a fear of falling."

Contemporary man, trapped in the frustrating situation of having achieved at last the capacity to create his own destiny but unable to discover normative commitments commensurate with that capacity, often reacts convulsively in one of two ways. On the one hand, he plunges into intense technological activism: If the activity towards some contingent end is intense enough it can create the illusion of purpose, it can cover up the normative anarchy which exists just beneath the surface. On the other hand, he may grasp at a reactionary idealism like those represented by the John Birth Society or Moral Re-Armament. The Birchers, unable to tolerate their new freedom and increasingly present neighbor, construct a paranoid model of reality which reduces their freedom to some form of counter-plot against the communists in behalf of a Jeffersonian political paradise, and they deny the demanding presence of the neighbor by reducing him either to a spy or patriot. The MRA pattern is similar, except that it is in the sphere of metaphysics and morals instead of politics. Religion is often a more passive form of this same regression: There are, it is argued, certain immutable "moral and spiritual values" to which we must "return," but these values are purposely left honorific and lifeless so that they cover the abyss of normlessness without having to submit to the test of action. The escapist route of higher education is often "the discipline": One can spend a life-time paying homage to its cabalistic intricacies and defending it against its detractors without ever seriously facing the question of the purpose of learning.

A further irony of our time is that the struggle for cultural purpose seems to be polarized around these two escapist positions: the idealist is called reactionary dreamer by the operationalist, and the operationalist is called a superficial manipulator by the idealist, and both are right.

What we need in order to achieve our cultural maturity is a sense of purpose passionate enough to overcome the anti-normative tendencies of our time and use our new freedom and awareness in behalf of man. This purpose must be weighty enough to escape both moralistic reductionism and calculated operationalism, and is best conveyed by the following story about St. Francis. It seems that Francis, with several of his friends, was walking down a road on a wintry afternoon; he was wearing a heavy cloak against the weather. Presently they approached a nearly naked beggar shivering with cold, and Francis promptly gave him his cloak. Some of his friends remonstrated with Francis, saying that if he caught cold and were incapacitated through lack of his cloak their movement would be without a leader, while the others congratulated him on the "goodness' of his act. Francis angrily silenced their squabble and told them that this man, like every man, was a brother and therefore the cloak was his by right if he needed it: It could not be kept from him by self-interest, however enlightened, nor could its giving be called "good" since it was already the beggar's by right. I take it that our new freedom and reality fulfill their "natural" ends when they are pressed into the service of such a vision of reality.

The passionate vision of a Francis, held expectantly within each present situation, has the power to use our new freedom and awareness for the shaping of our culture to human and humane form. But where do we find such power? Is there a cultural sub-community, like the roaming bands of prophets in early Biblical times, where this power is operative? I think such a way of life is evolving among the college and university students who, together with a few faculty, clergymen, and ghetto-dwellers, make up the so-called "freedom movement."

It is hard to realize that just two years ago, in *The Uncommitted*—a book on the American "youth culture"—Kenneth Keniston could argue that college students were marked by a lack of rebelliousness, by social powerlessness, and by privatized values. Erik Erikson was making the same sort of judgment when he described life for the college student as a "psycho-social moratorium." Slick journals were mildly deploring the passivity and the lack of good old-fashioned grit and gumption in the American college student. After quoting statistics which indicated that college seniors entering the work world were primarily concerned with pension plans and job security, editorials urged them to accept instead the risks of economic and professional life which have been the time-honored lot of the successful American entrepreneur.

But now the teach-ins are taking the place of panty-raids, voter registration campaigns in the deep South and northern ghettoes are replacing the spring bacchanals on the Florida beaches, tutorial programs are succeeding religious clubs, and after graduation—the Peace Corps and politics are taking the place of company apprenticeship programs, editorialists are telling students to "stick to their books" and to remember that time-serving apprenticeships are a necessary prelude to success in life. What, in fact, has happened, and why?

The passive, powerless, privatized "youth culture" which Keniston described had its more immediate roots in World War II: Veterans returning to college had taken all the risks they wanted, and now they were ready for college as an idyllic recompense for those harsh years. Although higher education was already participating much more than it had been in the research and development needs of the nation, the prevailing undergraduate assumption was that college represented a socially detached enclave, replete with its own romantic mythologies, designed to prepare persons in due and leisurely course for following a personal career trajectory. Society at large was simply the shadowy, yet stable and secure, background against which this trajectory was to be traced.

As the Cold War deepened, this model of higher education was clung to by students with something like desperation. University administrations, by and large, happily sanctioned the youth culture because it tacitly supported their *in loco parentis* power, and more or less confined their problem areas to the campus and to issues of private morality. Faculty were not inclined to disturb the youth culture because student passivity allowed them the time and energy to indulge more fully in the exciting, status-making research and development possibilities which the government was opening up for them.

The revolt against the youth culture was occasioned by the Negro struggle for civil rights in the South and has so far moved through three stages corresponding to three different levels of insight. The intense moral pressure caused by the southern civil rights struggle, led in many areas by the Student Non-Violent Coordinating Committee which was composed largely of Negro students, was enough to lead a few northern, white students into the struggle themselves. In the course of this participation, they began to discover that American culture and society, far from being the stable, secure world assumed by the youth culture, was wracked with such cruel and systematic injustice that commitment to revolutionary change was the only reasonable stance to take.

Furthermore, they discovered—through participation in direct action projects such as freedom rides and sit-ins—that their action could count for something in the world of political and social affairs. When these students returned to the campus, their achievement of moral commitment (of "the point beyond which there is no turning back," which is a crucial discovery for the maturity of each man) and of a real share in social power had a contagious quality for many other students who hadn't hitherto realized what a spiritual deprivation the youth culture was.

The second stage emerged when students began to realize that dramatic, random forays into the South was not a serious enough response to the problem, and that they had to regularize and localize (in the North) their activities. The Northern Student Movement was born out of this realization. Turning their own talents to use, they organized far-reaching tutorial programs for children in the social and racial ghettoes near their campuses. Some students moved into the ghettoes. They helped organize the disestablished to fight against local establishments, and they kept up a drum-fire of pressure to participate in the struggle on their uninvolved fellow-students. In this stage of development, students in the movement lived a kind of intellectual and spiritual schizophrenia, passively accepting and fulfilling the demands of a remote and static academic establishment while passionately involved in the struggle to change the character of an unjust society.

But when the students began to learn that intellectual and spiritual understanding, as well as simple political activism, would be necessary to bring about the social change they envisaged, it was inevitable that the academic establishments themselves should come under their fire, since these establishments were, by and large, unequipped to respond to the students' urgent needs. The "free-speech movement" at Berkeley was, for many students, the beginning of this discovery, and although most adults could echo President Eisenhower and see it only as "disgraceful riots," we must look beneath the turmoil to see what was really at stake. Mario Savio spoke for morally and socially sensitive students everywhere when he said, during that famous sit-in in Sproul Hall,

> Many students here at the university, many people in society are wandering aimlessly about. Strangers in their own lives, there is no place for them. They are people who have not learned to comprise, who for example have come to the university to learn to question, to grow, to learn—all the standard things that sound like clichés because no one takes them seriously. And they find at one point or other that for them to become part of society, to become lawyers, ministers, businessmen, people in government, that very often they must compromise those principles which were most dear to them. They must suppress the most creative impulses that they have; this is a prior condition for being part of the system. The university is well structured, well tooled, to turn out people with all the sharp edges worn off, the well-rounded person. The university is well-equipped to produce that sort of person, and this means that the best among the people who enter for four years wander aimlessly much of the time questioning why they are on campus at all, doubting whether there is any point in what they are doing, and looking toward a very bleak existence afterward in a game in which all of the rules have been made up, which one cannot really amend. [3]

Although it is easy to criticize the youthful naiveté of Savio's rhetoric, it is impossible to avoid the essential rightness of his analysis.

These students, finally driven by their moral passion to a new intellectual concern (How do societies and cultures work and how can they be changed?) and a spiritual quest (What is man's *real* life and what is the meaning of history?) are challenging the educational establishments in two ways. First, as in Berkeley, they are bringing direct pressure for specific reforms: Student rallies support good teachers who have been fired because they failed to measure up to research-oriented tenure criteria; philosophy departments which, in their preoccupation with esoteric linguistic and mathematical games, are unequipped and uninterested in helping students to assimilate intellectually their new social and moral experience, are under attack in student editorial columns; teach-ins pressure curricula committees to respond more relevantly to students' real questions. Secondly, the so-called "free university" movement is becoming an embarrassing challenge to the academic establishments. These

3. Mario Savio, "An End to History." Edited from a tape. Collected in *The New Student Left,* Cohen and Hale, eds. (Beacon Press, 1966).

independent, student-led "communities of learning" are springing up near every large academic center because, as the "Proposal for a Free University in Boston" points out, "Students returning from civil rights activity and community organizing projects found little of relevance in academia to the problems central to their concerns." The new free university in Boston aims, in the words of its prospectus, to involve the following groups:

> students and organizers seeking both the theoretical and empirical bases for ideology; people in community organizations who want to learn organizing skills, participate in political discussion, and gain or regain some aspects of general education; suburban opponents of the war who will want information as well as tools and perspectives for organizing in middle-class neighborhoods; professionals who are trying to redefine their roles in terms of social objectives; teachers and students dissatisfied with the content of their previous educational experience, as well as by the university's approach to learning and to the social relevance of intellectual activity; artists, writers, and actors who seek to explore new dimensions in their work, or to relate their work to the movement.

Thus, the students' exhilarating discovery of moral commitment and social power, occasioned by the civil rights movement in the South and solidified and deepened in the ghettoes of the North, has opened their eyes to some of the inadequacies of higher education and led them to attempt reforms.

Now let us consider the life-style of the participants in the movement: It is from this life-style that we have most to learn, I believe, since it is borne out of the Franciscan vision of the neighbor as brother and the perspective of historical realism. One way of seeing this life-style is through the eyes of its critics. On the one hand, operationally oriented change-agents like Saul Alinsky and many liberal politicians, who might be expected to be allies of the students, are very critical of them for not being pragmatic enough in their approach to social problems. I remember one Alinsky man saying to an S.D.S. (Students for Democratic Society) member, "What we want to know about you people is whether or not you're for real: You act like a bunch of poets!" But the students feel that Alinsky and the politicians, by playing the game in the terms and for the stakes essentially established by the *status quo,* don't raise the necessarily radical questions about the character of our society. They have read Silone's *Bread and Wine,* and they identify with Pietro Spina. On the other hand, they are criticized by academics, idealists and conservatives generally for compromising principle for political ends and for participating in political action when they ought to be sticking to their books. But the students argue that no serious involvement in social change is possible without being changed oneself, and that education without direct, concrete involvement in cultural issues is not education at all. In short, because they remain open to both the depths and the surface of events they are scored by idealists and operationalists from opposite directions.

But, to turn to a direct description of them I believe that their suffering —their intellectual, moral, and spiritual suffering—is what is most authenticating about their prophetic identity. I have spent long evenings with student radicals while they struggled to discover an "ideology." Most of them have studied Marxism but now feel that although Marx's analysis of society is useful, his anthropology of the individual is faulty. Existentialist philosophers, on the other hand, are usually far too simple-minded about political and economic realities to be accepted. As a matter of fact, it is doubtful, for all their desperate effort to find a secure ideological niche from which to see themselves and their history, that they will ever arrive in this promised land. Their honesty and openness to the myriad intellectual and emotional claims of experience will deny them the neat ideologies for which they search. But they find it equally difficult to be operationalists. Nothing is more frustrating to a political wheeler and dealer than to participate in an S.D.S. organizational meeting because questions of substance, of goals and essential meaning, are always in order and continually break down the smooth mosaic of cause and effect that is ostensibly being put together. The "really real" for them is to be discovered only by accepting the full weight, both the insistent depth and the empirical surface, of the here and now in their drive to keep faith with their vision of every neighbor as a brother. Only this precarious and passionate commitment, they feel, lets them into the action of life and allows them to be the shapers, not just the victims, of history.

Now, I am not here arguing that the student movement will, in fact, be able to shape our emergent technological culture to human and humane form. As I have pointed out before, the perspective of historical realism demands spiritual toughness, and many students find themselves sliding off into operationalism or idealism, or blunting the edge of their thrust by indulging in romantic self-pity. Furthermore, these students have the same psychological fish to fry in their personal evolution as any late adolescent in our society, and this task is often at cross-purposes with their cultural task. But their life-style reveals the ontological vision and epistemological stance which can use our new freedom and awareness to create a more mature culture. In our time of normative chaos their prophetic life is a sign of hope and an important model for society in general.

By now my not-so-hidden theological agenda should be obvious. Our new freedom and awareness and concommitant inability to assume an easy normative focus for the emergent culture really adds up to "the world come of age" which Bonhoeffer described. God is, in effect, kicking us in the pants and telling us that it is time to grow up. We are given the tools needed to shape a new culture and allowed to use them effectively only in the service of a prophetic commitment. We are even provided, in the student movement, one significant model of how that commitment can be assumed. Of course, there is no assurance that society will accept this challenge rather than hide in increasingly frenzied operationalism or increasingly brittle idealisms until we are overwhelmed by chaos, but these are our only two options.

The Church, as that community whose formal function is to bear witness in its life to the intentions of God for mankind, and to support and celebrate God's action wherever it is discovered, does not fulfill its role effectively in our present situation. The Church might learn a good deal from the student movement about what it means to be men of faith in our technological culture. Tillich in fact argues that the mature form of historical realism, which he calls "self-transcending realism," can only be held through faith; he says,

> Self-transcending realism is based on the consciousness of the "here and now." The ultimate power of being, the ground of reality, appears in a special moment, in a concrete situation, revealing the infinite depth and the eternal significance of the present. But this is possible only in terms of a paradox, *i.e.,* by faith, for, in itself, the present is neither infinite nor eternal. The more it is seen in the light of the ultimate power, the more it appears as questionable and void of lasting significance. So the power of a thing is, at the same time, affirmed and negated when it becomes transparent for the ground of its power, the ultimately real. It is as in a thunderstorm at night, when the lightning throws a blinding clarity over all things, leaving them in complete darkness the next moment. When reality is seen in this way with the eye of self-transcending realism, it has become something new. Its ground has become visible in an "ecstatic" experience, called "faith." It is no longer merely self-subsistent as it seemed to be before; it has become transparent. [4]

Is not the commitment to the neighbor as brother, revealed ever freshly through the "eye of self-transcending realism," precisely the commitment that the Church is called to live out of?

But, if the Church has much to learn about its own calling from the student movement, the movement stands in need of a reawakened Church. Students reject the Church because they see that its "faith" is often really an escape from, rather than a commitment to, history. As Tillich points out, "The man of today, who feels separated by a gulf from the theistic believer often knows more about the 'ultimate' than the self-assured Christian who thinks that through his faith he has God in his possession."[5]   The problem of the movement, however, is that it has no way to perceive that its apparently anomolous stance between operationalism and idealism, its inability to find a secure resting place in some Platonic form of Truth, is precisely its calling. The students do not have conscious access to the Judeo-Christian tradition in which their commitment is recognized as *the* calling of man. Their intellectual and spiritual suffering is often self-destructive because the tradition for *celebrating* their commitment has been rendered unavailable to them by the keepers of that tradition, i.e., by the Church. Thus, the Church, by recovering its own ability to live and celebrate the life of prophetic commitment as the authentic life of man, would also stand as an encouraging sign of the authenticity of every manifestation of that commitment which emerges in society. This is precisely the calling of the Church in our technological culture.

---

4. Paul Tillich, "Realism and Faith" in *The Protestant Era* (University of Chicago Press, 1948), p. 78.
5. *Ibid.,* p. 82.

But Then Came Man:
The Whole Picture

Stewart L. Udall

In this fast-spinning, fast-growing, and fast-changing world, man has achieved for the first—and unless he acts, perhaps for the last—time, the capacity to become human, truly human.

The door to this vision is held open by cybernetics and automation and computers. Used with intelligence and feeling and foresight, employed with compassion and humanity, programmed not just for goods, but for human goals, they have the potential to bring us—this entire world—into a new era where competition will be between cultures, not nations; where man will wish to be respected, not rich; where—as Norbert Wiener so simply and eloquently put it—there can be a "human use of human beings."

There is only one major impediment to all this. There is only one force which would close the door. That force is procreation: limitless unbridled, unthinking, and unnecessary procreation.

I am not simplistic enough to say that there are no other factors.

Of course, there are!

I am just realistic enough, however, to state that except for volcanoes and typhoons and other acts of God or nature, there are no problems on this planet but people problems. I speak of war . . . and of waste. I speak of hunger and of the appalling despoilation of the environment.

I speak of the very quality of life . . . human life, animal life, plant life. I speak of man and nature . . . and of man and nonsense. I speak of the nonsense in not facing up to the fact that—in short—all of our numbers can negate all of our dreams!

Five billion or so years ago this, our planet, was born. If all goes according to Halley—if not Hoyle—this, our earth, has quite a few million—perhaps billions—of years to go before the sun grows cool . . . before life on earth is deep-frozen forever.

So the question is not whether life in *our* time will vanish, but rather whether in our time *we* will vanquish life.

What is the situation now . . . globally and locally?

First, on a global scale:

There are more new stomachs than food supplies.

The new lives, the highest birth rates, are in the most underdeveloped nations where literacy and productivity are low, where the shortage of medical care is abysmally great.

On a global scale, entire continents are consigned to inveterate impoverishment, their peoples to the blandishments and bludgeonings as history attests of oppressors.

While we—and other nations able to do so—must continue to assist these nations, we now know that food is not enough. A bushel of wheat may keep a family from starving. But only momentarily. It does nothing to the birth rate. Looking ahead to 1985, we find that the food-aid needs of 66 developing countries will exceed by 12 million metric tons the grain that could be produced in the United States in excess of amounts required for domestic purposes and commercial exports.

No! A bushel of wheat is as nothing to the quicksand of a quickened birth-rate.

The answer to the global population crisis is multifaceted. It goes to education: 45% of the world's population can neither read nor write. How does one communicate with this vast and hungry audience? How communicate about farming, or fertility? How communicate about family or fecundity? How communicate about productivity and procreation?

Here is perhaps the greatest challenge! How do we educate whole continents . . . while there yet is time?

Here is a challenge to television, to satellites, to foreign aid, to a Ford Foundation. Here is a challenge to conscience and creativity!

I can think of no more painfully dramatic illustration of these facts than India. Nor can I think of a greater opportunity. And I believe the leadership of India, today, is receptive.

Since the first of this year, we have added about thirty million human lives to this planet. That is like the population—combined—of California and New York State. Imagine, now, the situation in those states . . . with that many more people . . . and with no prospect of dramatic increases in food production.

It is that critical.

It is that simple.

It is that tragic—for the world . . . and for us!

Meanwhile, in the United States, ours is a different sort of population crisis. Ours is a problem not of quantity, but of quality.

We add new lungs even as our deterioration of the air continues. And the water from which all life evolved is more foul than fresh.

Herman Melville, were he with us and working on *Moby Dick,* would shudder. His great white; whale no matter how elusive and mystical, would —if alive—be blackened by the oil of the *Torrey Canyon* were Ahab to pursue him towards the North Sea along the English coast.

Paul Gallico, whose *The Snow Goose* has become something of a major minor classic, would write that story with reservations, for the feathers of his Canada Goose might well be pinioned with oil slicks washed into the marshy nesting place where the author set her down.

There was a popular song titled "Blue Moon." It is time, I am afraid, for a ballad called "Smog Moon." Indeed, I wonder whether Mr. Poe could again pen these lovely lines:

> "For the moon never beams without bringing me dreams
>   Of the beautiful Annabel Lee,
> And the stars never rise but I feel the bright eyes
>   Of the beautiful Annabel Lee."

This is not a pretty picture, and it all portends a change in our literature and music. The new author is indifference, the new composer, callousness. And it is difficult to say whether we have been more industrious with our indifference or more inventive with our callousness.

One must construct a sound-proof room to remain aloof from the sonic boom. Everywhere in our metropolitan areas, noise assails the ears. The honking horn, unlike the rustling wind, soothes no souls!

There is more neon than sunset in our lives.

Aromas beset the nostrils. Along many of our highways there are more hamburger stands than trees. Discarded containers litter the landscape.

Crass developers and speculators defending themselves with protestations that they are merely building homes for the homeless, have bulldozed away trees and topsoil. They have cast nature to oblivion. They have left more than shoddy construction—what some architects call "cheapies"—in their wake. They have left eroded soil, silted rivers, and traffic-snarled shopping centers which are monuments to tastelessness.

We, the going-est people on earth, have caught ourselves in a transportation mess because our rugged individualism would not permit us to build unified mass transportation facilities. No, we—all of us—had to drive our own selves in our very own cars, accelerating—in some instances—as many horses as pulled an entire wagon train across the continent.

Strange, the elephants have their graveyards, places of silence and secrecy. But we had to contribute an elephantine mess to the open fields and riverbanks. Having exhausted our vehicles or tired of their styling, we began to discard them at the junkyards which so conveniently grace our countryside. These eyesores are so public as to be unavoidable, so reasonant with the discord of dismantling as to be obvious to the nearly deaf. Nature had reason when she caused the snake to shed its skin. I am unconvinced nature—in giving us reason—intended we should go through the annual rites of motor-car-moulting.

Before too long, some men will commute to the moon and beyond, but those of us on earth will have scant and scarce view of the stars. There is little sky left between the skyscrapers, and our urban congestion—with all the attendant problems—is becoming more acute. In just ten years, eighty percent of our people will be living on only one percent of our land. This means more megalopolis. This means more high-rise living and working. This may mean —ultimately, if one architect has his way—domed cities where the environment may be regulated ... where we will try, as best we can, to unnaturally keep things natural.

All of this is very interesting. Some years ago, Norman Cousins asked—after Hiroshima—whether modern man had become obsolete.

Now I would share with you another question and concern: Has he—modern man—become too omnipresent. And is his very omnipresence threatening his very survival as man, born of the seas and rooted to the soil, inspired by the stars and mesmerized into moments of contemplation by the Heavens we already are cluttering with burned-out rockets and cast-out debris?

We are at once the most resilient, most resourceful, most restive, most receptive, most radical, most reactionary people who ever lived. We have had time and the tide for everything but those moments of thought necessary to reverse the priorities ... to cause us occasionally to look before leaping.

This, in a sense, is illustrated by the following: of the five largest business enterprises in the United States, three are life insurance companies—for we believe in family. One is General Motors: we believe in going. One is American Telephone and Telegraph: for we believe in communicating.

There is only one problem. We have gone too far in some things. We have failed to communicate to ourselves and to the world the facts about our environmental and population crises. And, lastly, we have, while building up an industry that leaves proceeds to families, failed to dramatize in meaningful ways and on a massive scale another important kind of life insurance: rational population planning.

A bit of historical perspective would not be amiss. The telephone, we know, was invented by Mr. Bell in 1876. It was the year of our Centennial. It was the year Colorado entered the Union. It was the year of Custer's Last Stand and of *Tom Sawyer*'s first publication.

The automobile appeared a bit later on the scene. Though invented in Europe, it was American ingenuity which gave wheels—or, should I say, wings —to the horseless carriage. Ford parlayed the basic principle of mass production—the interchangeability of parts—into a prodigious output of cars of any color so long as it was black.

Kettering then gave us the self-starter which removed the last vestige of muscle from motoring. Suddenly if you avoided the horse because you didn't want to be thrown, you had no longer to risk a thrown sacroiliac while cranking the engine. *This* was progess!

Well, we have been ringing bells ever since Bell. We have than one-half of the world's telephones. Our words, our wisdom, our data and our date-making rely on the ten-cent call and the long-distance line.

We have been equally reliant upon—and far too aggressive with—the automobile and other vehicles powered by the internal combustion engine. Today we have ninety million cars, trucks, buses, and motorcycles on our roads. And to date, to our everlasting shame, we have killed more than one million of our fellow citizens on our highways, and injured millions more.

Now we have the facts about the phone. We have the facts about the auto. But yet to be discovered is the name of the salesman who gave impetus and inspiration to the insurance business. He was sick and tired of doors being slammed in his face and on his foot. And no wonder *that* was happening: he was selling "Death Insurance." The term frightened people.

So he decided, quite simply, he would become a salesman, instead, of "Life Insurance."

What's in a name?

Let the actuaries tell! And the Institute of Life Insurance!

This is germane to our discussion because in considering the relationships between the environment and population, and in espousing our points of view to our nation and to the world, it is reasonable to begin with a sense of the *target,* and a sense of *focus,* and a degree of *emphasis.*

Our *target* abroad is the uneducated and underdeveloped.

Our *target* at home is the indifferent and unaware.

Our *focus* must be three-fold: It must be on science. It must be on our social and political institutions and organizations. It must be on education. With this three-pronged focus in mind, we also will attain the fucrum . . . and the lever.

Our degree of *emphasis* must be great, but it can only achieve greatness if we are positive. And I regret that in our discussions and programs and pronouncements we often have been negative. Terms like "Planned Parenthood" and "birth control" and "family planning" bespeak restraint. They are, to that extent, self-defeating.

In terms as unequivocal as I can muster, I want to make it plain that from the smallest cell to the largest star, old dialogue inhibits new dreams, new ideas suffer from want of new idiom, and cliches stand in the way of creativity.

I mean this—seriously: we must cease to talk environmental failures and population or birth control, and begin to talk, instead, life insurance and life betterment.

Secondly, I want to make it unmistakably clear that the lamentations over the so-called gap between science and society are generally as vapid as they are victimizing. The critics of science have entered a quicksand of their own creation. It is in the laboratory, not in our forensics, that the best hope for attaining rational population growth has been found.

The fault is not at all in the license or attainments of science, but in the structures and strictures of society. The fault is not in the rigors of the laboratory discipline, but in the rigidities of our social and political institutions. The fault is not in the research, but in the translation of its fruits into viable social planning. The fault, in short, is not in the discovery, but in our applications or failure to apply new knowledge.

This coming June will be the 7th anniversary of our government's approval, for prescription use, of the "pill" in the United States. We have yet to accomplish the social planning and legislative framework through—and in—which that discovery—perhaps the most important discovery of the Twentieth Century—will reach the greatest number of people here and abroad . . . as the best available form of "quality-of-life-insurance."

No, I don't think at all our trouble has been that science has outraced society, but only that society has been laggard. Should we social planners tell the researcher to stop his research, or should we, too, begin to burn a bit more midnight oil, and to re-direct and in every possible way humanize the goals of all research and planning?

The reason we have air pollution is not basically Detroit. It is, rather, the social side of man—his desire to drive his own car, his general reluctance to pay the cost of mass transportation. The fault is in man's anxiousness to drive four-hundred horses when one hundred or even fewer would have gotten him to his destination just the same, but left the environment quite a bit less despoiled by fumes.

When Dr. Wiener made his plea for the human use of human beings, he was suggesting a new era made possible, we must remember, not by a revolutionary change in social attitude, but by science and invention—by automation, by cybernetics, by computers.

The reason we have water pollution is not basically the paper or pulp mills. It is, rather, the social side of man—his unwillingness to support reform government, to place in office the best qualified candidates, to keep in office the best talent, to pay it the best of wages, and to see to it that legislation both evolves from and inspires social planning.

The reason only ten percent of the annual crop of Ph.D.'s mathematics goes into teaching it *not* that industry pays more, but that the taxpayer wants to pay less.

Those who criticize science, or our emphasis on science, wallow about hoping that mere words, not fresh wisdom, will provide focus, fulcrum, and fight. And I think history will prove them wrong. I want to state this: There is *no* gap between science and society. There is only the gulf of fragmentation. We spend too much time on generating protest and not enough in shaping program. We expend too much energy on enunciating negatives, too little in promulgating positives. We give more time to our fight for fragmentation than to discovering how to coordinate efforts and to benefit through cross-disciplinary fertilization.

Can any discussion of environment take place today without reference to population? Can any discussion of population occur without reference to transportation? Can we leave the architect alone . . . without the city-planner? Can the union fight for a reduced workweek and remain indifferent to the facilities available for leisure?

The Sierra Club must be concerned over population curves, when it is population growth that has driven wildlife and wilderness to the wall. Planned Parenthood must express interest in the redwoods when we know that many of our greatest poets and social dialecticians have been deeply moved by nature: Whitman, Thoreau, Teddy Roosevelt, John Muir, Faulkner, Hemingway? These and others reflected on the quality of life, and what else but a quest for a better quality for life is the mission of Planned Parenthood?

Can any thinking human being, no matter how specialized his interest, can he be indifferent to the adequacy of our educational plant, to the level of teacher salaries, to the broadness of the curriculum? We *know* that it is from education is the recreator of the state and of society. We *know*—or should know—that, as Emerson stated:

"The true test of civilization is,

not the census,

nor the size of the cities,

nor the crops . . .

no,

but the kind of man the country turns out."

About this man we are turning out, we observe certain things. First, that he will be better educated than his parents. In spite of the division of departments into compartments, he will understand the need for interdisciplinary cooperation, and I think he will understand today's issues because his education will have crossed lines. At M.I.T., for example, it is not taught that urban history began when Fort Pitt became Pittsburgh but when agriculture replaced food-gathering . . . for it was at that point people became attached to the land and their installations took on permanent character.

Second, we know that our man of tomorrow will be more educated to think, and not merely trained to do. If he is a sociologist, he will have some understanding of mathematics that he can better assist the engineer or statistician who is programming the computer at the Department of Commerce where the Census is attempting to analyze population curves.

If he is a theologian, he will have some grasp of agricultural matters, for he will know the human soul is indeed more human when better fed, and that the ill-nourished must remain impoverished of spirit.

Third, our man of tomorrow will seek less after mammom that reason. In this nation, gradually, pockets of poverty will be eliminated. Youth will not be driven from the ghetto by the motivation to spend a lifetime making up for a childhood of material denials.

Fourth, our man will have more leisure and will use it better. It already is encouraging to note that the college graduates are watching less and less television, that amateur theatrical groups have so multiplied that the Brothers Klieg have been hard-pressed to supply enough lights, that piano manufacturers are working and selling at capacity.

Well, I did not come here to predict the future. I did come here to suggest —with both fervor and conviction—that our work . . . of improving life—for that is your work and mine—is coming into its own.

But we must talk a new idiom . . . not of population control, but of life insurance, and life improvement, and life betterment, and life enrichment.

And this is the time to begin. We have a "breather." We owe it, I think, to the pill, on the one hand, to pollution on the other; to contraception, if you will, and to congestion. In short, as people have become more and more crowded in urban areas, in a fouled environment, they have been less willing to have children, less anxious to expand families, and the timing of science and the Food and Drug Administration was perfect.

In 1966 the U.S. birth rate dropped to the lowest level since the mid-depression year of 1936. For all of last year, the rate was 18.5 births per 1,000 population. The figure for January, of this year, shows a further decline, to 17.7 (January 1966, was 18.2; January 1965, was 19.2). And the February figure for this year—just released—indicates another drop, to 17.4 (as against 18.8 for February 1966, and 19.7 for February 1965).

We must use our "breather" for social planning, for environmental restoration, for restoring the balance between man and nature, for learning to build beautiful and balanced cities.

Next, we must interject into our overall population-environmental picture some consideration of marriage, for, after all, most of us come from marriage.

By consideration of marriage, I mean, for example, this:

> That at the turn of the century, our life expectancy at birth was about forty-five years, and therefore we had to have our families young and our children close together.

Now, however, our life expectancy at birth is past the Biblical promise of Three Score and Ten, but our young people still show little appreciation of the number of their tomorrows. And I think one possible approach to further slowing down our population growth is to educate young people to the facts:

> That they *do* have time . . . to marry and to have families.

> That later marriages produce fewer divorces.

By consideration of marriage, I mean, for example, this:

> That there is enormous social pressure in this country in favor of marriage . . . and in favor of family. It is not coincidence that of the seven original astronauts, all were married, all had children.

> Nor is it coincidence that bank lending officers will be easier with the applicant who is married than with another, of similar background and circumstance, who is single, or married but childless.

> Nor is it coincidence that corporations will show preference to the em-

ployment application—other things being equal—where marriage and family are indicated.

Next, we had better set the best minds in the country to work on determining optimum population levels consistent with the quality of life we want to establish. We must enlarge our social and legislative planning to make these levels, increases, and goals, attainable and acceptable. We must, where necessary, change our institutional and organizational structures, liberating ourselves from the shackles of bureaucracy.

As of now, there is nobody who can tell us whether we can afford to have another four million people next year in terms of our diminishing—and not inexhaustible—supply of clean air.

As of now, there is no source we can turn to if we want to discover the hazards inherent in having eighty percent of our people living on one percent of our land, on our bicentennial . . . while our water supplies remain what they are, while our transportation remains what it is, while our recreational and health and educational facilities are as we *know* them to be—not merely *hope* they may be.

We continue to invite the automobile into our cities with each new garage we construct.

We continue to attempt the rebuilding of the cities, and provide too little funding for studies that might ultimately yield new cities and towns.

We continue to establish wilderness areas where those who believe with Thoreau that "There is no companionship so companionable as solitude"— where they can go. But, because of public pressure in the urban areas, the Air Force has reduced its testing of aircraft and the accompanying sonic booms over the cities. Instead, their tests now are held in the skies over the wilderness areas, where the congestion isn't, but where, with each boom, the solitude is broken, the sanctity disturbed.

Some treat! Some travail! Why not test over the oceans . . . and only there!

Perhaps in the not too distant future our social planning will include some dialogue about the impact of the tax structures on family size. Will it be worthwhile to grant tax incentives to those who keep the family small? Will it be advantageous to do the opposite with those whose additions to the family —after all—do ultimately cause us to invest more public funds in classrooms, recreational facilities, and hospital beds? As a Yale Professor of Public Health and Sociology, Dr. Lincoln Day, recently put it:

"Reproduction is a private act, but it is not
a private affair. It has far-reaching social
consequences."

So—unquestionably—does taxation and the rationale behind it!

We are at a moment in our history—and of the world's—when great things can happen . . . for man everywhere. We have the tools and many of the techniques required to launch and sustain a human revolution dedicated to the human use of the human beings, dedicated to the abolition of war and of waste.

We soon will be able to modify weather in meaningful—not isolated—areas, to bring rainfall to the parched lands and to make arid valleys verdant.

In the next few years, the Interior Department and the National Aeronautics and Space Administration will launch EROS—Earth Resources Observation Satellite. This will orbit the planet, giving us—for the first time—a quick and continuing look at the total environment. We will be able to detect new supplies of fresh water coming up from the ocean floor, as already, from the air, we have found several off Hawaii.

We will be able to detect sources of water pollution.

We will discover new mineral deposits.

We will discover defoliation due to plant disease and insects.

We will be better able to keep track of volcanoes.

We will be able to revise our maps with accuracy and speed never before possible.

In seventeen days we will collect data from space that it would take us twenty years to assemble from aircraft, and our information will never obsolesce. It will be fresh . . . like milk and eggs.

For the future, therefore, the question is not what we can do, for we will increasingly be able to create miracles and marvels and to humanize man and his environment.

The question has not to do with our capacity, but with our will; not with our imagination, but with our restraint; not with our acumen, but with our aspiration.

The question is this: Will society slow its growth, revitalize itself in the process, and begin to talk with science?

The question is this: Will we put an end to fragmentation and view the whole picture?

The question is this: *Will-we-use-our-knowledge?*

Are We Beginning to End Pollution?

Irwin Hersey

Air, water, earth—which along with fire were erroneously thought by Aristotle to be basic elements from which all matter derived—are under scrutiny from modern technology as they have never been before.

Unlike the ancient alchemists and philosophers, today's technologist is not concerned with analysis of these "elements" but rather in maintaining them as they should be: The natural environment for man's survival on earth.

Furthermore, environmental quality research and implementation are getting more support from spokesmen representing government, politics, sociology, psychology, education and the public at large—in an effort to keep the earth from becoming a dead satellite of the sun.

Ironically, the United States, as the most technologically advanced society on earth, is at the moment, the most acutely affected and at the same time most equipped to solve the problems of air, water, land and "noise" (or ear) pollution.

Pollution has been defined by the Committee on Pollution of the National Academy of Sciences—National Research Council as "an undesirable change in the physical, chemical, or biological characteristics of our air, water and land that may or will harmfully affect human life or that of any other desirable species, or industrial processes, living conditions, or cultural assets; or that may or will waste or deteriorate our raw material resources."

While "noise" pollution may not have the same origins as those which result in what we commonly think of as pollution, it is equally as harmful to "human life or that of any other desirable species," and is included here in the total view of the quality of our environment.

I—Our Air

That air pollution is now a worldwide problem is obvious to any visitor to any major industrial area almost anywhere on the face of the earth. It is equally obvious to the world's scientists, who are now quite concerned about the global effects of pollution. In fact, the American Association for the Advancement of Science held a two-day meeting on Global Effects of Environmental Pollution during its annual meeting in Houston last December.

Pollution usually stems from the fact that, under certain circumstances, natural processes are unable to remove from the air, land or water the increase in pollutants resulting from an increased scale of human activity. While in most instances the problems that then may arise are on a local scale, it is not at all unusual to find pollution effects lingering on long enough for the atmosphere or ocean circulation to spread them over the entire earth.

One example, of course, is radioactive fallout. Another is the rise in the concentration of atmospheric carbon dioxide produced by the large amounts of fossil fuels burned during the past few decades. The ocean's buffering action has just not been able to keep up with it, so we now find the carbon dioxide content of the atmosphere up by about 10 percent and still increasing. Other examples of pollutants resulting in worldwide effects would be pesticides and lead in the world's oceans, and lead compounds in the atmosphere.

However, while there is little argument over the fact that in some cases we already have what might be called worldwide pollution, there is virtually no agreement on the consequences of such pollution. Will the increase in the CO_2 content of the atmosphere upset the radiation balance of the earth and therefore, the world climate, or not? How long will it be before the increased amount of nitrogen compounds in the soil and water begins to affect the ecological balance?

Questions of this kind are becoming of increasing concern to many people both within and outside the government. In the government, for example, the Federal Council of Science and Technology has established a Committee on Environmental Quality, while the President's Scientific Advisory Committee is setting up a Panel on the Environment. The National Academy of Sciences has an Environmental Studies Board, and last year Congress organized a colloquium on the establishment of a National Policy for the Environment. AAAS has established a Committee on Environmental Alteration, and there are many private organizations concerned with pollution problems.

There has even been an attempt to organize some sort of anti-pollution effort on an international basis. Sweden has called on the UN to hold an international meeting on problems of the human environment, while the need for a global network to monitor environmental quality parameters has been emphasized by the International Biological Program.

The extent of the air pollution problem should not be underestimated. Last year, LaMont C. Cole, a professor of Ecology at Cornell University, delivered an address at a AAAS meeting entitled "Can the World Be Saved?" He opened his remarks by noting that the title was not his first choice, but the one he wanted to use, "Is There Intelligent Life on Earth?" had been preempted by a physicist discussing some of the same subjects. Then he added the wry comment that "there is evidence that the answer to both questions is in the negative."

Prof. Cole was blunt and to the point. Man, in the process of seeking a better way of life, is destroying the natural environment essential to any kind of human life, he stated. During his time on earth, he went on, "man has made giant strides in the direction of ruining the arable lands upon which his fuel supply depends, fouling the air he must breathe and the water he must drink, and upsetting the delicate chemical and climatic balances upon which his very existence depends." And, he concluded sadly, "there is all too little indication that man has any intention of mending his ways."

The evidence is there for all to see. Let's take the automobile exhaust fume problem, as an example. As noted, the august *New York Times* has already indicated that it feels little progress has been made to date in this area, and it is not optimistic about the future. Despite tougher pollution controls for automobiles, air pollution experts feel we may even be slipping in our efforts to cut down on automobile pollutants.

Thus, Austin N. Heller, New York's Commissioner of Air Pollution Control, said at a scientific meeting earlier this year that he didn't think the requirement for fuel controls on new cars would solve the pollution problem, since such devices would not be required on cars already on the road and there was nothing to insure that the controls would be kept in operating condition. Even more important, any reduction in the amount of pollution from each car would be more than offset by the continued increase in the total number of cars on the road.

On 1970-model vehicles, anti-pollution devices are expected to control 77 percent of the hydrocarbons and 68 percent of the carbon monoxide from automobiles, and 35 percent of the hydrocarbons and 37 percent of the carbon monoxide from gasoline-powered trucks and buses, and to limit smoke from diesel-powered vehicles to a faint plume. These standards are expected to be met by engine modifications designed to burn the fuel more thoroughly.

Unfortunately, no one feels that this is going to turn the trick. In fact, some people feel that the internal combustion engine may never reach the point in its development where pollution can be adequately controlled, particularly in crowded urban areas. This feeling has spurred the search for a replacement for the engine, with steam-driven and electrical vehicles coming under especially intensive investigation.

Thus, the Department of Transportation recently announced a $450,000 grant to test four steam-powered buses in San Francisco and a $300,000 grant to test buses using freon-powered external combustion engines in Dallas. Fuel cells and hybrid systems which would be part electric and part internal combustion are also under study.

However, for the foreseeable future at least, unless the steam car now being designed by Bill Lear of Learjet fame becomes an overnight sensation, we can expect to continue to be driving cars powered by internal combustion engines, since both the automobile and petroleum industries continue to stand squarely behind the engine in which they have so large a stake.

Interestingly enough, a crack has developed in the hitherto solid front presented by the two industries. When pollution control was merely a matter of making a few inexpensive adjustments, there was little disagreement between the automobile manufacturer and the petroleum company. Now that some rather expensive changes are being made on the 1970 models—estimates are that these could cost some $35 per car—auto makers have begun to take some pot-shots at the gasoline companies.

They claim, for example, that the lead added to gasoline to reduce engine knock results in a small but poisonous particulate emission of lead, which in addition to its health effects, would poison the suggested catalytic muffler, an alternative to expensive engine modifications.

The auto makers have also noted that hydrocarbons in the exhaust—a major cause of photochemical smog—would be reduced by cutting down on the butane content (which affects the volatility) of gasoline. The gasoline manufacturers respond that cutting down on the butane content would have drawbacks to the driver and would increase the cost of gas to a motorist by as much as 2 cents per gallon.

Much the same might be said of what is still a relatively minor source of air pollution—the airplane. However, scientists are becoming increasingly concerned over the fact that the airplane may have disproportionate importance because much of the CO_2 and water vapor released as engine combustion products enter the atmosphere at high altitudes and are only slowly removed from it.

As Prof. Cole has noted, when you burn a ton of petroleum hydrocarbons, you obtain as by-products about 1-1/3 tons of water and about twice this amount of CO_2. A Boeing 707 in flight turns this little trick in about 10 minutes. If you add up all the time spent in the air by the nation's air fleet in the course of a year, you can see that quite sizable amounts of CO_2 are released into the atmosphere in the course of a year. Prof. Cole estimated that it was 36 million tons just for flights terminating in New York.

Since the carbon-oxygen relationship which has existed as long as man has been on earth is essential to photosynthesis and hence to the maintenance of all life, upsetting this relationship could make life as we know it impossible. What man is doing today is unfortunately bringing this imbalance about, in large part by burning fossil fuels, which are hydrocarbons.

It is paradoxical that the exploitation of the fossil fuels—coal and oil and natural gas—have made it possible for more people to exist on earth simultaneously than has ever been possible before. It has also produced our greatest dilemma. The oceans hold our largest carbon deposits, since they take CO_2 from the atmosphere and precipitate it as limestone. Unfortunately, we are now burning fossil fuels so fast that the oceans are simply incapable of assimilating the CO_2 being released into the atmosphere.

Thus, while it is the hope of every developing country to become industrialized, and it now takes a far shorter time for this to come about than 50 or 100 years ago, we must be concerned over the fact that we are dumping more and more pollutants into the atmosphere. And, when the rate of combustion exceeds the rate of photosynthesis, we shall simply start running out of oxygen. And then we're in real trouble.

It should be noted at this point that there are different kinds of air pollution, caused by different things. The best-known, of course, is the photochemical smog so common to the Los Angeles area, but now noted with increasing frequency in many other areas. This is the result of hydrocarbons and oxides of nitrogen. The leading menaces from automotive exhaust are the hydrocarbons, which leads to smog, and carbon monoxide. Nitrogen oxides are also produced in engine exhausts, and contribute to the formation of smog, although many experts consider the hydrocarbons the most polluting element. Then there's the kind of pollution New York normally suffers from which results from interactions between particulate matter and sulfur dioxide. Primary sources of SO_2 are coal and residual oil, both of which have a relatively high sulfur content.

The nitrogen oxides problem is especially interesting, since, unlike the hydrocarbons and carbon monoxide formed when an internal combustion engine burns fuel, they are formed because of the very high temperatures created in the engine's combustion chamber. Pollution experts working for the automobile and petroleum companies tackled the hydrocarbon problem first because legislation required it, but the legislation appeared to be based on the knowledge that hydrocarbons are easier to control, and are believed by the experts to be the greatest source of smog.

However, spurred by the fact that the California Motor Vehicle Pollution Control Board is seeking to reduce nitrogen oxide emissions from the present 1250-1500 parts per million of exhaust gases to a desired 350 parts per million, the researchers are starting to come up with experimental devices that may turn the trick. Atlantic Richfield, for example, is working on a device which recycles some of the exhaust gases back into the engine and is said to be capable of reducing nitrogen oxide emissions by 85 percent.

It is interesting to note in this regard that portion of a statement on "Cleaner Air" by Standard Oil Co. of New Jersey dealing with nitrogen oxides. The statement, published as a three-page ad in a number of magazines last year, was taken from the testimony of Harold W. Fisher, a company director and vice-president, before a Senate Air and Water Pollution Subcommittee. All that Fisher had to say on the subject was: "There is no control as yet for nitrogen oxides, and the need for nitrogen oxide is not clearly defined. The knowledge in this area is not as advanced as that for hydrocarbons and carbon monoxide. However, we believe that reduction of nitrogen oxide emissions will be technically feasible if it is required." Perhaps he meant that control won't be required until it is technically feasible, a remarkably laconic paragraph in view of the subsequent California requirements, which are waiting on technical developments for implementation.

The high sulfur content of industrial fuel oil poses a different problem. Small-scale industrial heating operations often use #2 fuel oil (the typical home-heating variety), which is low in sulfur, and also more expensive. But engineers are still struggling with the problem of how to control the SO_2 from incinerating operations and big installations such as coal- and oil-burning power plants, which frequently emit boiler gasses at rates of the order of 500,000 cu. ft. per minute. About a dozen different approaches have been taken to the problem, but none has been particularly successful to date.

Obviously, containing such large volumes of gases even briefly for any kind of treatment is no easy matter. Consequently, a simpler solution would seem to be to use low-sulfur oil or natural gas. Unfortunately, the Federal Power Commission has refused some industrial applications to use natural gas because of the limited supply, and naturally low-sulfur oil is simply not so easy to come by. In addition, oil company representatives say that the widespread use of such oil, which has already started, would require millions of dollars worth of additional refining facilities and marked changes in the economic patterns of the oil industry. The National Air Pollution Control Administration believes that regular #6 (high-sulfur) fuel oil can still be used in areas where sulfur is not an air pollution problem.

Meanwhile, several states have moved to limit the sulfur content of fuels used by industrial concerns in their areas. In New Jersey, for example, despite many qualifying statements, fuel oil containing more than 1.0 percent of sulfur cannot be purchased for use after May 1968, and the allowable content will drop to 0.3 percent by October 1971.

The new standards already in effect will have some sizable economic consequences. There are three methods of preventing air pollution by fuel oil: finding new sources of oil with a natural low-sulfur content; refining the sulfur out of the oil, and desulfurizing flue gas. At the present time there is no economic or efficient method for the latter solution. New York has already imposed stringent regulations on the use of high-sulfur fuels, so unless increased quantities of the naturally low-sulfur oils are found, the refined oils, at an increased cost, will be the sole available commodity.

On the medical front, there is little question as to how dangerous pollution is, or is not. Certainly, in the notorious smog period in New York City in November 1966, the excessive morbidity and mortality that resulted was attributable in part to pollution.

Most pollution experts will tell you that if the present rate of population and industrial growth, automobile and air travel is maintained for another decade, with the pollution problem under attack at the present rate, lives are being shortened, morbidity rates are being increased, and the elderly and sickly will be living in a most unfit environment—a viewpoint with which many people who have lived through a bad day in Los Angeles or New York will readily agree. And a few doctors may even catagorize most of the large cities in this country as uninhabitable.

What can be done about it? Certainly the cities should be increasing their activities so as to be eligible for greater levels of federal funding. The federal government expects to see results from pollution efforts before granting higher levels of aid to local communities. Technology is available to control most sources of pollution, although developmental research is still needed. A representative of the federal government feels that social and political issues form the major roadblocks to implementation of this technology.

Finally, and perhaps most important of all, the federal government should be directly involved in the establishment and enforcement of pollution standards. Unfortunately, this will not be too easy to do, since it requires authorizing legislation from Congress, and industry is adamantly opposed to getting Uncle Sam into the pollution act. Quite frankly, they'd rather do it themselves or, at worst, suffer along with local or state regulations. It's interesting to note in this regard that, while industry tends to deplore non-uniform pollution standards, it shies away from nationally imposed standards in the belief that state regulations can be more easily tailored to local industry needs.

The Air Quality Act of 1967 established a timetable for planning regional air pollution control, giving the federal government the option of stepping in if the states failed to meet the deadlines. The timetable requires the Public Health Service to publish reports of the harmful effects of various kinds of air pollutants ("criteria"), based on the best and latest scientific information, and a companion report on the available methods for controlling the sources of those pollutants.

Another step is the designation by the P.H.S. of air quality control regions. The states involved in each region have nine months after all three actions are completed by the P.H.S. to submit regionally approved standards, based on public hearings, for air pollution control. The Secretary of Health, Education and Welfare approves or rejects these standards after comparison to the government's "criteria." Eight regions have officially been designated to date, with an estimated 57 regions to be defined by mid-1970.

The third step in the timetable gives the states in each air quality region another six months after approval of their standards to draw up plans for implementing them, including enabling legislation, local regulations, and target pollutants. Perhaps the biggest loophole in the legislation enters here, though, for there is no time limit for the actual implementation of the standards—only the implied threat of federal intervention if it is not carried out in a reasonable time.

The type of pressure the government can exert is exemplified by a recent statement by Dr. John T. Middleton, director of the National Air Pollution Control Administration, to the effect that the research that is just now beginning will "inevitably" lead to more stringent pollution controls. Dr. Middleton noted that pollution research was moving out of the laboratory, where healthy animals were studied in controlled situations, to comparisons of death rates for various illnesses in polluted and nonpolluted areas; investigations of deaths in heavy pollution incidents; and attempts to find out how air pollution multiplies the effects of other stresses on the body.

Preliminary indications, Dr. Middleton stated, are that air pollution is even more dangerous in life than in the laboratory. "It is likely," he added, "that accumulating evidence ... will show that there are biological risks associated with levels of exposure far below those now known to produce adverse effects."

At the same meeting at which Dr. Middleton spoke, Richard Sullivan, director of New Jersey's Division of Clean Air and Water, said that industry was already balking at standards that are being set as the result of currently realized dangers, and was pressing for pollution controls arrived at by a "consensus" of a commission representing all interests in the state, rather than by the state agency.

Industry is kidding itself. Educated guesses, including *Fortune's* and the government's, are that $3 billion per year for the next decade would put air pollution on a manageable basis. This is not an exorbitant figure, but let's be realistic and stress the timetable—the NEXT decade.

It's a job that obviously will have to be done through a major industry-government effort—and we're long past the point where it should get started.

II—Our Water

Water pollution has been defined by Robert T. Eckenrode, President of the System Sciences Division of Dunlap & Associates, Inc., as "the presence of toxic or noxious substances or forms of energy in natural water resources. Water pollutants may be noxious or toxic to people where the water source is used for drinking water, transportation or recreation; to industry where the source is used for industrial processing or cooling; or to nature where the pollutants disrupt the desired balance. Heat manifests itself as a water pollutant where waste water is returned to the source with thermal characteristics so different as to upset the natural or desired fauna and flora balance of the source."

So much has been written during the past few years on the extent of water pollution today that little useful purpose would be served by reviewing the matter in detail. However, the National Academy of Sciences—National Research Council report on "Waste Management and Control," published in 1966, gives some idea of the scope of the problem, and particularly how rapidly it is growing.

The report notes that any assessment of the magnitude of water pollution must consider the relationship between the total available supply of fresh water and the quantity of waste-carrying water. The former—the average annual stream flow that discharges into the oceans from the continental U.S.—is essentially fixed and amounts to about 1,100 billion gallons a day. The latter is the quantity returned to the stream flow after use by man, with its quality altered in one way or another.

A few past and projected ideas point up the scope of the problem. In 1954, some 300 billion gallons of the total were withdrawn daily, of which 100 billion gallons were consumed and thus represented depletion through use, and 190 billion gallons were returned to the streams. Corresponding values for the year 2000 are *889 billion gallons returned.* Thus, in 1954, withdrawals amounted to less than a third of the total and waste-ridden returns less than a fifth, in 2000, withdrawals will be a little over four-fifths and polluted returns about two-thirds of the total stream flow.

Water Use Is Increasing

Five categories of use were included in these data: Irrigation, municipal consumption, manufacturing, mining and steam-electric power generation. The projected increase for irrigation is very small, the principal change in returned irrigation water being an increase in the amount of dissolved mineral salts and some added contamination from chemicals used in fertilizers, pesticides and herbicides. Municipal sewage returns are expected to more than double, since more than 95 percent of the estimated U.S. population of 280 million will be living in urban areas. A sevenfold increase is expected in industrial wastes and the residues will be varied in character, containing oxygen-consuming ingredients as well as industrial chemicals of every kind.

The major withdrawal of water will be for power generation, and this water will be returned almost undiminished in quantity but at appreciably higher temperatures. As already noted, this "thermal pollution" can have a shattering effect on the ecology of streams into which it is introduced.

Pollutants entering water sources have been broadly classified into eight categories: Domestic sewage and other oxygen-demanding wastes; infectious agents; plant nutrients; organic chemicals such as insecticides, pesticides and detergents; other minerals and chemicals; sediment from land erosion; radioactive substances; and heat from power and industrial plants.

All pose different problems. For example, by 1980 it is estimated that the oxygen required to reduce sewage to stable compounds through the action of arobic bacteria will be large enough to consume the entire oxygen content of a volume of water equal to the dry-weather flow of all 22 of the U.S. river basins. A health hazard is posed by the incomplete elimination of infectious agents, while problems from excessive growth of algae blooms and plants have cropped up in many large lakes and remedial measures are largely lacking. Organic chemicals have already caused spectacular kills of fish and wildlife, and we know very little about the effects of long-term sub-lethal exposure. Methods of removing other minerals and chemicals, many of which are toxic, are poorly developed. Sedimentation fills stream channels and reservoirs, necessitating expensive additional purification measures. While intense public concern has led to the development of techniques to prevent radiation contamination under present conditions, the increase in nuclear power reactors by the year 2000 poses a serious challenge. And, since the amount of oxygen water can contain is reduced when the water is heated, the consequences for fish and aquatic life may be quite serious.

So much for the extent of the problem. The immediate reaction to all this is to throw up one's hands and give up. However, there are indications that something can be done about it, the prime example probably being Sweden, which is making a concerted effort to solve its water-pollution problems.

One particular problem that has concerned the Swedes is what has been done to many of the lakes near population centers into which unprocessed or incompletely treated sewage has been released over the years. While this is a temptingly cheap waste disposal method, it eventually destroys the lakes by adding large quantities of phosphorus and other nutrients, encouraging the growth of algae, which reduce the amount of oxygen in the water and eventually kill all fresh-water life. Sweden is today dotted with such "dead" lakes, many formerly prized recreational areas and now deserted.

A number of projects are now under way aimed at reviving these lakes. One is at Lake Trummen, near Vaxjo, which suffered from the injection not only of raw sewage, but also of waste from a local textile plant. While the pollution was stopped, the damage had long since been done. The lake is strangled with plant growth, making swimming impossible, and all the fish have died. A 16-ft-thick blanket of dead matter coats the lake bottom and the maximum water depth is now only 6 ft. Left alone, the polluted lake would eventually disappear.

Scientists from the Limnological Institution at Lund University, however, have launched a long-term project to bring the lake back to life and at the same time to gather data which may help in other such cases. The first stage of the project, a one-year research period, was just concluded, and the second stage, to get under way in the Fall, will consist of pumping up the relatively loose top layer of the sediment blanket, which contains most of the pollutants. Prof. Sven Bjork of the Limnological Institution says this layer contains from 10 to 20 times more phosphorus than the lower, more densely packed sediment. Various ideas for disposing of the rich mud will be tested—spreading it in forest areas and on farm land, and using it as filler in old gravel quarries in the area.

Low Budget—but Results

The pumping will take several months and, after it is completed, studies will be made of the lake's progress over the next nine or ten years. While this study period may appear to be unduly long for a small lake with less than 10,000 sq. ft. of surface area, it is intended primarily to gather experience that will be useful elsewhere, and the $200,000 price tag on the entire project does not seem high. As Prof. Bjork has noted, "it is still possible to do rather much with small investments" in the water pollution fight.

Other approaches are being tried to the same problem. One is a variant on the bubble technique used to pump air into submerged, perforated plastic hoses to keep harbors free of ice during the winter months. The method is being studied as a possible means of bringing lakes back to life. The injected air adds oxygen to the water near the bottom of the lake and also agitates it in such a manner that the water near the bottom of the lake comes into contact with air at the surface, thus providing a double dose of oxygen. There are some problems involved in using the technique. If, for example, the air is injected too rapidly, all the "dead" water may rise to the surface at the same time and possibly suffocate surface life. One aim of the project is to work out precise methods for applying the technique.

Another phase of the same project involves pumping up water from the bottom of a dead lake, running it through an artificial stream to aerate it, and then pumping it back to where it came from, thus adding oxygen and evacuating sulphurated hydrogen, which kills off fish life. These studies, covering four lakes and scheduled to last four years, will cost about $220,000.

A third project is focused on the potential of aluminum sulphate for curing lakes which have an excessive nutrient content. A year ago last Spring, two badly polluted lakes near Stockholm were sprayed with the substance, which combines with phosphorus in the water, flocculates and then settles to the floor of the lake, producing a harmless blanket on the bottom. A preliminary check last Fall showed that algae formation, usually quite heavy in the summer, had dropped to nearly zero and pollution had been cut down considerably. However, there is still some question as to whether the improvement is permanent, and only time can answer this question.

A Lake Drained of Life

Another project involves the famous Lake Hornborga in south central Sweden, at one time a world-renowned bird sanctuary which formed part of a chain of rest stops along the north-south route traveled by many migratory birds. Most birds now avoid the area, although this is not because the lake has been polluted, but rather because it has almost disappeared due to attempts by local farmers to create more arable land by sinking its water level five times, beginning in the 19th Century. The lake's area has shrunk from about 15 sq. mi. to the present 4-1/2 sq. mi. in the process and, as Lennart Vilborg of the Swedish National Nature Conservancy Office says, "to call it a lake today is somewhat pretentious."

The accuracy of Vilborg's comment is indicated by the fact that only about one-fifteenth of the area was open water when the field investigation was started in 1967. The rest was a vast sea of reeds, birch trees and a relatively small amount of other forms of plant life. Maximum depth is only 2-1/2 ft. and in the summer most of the "lake" is completely dry. In winter, the lack of oxygen is so great after the first freeze that the fish remaining in the lake die. The current attempt to restore Hornborga involves cutting the reeds and raising the water level.

That Sweden is particularly conscious of the pollution problem is indicated by the fact the country has formally asked the United Nations to sponsor an international conference in 1972 to combat the increasing tide of pollutants. The proposal drew 54 co-sponsors in its original form, and was adopted without opposition. In fact, it was, as one cynical UN observer put it, "the most acceptable international topic this side of motherhood."

Controlling the Polluters

The Swedes hope the conference will not only dispel public ignorance and apathy, but also enable researchers in this area to persuade their governments to take whatever measures are necessary to combat pollution, and produce a common outlook and direction in the consideration of environmental controls. Meanwhile, Sweden has passed a new law requiring all pollution-causing industries, as well as local councils, to conform to specified pollutant levels. Companies must apply for concessions to operate so that the harmful effects may be weighed against the economic benefits.

Evidence is beginning to appear that the U.S. is also becoming concerned over water pollution. For example, the Food and Drug Administration, reacting to growing public pressure for stricter enforcement of pesticide control laws, in April seized some 28,000 lb. of processed Lake Michigan coho salmon infected by pesticide residues. It's interesting to note that no one really knows today how many of the fish caught in U.S. lakes by commercial and sport fishermen every day are contaminated. A classic example is Clear Lake, Calif., where DDT, at the minuscule concentration of two one-hundredths part per million, was used to kill off a troublesome insect that hatched its eggs in the lake. As a result, plankton accumulated DDT residues at the rate of five parts per million, fatty tissue of fish feeding on lake-bottom life was found to contain several hundred to 2000 parts per million of DDT, and grebes and other diving birds died from eating the fish.

The New York State Health Department reports high concentrations of DDT in the state's central and northern lakes, and frankly admits its concern. As John Gottschalk, director of the U.S. Bureau of Sport Fisheries and Wildlife, puts it, "What is happening in Lake Michigan is an indication of what to expect elsewhere. There will be a day, and it may not be until the year 2000, when we are the coho salmon."

Concern over DDT led Senator Gaylord Nelson (D., Wis.) to commemorate the fifth anniversary of the death of Rachel Carson (who in the book "Silent Spring" seven years ago first exposed the pesticide threat) by introducing a bill to create a national commission on pesticides. Although federal regulatory legislation is on the books, it is rarely enforced and, in fact, there has been no criminal prosecution under the statute for some 13 years. As a result, the chemical industry is currently producing pesticides at the rate of better than a billion pounds annually, and their use is virtually uncontrolled.

It has been estimated that water pollution control is now costing about $6 billion annually, and this figure is expected to double by 1980. While it is difficult to come up with a figure on how much it would cost to do away completely with water pollution (and that's probably impossible), the NAS-NRC report on "Waste Management and Control" indicated that "the total cost of meeting all projected needs will undoubtedly run into tens of billions of dollars."

Remedies Are Costly

The report pointed out that one proposed system to separate all present combined sewer systems would cost $20 to $30 billion, and alternate solutions would also be very expensive. "An investment of tens of billions of dollars will be needed just to eliminate the backlog," the report went on. "For the future, it becomes clear that a major investment will be required to keep the quality of the waters in our streams, lakes and estuaries at reasonable levels."

What exactly can be done? The report suggested some possible areas of future technology which could contribute to water pollution control. One hope is that industrial techniques will be developed which would completely eliminate water transportation as a means of waste disposal. This would involve the replacement of liquid-based cleaning or plating techniques by mechanical or other methods, and by changing the raw materials used for certain processes in order to eliminate pollutant discharges.

In addition, synthesis, typified by the present conversion of some "hard" detergents to "soft," biodegradable detergents, could be made into a design criterion for new products. Although it is impractical to expect all products destined for eventual water disposal to be non-polluting or easily treatable, this general technology is deserving of major emphasis, particularly in the control of non-point sources.

Perhaps the most prominent area for a major breakthrough is the control of pollution from point sources by means of new waste treatment techniques. It could conceivably produce effluents as good as, or even better than, the original water supply. If treatment techniques could be developed which would permit the recycling of treated waste water for reuse, and the ultimate disposal of the waste concentrates, it might be possible to completely control pollution from municipal and industrial wastes.

The distribution of waste over a larger area or into a larger volume of water has some utility but can not effect major improvements in the total pollution picture. Detention or dispersion with time similarly hold no great promise for the future, although all these techniques could provide an immediate solution in some areas.

Diverting Waste Materials

Diversion of waste out of areas of use is somewhat similar to elimination, at least with respect to the use region. Although there is some question as to whether this technique would be competitive from the cost standpoint with other methods, the approach seems deserving of more attention than it has had in the past. In particular, the practicability of a two-pipe system for water supply and waste disposal appears worthy of study. Another technique, dilution, or low-flow augmentation, has broad applicability for both point and non-point pollution sources, and also appears deserving of emphasis.

Perhaps most important, the report concludes, the complementary application of several or all of these processes, together with increased water conservation and reuse of treated water, warrant review. Reengineering the home, and use of knowledge gained in developing the closed-cycle system found on the Gemini and Apollo spacecraft, could conceivably minimize the amount of waste leaving the home. Collection of non-point source pollution, or blending two streams of different qualities to provide a product of intermediate quality, could be helpful in some areas.

In addition, pre-concentration of waste concentrate could lead to the use of more economic disposal methods. The concept of dual use of existing facilities, such as adapting existing sewer systems for new treatment techniques, or grouping industries or municipalities to achieve the economies inherent in large-scale treatment, and even combining facilities for treating both waste water and solid waste, all may hold a practical answer to pollution abatement, and all are worthy of serious attention.

One thing is becoming more and more clear as time goes on and that is that the overall pollution problem can not be attacked piecemeal. Unfortunately, that's what we have been doing. The present war on pollution (in this country, at least) has been and is being hampered by its fractionation, not only at the local and state levels, but also at the federal level, where more than 80 different commissions, offices, agencies, administrations, corps and services are concerned with, and spend money on, pollution control.

This is especially frustrating because pollution control, even more than most urban problems, is a systems problem, since disposal of solid waste can cause air and land pollution, reduction of air pollution can lead to land pollution, and sewage disposal can cause water and air pollution. Systems techniques can be used to analyze the overall problem and yield insights to solutions, while preventing the use of specific solutions which could generate new and unexpected problems.

This systems approach has frequently been advocated before Congress, and has been used in a few projects by Aerojet-General Corp. for the state of California and by Dunlap & Associates. The NAS-NRC report also indicates that the systems approach offers the best possibility for devising a successful pollution-control program.

III—Our Land

A recent editorial in *Science* has pointed out that, while DDT has been one of the most important of all man-made chemicals, saving millions of lives through control of insect vectors and increased food production, it is a persistent and rather volatile chemical, and is frequently carried far from the point of its application.

The Choice Is Critical

It's fairly certain that, if the United States were to ban all pesticides, food prices would rise sharply and there would even be shortages of certain foods. And some people would like to see the use of all pesticides halted. The real issues, however, are the *choice* of pesticides (persistent vs. non-persistent, for example) and the conditions under which they are to be used. In this controversy, some agricultural experts and chemists defend the use of persistent chemicals like the chlorinated hydrocarbons (DDT, dieldrin, endrin, etc.), while conservationists advocate the use of nonpersistent chemicals such as the carbamates and phospho-organics, like malathion.

The defenders of the persistent chemicals point to their effectiveness and low cost. They note that DDT is relatively inexpensive and one or two applications may be sufficient for a season. Since much of the cost of using pesticides comes from labor, they argue that completely abandoning the chlorinated hydrocarbons would add appreciably to the nation's food bill. They also point out that, while DDT has been used for three decades, there has never been a documented instance in which human deaths have been traced to the chemical when it was properly employed, and relatively few have occurred even when it was improperly used. They also claim DDT and the other chlorinated hydrocarbons are not all that persistent and are slowly destroyed in the soil. Also, DDT slowly degrades in man and is excreted, so concentrations do not build up indefinitely.

Opponents of DDT point out that, along with its persistence, it may be carried literally thousands of miles from the place where it is being used. Thus residues of DDT have been found in seals and penguins in the Antarctic and, when a farmer uses it on his crops, some of it is destined to accumulate in you and me.

Some of the recent concern about DDT is due to new findings about its effects on animals other than man. A controversial report released recently described the carcinogenic effects of large amounts of DDT in tumor-susceptible mice. Interference with shell deposition in some birds has been traced to the pesticide, and it has also served as an estrogenic stimulus in rats. Most scientists agree that DDT use should be curtailed, and its consumption in the United States has been falling in recent years, although its use in other countries is still increasing.

However, pesticides are obviously not the only land pollutant. There are also the long-lasting radioactive elements from nuclear explosions. While the present evidence indicates that pollution to date from this source has not been great, existing experimental evidence regarding levels of radiation from fission products in soil is not as extensive as it should be, nor does it cover as long a period of time as might be desirable.

The solid wastes of agriculture are also frequently overlooked. Two such wastes—livestock and poultry manure, and field crop residues—are already produced in such quantities and are of such degradable organic material that their management poses serious problems. However, these materials could be a valuable national resource if they could be used as fertilizers and soil additives. At present they are not used because of the cost of transporting them from the places where they are produced to the areas where they are needed.

There is also the problem of disposing of solid wastes in metropolitan areas. Per capita production of refuse in the United States grew from 2.75 pounds per day in 1920 to 4.5 pounds per day in 1965 and is now growing at the rate of better than 4 percent per annum (not coincidentally, perhaps, the rate of growth in national product). Studies in the San Francisco Bay area indicate that, if all solid wastes are included, the rate of production is better than 8 pounds per person. The current cost of refuse collection and disposal is well above $3 billion, or more than for any public services except schools and roads.

Recovering Solid Wastes

Because of the increasing extent to which urban, suburban and rural areas are becoming crowded together, leaving little or no land for waste disposition, the storage, collection and disposition of solid wastes from both metropolitan residential areas and industry is getting more and more attention. Many industrial firms have for years treated waste products at the source where they can be recovered profitably before they become a public nuisance, and new legislation is forcing them to do so even when it is not economical.

Harold Rafton, a chemical engineer who has spent most of his career with the paper industry developing processes for the recovery of waste products from pulp manufacture, thinks future regional reclamation plans will make it possible to recover and re-use many valuable materials now being thrown away. They would work something like this:

In a reclamation plant, metal, lumber, tires or other large objects would be sorted by hand and sold to shredding plants. There, the metal pieces would be broken up and sold as scrap to steel companies for re-use as steel; lumber would be chipped and used as mulch to enrich the soil for highway beautification projects; and tires would go to scrap rubber plants which would separate the fabric, metal and wire components for re-use.

The rubbish remaining on the conveyor belt would be sent to a grinding machine, after which the lighter portions would drop onto a second belt with holes in it. Compressed air would blow the shredded paper and plastic fibers

—which would constitute a major portion of this type of waste—into a tank, while giant magnets would pick metallic materials off the belt. Glass, brick or cement fragments would run off the end of the belt into a container.

If the fiber and plastic materials blended, they could be used to make wallboard. If not, the fibers could be screened and separated from the plastics and then used for making paper or imitation wood. The hard plastics would go into another grinder, where they would be blended with a solvent to make a molasses-like dough which could be extruded as a crude but usable plastic. Small metal objects like tin cans would be melted into 100-pound pig-iron slabs, while the glass, brick and cement fragments would be crushed in another machine and used as a substitute for gravel in fill or as a base in road-building.

Unfortunately, Rafton's future reclamation plant would not accept garbage. However, other engineers are figuring out what to do about garbage. There is one European process, for example, which uses garbage to fuel steam turbine engines. It is much more likely, however, that tomorrow's garbage will be incinerated in special plants developed for that purpose. Incidentally, it is estimated that incineration of solid urban waste at the source could reduce the present work-load by 10 percent.

A pilot plant in Whitman, Massachusetts, developed by American Design & Development Corp. offers an example of how tomorrow's incinerator may work. The Melt-Zit Destructor operates at a temperature of 3,000 degrees F, completely destroying combustible materials such as paper and refuse. Non-combustible materials are melted and fused, with the 3 to 5 percent residue coming out of the incinerator like a stream of molten lava. This is cooled to a sterile metallic silicate resembling crushed cinders, instead of ending up as ash as happens in incinerators that operate at much lower temperatures. This residue can be used for shingles, fiberboard, road foundations or as sub-soil for gardens or recreation areas.

ADD claims a direct operating cost of no more than $3 per ton for rubbish disposal, which would be much lower than that of other incinerators. Heat from the Destructor would also be sufficient to purify thousands of gallons of salt or brackish water daily. Gases from the combustion process pass through a boiler and are converted into steam at the rate of 3 pounds of steam for 1 pound of refuse, and the steam passes through a conversion unit to purify the water.

Pipelines for Refuse

Harvard University has also studied the possible use of surplus World War II Liberty ships as offshore incinerators. The project was originated by the late Dr. Leslie Silverman, professor of engineering and environmental hygiene at the Harvard School of Public Health and an internationally recognized expert in air pollution control.

There is little question that the day of door-to-door pickup of refuse is rapidly drawing to a close, and that it will some time in the not-too-distant future be replaced by pipelines of some sort that can be connected to each house and will use air to suck the wastes from a fixed structure and convey it to either a disposal point or a transfer location. Such installations already exist in small numbers in housing developments in Europe, but have not yet been employed on a really large scale.

While a system like this would be almost ideal, the technical problems and costs involved in building a municipal system of this kind make it likely that it will be many years before we see even a modest trial system for evaluation purposes. The refuse truck is likely to be with us for some time to come.

Various other techniques are under study for handling solid wastes. Two of them, composting and chemical reduction, are worthy of attention, although it should be noted at this point that none of the processing methods proposed to date has proved economically feasible and most of them leave a residue which is often as difficult to eliminate as the original wastes.

Composting can convert about half of the solid waste of a community into a soil conditioner if it is subjected to aerobic decomposition at high temperatures. It is an old process used all over the world in areas of low labor costs and low soil fertility, but it has failed completely in this country because there is as yet no economical way of doing it and, perhaps even more important, there does not appear to be any sales market for it or even space where it could be stocked for distribution at no cost. Unfortunately, no one really wants a large compost heap today.

Chemical reduction of refuse has been under study and in some instances in operation for many years. For example, a steam treatment to release fats and other material from refuse was tried extensively some 50 years ago, but did not last very long. More recently, there have been attempts to effect the hydrolysis of cellulose, which constitutes a large portion of modern refuse. It can be hydrolized down to sugar, which can then be recovered and put to various industrial uses, such as the making of alcohol. However, the economics do not look too promising at this point.

As the NAS-NRC report on "Waste Management and Control" points out, an area of growing concern to scientists is the effect of sanitary land-filling on the ecology of tidal and swamp lands. The answer is undoubtedly one of degree, but there is a need for a study of the long-term requirements of our population, assuming no new means are found of disposing of solid wastes. There is little question about the fact that, as less and less land becomes available for solid waste disposal, incineration and other techniques will come into greater use.

Another kind of pollution results from over-fertilization of land under our land-bank programs. The unused nitrogen and phosphorus collects in the runoff from over-fertilized lands and enriches biological growth in adjacent lakes and streams to an excessive degree, making them uncontrollable and sometimes unusable as a source of drinking water, or even for recreation purposes.

This type of pollution can be abated through less intensive use of farm land and better control of fertilizer use. More frequent soil analysis and an expanded education program in this area are indicated, since the technology exists to correct pollution arising from over-fertilization.

Synergistic Effects

The same cannot be said about pesticides and herbicides, since they can accumulate in plants to a degree that may make the plants toxic to animals and human beings. Quantities of a few parts per billion in the soil may concentrate a thousandfold in plants and animals which serve as man's food. In addition, minute quantities of two or more chemicals may produce synergistic effects that greatly exceed the damage of one of them.

The NAS-NRC report lists as examples of this type of soil contamination the accumulation of chemicals such as benzene hexachloride in root crops; changes in the chemical composition of soils due to the killing of soil bacteria; increases in nitrogen levels due to the fact that nitrogen-killing bacteria have become more active; increased carbon dioxide and chlorine content in soil due to the digestion of chemicals by bacteria; and plant growth retardation due to the toxic effects of compounds.

As the report notes, all of this can result in lower crop yields, inferior quality of agricultural products and interference with farm programming. Even more important, no complete solution to the problem is presently in sight, although more careful selection and use of chemicals could help. Unfortunately, present technology probably is not sufficient to protect crops without a certain degree of soil contamination.

While substantial research is being carried out in this area, with particular emphasis on the development of less persistent and less hazardous chemicals, the report suggests that more research should be directed toward new approaches. It notes that many of the chemical poisons are absorbed and effectively destroyed by healthy soil rich in organic matter. This is an area in which more research is indicated. In addition, fresh approaches, including the wider use of artificial lures, insect sterilization and better use of natural predators appear promising as effective and inexpensive long-range solutions.

While more efficient incineration will no doubt provide a relatively effective short-range solution to one facet of the land pollution problem, the magnitude of the problem by the turn of the century will demand new answers, in the form of solid waste disposal systems which have not yet been designed. The NAS-NRC report suggests that these solutions will probably take the form of systems very similar to that described by Harold Rafton—that is, they will salvage and re-use all components of what is now waste, putting back into the soil those components that will benefit the soil, and returning to industry and commerce those components that are industrially useful. The systems, designed to conserve manpower, will incorporate features that greatly reduce or eliminate the costly collection practices presently in use.

The NAS-NRC report offered an interesting cost comparison of the three solid-waste disposal methods used today: Sanitary land filling, central incineration and composting. Capital costs, with land excluded, for each practice are $1,000-2,000, $3,500-7,000 and $1,500-10,000 per ton per day, and operating costs, $1.25-2.25, $3.50-5.00 and $2.00-7.00 per ton per day, respectively.

As these figures indicate, sanitary land filling has a definite cost advantage and thus is widely used. It is simple, disposing of all solid waste without preliminary preparation or treatment. Its disadvantages are that it requires considerable land area, and land is rapidly disappearing, and it has a potential for aggravating water-pollution problems.

Incineration has the advantages of rapid disposal and low nuisance value, partly because it eliminates putrifiables. Its disadvantages are high air and water pollution potential, large capital investment and high operating cost because it requires relatively skilled operators. Also, today's incinerators still leave a residue which in turn has to be buried or otherwise disposed of.

Composting is not widely practiced for reasons given earlier. However, a 1961 British study offered some evidence that controlled composting in sanitary land fills, and forcing air through the fill, could make polluted ground water adjacent to the fill self-purifying. If additional research confirms this, composting could provide a solution to one of the major problems associated with sanitary land filling.

Packaging Is Villain

How serious is the city solid-waste problem? It can be judged by examining just the major contributor to solid waste—discarded packaging. At present, almost 35 million tons of packaging are used in the United States each year. Of this, more than 60 percent is paper, 20 percent is glass, 16 percent metal, and the balance plastic. The metal group includes steel, tin, lead and aluminum, all with potential salvage value. These figures indicate that an average community of 10,000 people discards about 1,000 tons of paper and 172 tons of metal per year from packaging materials alone. Even more important, packaging trends indicate that the number of discarded containers will continue to increase for years to come.

The United States has been described by a wag as a nation standing knee-deep in refuse and hurling rockets at the moon. It is, unfortunately, a relatively accurate description. However, there is growing evidence that the land-pollution problem is being attacked at long last and, with luck, may even be solved before we are buried by it.

IV—Our Hearing

Some years ago, science fiction writer J. G. Ballard wrote a short story called "The Sound Sweep," in which he envisioned a machine called a "Sonovac," which was used to sweep up the noises of the past and take them to sonic dumps, where they were compacted and reduced to silence for all time.

There are a lot of people around today who feel that, if something like a Sonovac isn't invented in the very near future, we may be in real trouble. As Dr. Vern O. Knudsen, a physicist and former Chancellor of the Univ. of California at Los Angeles, put it, "Noise, like smog, is a slow agent of death. If it continues to increase for the next 30 years as it has for the past 30, it could become lethal." And Dr. Knudsen knows what he's talking about, since he's spent some 40 years on the study of sound and is one of the world's leading experts in this area.

Nor is Dr. Knudsen alone in his viewpoint. The Federal Council for Science and Technology, for example, last year issued a report which called noise a major health hazard in American industry and an increasingly serious nuisance throughout the world. "A new world of whining air conditioners, whirring engines, pneumatic hammers and the neighbor's radio have intruded on our peaceful world in a rising tide of unwanted sound," the report stated. It concluded that the noise problem had reached "a level of national importance and public concern," and recommended a series of sweeping research programs by Federal agencies to investigate the fundamental nature of noise in an effort to better understand its effects on human health and efficiency.

The report indicated that the overall loudness of the noise environment was doubling every ten years, and that it was becoming more and more difficult to escape from it. Dr. Knudsen estimates that the general level of urban noise has increased about 1 db per year for the past 30 years.

"Prolonged exposure to intense noise," the report indicated, "produces permanent hearing loss. Increasing numbers of competent investigators believe that such exposure may adversely affect other organic, sensory, and physiognomical functions of the human body." As a result, it recommended that the government develop guidelines for noise level standards in the urban environment and provide leadership for noise abatement and control programs.

The deleterious effects of noise on human beings have been indicated in a number of studies, not the least interesting of which was a project in which a team of McGill University epidemiologists studied impairment of hearing among the members of 12 Montreal rock groups. The study included both measurements of the noise produced by the groups and interviews with the musicians. The conclusion? That rock musicians would be wise to wear muffs or ear plugs if they wanted to prevent temporary, or even permanent, ear damage.

The Danger from Music

Several band members felt that their hearing had been impaired and one guitarist reported that he had lost all hearing in one of his ears. Like other studies, this one reported that the average rock music in Montreal produced a noise level of 120 db, compared with noise ratings of 140 db for jet planes, 130 for pneumatic riveters, and less than 110 for power mowers. The pain threshold is about 140 db.

While hearing loss due to noise is related to intensity, frequency and length of exposure, with the effects varying widely among individuals, the McGill study stated flatly that "noise exposure of the intensity and frequency encountered in this study may cause auditory impairment." It went on to note that, "if the current trend in musical entertainment and social activity (in discotheques) continues, these observed deleterious effects should become progressively more important in medical practice." Moreover, the study indicated that amplification systems used by the Montreal rock groups were of uniform capacity but were usually played at 50 percent of maximum output because of fear of burning out expensive electronic equipment. Obviously, the more affluent rock groups don't have to worry about this and can push their amplifiers to the limit.

In recent months, Ralph Nader has turned his attention to the same problem and, fearing for the hearing of a new generation of Americans, is trying to get state and governmental regulation of rock noise. In a letter to two Senate subcommittees, he noted that hard rock groups in the Washington, D.C. area were producing 100–116 db of noise, and that readings as high as 138 db had been recorded in go-go joints elsewhere. In seeking noise limitations, Nader referred to a University of Florida study which indicated that continuous dance music at 90 db led to teen-age hearing losses in certain high ranges of spoken sounds.

But the noise produced by rock groups is, of course, only a part of the total problem. A German study, for example, has shown an unusually high number of abnormal heart rhythms among steelworkers continuously exposed to a severe noise environment. In addition, Italian weavers, also exposed to intense noise, have shown abnormal brainwave patterns, some of which are suggestive of personality disorders.

Even outside of heavy industry, the Federal Council for Science and Technology has found evidence that the American environment is getting noisier, with the noise level in kitchens, for example, beginning to approach that found in factories. Its report indicated that in general, the traffic noise from freeways and downtown areas probably disturbs more people than any other kind of outdoor noise, with the trailer truck probably the "most notorious" noise producer. It's interesting to note that little has been done to date by truck manufacturers to improve this situation.

Sounds in the Home

Modern building designs and construction materials were also criticized for not protecting against noise from both within and outside the building, although this is an area in which progress is being made. Nevertheless, the report noted that the lightweight construction and open-plan design of most modern buildings provides very little in the way of noise protection.

Dr. Knudsen feels that, even if the U.S. (and the entire world, for that matter) is getting noisier, we are also devising better ways of muffling noise. He believes that most, if not all, noise sources can be controlled—or rather could be, if people were really interested. For example, he feels traffic noise could very easily be controlled, since combustion engines could be far quieter than they are today.

"When automobiles come from the factory," Dr. Knudsen has said, "they're equipped with good mufflers . . . You're not bothered by the noise from a Cadillac, for example, and there's really no reason why a bus or truck should be noisier than three or four Cadillacs." He suggests that, as a starter in the attack on traffic noises, traffic policemen should be equipped with sound level meters to detect noisemakers, as is done in Sweden.

Soundproofing Is Effective

The sound specialist also believes it is possible with modern technology to keep outdoor noises outside of buildings. Adding from 5 to 10 percent to the cost of a building, he points out, can provide the equivalent of 10 in. of concrete shielding against noise. The extra investment buys thick, tight, solid doors; sound-absorbent entrance halls; insulated walls and ceilings; heavy walls for bathrooms; and double windows with sound-deadening air space between the panes. These measures could reduce interior noise by 50 db, and most of them are common in European countries, in some of which building codes already require a 50-db reduction in interior noise levels.

As Dr. Knudsen has noted, the worst is yet to come, with the sonic boom of the supreme transport. He sees little hope of doing anything about the boom for the foreseeable future, at least, and other experts tend to agree with him on this.

Dr. Richard Seebass of Cornell University, for example, concluded a recent paper on the subject by noting that "we do not foresee any revolutionary concept that will totally eliminate the sonic boom," although he prognosticated a continual evolution of SST designs with improved boom characteristics. And Dr. Garrett Hardin of the University of California at Santa Barbara, for example, has referred to the boom as "something much worse than noise," adding that experiencing it is like "living inside a drum beaten by an idiot at insane intervals."

That a lot of people are now worrying about sonic boom indicates we have come a long way from the days when the Air Force tried to sell it as "the sound of freedom" and aircraft company public relations experts told us we'd get used to it, just as we had gotten used to the noise from jet engines. Unfortunately, a sonic boom is not quite the same thing as jet engine noise. A report made by a group of scientists to the Secretary of the Interior last year, for example, noted that "each boom would be perceived by its hearers as equivalent in annoyance to the noise of a large truck traveling at 60 mph at a distance of about 30 ft."

SSTs Are LOUD

The report went on to say that the number of supersonic transports expected in the late 1970s would subject between 20 and 40 million Americans living under a path 12-1/2 mi. on either side of the expected flight tracks to five to 50 booms per day, while an additional 35 to 65 million within 12-1/2 to 25 mi. of the flight path would be subjected to one to 50 booms per day of somewhat lower intensity, and 13 to 25 million more would experience one to four high-intensity booms.

Those are pretty big numbers, and they loom even larger when one realizes that no real solution to the boom problem appears imminent. It is now almost universally acknowledged that, until a solution to the boom problem is found, no SST will fly supersonically over inhabited areas. It is interesting to note in this regard that even the Russians do not expect to fly their Tupolev-144 SST at supersonic speeds over populous areas of the USSR, indicating that, even in a dictatorship, the sonic boom has caused public relations problems.

We are still not yet clear as to the full consequences of repeated exposure to sonic booms, despite tests made in this country in Southern California and Oklahoma City, and in France, although there seems to be a general feeling that crisscrossing this country with sonic-boom targets could cause psychological, aesthetic and possibly even physical damage. We do know, however, that 30 percent of the people questioned in sonic-boom tests consider the noise either intolerable or unacceptable, while another 50 percent find it objectionable to some degree.

Most of the effects of the boom are associated with the general increase in noise levels to which we are already being exposed. However, sudden shock noises like those produced in supersonic flight can be psychologically upsetting. Local politicians in virtually all metropolitan areas, for example, report that they receive more complaints from people disturbed by sudden noises than by any other form of environmental pollution.

There can be little doubt that an anti-noise movement is developing. During the past year, for example, several bills have been introduced in Congress to establish noise standards, encourage research, offer grants and technical aid, or otherwise assist in reducing the hazard of noise. Many manufacturers, under public pressure and guided by laws establishing controls, are vigorously seeking ways and means of suppressing noise or designing around it. Local governments have established anti-noise ordinances, but these are primarily concerned with localized and conventional kinds of noise. As yet, there is little evidence that metropolitan areas can get together to establish ordinances dealing with more generalized noise.

There is a feeling that dealing with the noise problem by legal means may prove inadequate, primarily because the police powers in the areas of health, safety and public welfare may not be specific enough to cover psychological or esthetic damage. Certainly there is little evidence to date of any intent to protect the public against noise damages which would be primarily of a psychological or esthetic character.

Esthetic Damages

One important study of the noise phenomenon is being carried out by a bio-engineering group at the Carnegie Mellon University in Pittsburgh. The problem is being analyzed from the systems standpoint, with alternative program objectives stated in terms of intensities and other characteristics of noise, and various control measures then considered in order to determine the least costly means of achieving particular noise level objectives.

This question of legal penalties for excessive noise is a very complicated one. As Dr. Knudsen has pointed out, noise can be harmful in several different ways. First, of course, it can impair hearing. One research team has found that Scandinavian shipyard workers on the job for more than 30 years were unable to hear a whisper a yard away. Second, it simply blurs or masks sounds you want to hear, or, in the words of a highly successful Broadway play of last season, "You Know I Can't Hear You When the Water's Running."

Noise also interferes with rest and impedes convalescence. In fact, Dr. Knudsen traces his lifelong interest in noise to a period during which he was being treated for duodenal bleeding in a hospital near a truck route. "I thought it would drive me out of my mind," he says. "When the trucks would roar by, I'd actually feel pain in my stomach." After he left the hospital, he developed the ear protectors U.S. servicemen have worn ever since WW II, and when his royalties ran out about 10 years ago, some 8 million pairs had been made.

Noise also prevents concentrated mental effort. The noise from a jet plane passing overhead, for example, can stop any school class in its tracks, and anyone who has ever attended an outdoor concert knows how annoying the sound of a plane or truck passing by can be. And, last but by no means least, noise frequently causes stress and nervousness, and therefore the diseases that result from tension.

The growth of scientific interest in the noise problem was indicated by the National Conference on Noise as a Public Health Hazard, sponsored by the American Speech and Hearing Assn. in Washington, D.C., in June 1968. This was the first such meeting ever held in this country and, as Dr. William H. Stewart, Surgeon General of the Public Health Service, said in his keynote address, it was symptomatic of the fact that the public was getting fed up with "the indignity of the sound barrage."

Papers submitted at the meeting demonstrated that noise can affect our psychological state in many different ways. It is no coincidence, for example, that the mining and quarrying industries, which use drills which produce noise far in excess of 120 db, are among those having the highest accident and injury rates. In office situations, noise can create inefficiency by decreasing the effectiveness of communication between workers, while the inability of factory workers to hear warning shouts over the noise of their equipment is unquestionably a factor in accidents and injuries.

It has been found that long-term exposure to 90-db noise can lessen vigilance, which is dangerous in jobs requiring watch-keeping, such as monitoring several pressure gauges at once. Noise at a sustained level of 120 db (which is less than that experienced by a riveter) has been found to impair equilibrium, particularly when there has been unequal stimulation of the ears. Another study has demonstrated that loud noise could reduce eye movement and the ability to focus.

In general, acoustic scientists have found that it is not the quantity of work which suffers when workers are exposed to sustained high-level noise, but the quality. Thus, an assembly line worker would not turn out fewer pieces of work in a given period of time, but is likely to turn out fewer pieces which will pass inspection. It is believed, in fact, that people in noisy locations tend to work faster, and consequently less carefully, in the hope that they can soon get rid of the stress.

Astronauts Were Affected

It's interesting to note in this regard that a group of astronauts subjected to the 145-db noise of a jet engine at full power found it difficult to do even the simplest mathematical problems and tended to write down any answer that came into their heads just to get the test over with. Other experiments have shown that noise has the greatest impact on the performance of tasks which demand the most effort from a worker.

Many factors can affect the manner or degree to which we are annoyed by noise. It has long been known, for example, that the ability of a sound to annoy varies with its intensity and pitch. Studies have also shown that variable sounds—those which occur randomly in time or change their loudness, pitch or point of origin—are more annoying than steady-state noises. Dr. Knudsen has found, for example, that, before he began wearing ear plugs to bed, three out of four times when he awoke during the night, a sudden noise was to blame. Also, a normal sleeper isn't bothered by the steady ticking of an alarm clock, but might have trouble if the ticking grew randomly louder and softer.

One laboratory study has shown that the performance of subjects exposed to noise varied with the type of pre-noise briefings they were given. Those who were told noise would impair their performance tended to suffer such impairment, while those who expected to perform better actually did so. Personality factors also tend to affect a person's reaction to noise. In general, those who reveal anxiety, introversion or neurotic tendencies on a personality test tend to show the poorest performance on a test of vigilance.

Consequently, it is extremely difficult, if not impossible, to create a meaningful scale for rating the potential annoyance of any particular sound with respect to a given subject. In every case, it's necessary to take into account the content and context of the sound, as well as the temperament of the subject.

There is little question that the most annoying kind of noise for most people in normal situations is the noise which affects sleep or relaxation. It is a well-known fact, for example, that, in communities where air traffic is a problem, the largest number of complaints arise from interruptions of sleep and rest.

The Context Is Important

The context in which the noise is heard is also important. Thus, the noise of vehicular traffic is generally far more annoying to people while they are in their own homes than when they are outdoors. About everyone knows that the soft patter of little feet in your own apartment becomes the noise of a thundering herd of elephants when it comes from the apartment overhead or next door.

Many people feel that it is really irrelevant to try to prove that noise is a real or potential hazard to public health, since it can more easily be defined as any sound that interferes with human activity. Thus, they argue, we should not be troubling ourselves by trying to set up objective standards which prove that noise is either dangerous or troublesome. Instead, we should decide that noise above a certain level, or of a certain kind, is objectionable, and then assert that we simply will not tolerate it.

This view was expressed by Wilber M. Ferry of the Fund for the Republic at the Washington conference. Ferry argued that the case against noise should be based as much on civilized standards as on the basis of public health or economy. He emphasized that a society must cling to certain civilized values which need emphasis "just as much as demonstrations of growing deafness, stress and inefficiencies (which arise) out of increasing noise. Quiet and privacy are positive values, not to be considered expendable except on showings of the utmost importance."

Ferry, who is not a scientist, also made one other very important point in his paper. He said that what we do about noise will reflect our relationship to technology. He noted that we have neither been interested nor successful in controlling noise because we have not been interested or successful in controlling technology. "Until we are willing to undertake to deal with the cause," he concluded, "fussing around with the symptoms is not likely to produce far-reaching results."

All of which may be true, but the question is: How do you go about controlling noise, or, more precisely, policing its control? For example, Dr. Aaron J. Teller of Cooper Union, an authority on pollution, suggested at the recent 62nd meeting of the Air Pollution Control Assn. in New York that taxes be levied on wastes spewed into the environment to pay for the damages they are causing now and to compensate future generations for the loss of natural resources. It's difficult to see how a scheme of this kind could be applied to the noise pollution problem.

However, it's interesting to note that a recent study made for builders and lenders which also contains tips for home buyers warns that anyone buying shelter could find himself owning obsolete housing before he moves in unless he makes sure that noise within a dwelling can be controlled, but that sound-control measures should be planned before construction, when they are relatively inexpensive. Corrective sound control after a dwelling is built is invariably expensive, it notes.

The study is particularly interesting because it was originally a supplement to the *Construction Lending Guide,* published by the U.S. Savings and Loan League to aid loan officers, but is now available to anyone interested in building or owning a home or apartment. It indicates quite clearly that there is a growing concern over the need to make homes and apartments soundproof.

The Determination to Control

What the future holds with regard to anti-noise legislation is extremely difficult to predict. The variety and kinds of noise which must be considered; the geographical areas that must be taken into account in connection with each type of noise; the difficulties involved in allocating the costs of noise reduction; and the ticklish problem of how to police noise-control legislation all require careful study and analysis. The situation is a very complicated one, and there is no doubt that entirely new laws and institutions will have come into being to insure implementation of the most effective control programs, once these have been determined.

Nevertheless, there is every indication that this will be done. Many of the means for ending noise pollution are already at hand; research will make whatever else is necessary available in the not-too-distant future. All that's required at this point is the firm resolve not to continue to live in a society where dangerous, unhealthy or uncomfortable noise is tolerated. And there is increasing evidence that we are rapidly reaching this stage.

Thermal Pollution:
Hot Issue for Industry

Raul Remirez

Prodded by control dictums, industry is seeking adequate measures to cool its process water prior to discharge. Regulatory agencies are moving ahead with enforcement action while studying the actual effect of thermal input on the biological life of a body of water.

Industry is now swimming in its own hot water—the water that it discharges into rivers, lakes and the ocean after using it for cooling and other process functions. Some firms call the release of hot water "thermal enrichment," but most conservationists consider it a formidable threat to aquatic life. And the government, which dubs it "thermal pollution," is determined to curb it with new legislation and federal standards for temperature. All this means huge expenditures for industry offenders, because they must satisfy the new regulations by developing new techniques and equipment to get rid of the waste heat.

Thus, thermal pollution, which was creating only a flurry of interest six months ago, is now on center stage. Reports of pollution trouble-spots (California, Florida, the Pacific Northwest, New England) are coming at a fast pace. Professional meetings—e.g., the Cooling Tower Institute meeting this January in New Orleans—have been scheduling special sessions on the subject. And the government, through Sen. Edmund S. Muskie's (D.-Me.) Public Works Pollution Subcommittee, held crucial hearings last month on most aspects of thermal pollution.

For all the current noise, however, none of the parties involved seems to know how much damage the heated effluent is now causing. Conservationists have reported cases of killed or diseased fish in high-discharge areas. But there are some instances when the effect of heated discharges may even have been beneficial. Both government and industry recognize this dichotomy, and are urging more research into all aspects of thermal pollution. The government, however, presaging a worsening situation in the near future, prefers to act now, along the line of pollution prevention instead of after-the-fact abatement.

Guidelines Yes, Standards No

Part of the controversy surrounding the subject of thermal pollution focuses on the temperature standards that are being adopted at a fast pace for U.S. rivers, lakes, estuaries and coastal waters. Electric utilities oppose them on the grounds that there aren't enough data on the effects of heated effluents on aquatic life. They would prefer to see interim guidelines rather than permanent standards. The Dept. of the Interior, on the other hand, says that the information now available is adequate for setting standards.

According to Interior, the cornerstone for present temperature limits is the Clean Water Act of 1965, which contains provisions to deal with any factor that contributes to pollution (including temperature). The Act required all states, through local pollution-control boards, to submit to Interior's Federal Water Pollution Control Administration by July 1, 1967 a program for combating pollution in interstate bodies of water.

Partly because present damage cannot be measured, private utilities stand solidly in their opposition to water temperature standards. At a recent seminar on thermal discharges held in Albany, N.Y., Howard D. Philipp, chief nuclear engineer for Niagara Mohawk Power Corp., voiced serious objections to the standards on behalf of seven major utilities that serve 98% of the state's power requirements.

In a plea for provisional, rather than permanent standards, Philipp said that temperature-tolerance levels and lethal temperatures are known only for a limited number of fish. These data, he added, are too scattered and indefinite to be of practical value. On this basis Philipp criticized the maximum-temperature-rise and rate-of-temperature-rise limits set by some states.

Adopted temperature criteria, says Philipp, should be applied to water at the end of the mixing zone—i.e., the area where the heated effluent mixes horizontally and vertically with the receiving water. Some states (Massachusetts, Maine, Pennsylvania and Wisconsin) have recognized this mixing effect that dilutes the discharged heat and incorporated the concept in their water-quality standards.

State Standards—The states met their deadline last year with programs including proposals for setting water-temperature limits. These, however, do not become federal standards until Interior gives its approval. As of last month, the department had given the nod to 17 only, but swift approvals are expected for the remaining states.

In deciding whether to accept or change proposed state standards, Interior considers temperature criteria based on the water's intended use—i.e., fish propagation, body con-

tact sports, industrial, municipal supply, etc. It is also stay-
ing close to the recommendations made in a report issued by
the National Technical Advisory Committee last summer,
which suggests specific temperature limits that would avoid
harming both warm-water and cold-water fish.

According to Interior, temperature limits set by already-
approved standards don't vary much from state to state.
There is generally a maximum temperature allowable (93 F.
is a typical limit), a maximum allowable increase over ambi-
ent temperature (a common figure is 5 deg. F.), and, frequently, a maximum rate of temperature increase.

The Finger of Guilt

If all this points to some confusion, at least all detractors seem to agree
on who bears the burden of guilt. It is the electric power industry, by far the
biggest user of fresh water for cooling and condensing purposes. Utilities
generally operate on the basis of once-through cooling, i.e., they draw water,
use it and then discharge it hot without cooling it first.

But that's only part of the problem. A real concern today is the close
relationship that exists between future power demands and thermal pollution.
As the Dept. of Interior sees it, power generation has doubled every decade
since 1945, and future demands are such that the time span needed for the
same increase may only be five years.

Speaking in present terms, about 70% of the industrial thermal-pollution
load is attributed to the steam electric-power industry, which discharges approximately 50 trillion gal./yr. of heated water. By 1980, it is estimated that
the power industry will use one-fifth of the total fresh water runoff in the U.S.
for cooling, thus, unless it is controlled, releasing 100 trillion gal./yr. of heated
water.

Future demands not only call for more generating plants, but also for
larger units that will release even more heat. At present, about 81% of the
electricity generated comes from thermal (fossil-fuel and nuclear) plants, and
most of the balance is hydroelectric power. But there are relatively few hydro
sites remaining that can be economically developed, and some of these cannot
be touched because of scenic and other aesthetic reasons.

On this observation, one industry source estimates that by the end of the
century, thermal power will represent 92% of all generating capacity in the
U.S. The booming nuclear-power industry will take a large share of this total
(see charts, page 427).

Present nuclear plants, which operate with water reactors, reject about
50% more heat than do fossil plants of equivalent capacity.

All told, the future will probably see a mix of water-nuclear, fossil and breeder-nuclear plants rejecting nine times more heat by the year 2,000 than they do now.

What should be done about it? The government has already anticipated the problem and acted accordingly by setting standards. Utilities now must find economic ways to cool waste water, or develop new ways of utilizing the valuable heat it contains.

How Much Harm?

With all this action in the works, one question keeps fueling the controversy. How much harm does hot water really cause? Senator Muskie himself, at the opening remarks of his Senate subcommittee hearings, said "there is a lack of specific information required to determine the ecological effects of a large volume discharge from particular installations."

In other words, it is difficult to tell to what extent aquatic life will be affected. This is partly because each individual case of heat rejection by utilities is framed by its own set of variables. For example, when a power plant returns condenser cooling water to a river, existing ecological conditions determine what part or parts of the river are affected, how high the temperature rises, and how long the effect lasts.

The rate at which the added heat is dissipated to the environment depends partly on conductive and evaporative heat losses from the water surface. The magnitude of these losses, in turn, is related to the temperature of the water

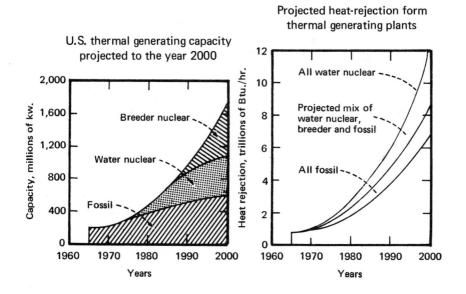

surface, the wind speed over the surface, and the turbulence of the body of water. Thus, the rate of heat transfer varies greatly with meteorological conditions, and with certain hydrodynamic characteristics of the water body.

Even if the river in question is substantially heated for an appreciable length of time, the precise effects on aquatic life are still unknown. Teams of ecologists are now studying these harmful aspects of thermal pollution, and Interior's Bureau of Commercial Fisheries has an unpublished report that points to the dangerous conditions for cold-water fish in the Columbia River system. The latter waterway, which forms the boundary between the states of Washington and Oregon for 200 miles, has been mentioned as the site for as many as 38 nuclear plants.

Thermal-Effects Probe

Interior has begun a $600,000 study on the effects of thermal inputs on the biological life of the Columbia River. Results of the two-year study, says Secretary of Interior Stewart W. Udall, "are certain to have a major bearing on the whole unsettled question of thermal effects."

The new study, to be directed by the Federal Water Pollution Control Administration's Northwest Regional Office, will have as one of its first jobs that of determining the relationship between nitrogen levels and temperature, and the effects of the combination on fish life.

The investigation, to take in the entire Columbia, will expand work already being done on part of the river. Work under way includes development of a mathematical model to forecast temperature changes brought about by various heat inputs.

An immediate aim of the study, says Interior, is to resolve inconsistencies in temperature standards contained in the water quality standards of the states of Washington and Oregon. It is hoped by Interior that a scientific basis can be found for determining how much variation in stream temperature above normal should be permitted.

What to do?

Utilities, too, are mobilizing their resources. Consolidated Edison Co. (New York) is making hydraulic model studies and aerial infrared surveys of the Hudson River in the vicinity of Indian Point, where additional nuclear units will be built. And a group of utility companies in Texas is funding a 3-1/2-yr. research project at Texas A&M University. The $200,000 study will determine the effects of heated discharges on fish in three of the state's fresh-water reservoirs.

Compliance with proposed standards, say the utilities, will seriously limit the number of sites available for power generating stations. But it is probably the staggering cost of complying without changing the power-plant's steam cycle that worries the industry most.

The use of reservoirs or mammoth cooling towers could alleviate the heat problem, but this is an expensive remedy. It could add up to a major share of the $1.8 billion that Federal Water Pollution Control Administration (FWPCA) believes industry will have to spend on thermal pollution control in the next five years. It is estimated that operating costs for these facilities may increase the cost of electricity from 1 to 5%.

Reuse of the now wasted heat is another possibility, and one that could eventually generate new heat-transfer technology. In this area, the chemical process industries (CPI) which rarely uses once-through cooling (one reason the CPI is considered a minor contributor to thermal pollution) has developed considerable know-how over the years.

There is also the possibility of looking into the beneficial effects of heated waste. Hot water can keep waterways open in winter, even benefit some fish and aquatic plants. The Long Island Lighting Co. is reportedly using a portion of its waste heat to assure an oyster hatchery an even water-temperature; this is said to lengthen the oyster's spawning period. And the Washington State Water Research Center is seeking federal approval and funding for a $3.7-million study to determine if thermally polluted water from AEC's Hanford operations at Richland, Wash. can be used for irrigation (*Chem. Eng.,* Feb. 12, p. 52).

Cooling Towers

Most utilities, however, are looking into ways to cool the used water before discharging it. One alternative is to build a cooling pond or reservoir. But the sheer size of future power plants, particularly those using nuclear power, seems to preclude this approach. Niagara Mohawk Power Co.'s chief nuclear engineer, Howard D. Philipp, estimates that one acre of pond surface is needed for every 1,000 kw., which means that thousands of acres would be required for plants in the 0.5 to 1-million-kw. range, such as many of the projected nuclear plants.

Much attention is being focused on giant (upwards of 400 ft. high) natural-draft cooling towers. This is apparently the only type of tower that can handle the enormous volumes of water needed by big power plants. Here, too, the cost is something to consider. Tor Kolflat of Sargent & Lundy (Chicago, Ill.) gives an additional-investment figure of $11/kw. for least expensive model of such natural-draft towers. This means that for a 1-million-kw. plant, which costs from $100 million to $130 million, an additional outlay of $11 million must be set aside for cooling towers.

A natural-draft tower is essentially an empty shell in the form of a chimney, which has the primary function of creating a draft. There is packing at the bottom, through which the water trickles, drops or flows in a predetermined manner so as to give up a portion of its heat to the air stream flowing past. The distinctive hyperbolic shape of the shell gives greater strength, and closely matches the pattern of air flow through the unit.

Because of their cost, natural-draft towers have never been popular in the U.S. The first U.S. one was built in the early 1960's, and only 19 more have been erected since then, all by utility companies. In Europe, however, they have been a common sight for the last 50 yr. Lower European costs for labor and reinforced concrete (the preferred construction material), and higher costs for land (hyperbolic towers take up less space than conventional units) explain the European's preference.

U.S. usage is bound to change as nuclear power generation gains momentum, according to W. J. Jones of Hamon-Cottrell, Inc. (Bound Brook, N.J.). Hamon-Cottrell is now building three hyperbolic towers for the Paradise generating plant of the Tennessee Valley Authority near Drakesboro, Ky. One of these, 437 ft. high, is considered the world's biggest natural-draft tower.

However, even large, efficient towers are expected to provide only a partial answer to thermal pollution problems. Some construction difficulties may be encountered in the U.S., e.g., hyperbolic units require a soil bearing capacity of about 4,000 lb./sq.ft., which is seldom found in certain areas such as the Gulf Coast. And because they handle large quantities of water, hyperbolic towers may create fog problems if placed close to highly traveled areas, thus creating traffic hazards.

Who Should Regulate?

The question of who has authority to regulate pollution was the central theme of Senator Muskie's subcommittee hearings. Target of the committee was nuclear power and thus, indirectly, the Atomic Energy Commission (AEC), which grants all licenses for nuclear power reactors. Although AEC may now be changing its original position, it has always contended that it had no jurisdiction over the heated effluents released by its licensees.

Some observers feel the AEC would reverse its views if it received some guarantee that there would also be control over fossil-fuel plants. But Senator Muskie believes the AEC is responsible for thermal discharges anyway under the provisions contained in executive Order 11288 (July 1966), which say that federal agencies should exercise responsibility over water pollution caused by their licensees.

The hearings, which attracted representatives from state agencies, engineering firms, private utilities, and scientists, were triggered by an AEC permit allowing a utility to build a 540-mw. nuclear plant on the Connecticut River, at Vernon, Vt. The states of Vermont, New Hampshire and Massachusetts protested the permit on the grounds that AEC had excluded requirements for thermal-pollution control.

Although AEC stated that its permit clearly left the door open for possible future changes in the nuclear plant to meet water quality standards, it took some interstate wrangling before Vermont Yankee Nuclear Corp., the builder, decided to earmark $6 million for the construction of cooling towers. Cost of these units must be added to the total capital investment figure of $100 million, and their operation, says Vermont Yankee, will boost total operating costs by $900,000/yr.

Results of the hearings could serve as guidelines for other areas where groups of nuclear installations are being planned. There is already talk about: a nuclear unit to be built in Florida, which will discharge heated water into Biscayne Bay; the proposed increase in Lake Michigan nuclear power plants from one to six; thermal pollution in the coastal waters off Hawaii; and the situation developing along the Columbia River.

The Garbage Explosion

Charles A. Schweighauser

The accumulation of solid waste in the United States is reaching alarming dimensions. Each person throws away more than half again as much waste as he discarded fifty years ago. A larger and more affluent population, buying an increasing quantity of goods designed to be discarded after temporary use, produces a gigantic disposal problem. Each one of us in a year throws away 188 pounds of paper, 250 metal cans, 135 bottles and jars, 338 caps and crowns, and $2.50 worth of miscellaneous packaging. And every year we amass 2 per cent more refuse which, coupled with a 2 per cent annual population growth, indicates a 4 per cent annual growth in the solid waste disposal problem.

In 1920, the daily per capita disposal was somewhat less than 3 pounds; in 1965 it was 4.5 pounds, not including industrial solid wastes, which account for an equal amount. In 1920 the citizens of this country were throwing away 100 billion pounds per year; today the amount is more than 720 billion pounds per year—not including 6 trillion pounds of mineral and agricultural solid wastes. By 1985, household wastes alone will amount to an estimated 1.25 trillion pounds per year.

The trend is illustrated by the history of glass containers. The first "no deposit, no return" beer bottle was made in 1938; in 1958 more than 1 billion bottles were made, and in 1965, nearly 5 billion bottles were distributed. The throwaway soft-drink bottle production was more than 1 billion. By 1970, the estimated combined beer and soft-drink use will exceed 12 billion nonreturnable bottles. That's 33 million bottles a day.

Nearly all major urban areas have run out of suitably inexpensive land for solid waste disposal, and are forced to dispose of solid waste by either long-distance removal or incineration. Incineration is more expensive than land disposal because of smoke pollution laws, and incinerator residue must be disposed of somewhere on the land.

The traditional method of disposing of solid waste was to put it either on the land, under the land, or down the side of a bank, where it decayed and was covered by vegetation. Very little thought was given to the pollution that resulted from using the land as a dump. Water, air and visual pollution weren't noticeable because their effects weren't very large. With an increasing population producing proportionately more waste, the physical insult to the land and to human sensibilities can no longer be tolerated. Urban areas must look for new methods and procedures to handle solid waste, as traditional techniques become unsatisfactory.

Few successful attempts have been made to re-use the paper, glass, plastics, rubber, rags and garbage that make up most of our domestic solid waste. About 35 per cent of total paper production, and about 10 per cent of plastics are recycled; glass not at all. The re-use of some metals is higher, as most major nonferrous metals can be economically salvaged. Copper re-use accounted for about 40 per cent of the supply in the United States in 1963, discarded lead was recovered at a rate that was more than double that produced from domestic mines, and scrap aluminum accounted for about 25 per cent of the total supply in the same year. Recovered scrap iron and steel currently account for about 50 per cent of total production. The recovery of rubber for chemicals, rubber and fibers is beginning to increase, and now stands at about 15 per cent.

The nearly 180 million tons of annual municipal refuse are estimated to contain ferrous and nonferrous metals valued at more than $1 billion. Each ton of residue from incineration contains 500 pounds of iron and 50 pounds of aluminum, copper, lead, tin and zinc. Fly ash from incinerators weighs about 20 pounds for every ton of refuse incinerated, and contains enough silver and gold to be comparable to a normal mine assay in the West.

Our solid waste disposal problems would be much worse if some materials were not recycled. A great deal more recycling could be done, but rising labor costs, uncertain markets, synthetics, and the mixing of refuse all make re-use costly and difficult.

Our industry is organized to use continuous input of new, rather than recovered, materials, and is sustained by constantly increasing consumer affluence and demands, built-in obsolescence, self-service merchandising, and competitive enterprise. We collect sometimes widely scattered resources, process and distribute them. But the responsibility of the private enterprise manufacturers stops at the shipping door. Neither distributors nor the retailers ever took responsibility for the disposal of the material they distribute (returnable bottles were an exception). Goods are used and then discarded; there are no consumers in the literal sense of the word. Responsibility thus passes from producer to user to a local government disposal agency.

So far it has been economically more feasible to build a new product out of new resources, because our industrial systems are so constituted, than to recover and re-use old products and their parts. Re-use systems could work in one of two ways: the material to be recycled could be collected by the producer, or the public agency or private individual keeps responsibility for returning the product to the producer/manufacturer (reverse distribution). The former system has worked, but in a rather disorganized and unsystematic manner; missing are economic incentives to encourage an efficient re-use system. Compounding the problem is our resistance to investment of money in an item that we will never use again. Solid waste disposal, therefore, has a low priority status with the general public and its agencies.

Recycling of garbage presents some interesting and difficult situations. Before urbanization, domestic garbage was recycled through pigs, chickens and other farm animals. Piggeries still use domestic garbage, but on a lesser scale. Garbage must be separated from nonorganic solid waste, usually in the home, which causes problems. Most states require domestic garbage to be cooked before it is fed to swine in order to stop the trichinosis cycle, a requirement that is necessary for public health but that is also expensive and time consuming for the piggery operator.

Another recycling method for garbage is composting, a process that involves biochemical degradation of the organic material. This material must be separated from ferrous metal, usually by magnet, and from all other metals, glass, paper, rubber and plastics by hand. The remainder is shredded, put through a short aerobic period to increase bacterial action and to hasten decomposition, and then allowed to cure for several months. The final product is a soil conditioner.

The product is of good quality, but the cost of producing it is greater than that of other types of soil conditioners. The cost of compost material from other sources is higher in Europe than in the United States, and thus composting of solid waste is much more widely practiced there. Contribution to the cost factor is labor (including pickers), rather expensive equipment, time for the product to mature, and the relatively low percentage of organic material in the total refuse. About 30 per cent of the original material must still be disposed of by other means. Vermin are also hard to control in a compost operation.

A number of schemes have been proposed, and a few are being developed, using other kinds of technology. For example, containers that dissolve in water or by the action of soil acids and sunlight are being investigated. Organic solid waste, mixed with sewage sludge, may give a high-grade composting material, provided that the pathogens can be removed. This process may also be valuable in curtailing excessive use of nitrogen, and thereby slowing down the tendency toward entropy in lakes and waterways.

Home solid waste grinders, analogous to contemporary garbage grinders, have been proposed. These larger and more rugged units might handle objects of paper, glass, plastics and light metal up to a cubic foot or more in size. After grinding, the material would be disposed of through sewer systems.

The Japanese have built large compactors to reduce solid waste to high-density blocks, which are then encased in an asphalt sealer and used for building foundations and other construction purposes. The city of Cleveland is also experimenting with a similar technique, using a mixture of solid waste, fly ash, dried sewage sludge, river and lake dredgings, and incinerator residue made into small, compact bricks for use as fill material to reclaim submerged lands adjacent to Lake Erie.

Paper products make up the largest percentage of most domestic solid waste. Since paper is nearly all cellulose, and since ruminants (cattle) can digest cellulose and turn it into protein, a number of public and private groups are experimenting with the use of paper products, supplemented with vitamins and minerals, as cattle feed. Today's newspaper may be tomorrow's steak.

Power production from incinerated solid waste has also been proposed. Milan, Italy, will soon be running all of its streetcars and subways by electricity generated by incinerated solid waste. Similar attempts in this country have been less successful, however, because of the inconsistent quality of the refuse to be incinerated and the production of electricity by other, hitherto less expensive means.

Other ideas have been put forward, such as adding hydrogen to wastepaper to make a high-grade fuel. Sanitary land-fill techniques along the sides of highways, power lines, sewer interceptors and other public rights of way have also been suggested. Such innovations have met with little enthusiasm, however, due to economic factors, entrenched procedures and lack of organized promotion campaigns.

The annual sum spent on solid waste collection and disposal in this country is large in comparison with other services. The annual cost of refuse collection and disposal, according to a recent federal document, is estimated at more than $4.5 billion, an amount that is exceeded only by schools and roads among public services. An estimated additional $750 million will have to be spent each year over the next five years to bring the collection and disposal systems of the nation to an acceptable health and aesthetic level.

The figures show only how much we have extracted from our natural environment, used briefly, and discarded permanently. They tell nothing about the supply of metals, pulp and other resources still left in the natural environment that we will use once and discard. Will we run out of resources before we choke on our own midden?

We can no longer treat solid waste as something to be shoved so far away that we'll never see it. There just isn't anywhere left to put it. We must learn to treat solid waste as a fact of life, and learn how to live with it, as we have learned to live with, or at least tolerate, sewers, automobiles, manufacturing plants and all other aspects of our modern age.

It should also be realized that our traditional attitude toward solid waste disposal is untenable in view of further environmental deterioration. This attitude can be characterized by the phrase "symptom-chasing." The disposal problem will not be solved by improving procedural techniques: it can only be postponed for a few years at best. The ultimate goal of refuse disposal—or any pollution control—is 100 per cent recycling of materials and energy. Advanced refuse recycling systems for urban areas are in the distant future; for the countryside they are even further away.

The problem can be viewed from another perspective. To control the effluent of our affluent society one must understand and rectify the causes in human behavior, and not just the symptoms. If a man has a brain tumor we would hardly expect his total treatment to be an aspirin. We must somehow convince the refuse makers—both the producers and the consumer-users (that means all of us)—that our overpackaged, overstuffed, throw-away life-style will bring long-term ecological and economic disaster. We probably will not much longer have the luxury of making decisions that will affect the ecosystem, as nearly all decisions ultimately do, based solely on economic and political expediency.

It is comparatively simple to write about solid waste practices and to recommend certain changes and improvements. It would be infinitely more difficult—but much more to the point—to study ways of drastically reducing solid waste effluent. And we must recognize that our solid waste situation, as well as our larger environment, can be improved only by self-imposed restraints.

Can Engineering Cope with the Debris of Affluence?

Anthony Hannavy

High-geared mass production pours out goods faster than dumping grounds can absorb the stuff that's made obsolete. So engineers look for better ways of rubbish disposal

Waste products of our affluent society are piling up in enormous heaps. Each year, the U.S. discards 6 million cars ... 50 billion food and beverage cans ... 25 billion bottles ... 65 billion metal and plastic jar and can caps ... in all, 152 million tons of solid wastes to be disposed of before the next load.

Around every big city, defacing the landscape and polluting the air, are reeking incinerators and smoldering dumps. The nation spends $3 billion of public money annually to dispose of garbage and refuse—without doing the job very well.

Big Assignment

The task is on a vast scale. How much is 152 million tons of rubbish? If you could spread it around the world at the rate of 100 tons per running mile, the band of rubbish would reach 60 times around the equator, with some to spare.

Solid wastes are being produced on an industrial-society scale but are being disposed of by methods from horse-and-buggy days. According to one expert, the U.S. ought to be spending at least $100 million a year on research toward better methods. It is spending about $4 million a year through the Depts. of Interior and Health, Education & Welfare. States and municipalities do little in the research field.

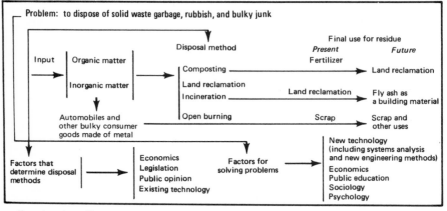

To analyze the problem, a systems engineer plots the factors like this, then weighs input and output considerations.

The disposal problem ranks low among public issues, even with rising indignation being expressed about air pollution and the ugliness of dumps. It won't be solved, experts say, until engineers are turned loose to design disposal systems with some tiny fraction of the effort they put into design of goods that will wind up on the scrap heap.

This won't happen until public opinion and economics dictate.

Compounding a Problem

Every so often, someone suggests that engineers should design products with greater awareness of the eventual need to dispose of them. But it's hard to get a product manager to take seriously an idea that leads, in extreme, to collapsible automobiles or soluble refrigerators.

In reality, many a product improvement adds to the disposal problem. As autos, for example, are made with a higher nonmetallic content, their scrap value diminishes. Aluminum cans for beer and soft drinks are proving more durable on the dump than the old tinned-steel ones. Beaches and picnic grounds are littered with pull tabs from beverage cans. On the street, cigarettes disintegrate but the filters don't.

As the American people buy more hardgoods, they scrap more. As they buy more food, more goes into the garbage pail. As the population continues to rise, there's more solid waste of every kind. By 1980, today's national total of 152 million tons of garbage and rubbish will leap to 260 million tons.

For practical reasons—social, political, and economic, as well as technological—the major effort of engineers will be devoted to finding better ways to dispose of solid wastes. Above all, the techniques of systems analysis must be applied.

How It's Done Now

Present disposal methods are aimed at compacting waste by burning or burying it after sorting out salvageable metal and rubble. Some garbage is still fed to pigs, though state health laws requiring sterilization of it by boiling have discouraged the practice.

The residue may be burned in the open at dumps or in incinerators, with or without recovery of heat energy as a byproduct. Or it may be buried in sanitary landfills, compacted and leveled by bulldozing and covered with earth. Microbial action of decay that causes the refuse to settle can be speeded and better controlled by composting, in the open or in mechanical systems.

Open Burning

Reducing the volume of refuse by open burning is a primitive but widely used method of disposal. It is so inefficient, so unsightly, and so hazardous to health that six states have banned this method and sanitation engineers strongly recommend that it be outlawed everywhere.

Fumes from the burning have caused paint to peel from houses near dumps. Researchers in New Orleans have found a relationship between asthma cases admitted to hospitals and the presence of a silica crystal in the patients' lungs that is associated with partial combustion. In dump fires, organic matter is incompletely burned, leaving a putrescent mass to attract rats and breed flies. Researchers have found that each cubic foot of uncovered garbage can produce 70,000 flies. And nearby water can be polluted by drainage from dump sites.

Landfill

Burying rubbish is among the oldest disposal methods, and it is the cheapest. Half of New York City's 15,000 tons of waste per day goes into filling marshes and extending shores. For sanitation, the refuse is covered with about 2 ft. of dirt. Nature does the rest as fungi and bacteria break down the organic matter. Typically, refuse is compacted ultimately to one-sixth of its original volume.

Landfill has two major drawbacks. It is vulnerable to many of the nuisances of dumps used for open burning: rats, flies, stench, and unsightliness even when earth is spread quickly. Also, urban areas are running out of sites for fill—a population of 1 million needs room for 5000 cu. yd. each day.

It is possible to create artificial hills of sanitary landfill, as in Frankfurt, Germany, where trees are planted on 25-deg. slopes. Last month, Du Page County, Ill., reported progress on a two-year project to create a 118-ft. ski slope on its flat countryside. The hill will be composed of 800,000 cu. yd. of garbage, covered with earth dug out to create a recreation lake.

Composting

Organic matter is also reduced by the microbial action in composting, but faster and under better control than in landfill. Biological activity is spurred by boosting the moisture content of the waste to a range of 40% to 70%.

Heat from this reaction sends the compost heap to 70 C. Then the pile is turned, aerating the refuse so the metabolic action remains aerobic. After several cycles, the temperature will no longer reach 70 C but will drop, indicating stability in the compost pile.

In the open, composting takes 6 to 10 weeks; in the mechanical system, three to six days. Researchers here and abroad are working on ways to cut this time. At New York Univ., Profs. Raoul Cardenas and Alan H. Molof have proposed a method of accelerating the decomposition of cellulosic waste.

The compost process reduces organic solids by 40% of their weight, through partial conversion to carbon dioxide and water. The residue is a humus material that has little value as fertilizer but some value as a soil conditioner. It is in demand in many countries, but not in the U.S.

Adding Value to Wastes

One application of systems analysis is producing ideas for adding value to solid wastes, putting the nation's refuse to more valuable uses than the mere filling of land on sites that are becoming scarce anyway. Incinerator ash, for example, has for years been used as an aggregate in concrete.

Japanese engineers have carried this idea a step further. In a giant press, they compress garbage under 2980 psi to reduce its volume by 75%. Then they coat the "rocks" of garbage with iron, so the blocks can be welded together for structural strength in building foundations and retaining walls.

Even without the iron cladding, the compressed garbage has value in civil engineering. The Japanese engineers claim that the compression destroys all microorganisms, eliminating any odor problem.

Health Safety

As part of its long-term Solid Wastes Program, the U.S. Public Health Service is financially supporting a new compost plant at Johnson City, Tenn. According to Ralph Black, deputy chief of the program for USPHS, one phase of the seven-year experiment is the adding of sewage sludge to compost for more value as fertilizer and soil conditioner.

The researchers are trying to determine the safety of such a process. The key question is how long pathogens, the agents that infect people, survive in compost. The Tennessee Valley Authority is operating the plant to show the worth of compost in improving the soil.

Adding Plant Foods

In another compost plant designed by Dr. Stephen Varro, vice-president of National Waste Conversion Corp., New York, large volumes of refuse are upgraded by the addition of plant foods. Varro has demonstrated the feasibility of his process in a prototype plant at Manhattan College, Riverdale, N.Y.

In Varro's design, incoming refuse passes a magnet that removes ferrous metals, and the rest is ground to a pulp. This pulp is fed into an aerobic digester, made up of several decks, with mechanical rakes to move the material from one deck to another in a continuous process. Moisture, temperature, and other conditions are automatically controlled. Compost emerging from the digester is dried, compacted, and granulated to farm fertilizer grade.

The whole process takes only 72 hr. Plant foods that are added during the treatment are chosen to provide slow-releasing nitrogen in the final fertilizer. Fast-releasing nitrogen imparts rapid initial growth to plants, but only slow-releasing nitrogen provides sustained growth.

Salvaging Value from Scrap

Bulky metal rubbish—automobiles, home appliances, and other junked machines—represent both a dumping problem and a salvage opportunity. Reclamation value is restricted largely by the economics of getting the junk to salvage centers and of extracting reusable metals.

According to the U.S. Bureau of Mines, well over 6 million cars are scrapped each year, representing a recoverable total of about 9 million tons of ferrous and nonferrous metals. But the $6-billion auto scrappage business calls for expensive, sophisticated machinery.

To the Bone Yard

An automobile usually goes into retirement at the used-parts dealer's yard, where it remains until all salable parts have been removed. This scavenging may take years. Then the car is either dumped or sold to a scrap processor.

At this point, the car's fate depends largely on how far away the scrap plant is. If it costs more than the scrap value to haul the carcass, obviously salvage doesn't pay. Yet scrap processors must concentrate their activities. Shredding equipment costs $1.5 million to $4 million; a crusher costs about $90,000.

At the salvage plant, a car is stripped of its engine and transmission. The body is flattened into a 12-in. slab, cut into sections and fed through a hammer mill, then baled, sheared, or shredded. Some shredders now in operation can process more than 350,000 auto bodies a year, and their finished product is excellent for making steel.

Crushers may prove inexpensive enough to be spotted around the country, easing the problem of transport to the scrap processors. They can compress a car into a block 2 x 6 ft. at the rate of 18 cars an hour. A truck that can carry only four or five cars can transport 28 such flattened bodies.

Sorting It Out

Engineers at Hugo Neu-Proler Co., Los Angeles, have had experience with shredding 300,000 to 350,000 cars per year. The auto body is shredded into fist-sized chunks. Magnetic separators pick out the ferrous parts, and the final product has an iron content of about 97%, suitable to be fed directly into the steel furnace. In the process, however, the company is left with 300 tons per year of nonmetallic residue. So it has its own dumping problem.

The company also notes that a lot of paperwork is involved in the scrapping of a car. Title transfers and other legal items make it impossible to send a car to the shredder in less than 7 days, and red tape often takes 40 days.

Smaller shredders, to cost around $300,000, are being designed to distribute this form of scrap reclamation more widely.

Melt-Down Technique

Hyman-Michaels Co., Chicago, plans to market a new high-speed Japanese auto salvager, made by Tezuka Kosan. A plant has been in operation at Funabashi, near Tokyo, for a year, crushing and cooking 50 cars per 8-hr. shift. Three more plants are due for installation in Japan this year, and three are in negotiation for sale in the U.S.

The system, called Carbecue, burns off nonmetallic material, melts non-ferrous metals out of the scrap, and bales the ferrous metal that remains. Each car passes through a furnace, three cars at a time. In 900 C, the cars rapidly heat up. First lead, then aluminum, and finally copper drip out for collection on a conveyer belt as their melting points are reached. All this takes only 8 minutes.

The remaining ferrous material is compressed into a bale with an iron content of 97% or 98%. Processing 50 Japanese cars a day yields the Funabashi plant about 30 metric tons of iron scrap. With bigger American cars, Tezuka engineers estimate a yield of a metric ton of iron and half a ton of nonferrous scrap from each automobile.

Upgrading Ore

The Bureau of Mines is spending about $5 million a year on research projects to turn scrap into useful material. In one project, low-grade taconite is upgraded by being roasted with auto scrap to produce a low-cost raw material for blast furnaces. In another, researchers at Illinois Institute of Technology are working on ways to convert residues of aluminum processing into building blocks.

Yet another scheme, being tried in Minnesota, uses scrap metal in a new process for converting nonmagnetic iron ore into magnetite for the iron and steel industry.

Burning It Up

Both in the U.S. and abroad, the trend seems to be toward incineration of solid wastes—with or without any attempt to recover any salvage or by-product value. The main point is to get rid of the mountains of refuse that every city piles up daily.

Incineration is the most expensive of the solid-waste disposal methods. But it appears to be, at the moment, the most practical. Moreover, its efficiency is being improved and its cost reduced by modern engineering.

One City's Solution

Two huge incinerators, costing more than $50 million apiece, are being designed right now for New York City. They will serve at central points, one in the south Bronx and the other on the former Navy Yard site in Brooklyn, to burn a major portion of the city's daily 15,000 tons of solid wastes. Each incinerator, fed by barge transport from several collection points, will reduce over 7000 tons of waste to 1500 tons of residue.

Incineration has its own problems, of course. One is air pollution, a problem with any kind of burning. The other is the disposal of the incinerator residue.

Maurice M. Feldman, chief engineer of New York's Sanitation Dept., is confident that designers can solve the air pollution problem. Electrostatic precipitators, sedimentation chambers, water sprays, and centrifugal separation of materials by density can control the emission of particulate matter.

The daily 3000 tons of residue from the two giant incinerators will be conveyed by barge to a landfill site some 20 mi. away. However, incinerator ash is not a wholly desirable material for fill.

In a study sponsored by the California Water Pollution Control Board, engineers found that water trickling through incinerator ash dumps will leach contaminating alkalies and salts from the dump. In one year, they measured from an acre of ash 12 in. thick: more than 1.5 tons of sodium and potassium, 0.9 ton of chloride, 0.24 ton of sulfate, and 4 tons of bicarbonate. This stuff runs off into nearby waters.

Water Pollution

As with open dumping and landfill with raw wastes, sites for burial of refuse must be available within economical barge or truck hauling distance. Such sites are becoming scarce around all major cities.

Moreover, when leaching from dumps pollutes streams and lakes, the public pays for the pollution in other ways: the need to clean a water supply, the loss of recreational facilities, an effect on health, the destruction of an industry that depends on water, such as fishing or even a paper mill.

Some city engineers say that when available land for dumping runs out, they will dump incinerator ash far at sea or in lakes. This type of thinking has already ruined the fishing industry in Lake Erie. Chemicals pumped into the lake produce algae that die, and the dead algae feed bacteria that absorb all or most of the oxygen in the water. Without oxygen, marine life can't survive.

Even the waste water from the fly-ash sprays and the quenching of ash residue in incinerators presents a pollution problem. Little is known of the characteristics of this water. Engineers at many incinerator plants in the U.S. are carrying out research to determine these characteristics and to work out ways to reclaim the water.

With efficient processes of treating waste water at the incinerators, the recycling of water promises substantial savings in operating costs of big incinerator plants.

Byproduct Power

Some designers suggest further economies by putting the incinerators' waste heat to work generating electric power, as is done in Europe. However, fuel costs are higher in Europe than in the U.S., so it pays to make steam for electric power a major aim in incinerator design. In the U.S., the steam will remain only a byproduct as long as natural fuels can be bought at reasonable prices.

Better Incinerator Design

Today's incinerators are being designed to take care of bulky refuse such as demolition lumber, furniture, home appliances, and junked automobiles. As a rising number of state and local authorities ban the dumping of such junk in open-burning sites, incinerators will have to cope with the disposal problem.

Dr. Elmer Kaiser of New York Univ. is now experimenting with an incinerator capable of handling bulky objects. He and Norman W. Wagner, supervisor of sanitation for Stamford, Conn., have also designed a multipurpose incinerator that can process 200 tons a day.

This incinerator, says Wagner, will handle many materials that an ordinary incinerator cannot. It will burn logs, stumps, brush, demolition lumber, furniture, tires, plastics, the nonmetallic parts of auto bodies, and nondegradable, highly volatile liquid wastes from industry. It will be fitted with centrifugal and electrostatic dust interceptors to guard against pollution of the air.

Not Enough Known

Design engineers are working under a handicap in planning more efficient incinerators. Despite the fact that refuse is burned by the millions of tons each year, no one can provide full and accurate chemical analysis of the combustion process. More work is now being done in this field.

However, research in turn is handicapped by a lack of uniformity in temperature and chemical measurement techniques. There has been little or no standardization of temperature-sensing instruments, for example, or of their placement in the combustion chamber.

Confusion results from this lack of standardization. Engineers at one plant may insist that a temperature range of 1600 to 1800 F is essential for best results, while engineers at an almost identical plant will say a range of 1800 to 2000 F must be maintained. And both groups may be right—the thermocouples that provide the readings may be at opposite ends of the combustion chambers.

Distillation

A variation on incineration of refuse is being tried both in the U.S. and overseas: the process called destructive distillation and gasification.

In this process, wastes are heated to between 1300 and 1500 F out of contact with air. After the moisture has been driven out, the organic matter is distilled to approximately equal percentages of gases, liquids, and solids.

The gases include carbon monoxide, carbon dioxide, methane and higher hydrocarbons, hydrogen, and nitrogen. The liquids range from alcohols to tars. The solids are a char made up of carbon and minerals. Total volume of the wastes is greatly reduced, easing the disposal problem.

Engineers are also working on gasification of wastes in a deep-bed gas producer supplied with air in less than half the normal combustion mixture. Plants are being run on a pilot scale to determine performance and operating cost. Experts say the residue should take up the same landfill space as the ash from ordinary incineration, but other factors may prove advantageous.

The Packaging Litter

The ultimate in design for disposal may be achieved in containers and wrappings. The packaging industry is working intensively on research in materials that will dissolve after being discarded.

In an age of processed foods, cans and plastic wrappers add enormously to the dumping problem. Unlike paper, plastic packaging doesn't quickly disintegrate in exposure at the dump. Unlike tinned-steel cans, aluminum containers don't break down in months.

Last year, the food industry spent more than $34 billion for plastic film to wrap foodstuffs. According to the President's Science Advisory Committee, which has studied the waste-disposal problem, each American every year throws away 135 bottles, 250 cans, and 340 metal or plastic caps.

To meet the disposal problem head-on, Dow Chemical's Packaging Sales Dept. is trying to perfect a bottle that will disintegrate after it has been emptied, and the U.S. Dept. of Agriculture is studying edible food wrappers. The packaging industry has formed a packaging materials research council to coordinate approaches to solving the disposal problem before it arises.

The Latest Ideas

The Agriculture Dept.'s novel approach to food packaging is to make the wrapper part of the food. Scientists are experimenting with a sprayed-on package made of a mixture of glyceryl triacetate and vegetable oils. Such a coating, they believe, will preserve the food and can either be washed off before the food is cooked or else eaten after it has melted during the cooking.

To reduce the litter of millions of long-lasting aluminum cans, Reynolds Metals Co. has launched a reclamation program on a pilot basis in Florida. It has set up collection points at several service stations in Miami Beach, and it offers 1/2 cent for each empty aluminum can turned in. The cans are collected and processed for reuse by secondary aluminum producers. The program is too new for significant results yet.

Sanitation engineers say a high percentage of packaging material is being reclaimed profitably. They also say plastics, especially the heavier materials, require special attention at disposal plants. Incineration gets rid of unsalvage-able plastics, but only if combustion is regulated for that purpose. Metropoli-tan Waste Conversion Corp. of Wheaton, Ill., is building a special plastics-burning plant in Florida due to go into operation shortly.

In the Broad Picture

Far from being a problem primarily for sanitation engineers, waste dis-posal today enlists the attention of a wide range of engineers, scientists, public health officials, city planners, and company managers. The lag in action on the problem is not traceable to a lack of interest but to delay in public and professional recognition of the stature of the problem and the complexity of its solutions.

"Sanitation" still doesn't have the allure of "aerospace" or "solid-state electronics." Yet the trend of public pressure toward the alleviation of air pollution (*PE*—Dec. 19 '66, p 32) and water pollution dictates as serious and as sophisticated an effort as is being made in those glamour fields.

Meanwhile, the politics and the economics of waste disposal will have to catch up with technology.

Criteria

According to William L. Plumb, a New York industrial designer, designers of all sorts of products will have to become more aware of what happens to a product after it has served its use. But company management will have to decide how much attention to this factor is economical.

Most packaging engineers, for example, feel that wrappings, cans, and bottles could be manufactured today or tomorrow so as to disintegrate after being emptied and discarded. But without insistent public demand, would it be economical?

Similarly, communities consider short-term economics in choosing methods of waste disposal: open burning, landfill, composting, or incineration. Yet, from the viewpoint of total environment, a different method may be demonstrably more economical. Again the question is: What does the public insist on?

Hidden Costs

For example, open burning may be selected as the least expensive method of dealing with a city's solid wastes. Engineers and public health officials are becoming aware of hidden costs that, in the long run, may make open burning one of the most expensive methods.

Energy in the form of heat is lost into the atmosphere. Plants and trees are killed by fumes—and the public is increasingly aware of esthetic values. Residents of nearby homes may have to paint their houses oftener and go to the doctor oftener for respiratory ailments. Leaching from the dump may pollute a stream or lake that is someone else's water supply. Thus, what someone saves by open burning, someone else has to pay for, perhaps twice over, in the long run.

Similarly, sanitary landfill ranks next to open burning as a cheap disposal method, provided sites are reasonably near the collecting points. Yet composting or incineration do a better job of reducing the volume of solid wastes and of minimizing the nuisances.

Applying Systems Analysis

In this country, composting is less economical than abroad. As a soil conditioner, even with additives as a fertilizer, the product of composting cannot compete with chemically produced fertilizers. But as land around cities gets scarcer, compost may be in greater demand in reclamation of marshes and shores, and its economics may become more attractive.

In the government's Solid Wastes Program and in other approaches to the disposal problem, systems analysis is being applied to an ever greater extent. Local community problems are being tied in with a national view of how they might be solved. The interests of other segments of the public are being brought into consideration.

Rail Haul

The American Public Works Assn., for example, is making a three-year study of rail haul as an economical means of transporting refuse to landfill areas. The association believes the railroads can give waste disposal a mass-production scope. Karl Wolf, of its staff, points out that one railroad car can carry 60 to 100 tons of refuse. Unit trains of 100 cars could be made up to shuttle wastes to disposal sites such as abandoned mines and quarries hundreds of miles away.

The study includes materials-handling equipment at loading points and disposal sites. One idea is to use specially designed pumps to remove refuse from the cars.

Seeking the Ideal

Ultimately, the goal is to recycle the processed solid wastes. Systems analysis is being used (1) to determine the best solution for each city's waste problems and (2) to seek engineering solutions that will provide the best opportunity for reuse of wastes.

Engineers who use systems analysis point out, however, that an ideal engineering solution may not be acceptable to the public, for economic, social, or political reasons. So they include as inputs for their models such considerations as public health, land-use patterns, technology, economic costs and benefits, and public acceptance.

As one city sanitation engineer sums up the problem: Most people think of economics solely in terms of saving money, of minimizing direct costs. They should be thinking of making the best use of their community's resources. Then the choice would be clear when it comes to a question of paying more in taxes for better waste-disposal methods or of impairing the community's health and attractiveness.

The Disposal Crisis:
Our Effluent Society

Eric B. Outwater

From now to the turn of the century the five boroughs of New York City will generate enough bottles, cans, paper and other solid waste to cover Central Park with a plateau the height of the Empire State Building.

No means for the disposal of most of it, short of shipping by train for deposit elsewhere, has as yet been conceived. It is no longer just a problem of "Keeping America Beautiful" by depositing trash in a litter basket; at a cost of from 60 to 90 cents to pick up one item of litter that has not been placed in the basket, we are now generating waste at the staggering rate of almost five pounds per day per person, and "where to put it" is growing into a crisis of monumental proportions.

Waste collection and disposal now cost the American public $6 billion per year, and the price keeps escalating. The word "disposal" is an unfortunate choice of words; "deposit" would be more apt, as more than 80 per cent of the solid waste in the United States goes into the ground pretty much in the form we discard it. The alternate methods of disposal are also far from ideal. The most common is to dump into an open pit, where combustion takes place which pollutes the air. Then leaching by rain pollutes the water table, and vermin infestation adds another unpleasant dimension to the problem.

The unsavory sight of such an area during its use and the uncertain qualities of its future use as a site for home building or as a recreational area leave only one redeeming feature—it's cheap. When waste is placed into the ground, there remains a residue from all methods of disposal that should be treated by a technique called sanitary land fill. By use of mechanical equipment the waste is not only compacted to reduce its volume, but is also daily covered by a layer of soil so that any burning is inhibited and exposure of the waste is at a minimum. By burying them we tend to forget the fact that many items of solid waste are virtually indestructible even after their deposit in the ground. A plastic bottle buried today will be in virtually the same condition ten or even a hundred centuries later.

Although most people believe that municipal incineration takes care of the problem, this is far from the case. The poor design of incinerators has made this technique not only costly but also a major contributor to air pollution. In New Jersey, of 32 incinerators built, only eleven are still operated. Uncontrolled incineration of the familiar open-burning or on-site type leaves bulky debris, and results in much self-defeating chaos as the 300,000 tons of solid matter that are discharged in the air of New York City every year, with the resultant vast cost of removal through cleaning and sweeping. As the composition of waste includes more and more fire-resistant items, such as glass and metal, these add to the residue that must be removed from the incinerator and deposited elsewhere.

A much-touted system of re-use and reclamation to recover part of the cost is known as "composting." This system does produce something of value to the farmer. But, unfortunately, the cost of production and the fact that our largest waste-generating areas are far from agricultural areas has made composting in most areas uneconomical when compared to its more efficient chemical equivalent.

Ideally, waste should be reclaimed and cycled into a re-use pattern such as the system built in the space capsules. Some effort has been made with the aluminum beer can, but economies of re-use and a trend in consumer demand has resulted in almost all consumer packaging being on a one-trip, single-service basis. Even with waste paper: as recently as fifteen years ago over 60 per cent was salvaged, but today, due to the economics of the paper industry, only 10 to 20 per cent is recycled. Regrettably, manual sorting is the only way known to separate plastic, aluminum and glass from waste. Ideally, manual sorting should be done in the home or at the point of initial disposal. To accomplish this, a vast consumer education program would be necessary, as well as a system of economic incentives for recycling.

The crisis in disposal is also complicated by the arbitrary political boundaries between communities. Where one area has high population density, another might have land available for disposal. Getting two communities to agree on a mutually acceptable arrangement has often proved impossible. Compounding the difficulty is the fact that physical boundaries of watersheds, airsheds and disposal areas are not likely to coincide.

This points up the competition for sites between the community need for living space, and the community need for a deposit area or disposal site. The inability of local governments and private groups to solve their own problems and the unwillingness of industry to acknowledge its role in the disposal crisis have raised the possibility of massive intervention of the Federal Government.

Federal Intervention, Or . . .

There are, of course, many sources of pollution other than solid waste that affect our environment. As the number of automobiles soars and the level of air pollution increases, the ability of the atmosphere to cleanse itself of exhaust falls farther and farther behind. Additionally, there is oil pollution in the ocean, pollution of water through dumping raw sewage and industrial waste, and even noise pollution.

It is interesting to note that many of the proponents of conservation talk in such phrases as "clean air," "pure water" and "litter free." These are, of course, no longer realistic goals for a conservationist—he can now only strive to keep the levels of waste and pollution below the community's threshold of non-acceptability. These levels of tolerance are closely related to what a community is willing to pay, in relation to what it is willing to put up with. This is normally the conditioning of people already hardened through continued exposure to varying degrees of pollution. A ghetto family is likely to be resigned to dirty streets, but draws the line at rat infestation.

In most areas we are willing to accept the fact that if things don't get worse, we can put up with them. But, of course, the situation everywhere will get worse, in geometrical proportion, in the near future. New York City is facing the problem of having no more deposit sites within seven years, at the present rate of generation, without resorting to using trains to ship the material to some distant site.

The hope has been often expressed that technological breakthroughs will come to alleviate the problem. These are usually a matter of incentives, which do not appear as yet to exist. A survey of the hundred largest manufacturers in the United States shows little or no involvement in the development of any methods to alleviate the waste problem, even that generated by the use of their own products. Rarely has clear-cut acknowledgement been made of their responsibility for adding to our pollution and waste-disposal problems. This is understandable, since the aims of the package manufacturer and the conservationist not only fail to coincide, but conflict. The conservationist would like to see packaging homogeneous, easily degradable and uniform in size. The packager strives to achieve marketability through more types of packages and combinations of materials that are the strongest possible.

What will it take to make industry assume its responsibility in this area? Perhaps a Ralph Nader or Rachel Carson will come along and act as a catalyst, or some new breakthrough in reduction and energy exchange may be developed. What is much more likely is belated governmental recognition of the problem, followed by severe and strict federal and state legislation and interference at every level.

The following excerpt from a bill introduced in the New York State Senate by Senator Seymour in February 1968 was one of at least seventeen acts introduced in as many state legislatures concerned with ultimate disposal of packaging materials: "To prescribe regulations governing the design and composition of disposable containers manufactured or sold within the state, to facilitate reclamation of waste materials contained therein or to aid the decomposition or other disposal thereof."

When we consider that pollution problems result not only from such tangible items as old cars and furniture, ships, packages, smoke and building rubble, but also from heat, noise, waste water, invisible odorless gas, and crop fertilizers, and affect health, crops and even the supply of oxygen in our atmosphere, we realize the staggering opportunities for the well-meaning government bureau to involve itself in every level of our existence.

... Voluntary Limitation?

Effective voluntary limitation of much waste generation would be possible if the economic advantages were fully understood and translated into tangible financial savings. Lower maintenance costs to home owners, the easing of discomfort and the reduction of doctor bills for asthma sufferers could be proven if, for instance, the sulfur content of fuel oil used for heating were lowered. The marketing convenience of one-way packages for beverages would not be as attractive if the disposal costs were reflected in tax savings by bottlers who furnish returnable deposit bottles.

Federal legislation enacted in a time of acute crisis, by necessity, is normally overly restrictive when applied nationally, regardless of degree of the crisis in a particular area. Waste and pollution problems vary markedly by area. Local ordinances could reflect the needs of particular communities and would tend to be more equitable. To pass worthwhile local ordinances and to encourage community cooperation, there must be an accurate source of unbiased information. Until recently no public service organization existed on a national level that could supply the necessary unbiased and accurate information on which a local community could base its planning, and which could factually mediate between political groups.

A new non-profit organization has been organized called FORCE, the Foundation for Responsible Conservation of our Environment, which will be programming a data bank to give up-to-date answers to such questions as cost and efficiency of various disposal methods, as they apply to a particular area. Also under study will be the level of public tolerance for waste and pollution. More important, FORCE will provide information assistance, such as accurate data regarding the addition or deletion of various sources of waste and pollution on a national level. Thus, if a soft-drink bottler switches to a particular package, this fact will be programmed and related to its effect on the disposal facilities of a specified community in that bottler's distribution area, so that long range planning and local legislation can be geared to fact, as it applies to that area. This will eliminate guesswork, or the costly and lengthy undertaking of a private study by a consulting firm on an area-by-area basis.

It must be apparent that James Reston's recent statement in the *New York Times* that "the old optimistic illusion that we can do anything we want is giving way to doubt, even to a new pessimism" is again confirmed by our seeming plunge into a conflicting chaos of cause and effect in the constant changes in our human environment. We should take heart however from the fact that modern man can adjust and thrive in an atmosphere of environmental pollution, crowding, dietary deficiency, monotony and ugliness, at the same time heeding the warning of René Dubos in his book *So Human an Animal* that "all too often the wisdom of the body is a short-sighted wisdom" and "evaluated over the entire lifespan, the homeostatic mechanisms through which adaption is achieved often fail in the long run because they result in delayed pathological effects."

We must hope that man's massive interventions will not, in William Faulkner's words, "turn the earth into a howling waste, from which he would be the first to vanish."

Can Anyone Run a City?

Gus Tyler

Can anyone run a city? For scores of candidates who have run for municipal office across the nation this week, the reply obviously is a rhetorical yes. But if we are to judge by the experiences of many mayors whose terms have brought nothing but failure and despair, the answer must be no. "Our association has had a tremendous casualty list in the past year," noted Terry D. Schrunk, mayor of Portland, Oregon, and president of the U.S. Conference of Mayors. "When we went home from Chicago in 1968, we had designated thirty-nine mayors to sit in places of leadership. . . . Today, nearly half of them are either out of office or going out . . . most of them by their own decision not to run again." Since that statement, two of the best mayors in the country —Jerome P. Cavanagh of Detroit and Richard C. Lee of New Haven—have chosen not to run again.

Why do mayors want out? Because, says Mayor Joseph M. Barr of Pittsburgh, "the problems are almost insurmountable. Any mayor who's not frustrated is not thinking." Thomas G. Currigan, former mayor of Denver, having chucked it all in mid-term, says he hopes "to heaven the cities are not ungovernable, [but] there are some frightening aspects that would lead one to at least think along these lines." The scholarly Mayor Arthur Naftalin of Minneapolis adds his testimony: "Increasingly, the central city is unable to meet its problems. The fragmentation of authority is such that there isn't much a city can decide anymore: it can't deal effectively with education or housing."

Above all, the city cannot handle race. Cavanagh, Naftalin, and Lee— dedicated liberal doers all—were riot victims. Mayor A. Sorensen of Omaha had to confess that after he'd "gone through three-and-a-half years in this racial business," he'd had it.

Although frictions over race relations often ignite urban explosives, the cities of America—and the world—are proving ungovernable even where they are ethnically homogeneous. Tokyo is in hara-kiri, though racially pure. U Thant, in a statement to the U.N.'s Economic and Social Council, presented the urban problem as world-wide: "In many countries the housing situation . . . verges on disaster. . . . Throughout the developing world, the city is failing badly."

What is the universal malady of cities? The disease is density. Where cities foresaw density and planned accordingly, the situation is bad but tolerable. Where exploding populations hit unready urban areas, they are in disaster. Where ethnic and political conflict add further disorder, the disease appears terminal.

Some naturalists, in the age of urban crisis, have begun to study density as a disease. Crowded rats grow bigger adrenals, pouring out their juices in fear and fury. Crammed cats go through a "Fascist" transformation, with a "despot" at the top, "pariahs" at the bottom, and a general malaise in the community, where the cats, according to P. Leyhausen, "seldom relax, they never look at ease, and there is continuous hissing, growling, and even fighting."

How dense are the cities? The seven out of every ten Americans who live in cities occupy only 1 per cent of the total land area of the country. In the central city the situation is tighter, and in the inner core it is tightest. If we all lived as crushed as the blacks in Harlem, the total population of America could be squeezed into three of the five boroughs of New York City.

This density is, in part, a product of total population explosion. At some point the whole Earth will be as crowded as Harlem—or worse—unless we control births. But, right now, our deformity is due less to overall population than to the lopsided way in which we grow. In the 1950s, half of all the counties in the U.S. actually lost population; in the 1960s, four states lost population. Where did these people go? Into cities and metropolitan states. By the year 2000, we will have an additional 100 million Americans, almost all of whom will end up in metropolitan areas.

The flow of the population from soil to city has been underway for more than a century, turning what was once a rural nation into an urban one by the early 1900s. Likewise, the flow from city to suburb has been underway for almost half a century. "We shall solve the city problem by leaving the city," advised Henry Ford in a high-minded blurb for his flivver. But, in the past decade, the flow has become a flood. Modern know-how dispossessed millions of farmers, setting in motion a mass migration of ten million Americans from rural, often backward, heavily black and Southern counties to the cities. They carried with them all the upset of the uprooted, with its inherent ethnic and economic conflict. American cities, like Roman civilization, were hit by tidal waves of modern Vandals. Under the impact of this new rural-push/urban-pull, distressed city dwellers started to move—then to run—out. Hence, the newest demographic dynamic: urban-push and suburban-pull. In the 1940s, half the metropolitan increase was in the suburbs; in the 1950s, it was two-thirds; in the 1960s, the central cities stopped growing while the suburbs boomed.

Not only people left the central city; but jobs, too, thereby creating a whole new set of economic and logistic problems. Industrial plants (the traditional economic ladder for new ethnic populations) began to flee the city in search of space for factories with modern horizontal layouts. Between 1945 and 1965, 63 per cent of all new industrial building took place outside the core. At present, 75 to 80 per cent of new jobs in trade and industry are situated on the metropolitan fringe. In the New York metropolitan area from 1951 to 1965, 127,753 new jobs were located in the city while more than three times that number (387,873) were located in the suburbs. In the Philadelphia metropolis, the city *lost* 49,461 jobs, while the suburbs gained 215,296. For the blue-collar worker who could afford to move to the suburbs or who could commute (usually by car) there were jobs. For those who were stuck in the city, the alternatives were work in small competitive plants hungry for cheap labor and no work at all.

Ironically, the worthwhile jobs that did locate in the cities were precisely those most unsuited for people of the inner core, namely, white-collar clerical, administrative, and executive positions. These jobs located in high-rise office buildings with their vertical complexes of cubicles, drawing to them the more affluent employees who live in the outskirts and suburbs.

This disallocation of employment, calling for daily commuter migrations, has helped turn the automobile from a solution into a problem, as central cities have become stricken with auto-immobility; in midtown New York, the vehicular pace has been reduced from 11.5 mph in 1907 to 6 mph in 1963. To break the traffic jam, cities have built highways, garages, and parking lots that eat up valuable (once taxable) space in their busy downtowns: 55 per cent of the land in central Los Angeles, 50 per cent in Atlanta, 40 per cent in Boston, 30 per cent in Denver. All these "improvements," however, encourage more cars to come and go, leaving the central city poorer, not better.

Autos produce auto-intoxication: poisoning of the air. While the car is not the only offender (industry causes about 18 per cent of pollution; electric generators, 12 per cent; space heaters, 6 per cent; refuse disposal, 2.5 per cent), it is the main menace spewing forth 60 per cent of all the atmospheric filth. In 1966, a temperature inversion in New York City—fatefully coinciding with a national conference on air pollution—brought on eighty deaths. In 1952, in London, 4,000 people died during a similar atmospheric phenomenon.

The auto also helped to kill mass transit, the rational solution to the commuter problem. The auto drained railroads of passengers; to make up the loss, the railroads boosted fares; as fares went up, more passengers turned to autos; faced with bankruptcy, lines fell behind in upkeep, driving passengers to anger and more autos. Between 1950 and 1963, a dozen lines quit the passenger business; of the 500 intercity trains still in operation, fifty have applied to the ICC for discontinuance. Meanwhile, many treat their passengers as if they were freight.

Regional planners saw this coming two generations ago and proposed networks of mass transportation. But the auto put together its own lobby to decide otherwise: auto manufacturers, oil companies, road builders, and politicians who depend heavily on the construction industry for campaign contributions.

The auto is even failing in its traditional weekend role as the means to get away. On a hot August weekend this year, Jones Beach had to close down for a full hour, because 60,000 cars tried to get into parking lots with a capacity of 24,000. The cars moved on to the Robert Moses State Park and so jammed the 6,000-car lot there as to force a two-hour shutdown.

Overcrowding of the recreation spots is due not only to more people with more cars but to the pollution of waters by the dumping of garbage—another by-product of metropolitan density.

Viewed in the overall, our larger metropolises with their urban and suburban areas are repeating the gloomy evolution of our larger cities. When Greater New York was composed of Manhattan (then New York) and the four surrounding boroughs, the idea was to establish a balanced city: a crowded center surrounded by villages and farms. In the end, all New York became citified. Likewise, the entire metropolitan area is becoming urbanized with the suburbanite increasingly caught up in the city tangle.

The flow from city to suburb does not, surprisingly, relieve crowding within the central city, even in those cases where the city population is no longer growing. The same number of people—especially in the poor areas— have fewer places to live. In recent years, some 12,000 buildings that once housed about 60,000 families in New York City have been abandoned, with tenants being dispossessed by derelicts and rats; 3,000 more buildings are expected to be abandoned this year. The story of these buildings, in a city such as New York, reads like a Kafkaesque comedy. For the city to tear down even one of these menaces involves two to four years of red tape; to get possession of the land takes another two to four years. Meanwhile, the wrecks are inhabited by human wrecks preparing their meals over Sterno cans that regularly set fire to the buildings. By law, the fire department is then charged with the responsibility of risking men's lives to put out the fire, which they usually can do. However, when the flames get out of hand, other worthy buildings are gutted, leaving whole blocks of charred skeletons—victims of the quiet riot.

Other dwellings are being torn down by private builders to make way for high-rise luxury apartments and commercial structures. *Public* action has destroyed more housing than has been built in all federally aided programs. As a result, the crowded are more crowded than ever. Rehabilitation instead of renewal doesn't work. New York City tried it only to discover that rehabilitation costs $38 a square foot—a little *more* than new luxury housing.

The result of all this housing decay and destruction (plus FHA money to encourage more affluent whites to move to the suburbs) has been, says the National Commission on Urban Problems, "to intensify racial and economic stratification of America's urban areas."

While ghetto cores turn into ghost towns, the ghetto fringes flare out. The crime that oozes through the sores of the diseased slum chases away old neighbors, a few of whom can make it to the suburbs; the rest seek refuge in the "urban villages" of the low-income whites. Cities become denser and tenser than they were. In the process, these populous centers of civilization become —like Europe during the Dark Ages—the bloody soil on which armed towns wage their inevitable wars over a street, a building, a hole in the wall. Amid this troubled terrain, the free-lance criminal adds to the anarchy.

All these problems (plus welfare, schooling, and militant unions of municipal employees) hit the mayors at a time when, according to the National Commission on Urban Problems, "there is a crisis of urban government finance . . . rooted in conditions that will not disappear but threaten to grow and spread rapidly." The "roots" of the "crisis"? The mayor starts with a historic heavy debt burden. His power to tax and borrow is often tethered by a rural-minded state legislature. He has lost many of the city's wealthy payers to the suburbs. His levies on property (small homes) and sales are prodding Mr. Middle to a tax revolt. The bigger (richer) the city is, the worse off it is. As population increases, per capita cost of running a city goes up—not down: density makes for frictions that demand expensive social lubricants. Municipalities of 100,000 to 299,000 spend $14.60 per person on police; those of 300,000 to 490,000 spend $18.33; and those of 500,000 to one million spend $21.88. New York City spends $39.83. On hospitalization, the first two categories spend $5 to $8 per person; those over 500,000 spend $12.54; New York spends $55.19.

Expanding the economy of a city does not solve the problem; it makes it worse. Several scholarly studies have come up with this piece of empiric pessimism: if the gross income of a city goes up 100 per cent, revenue rises only 90 per cent, and expenditures rise 110 per cent. Consequently, when a city's economy grows, the city's budget is in a worse fix than before. This diseconomy of bigness and richness applies even when cities merely limit themselves to prior levels of services. But cities, unable to cling to this inadequate past, have had to step up services to meet the rising expectations of city dwellers.

The easy out for a mayor is to demand that the federal coffers take over cost or hand over money. But is that the real answer? The federal income tax as presently levied falls most heavily on an already embittered middle class— our alienated majority. Unable to push this group any harder and unwilling to "soak the rich," an administration, such as President Nixon's, comes up with revenue-sharing toothpicks with which to shore up mountains. Nixon has proposed half a billion for next year and $5-billion by 1975, while urban experts see a need for $20- to $50-billion each year for the next decade. A Senate committee headed by Senator Abe Ribicoff calls for a cool trillion.

But even if a trillion were forthcoming, it might be unable to do the job. To build, a city must rebuild: bulldoze buildings, redirect highways, clear for mass transportation, remake streets—a tough task. But even tougher, a city must bulldoze people who are rigidified in resistant economic and political enclaves. The total undertaking could be more difficult than resurrecting a Phoenix that was already nothing but a heap of ashes.

What powers does a mayor bring to these complex problems? Very few. Many cities have a weak mayor setup, making him little more than a figure-head. If he has power, he lacks money. If he has power and money, he must find real—not symbolic—solutions to problems in the context of a density that turns "successes" into failures. If a mayor can, miraculously, come up with comprehensive plans, they will have to include a region far greater than the central city where he reigns.

A mayor must try to do all this in an era of political retribalism, when communities are demanding more, not less, say over the governance of their little neighborhoods. In this hour, when regional government is needed to cope with the many problems of the metropolitan area as a unity, the popular mood is to break up and return power to those warring factions—racial, economic, religious, geographic—that have in numerous cases turned a city into a no man's land.

Is there then no hope? There is—if we putter less within present cities and start planning a national push-pull to decongest urban America. Our answer is not in new mayors but in new cities; not in urban renewal but in urban "newal," to use planner Charles Abrams's felicitous word.

We cannot juggle the 70 per cent of the American people around on 1 per cent of the land area to solve the urban mess. We are compelled to think in terms of new towns and new cities planned for placement and structure by public action with public funds. "All of the urbanologists agree," reported Time amidst the 1967 riot months, "that one of the most important ways of saving cities is simply to have more cities." The National Committee on Urban Growth Policy proposed this summer that the federal government embark on a program to create 110 new cities (100 having a population of 100,000, and ten even larger) over the next three decades. At an earlier time, the Advisory Commission on Intergovernmental Relations proposed a national policy on urban growth, to use our vast untouched stock of land to "increase, rather than diminish, Americans' choices of places and environments," to counteract our present "diseconomies of scale involved in continuing urban concentration, the locational mismatch of jobs and people, the connection between urban and rural poverty problems, and urban sprawl."

New towns would set up a new dynamic. In the central cities, deconges-
tion could lead to real urban renewal, starting with the clearing of the ghost
blocks where nobody lives and ending with open spaces or even some of those
dreamy "cities within a city." The new settlements could be proving grounds
for all those exciting ideas of city planners whose proposals have been frustrated by present structures—physical and political. "Obsolete practices such
as standard zoning, parking on the street, school bussing, on-street loading,
and highway clutter could all be planned out of a new city," notes William E.
Finley in the Urban Growth report. These new towns (cities) could bring jobs,
medicine, education, and culture to the ghost towns in rural America, located
in the counties that have lost population—and income—in the past decades.
Finally, a half-century project for new urban areas would pick up the slack in
employment when America, hopefully, runs out of wars to fight.

The cost would be great, but no greater than haphazard private developments that will pop up Topsy-like to accommodate the added 100 million
people who will crowd America by the year 2000. Right now we grow expen-
sively by horizontal or vertical accretion. We sprawl onto costly ground,
bought up by speculators and builders looking for a fast buck. Under a national
plan, the federal government could buy up a store of ground in removed places
at low cost or use present government lands. Where private developers reach
out for vertical space, they erect towers whose building costs go up geometri-
cally with every additional story. On the other hand, as city planners have been
pointing out for a couple of decades, "it has been proved over and over again
by such builders as Levitt, Burns, and Bohannon" that efficient mass production of low-risers "can and do produce better and cheaper houses." Cliff
dwellings cost more than split-levels.

The idea of new towns is not untested. "There is little precedent in this
country, but ample precedent abroad," notes the Committee on Urban
Growth. "Great Britain, France, the Netherlands, the Scandinavian countries
—all have taken a direct hand in land and population development in the face
of urbanization, and all can point to examples of orderly growth that contrast
sharply with the American metropolitan ooze." To the extent that the U.S. has
created new communities it has done so as by-products: Norris, Tennessee, was
built for TVA to house men working on a dam; Los Alamos, Oak Ridge, and
Hanford were built for the Atomic Energy Commission "to isolate its highly
secret operations."

What then is the obstacle to this new-cities idea? It runs contrary to the
traditional wisdom that (a) where cities are located, they should be located, and
(b) that the future ought to be left to private enterprise. Both thoughts are a
hangover from a hang-up with laissez faire, a Panglossian notion that what is,
is best.

The fact is, however, that past reasons for locating cities no longer hold
—at least, not to the same extent. Once cities grew up at rural crossroads; later

at the meeting of waters; still later at railroad junctions; then near sources of raw material. But today, as city planner Edgardo Contini testified before a Congressional committee, these reasons are obsolete. "Recent technological and transportation trends—synthesis rather than extraction of materials, atomic rather than hydroelectric or thermoelectric power, air rather than rail transportation—all tend to expand the opportunities for location of urban settlements." Despite this, the old cities, by sheer weight of existence, become a magnetic force drawing deadly densities.

Furthermore, concluded Mr. Contini and a host of others, "the scale of the new cities program is too overwhelming for private initiative alone to sustain, and its purposes and implications are too relevant to the country's future to be relinquished to the profit motive alone." The report of the Urban Growth Committee stresses the limited impact of new towns put up by private developers such as Columbia, Maryland and Reston, Virginia. "They are, and will be, in the first place, few in number, serving only a tiny fraction of total population growth. A new town is a 'patient' investment, requiring large outlays long before returns begin; it is thus a non-competitive investment in a tight money market. Land in town-size amounts is hard to find and assemble without public powers of eminent domain. Privately developed new towns, moreover, by definition must serve the market, which tends to fill them with housing for middle- to upper-income families rather than the poor."

The choice before America is really not between new cities and old. Population pressure will force outward expansion. But by present drift, this will be unplanned accretion—plotted for quick profit rather than public need. What is needed is national concern for the commonweal in the location and design of new cities; a kind of inner space program.

PART IV

21st Century

INTRODUCTION

There appears to be general agreement that, when attempting to predict what the future holds, there is a continuum from unprecedented disaster to potentialities never before dreamed of. The pessimist will choose that portion of the spectrum which portends disaster through such uncontrolled happenings as environmental decay or nuclear holocaust. The optimist contends that the world will continue to progress technologically, socially, culturally, and economically through continued advance. The eclectic believes that the future lies in the control and/or steering of technological progress.

The issues and quandaries presently being faced are illustrated in the writings of the authors of this section. The range is diverse and further questions emerge as allied thoughts for contemplation.

Through these readings it is apparent that the future appears to hold unlimited potential and that the lives of all will be touched, at least to some degree, by technology.

Science and its lusty offspring, technology, are more vigorous and dynamic than in any other previous era or cultural system. They have provided this age with more of its unique characteristics, both those with merit and those deteriorating society, than any other age.

The keys to the future are "change" and "adjust." In the near past man tended to think in terms of permanence and stability. However, if the current trends are examined, it will be realized that relatively few traditions and/or institutions exist in the same form they did only a few years ago. The United States and the world must adjust to this change if it is to receive the promising benefits of the future.

Possibly the future lies in the answer to a profound question raised by John Diebold of the Diebold Corporation—to decide what on earth you want to do, just because you can do it.

Must Technology and Humanity Conflict?

Joseph Wood Krutch

Advances in science and technology are no guarantee of the "good life." Perhaps a set of national criteria can help prevent further threats to human well-being.

The quarter century just past has seen advances in science and technology unprecedented in human history. Atomic fission, space travel and, most recently, the discoveries relating to the innermost secrets of life and heredity would hardly have been predicted even a generation ago except, perhaps, for some dim future as far ahead as thoughts can reach. They have come upon us almost unaware, and we hope, of course, that somehow good will be the result of all three. But even the most optimistic are bound to admit that each is also a potential threat. The growing question is: Does our nation need some central advisory body or criteria to judge whether advances in science and technology contribute to human welfare or create still other potential threats?

Atomic fission raises the possibility that civilization may be destroyed. A war of the worlds made possible by space travel is not so likely as the possibility that its appeal to the imagination may tempt us to spend the money, energy and brain power on it which might more profitably be employed in solving human problems here on earth. The threat from macromolecular biology may be more remote but is, ultimately, at least equally frightening. If we believe the claims of those in the best position to know, it promises ultimately the power to determine the intelligence, character and temperament of unborn generations. And that suggests the possibility of a totalitarianism more complete than we have previously dreamed of.

Up until now, there always seemed the possibility that human nature could not be completely controlled either for good or for ill. But what if human nature itself can be changed or abolished?

None of these threats will necessarily become a reality. And in the past, science has bestowed upon us enormous increases in comfort, health and affluence. Yet during the quarter century which gave us atomic fission, space travel and the achievements of biochemistry, some aspects of the human condition have deteriorated. Ours is much more conspicuously an age of anxiety than it was. Specifically, overpopulation, environmental pollution and epidemic violence are all new. They are the most pressing problems of today and are likely to continue to be for a long time to come despite atomic fission, trips to the moon and the sensational discoveries in biochemistry.

It is not necessary to believe that advances in science and technology are the cause of these deteriorations. But one has accompanied the other and the minimum conclusion to be drawn is that advances in science and technology alone are no guarantee of accompanying improvement in the human condition. What is new today is the fact that warnings are now beginning to be heard from within the very professions and organizations which have formerly been most likely to assume that progress is inevitable and that progress in technology inevitably means improvement of the human condition.

As Dr. Laurence Gould, former president of the American Association for the Advancement of Science, pointed out, most great civilizations of the past have succumbed to deterioration from within rather than to attacks from without (see *Bell Telephone Magazine,* January/February 1969).

Dr. Gould's thoughts caused the Institute of Life Insurance to take full-page advertisements in various publications to ask: "Could this happen to us? To our families? To our way of life? Could this happen to America the Beautiful? Well, look around. You can see signs of it at this very moment in every major city of this country. You can see it in the slums, in the jobless, in the crime rates, in our polluted air, in our fouled rivers and harbors and lakes. You can see it in our roads strangled with traffic. . . . We must all do something about it. While there is still time. Before our cities become unfit places in which to live."

Many Scientists Question Fundamental Assumptions

To be sure, there have always been mavericks (usually men of letters) who expressed doubts about "progress." And there was also the philosopher in Samuel Johnson's moral romance *Rasselas* who had invented a flying machine but refused to reveal its secret because, so he said, men should not be permitted to fly until after they had become virtuous. But despite such grumblers, the generally accepted assumptions ever since the sixteenth century had been that every increase in knowledge, power or technical ingenuity would in the end contribute to the improvement of the human condition; that the extent to which science and technology had developed was indeed a measure of the extent to which a good life was being led in any community; and finally, that, as the Marxists said, all changes in the human condition are simply the inevitable consequences of evolving technology. From these propositions, and especially from the last, there was a tendency to conclude that one need not concern oneself with any other aspect of a good life for the simple reason that the propositions will all develop as byproducts of the advancing knowledge, power and affluence.

Because technological changes have been so accelerated during the past quarter century, responsible scientists and technicians have begun to question fundamental assumptions. And they are suggesting that the time has come when we must realize that certain powers are dangerous, certain inventions are threats rather than promises, and that, in a word, we must begin to ask not simply *can* we do this or that but *should* we do this or that. We may not be able to wait for men to become virtuous before sending them to the moon, but we realize that sending them there will not necessarily make them more so.

Neither Anthony Weiner, formerly of the Massachusetts Institute of Technology, nor Herman Kahn, formerly of the Rand Corporation, are men whom we would expect to be enemies of science and technology, yet in their recent book *The Year 2000* they issue a solemn warning: "Practically all the major technological changes since the beginning of industrialization have resulted in unforeseen consequences. . . . Our very power over nature threatens to become itself a source of power that is out of control. . . . Choices are posed that are too large, too complex, too important and comprehensive to be safely left to fallible human beings."

New Drugs Put Godlike Powers in Medical Scientists' Hands

Some threats are indirect in the sense that they affect the environment which in turn affects the human being. Those created by biochemistry threaten directly human intelligence, personality and character. They put into man's hands godlike powers he himself is not sufficiently godlike to be trusted with. In April 1968 the University of California psychologist, Dr. David Krech, told a Senate subcommittee that within the next decade medical scientists will be able to exert a significant degree of control over man's mind.

Different drugs, said Dr. Krech, affect different kinds of intellectual activity. Thus an antibiotic called Puromycin prevents long-time, but not short-time, memories. Injected into an animal, it "permits it to put in an ambitious day's work although it will not build up a permanent body of experiences or memories or abilities." It might be used to produce, for instance, a body of subhuman but docile workers much like those who compose one of the biologically established social classes in Aldous Huxley's *Brave New World,* considered wild fantasy only 37 years ago.

What Humans Will Determine How New Powers are Used?

Dr. Krech's statement is cautious because it looks a mere decade ahead. His colleague at the University of California, Prof. Robert Sinsheimer, professor of biophysics, raises more alarming possibilities without setting a definite deadline: "Eventually we will surely come to the time when man will have the power to alter, specifically and consciously, his very genes. This will be a new event in the universe. No longer need nature wait for the chance mutation and the slow process of selection. Intelligence can be applied to evolution. How might we like to change his genes? Perhaps we would like to alter the uneasy balance of our emotions. Should we be less warlike, more self-confident, more serene? Perhaps. Perhaps we shall finally achieve these long-sought goals with techniques far superior to those with which we have had to make do for many centuries."

Is such a power too great to be trusted to "fallible human beings?" Will it be used to make us more or less warlike, self-confident and serene? The answer, no doubt, is that this will depend upon whose hands the power falls into. And as things now stand, we who already control so much have developed no way of determining into whose hands any of our new powers will pass.

In its simplest form, the question of proper use arises in connection with what has come to be called "the responsibility of the scientist." Is it his duty, before giving a discovery or an invention to the public, to ask the simple question, "Will this knowledge (or indeed *can* this knowledge) be well used?" What would have been the nearly unanimous reply until very recently was given a few years ago by Edward Teller in connection with atomic fission and its possible catastrophic effects: "I believe that we would be unfaithful to the tradition of western civilization if we were to shy away from exploring the limits of human achievement. It is our specific duty as scientists to explore and explain. Beyond that our responsibilities cannot be greater than those of any other citizen of our democratic society." At an opposite extreme, Robert Oppenheimer has confessed to the feeling that he has "known sin" as a result of his involvement in the creation of the bomb.

We Can Do More Than Hope!

One need not question the reasonableness of either one or the other of these attitudes to see that, justifiable or not, neither is really helpful. Neither washing the hands nor wringing the hands solves the problems created by the fact, now plainly evident for the first time in history, that we have begun to assume powers which need to be channeled and not simply turned loose upon the world to see what will happen. Knowledge is power, but it is not equally evident that power is always good. Science is knowledge; technology is doing. But it is no longer safe to say that whatever we can do we should do.

Prof. Sinsheimer is an optimist: "After two billion years, this is the end of the beginning. It would seem clear, to some achingly clear, that the world, the society, and the man of the future will be far different from that we know. . . . We must hope for the responsibility and the wisdom and the nobility of spirit to match this ultimate freedom."

There remains the question of whether we can indeed do nothing more than hope (and perhaps fear). Prof. Sinsheimer seems willing to propose no alternative, but neither Prof. Krech nor the Messrs. Weiner and Kahn are willing to leave to chance the use which will be made of powers still difficult for most of us to imagine. "The issues I have raised," said Prof. Krech, "are much too pervasive and too profound to permit the physician to scribble the necessary social policy on his prescription pad," and he proposes a national commission to determine how the new drugs shall be used. Messrs. Weiner and Kahn go even a step further than that. For the future they suggest that neither science nor technology shall pursue new discoveries or develop new applications of scientific knowledge without first investigating their possible consequences for evil as well as for good.

In saying that, they go far beyond the mere rejection of Dr. Teller's contention that the scientist should assume no responsibility for the use made of his discovery. They propose the attitude that the philosopher in *Rasselas* assumed when he believed that an invention should not be given to the world until men had become virtuous enough to make only good use of it. It is not likely that Dr. Johnson himself would regard his philosopher's proposal as practicable, and one begins to wonder if any other really is.

It is very well to suggest "a national council to evaluate how mind-influencing drugs shall be used" or to say with Messrs. Weiner and Kahn that the choices are too fateful "to be safely left to fallible human beings." But where are we to find infallible human beings to whom we may safely trust the control of the uncontrollable? A committee invested with the powers to determine just how virtuous mankind had become, how the human mind and character should be changed, or even what technological uses should be made of our increased power over nature would be possible only in a totalitarian society so absolute that its decrees would, in the end, probably be more stupid than those made today by pure chance. The brave new world would already be here.

Criteria for Decisions Needed

Obviously we will have to make do with something less fruitful than tight bureaucratic control of science and inventions. On the other hand, there is a desperate need for something more effective than now exists. We need some body or bodies which would at least suggest, influence and direct the ends toward which research and its applications are directed and encourage certain tendencies even if the bodies did not go so far as to forbid others.

The beginnings of the power to do at least this much already exist to a significant extent by virtue of the fact that a large proportion of all scientific and technological projects now being actively pursued owe their existence to grants from the national government, various foundations, or industry itself. To a considerable extent these institutions determine what enterprises in either pure or applied science will be undertaken. They must have certain criteria on which they make their decisions.

In the case of the government, the criteria are largely those relating to supposed military necessity. In the case of industry, they relate largely to the possibility of ultimate profits. But neither government nor industry follow rigidly these criteria. Both support projects which seem to be simply "in the public interest."

Central Group Could Advise on Threats of Technology

Nevertheless and insofar as one can judge, the enterprises which government, industry and the foundations support are selected on the basis of criteria usually not up to date. They are based upon the nineteenth century assumptions: all knowledge is good; even the most useless scientific fact may turn out to be important; every technological advance is beneficial; progress is inevitable, and so forth and so forth. But none of these things is true any longer. There are by now so many things which could be found out and so many things which could be done that we must pick and choose whether we do them on the basis of conscious or unconscious preferences. Unless everything quoted in this article from distinguished contemporary scientists is untrue, there is a crying need for revised criteria to help those who support research and development decide which project should be supported.

Perhaps the time will come soon—perhaps it is already here—when some central advisory committee will both advise and, without exercising any absolute authority, be able to do something more than merely advise, since it will control a good many of the purse strings. But even without the existence of a central body, those existing bodies which control important purse strings and apply certain criteria could accomplish a great deal by revising the criteria.

The implications of the need for criteria or some controls are strong for the United States and the rest of the world because all mankind faces the same threats. But international controls are almost unthinkable in a time when nations squabble over matters much more petty than the destiny of man. So a start could best be made right here at home.

There are several steps to be taken before either the existing bodies come up with acceptable criteria or a national board is created. These steps consist in recognizing the following facts:

That science and technology can be catastrophic as well as beneficial.

That we cannot afford to wait and see what the effects of any specific application of new powers will be because once they have been acquired they cannot be eliminated. Whether atomic fission and the ability to control the mind are regarded as promises or threats, we are saddled with them, and they cannot be abolished now.

That science for science's sake, the pursuit of knowledge without thought of how it may be applied, are no longer tenable aims.

That knowledge and power are good only insofar as they contribute to human welfare.

In a sense Congress determines whether certain projects shall be undertaken. It votes money for the space program, the National Science Foundation and so forth. But it does not really establish any priorities except insofar as it votes or refuses to vote funds. Nobody in a position to exercise any real influence ever asks what is, for example, the relative importance of getting to the moon and abolishing pollution on earth. Is it more important to build supersonic transport planes than to clear the slums? Those are questions which should be examined in the light of meaningful criteria. The criteria, however, have never been formulated. A body charged with formulating them and composed of men in the humanistic and the physical sciences could at least make a beginning.

The Control and Use of Technology

Nigel Calder

Suppose a politician sought election on the following platform:

I offer you a richer and more hectic world, with more people, more noise, more automobile accidents, more neuroses and stress diseases than your wildest dreams evoke. Vote for me and I shall waken you each night with the exciting bang of passing supersonic air liners. I shall foment yet harsher economic and political nationalism. I shall lead you to glory in shooting up unruly people of lands whose standard of living is rising far more slowly than yours; we shall fight, too, for possession of the riches of the deep ocean. Together we shall banish wild animals to the museums where they belong and create an entirely man-made world. I offer you prostheses so potent that the day will come when we cannot tell man from machine. Our technocrats will continue to catch you unawares with new inventions. By such means we shall strengthen the power of central government and, through our national computer network, we shall be watching your every action, transaction and misdemeanor with a brotherly eye.

If anyone were bold enough (or honest enough) to declare such intentions he would doubtless be supposed to have committed political suicide. Certainly any tolerably well-intentioned politician would be honestly shocked to have it suggested that such was *his* program. And yet we are heading now to approximately the world I have described—under our benign leaders. It is the unwitting program of the technically advanced nations of the world, because politicians show a remarkable fatalism, or blind enthusiasm, for the application of new scientific and technological knowledge—attitudes that are deplorable, but not beyond remedy.

It is true that politicians have (belatedly) come to recognize the importance of science, but they have carefully enshrined it as a peripheral activity for government. Those who are readiest with the cliché par excellence of the 20th century—the one about science and technology offering unlimited opportunities for good or ill—are often the last to get down to cases and answer the question: "What do you think is good, and what ill, in the present tendencies in applied science?" (Antibiotics and the H-bomb won't do as answers—they are part of the cliché.)

Sometimes, the politician counterattacks by saying that people are trying to drag science into party politics—to get politicians and parties to favor this and disfavor that sphere of scientific activity. That is precisely what I am doing. Ho, ho, comes the rejoinder, the spirit of science must not be enslaved; that was the error of the dictators. . . .

The obvious distinction that must be drawn is between scientific inquiry and the uses made of scientific knowledge so acquired. The former, I agree, must not be enslaved by society; but society is already in the thrall of the latter —not of any person or group of people (e.g., "the scientists"), but of the principle that says, if we can do it, we shall.

As one British Nobel Prize-winner has observed, we can regret nuclear weapons without regretting the knowledge that made them possible. The inability to resist a bright idea is often justified by the argument, "If we don't do it, somebody else will." There are fears for the prestige, military power and economic strength that underpin a nation's way of life. My fear is for the way of life, which may itself succumb to attempts to defend it by sacrifices to the god of efficiency and by remorseless and unthinking application of science in a sort of technological nationalism. Things that radically change everyday life, like television and automation to name but two, are brought casually into being. We have allowed automobiles to ruin the centers of our cities; we have turned our countryside into a two-dimensional chemical plant. The fact that all nations are following the same path does not prove it to be the right one.

Perhaps at this point I should affirm my own enthusiasm for human ingenuity, lest any reader suppose I am in some way "anti-science." Some of the impending applications of science fill me with delight; these include the desalination of water, the revolution in information-handling and communication promised by the big new computers, and the ability to treat mental disease and virus disease chemically. About others—air-cushion vehicles which can overcome the lack of roads in underdeveloped countries and provide 300 mph railways for advanced countries, or the automation of such demeaning work as clerking and mining—I complain that programs to apply them are going too slowly. I look forward hopefully to the day when food can be made artificially, when we shall have energy without limit by controlled nuclear fusion, when we shall have economic ways of winning minerals from low-grade ores. What I abjure is my right (or anyone's right) to say, "Let these things be done," without their being first subjected to the proper democratic tests of public approval and to a review of priorities. If they come unbidden, then democracy is powerless in matters touching the lives of everyone.

But how can we relate such matters to party politics? In principle it is very simple; in practice it will require unaccustomed thought on the part of politicians. To begin with, all technological proposals should, as a matter of course, be energetically opposed by someone, if only so that the dilemmas or uncertainties surrounding them can be exposed, and the complex technical and social issues aired. It is astonishing, for example, that in the United States "national goals in space" were not adequately contested before irreversible decisions had been taken and a substantial part of the nation's resources and scientific manpower committed to landing a few men on our barren satellite. Similarly, the Anglo-French supersonic air-liner project was started without any public debate.

That first requirement is minimal, and rather negative. The second and bigger step that party politicians must take will come from their realization that technological change is at least as full of consequence for the character of society as are the traditional preoccupations of politics. Different combinations of technology will produce different kinds of society. I have already outlined, in the fictional manifesto at the outset, the kind of society toward which we appear to be heading, out of control. A moment's reflection shows that we could do much better, and that in fact we have choices.

If politicians seriously attempted to evaluate current technological opportunities in the light of their political beliefs, they would find that different social views would tend to favor different items, and that they could, in effect, translate a large part of their policy into technological terms. If, for example, you believe in a strong, vigorous central government dedicated to a great technological leap into the future, you should favor big technical projects, big computers and power units, and the development of ocean resources; you should demand a large investment in preventive medicine, building and urban research, and the social sciences.

If, on the other hand, you are concerned with support for the "small man," local private enterprise and the well-being of the individual at home and abroad, you ought to prefer, for example, small-scale projects, energy sources and computers, better transport and communications, versatile production methods, natural products research and rural development, new technical methods of fighting crime, and a general bias toward improving the old technologies, rather than superseding them with new ones.

There are many ways of shuffling the pack to produce combinations of technologies that would match the particular set of beliefs of a given party. I leave it as an exercise for the reader to consider what, in the light of his own political beliefs, he favors and disfavors in current scientific possibilities. There are, of course, many more elements in the scene than I have mentioned; and where there are apparent gaps, the program maker is quite entitled to dream up new research projects.

If I may say so as a foreigner, a special responsibility falls on Americans. The United States is the pace-setter in technology, and latent in the present work of its scientists and engineers are many new wonders which will transform the world at an ever-accelerating rate. America has—in comparison with Britain, for example—the incalculable advantage of well-tried executive and legislative machinery for reviewing scientific matters, and an academic community well accustomed to serving the nation. If the United States should decide to exercise some choice, in what it encourages and what it rejects, the common man in every land who feels that he is being swept along helplessly in a torrent of innovation would be forever grateful.

Time is man's sternest discipline, unless he chooses to ignore it. In the present there are many possible futures, good and bad, for each of us as an individual and for the world at large. But when time passes and the future becomes the present, it is unique and there is no choice. *The* future, the one we or our children will actually inhabit, is too serious a matter to be left to science fiction, too much a concern of humane politics to be left to the blind drive of technology.

A Crucial Time for Technology

T. A. Wilson

What I'm going to say may strike you at first as being slightly redundant before this audience, and that is—that I'm in favor of technology. I realize that this is like being in favor of motherhood, which is literally true in the case of Cal Tech and technology. I share what I'm sure is a common feeling of attachment to the alma mater and what it stands for, and the events of today have heightened this.

But I think the subject of our present situation in the advance of technology is a pertinent one to talk about. At the very time when the case for technology seems obvious, we are in danger of seeing a de-emphasis of activity in this realm.

We are in a period when the focus is on the dollar—for very good reason. People are becoming aware—or Congressmen are at least—that some $137 billion have gone into research and development in the past ten years and that over half of those dollars have been public funds. A few years ago, when the focus was more on our technological race with the Soviet Union, that fact would have been generally applauded.

Now it is viewed in the environment of other problems like minority unrest, equal opportunity, crime in our streets, population pressures, disease, poverty, hunger, urban blight, pollution.

These problems are real. They cry for attention. I'm thoroughly in favor of intercepting these problems—or retrieving the ball if we've lost it—but I think a further advance of technology must be a part of the process. Certainly technology should not be cast in the role of a competitor to the main stream of this effort. One of the fundamental factors that has made us strong as a nation has been our constantly advancing technology, which has fed our growth. It is the yeast in the economy that expands in all directions—and now it has an urgent new direction in which to expand.

However, the possibility exists that others may seek the remedy in another direction in the form of a cutback in research and development—in scientific exploration and invention. There is a certain amount of talk in Washington about a moratorium on science until society gains the wisdom to use technology safely. In the words of Congressman Emilio Daddario, reporting on the findings of his science, research and development sub-committee, "We are warned that mankind can know too much for its own good. A catch up period is proposed—so that mores and national conduct can develop which are equal to the choices forced by science."

Congressman Daddario and his committee disagreed with this viewpoint, but the threat of serious cutback exists. On the other hand if the appropriate relationship between technology and the solution to the pressing problems of the day can be seen, there is an unprecedented opportunity before us.

The really basic solutions in the realm of power, food, water, and increase of wealth versus population growth, depend on continued advance of technology. Atomic power offers a remedy for parts of the world that are without the natural resources for power generation. A new strain of rice is proving to be a partial answer to food production in the Philippines. Large scale desalination of seawater is being proposed for the desert Middle East. New forms of locomotion and traffic control are being forced by urban transportation tie-ups, and so on. The demands for change and the expectations of change are so great that just pecking at the problems of society by conventional methods is not going to get us to the post office.

But I'm sure the all-important reason for continuing to move forward technologically is economic. Applied science is basic to the advance of our standard of living—both in the generation of wealth and the distribution of it. Our economic growth is not brought about by everyone working that much harder; in fact man works less—it comes from accomplishing more with the work that each man does.

Economists attribute from 50 to 80 per cent of the growth of our gross national product to increased productivity stemming principally from investment in applications of new technology. It is activity generated by technology that employs people. So, in seeking out remedies for our social problems, a great many of which have an economic base, we need to keep technology *in* the equation. The creation of new jobs and the acquiring of the funds needed to solve problems *both* depend on a strong economy, well supported by research investment.

Actually we have become accustomed to talking in terms of spin-offs from scientific discovery and experimentation which result in new employment and business activity. The television industry was spun off the microwave radar developments prior to World War II. Lesser spin-offs like Teflon fry-pans are constantly occurring from science and engineering, and they will continue to occur, along with more substantial ones.

Our space people can get very enthusiastic about the long range prospects for bigger effects. They envision a whole new order of communication in which the individual has his pictophone console that will connect him with anyone anywhere via the satellites, and will bring in any type of needed information whether library research or an orbital view of fish migration in the Caribbean or crop growth in Kansas.

But politically speaking, the spin-off theory is not a tangible package to sell. You can't tell in advance in just what direction spin-offs are going to occur. You have to accept the basic proposition that they do occur.

There is another compelling reason for maintaining at least the present level of technological effort in this country. I think it is categorical that as you increase the performance of an article or a system to meet the demands of society, each increment of improvement requires progressively more effort, until finally you reach a point where the constraints fence you in. Then it takes a breakthrough to get outside the fence.

The development of transportation, for example, has progressed to the point where now we encounter the constraint of traffic congestion—in our city streets, at our airports, even on the airways. Yet the requirement for growth continues—the population grows, the GNP grows, the individual's aspirations grow, the market grows. The pressures won't let you shut off progress. To try to meet one aspect of this challenge we have a five-year program at Boeing— aimed at automatic flight management of commercial air transports from take-off to landing, including provision for zero-zero weather. This is being accomplished with an assist from missile and space age technology. When operational it should relieve the airways portion of the constraint.

But new breakthroughs, like spin-offs, defy programming. In all fields there is still a great need for open-ended research, without an application in sight—the type in which many universities participate and where Cal Tech excels.

We carry it on in our own company in an organization called The Boeing Scientific Research Laboratories. One of the ground rules there, is that when the use of a piece of research becomes clear, you move it to an engineering project and get back to your research. The lab is staffed insofar as possible with outstanding scientists in their fields and they are required to fish in new waters. Another ground rule is that nothing is classified. There must be freedom to exchange experiences with any other member of the scientific community anywhere.

I think it is highly important that this type activity continue in as many places as possible. Peter Drucker contends incidentally that there is now a conscious discipline for the imaginative leap into the unknown. As he puts it, "We are developing rigorous methods for creative perception. Unlike the science of yesterday, it is not based on organizing our knowledge. It is based on organizing our ignorance." Some of the same freedom from boundaries that exists in basic research is required in applied research and development. You can get trapped by the planning process, although I am categorically in favor of that process as a means of conserving funds and directing energy toward achievable goals. The problem comes when you make 100 per cent assurance of a successful conclusion the prerequisite to a start. This puts unreasonable constraints on development and does not recognize the process of growth and learning along the way.

All government R&D should not be oriented to provable ends. The military is not advancing technically as it should. There is still a place for the experimental model. A national security policy that does not insure technological leadership is not security—it is insecurity.

In the realm of truly open-ended endeavors, we have the space program as a classic example. We ought to make sure that its ends stay open, at least in part, after the current goal of the moon is reached. Quite apart from the direct returns in scientific discovery and the continuing technological spin-offs, I have referred to, I think there are unseen revenues from the space effort in the stimulus it provides to youth, to education and to the enlarging of all our perspectives. It tends to push away the limitations to thinking, at a time when society is crowded, frustrated and depressed by limitation. It provides a fresh outlook which is precisely the type of outlook that must be taken toward solving our problems on the surface of the earth.

If you proceed on the assumption that it is going to take a generation of effort to solve our most urgent social problems, then today's youth is going to play a large part in arriving at the solutions. Therefore, I think the attitude of youth is important. Here we do have something to build on.

We can't (at least I can't) get my kids to cut their hair but youth can be influenced where there is credibility—where there is something they can believe in. The advance of technology is something they can understand and tie to. Its whole starting point is not one of despondency and frustration but an attitude of, "Yes, we can do this thing if we'll apply ourselves to it." We need to make the transfer of that attitude from technology to social affairs—or marry the two.

Of course social and urban problems are more difficult to solve than those such as food production, communication, transportation, etc., that industry has more successfully dealt with in the past. The present urban situation presents a very muddy picture, with a fantastic array of interrelated problems and pressures. Someone has to perform a clarifying role if we are to have novel and effective approaches to these problems. It may be that the key universities can serve in this capacity and become the catalysts to action. Cal Tech's planned approach to humanities is certainly a step in the right direction. I commend the institute for its "Science for Mankind" program.

I'd like to get back to the proposition that technological research and development is fundamental to the process of moving ahead on all fronts. The question arises, whose responsibility is this basic advance? The government's? The academic community? Industry's? I think it is inescapable that all these must be involved. The government must certainly be involved, and in a strong supporting role. National goals are at stake. These goals have to be a few marks higher than can be met with known technology. Therefore, the element of venture or risk cannot be completely removed from government policy any more than it can from corporate policy if we are to remain viable and dynamic as a nation, avoiding stagnation and decline.

Basic research is a venture, space exploration is a venture. The SST is a venture. If we wait until we fully see the end rewards before starting, we never start. But we might ask ourselves how much credit we deserve as a nation for the original decision to establish moon landings as a national goal. Or did we do it mainly under the goad of Soviet competition? The Soviet Union established it as a national objective and we responded. We are dependent on public opinion for national objectives, and the Soviet challenge gave public opinion its motivation. However, if a move is not pressed on us as an answer to a threat, does the public appreciate the rewards enough to embark on technology advance as a policy in its own right and be willing to pay the bill? I think the determination of whether or not we have a science moratorium may lie in the answer to this question.

Actually the challenge presented by our urban crises and the effects of poverty at home and in other countries can be taken as a prod to action just as Soviet military and space rivalry has been. But in those cases, technology was accepted as the method of meeting the challenges, whereas with urban emergencies we are in effect being told we must draw back from technology. I have faith that an informed and aware citizenry will make the right decision, but we all share the obligation of informing the public.

The general advancement of technology should not depend just on the expenditure of government money—I certainly don't want to imply that. But I'm distressed at the criticism of the use of government money. It is very appropriate that industry should share in the costs of R&D, to provide an incentive for efficiency, but it must be appreciated also that industry puts a lot of money into research in commercial areas that subsequently benefits the government sector.

I am not putting a halo around the large company as the technological fountainhead. Very often the larger corporation's interests will run along certain lines and it must focus its main stream of effort in the directions that appear most promising. Consequently it is most often the small group or the dedicated individual that does the real pioneering—much of this on the university campus.

But overall, the tendency is to place more and more of the research cost burden on the private sector. The government, for example, is increasingly disallowing the costs of independent research and development on the part of government contractors. The effect is to restrain R&D.

The government's own R&D programs are more and more difficult for their sponsors to justify. This affects universities as much as industry, if not more. When a NASA program is cut, among the first things to go are the support of fellowships and grants to schools for special work. This places the burden on some other sector of the economy and this is a burden which industry must recognize, especially with respect to private institutions.

In some countries the whole burden of research and development is public —or national. We would not want that arrangement. But I think public and private interests do run parallel. I think aerospace is an excellent example of a business where technology advance is vital to our international competitive position. This industry is not asking for barriers against foreign imports; it generates large exports and large favorable balance of payments by application of technology.

Past experience has shown that where significant long-term needs exist, industry eventually finds a way to provide solutions, and at a profit. Often it is with government support, at least initially, as has been the case in all major forms of transportation development—rails, roads, ships, airports—and in the development of our petroleum supply.

It is not hard to conceive of industry solutions to things like low-cost housing, cost effective urban transportation, and even vocational training on a broad scale. If this results from the profit motive, I don't think that requires an apology. If industry gets a job done well, it is recognized by society as performing a useful, legitimate role, and society is willing to pay it a reasonable fee to make this result possible.

But the market for what we might call "improvement" or "problem solving" in the urban area is not very clearly defined, to say the least. It is diffuse and multi-customer. It is not well researched. And the customer—the urban entity if you can identify it—is not skilled in specifying requirements.

To get at the answers will no doubt require the combined efforts of government, universities and industry in defining the jobs to be done and creating the market that mobilizes the required effort. Government may have to be the enabler, the university the thinker, industry the doer, although these are certainly not mutually exclusive roles.

Here at Cal Tech elements of the organization and individuals are already involved in the urban problems realm and exploring a wide range of possibilities. MIT and Harvard are active in the area.

Quite possibly the universities can provide the independent non-political "data banks" of pertinent information and "know-how" for local governments and citizen groups to consult. In Seattle our Forward Thrust program planners found that their most time-consuming problem was in getting good data and analyzing them. Forward Thrust was, and is, a combined community effort to put together and fund an integrated program to take care of several of the Puget Sound area's needs such as urban transportation, sewage disposal, reduction of juvenile delinquency, etc. The program has made it possible to plan projects such as a multi-purpose stadium on the one hand while insuring that funds are also available for youth centers and other desperately needed facilities.

This effort is just one example of the aroused public interest which exists throughout the country. There are some jillion committees concerned with our urban situation; in fact—several in each community and at every level of both government and industry. I would hesitate to call this effort chaotic but it is far from integrated.

It would be regrettable if these groups leave an aggressive, purposeful national technology advancement effort out of the process of finding solutions to today's problems. As I may have at least faintly implied already, I think this effort is mandatory.

Consequently, ladies and gentlemen, my message is—forward technology! I think we have a head of steam up in this country and it would be disastrous to turn off the engine just as we're coming to the grade.

Thank you.

The Next Industrial Revolution

Athelstan Spilhaus

We must have a new industrial revolution even if a few of us have to generate it. Other industrial revolutions have come about unplanned. The first was hailed as a way of ennobling human beings by substituting steam and electrical power for their muscles. This it undoubtedly did, but the generation of power brought with it side effects—including air pollution—which, far from being ennobling, were and continue to be degrading to human existence. In the second revolution the multiplication of "things" came about—"things" that at last could be mass-produced, so that people could have more and more of them. Thus was generated the solid-waste problem.

A third revolution was the tremendous growth in industrial chemistry, and the ability to tailor-make chemicals in vast quantities very cheaply, for all kinds of purposes—for example, pesticides intended to selectively destroy forms of life inimical to various groups of human beings. But these turned out not to be so selective; they have upset the little-understood ecological balance, and have polluted and poisoned the waters.

In preparation for the next industrial revolution, I suggest that we revise our vocabulary. For instance, there is no such thing, no such person, as a consumer. We merely *use* "things"; and, according to the law of the conservation of matter, exactly the same mass of material is discarded after use. Thus, as the standard of living goes up, the amount of waste and consequent pollution must go up.

I believe we must base the next industrial revolution—a planned one—on the thesis that there is no such thing as waste, that waste is simply some useful substance that we do not yet have the wit to use. Industry so far is doing only half its job. It performs magnificent feats of scientific, technological, and managerial skill to take things from the land, refine them, and mass-manufacture, mass-market, and mass-distribute them to the so-called consumer; then the same mass of material is left, after use, to the so-called public sector, to be "disposed of." By and large, in our society, the private sector makes the things *before* use and the public sector disposes of them *after* use.

In the next industrial revolution, there must be a loop back from the user to the factory, which industry must close. If American industrial genius can mass-assemble and mass-distribute, why cannot the same genius mass-collect, mass-disassemble, and massively reuse the materials? If American industry should take upon itself the task of closing this loop, then its original design of the articles would include features facilitating their return and remaking. If, on the other hand, we continue to have the private sector make things and the public sector dispose of them, designs for reuse will not easily come about.

We industrial revolutionaries must plan to move more and more into the fields of human service, and not leave such concerns to the so-called public sector. We have seen our food supply grow to abundance in the United States, with fewer and fewer people needed to grow it. We are seeing the automation of factories, with an abundance of "things" provided by fewer and fewer people. On the other hand, we have a shortage of human services and a shortage of people providing these services. It follows quite simply that, if private enterprise is not to dwindle, while the public sector grows to be an all-embracing octopus, then private enterprise must go into the fields of human service.

The next industrial revolution is on our doorstep. Let us be the revolutionaries who shape it, rather than have it happen—and shape us.

Will the Computer Outwit Man?

Gilbert Burck

Ever since it emerged from the mists of time, the human race has been haunted by the notion that man-made devices might overwhelm and even destroy man himself. The sorcerer's apprentice who almost drowned his world, Frankenstein's frustrated monster who tortured and destroyed his creator, the androids that mimic human beings in the frenzied pages of today's science fiction magazines—all play upon the age-old fear that man's arrogant mind will overleap itself. And now comes the electronic computer, the first invention to exhibit something of what in human beings is called intelligence. Not only is the computer expanding man's brainpower, but its own faculties are being expanded by so-called artificial intelligence; and the machine is accordingly endowing man's ancient fears with a reality and immediacy no other invention ever has.

The fears are several and intricately related, but three major ones encompass the lot. The one that worries the columnists and commentators is that the computer will hoist unemployment so intolerably that the free-enterprise system will be unable to cope with the problem, and that the government will have to intervene on a massive scale. It is enough to repeat here that the computer will doubtless go down in history not as the explosion that blew unemployment through the roof, but as the technological triumph that enabled the U.S. economy to maintain the secular growth rate on which its greatness depends.

The second fear is that the computer will eventually become so intelligent and even creative that it will relegate man, or most men, to a humiliating and intolerably inferior role in the world. This notion is based on the fact that the computer already can learn (after a fashion), can show purposeful behavior (narrowly defined), can sometimes act "creatively" or in a way its programmer does not expect it to—and on the probability that artificial-intelligence research will improve it enormously on all three counts. Meanwhile there is the third fear, which is that the computer's ability to make certain neat, clean decisions will beguile men into abdicating their capacity and obligation to make the important decisions, including moral and social ones. This fear as such would be academic if the second one were realized; for if the computer ever betters man's brainpower (broadly defined), then its judgments will be superior too and men finally will be outwitted. To appraise both fears, therefore, we must examine artificial-intelligence research, the formidable new science that is striving so industriously to make the computer behave like a human being.

The Routes to Judgment

The goal of artificial-intelligence research is to write programs or sets of instructions showing the computer how to behave in a way that in human beings would be called intelligent.[1] The workers proceed on the assumption that human nervous systems process information in the act of thinking; and that given enough observation, experiment, analysis, and modeling, they can instruct a digital computer to process information as humans do. Broadly speaking, they simulate human intelligence in two ways. One is to build actual counterparts of the brain or the nervous systems with computer-controlled models of neural networks. The other and more productive approach (so far) is to analyze problems that can be solved by human intelligence and to write a computer program that will solve the problems. Most if not all their toil involves programming the computer "heuristically"—that is, showing it how to use rules of thumb, to explore the likeliest paths, and to make educated guesses in coming to a conclusion, rather than running through all the alternatives to find the right one. Looking closely at their ingenious achievements, which appear so marvelous to the layman, one begins to understand why it is easy to make sweeping projections of the machine's future as an intelligent mechanism.

But these research workers have a hard if exhilarating time ahead of them. When one turns to living intelligence, one is struck with the colossal job that remains to be done—if the word "job" is not too presumptuous to be used in this context at all. If one defines intelligence merely as the ability to adjust to environment, the world is positively quivering with what might be called extracomputer intelligence. Even the lowest species can reproduce and live without instruction by man, something no computer can do. Moreover, the exercise of intelligence in animals, and particularly in higher animals, is a stupendously complex process. As Oliver Selfridge and Ulric Neisser of M.I.T. have put it in a discussion of human pattern recognition, man is continually confronted with a welter of data from which he must pick patterns relevant to whatever he is doing at the time. "A man who abstracts a pattern from a complex of stimuli has essentially classified the possible inputs," they write. "But very often the basis of his classification is unknown even to himself; it is too complex to be specified explicitly."

Yet specify the researcher must if he is to simulate human behavior in the machine. For the electronic computer is a device that processes or improves data according to a program or set of instructions. Since it is equipped with a storage or memory, a stored program, and a so-called conditional transfer, which permits it to make choices, it can be instructed to compare and assess

---

1. Many researchers shudder at the phrase "artificial intelligence." Its anthropomorphic overtones, they say, often arouse irrelevant emotional responses—i.e., in people who think it sacrilegious to try to imitate the brain.

and then judge. Nevertheless its conclusions are the logical consequences of the data and program fed into it; and it still must be told not merely what to do, but how, and the rules must be written in minute and comprehensive detail. Ordinary programs contain thousands of instructions. The research worker cannot make the machine do anything "original" until he has painstakingly instructed it how to be original, and thus its performance, at bottom, depends on the intelligence or even genius of its preceptor.

Although the men doing artificial-intelligence research are themselves extraordinarily intelligent, they disagree widely and acrimoniously not only about what exactly they expect to do but even about what exactly they have done. In a recent paper John Kelly of Bell Laboratories and M.I.T.'s Selfridge, who between themselves disagree about almost everything, agreed that the most controversial subject in this work is whether man's ability to form concepts and to generalize, an age-old concern of philosophers, can ever be imitated. Some conservative extremists think no machines can ever emulate this faculty, while their opposites believe the problem is about to be solved. The moderates take various stands in between.

Everybody in the intelligence business would probably agree with Kelly and Selfridge that the task of simulating human intelligence has a long way to go. "Our position may be compared to pioneers on the edge of a new continent," they write. "We have tested our axes on twigs, and made ladders and boats of paper. In principal we can cut down any tree, but obviously trees several miles in girth will take discouragingly long. We can span any river with bridges or boats in principle, but if the river is an Amazon with a thirty-knot current we may not be able to do it in fact. Then again, the continent may be two light years across. However, as pioneers, what we do *not* see is a river of molten lava, which at one sight would make us admit the inapplicability of our tools."

The Machine that Learns

One man who has chopped more than a few twigs and crossed a stream or two, but has no illusions about the rivers and forests ahead, is Arthur Samuel, consultant to I.B.M.'s director of research. Samuel, who has been classified as close to center in the field of intelligence research, pioneered in machine learning in the late 1950s by teaching a computer to play checkers so well that it now consistently beats him. To program the machine to play the game, Samuel in effect stored a model of the checkerboard in the computer. Then he instructed the machine to look ahead as a person does, and to consider each move by analyzing the opponent's possible replies and counter-replies. Although the machine theoretically could search *all* possible choices, there are 10^{40} such choices in every game; and even the fastest computer would take longer than many think the universe is old to play a game by ticking off them all.

So Samuel instructed the machine to proceed heuristically by feeding it two "routines" that showed it how to learn from experience. One routine told it how to remember past positions and their outcomes, and the other told it how to improve the way it appraises positions. The machine got steadily better, and was soon superior to its master. But this does not mean, Samuel insists, that the computer is categorically superior to man. It beats him not because it uses information or techniques unknown or unknowable to man, but simply because its infallible memory, fantastic accuracy, and prodigious speed allow it to make a detailed but unimaginative analysis in a few seconds that would take man years to make. When, if, and as a championship chess program is constructed, says Samuel, the same generalization will hold. No chess program can yet play much better than an advanced novice; the reason is that chess is vastly more complex than checkers, and nobody has yet devoted enough time or thought to the task.

The limitations of the computer, Samuel likes to point out, are not in the machine but in man. To make machines that appear to be smarter than man, man himself must be smarter than the machine; more explicitly, "A higher order of intelligence, or at least of understanding, seems required to instruct a machine in the art of being intelligent than is required to duplicate the intelligence the machine is to simulate."

Samuel scoffs at the notion that the great things in store for the machine will give it a will, properly defined, of its own. In relatively few years machines that learn from experience will be common. The development of input and output devices has some way to go, but in twenty years or so businessmen may be able to discuss tasks and problems with computers almost as they now discuss them with other employees; and televideo-phones, for companies that cannot afford their own computer systems, will make central machines as convenient as the telephone itself. Programs will be written to instruct machines to modify their own rules, or to modify the *way* they modify rules, and so on; and programs will also instruct one machine to program another, and even how to design and construct a second and more powerful machine. However, Samuel insists, the machine will not and cannot do any of these things until it has been shown how to proceed; and to understand how to do this, man will have to develop still greater understanding. As the great originator, he will necessarily be on top.

The Artificial Brain

An *apparent* exception to such a conclusion, Samuel acknowledges, is that man may eventually build a machine as complex as the brain, and one that will act independently of him. The human brain, which contains roughly 10 billion cells called neurons, works strikingly like a computer because its neurons react or "fire" by transmitting "excitatory" or "inhibitory" electrical impulses to other neurons. Each neuron is connected to 100 others on the average, and sometimes to as many as 10,000, and each presumably gets more than one signal before "firing." Whether these neurons are at birth connected at random or in a kind of pattern nobody knows, but learning, most research workers think, has something to do with changes in the strength and number of the connections. Some hold that creativity in human beings consists of "an unextinguishable core of randomness," i.e., creative people possess a lot of random connections.

To imitate the behavior of neurons, several researchers have built "self-learning" or "adaptive" machines using mechanical, electrical, and even chemical circuits, whose connections are automatically strengthened, as they presumably are in the brain, by successful responses. The Perceptron, built by Frank Rosenblatt, a Cornell psychologist, has for years been demonstrating that it can improve its ability to recognize numbers and letters of the alphabet by such a routine. Because such devices are proving useful in pattern recognition, they doubtless will be improved.

Theoretically, these techniques might create a monster capable of acting independently of man; but Samuel argues that the brain's complexity makes this highly unlikely. Even if great progress is made in imitating this complexity, not enough is known about the interconnections of the brain to construct a reasonable facsimile thereof; the chance of doing so, Samuel guesses, is about the same as the chance that every American will be stricken with a coronary on the same night. As others have estimated it, moreover, the total cost of duplicating all the brain's cells and connections, even at the ludicrously low cost of only 5 cents per cell and 1 cent per connection, would come to more than $1 quintillion, or $1 billion billion. Some of the brain's functions can probably be reproduced with vastly fewer cells than the brain contains, however, and the odds on building a tolerably "human" model brain will doubtless improve. But they will improve, Samuel reiterates, only as man improves his understanding of the thinking process; and his ability to control the mechanical brain will increase to the extent he increases his own understanding.

The Optimistic Extremists

Some enthusiastic and optimistic research scientists feel that such judgments tend to understate the potentialities of the machine. Allen Newell and Herbert Simon of the Carnegie Institute of Technology argue that man is no more or less determinate than the computer; he is programmed at birth by his genes, and thenceforth his talents and other traits depend on the way he absorbs and uses life's inputs. The day will come, they prophesy, when a program will enable the machine to do everything, or practically everything, that a man's brain can do. Such a program will not call for "stereotyped and repetitive behavior," but will make the machine's activity "highly conditional on the task environment"—i.e., on the goal set for it, and on the machine's ability to assess its own progress toward that goal.

Meanwhile, Newell and Simon insist, the computer will surpass man in some ways. Back in 1957, Simon formally predicted that within ten years a computer would be crowned the world's chess champion, that it would discover an important new mathematical theorem, that it would write music of esthetic value, and that most theories in psychology would be expressed in computer programs or would take the form of qualitative statements about the properties of such programs. Simon's forecast has only three years to go, but he thinks it still is justified, and he sticks to it. Not only that, he confidently looks forward to the day when computers will be tossing off countless problems too ill-structured for men to solve, and when the machine will even be able to generalize from experience.

Together with Newell and J. C. Shaw of Rand Corporation, Simon has done a great deal of pioneering in artificial intelligence. The trio was one of the first to introduce heuristic methods; and their first notable achievement, in 1956, was the Logic Theory Program, which among other things conjures up proofs for certain types of mathematical and symbolic theorems. It was the Logic Theory Program that independently discovered proofs for some of the theorems in Russell and Whitehead's *Principia Mathematica,* and in at least one instance it provided a shorter and more "elegant" proof than Russell and Whitehead themselves.

Simon and Newell's General Problem Solver, mentioned in the first chapter, is an ambitious feat that instructs the computer to behave adaptively by solving sub-problems before going on to knock off bigger ones, and to reason in terms of means and ends. Using the General Problem Solver, Geoffrey P. E. Clarkson of M.I.T. has successfully instructed a computer to do what an investment trust officer does when he chooses securities for a portfolio. Clarkson analyzed the steps the officer takes, such as appraising the state of the market, and also analyzed these steps according to the postulates of the General Problem Solver. Thereupon Clarkson constructed a program that in several actual tests predicted with astonishing accuracy the trust officer's behavior, down to the names and number of shares chosen for each portfolio. The General Problem Solver, furthermore, aspires to endow computers with more than such problem-solving faculties. It tries to show how people solve problems, and so provide a tool for constructing theories of human thinking. Its techniques, says Simon, "reveal that the free behavior of a reasonably intelligent human can be understood as the product of a complex but finite and determinate set of laws."

One of the most assiduous of the optimists is Marvin Minsky of M.I.T., who believes we are on the threshold of an era that "will quite possibly be dominated by intelligent problem solving machines." Minsky has divided the intelligence research scientists' achievements and problems into five main groups: (1) search, (2) pattern recognition, (3) learning, (4) planning, and (5) induction. To illustrate: in solving a problem, a worker can program a computer to (1) search through any number of possibilities, but because this usually takes too long, it is enormously inefficient. With (2) pattern recognition techniques, however, the worker instructs the machine to restrict itself to important problems; and with (3) learning techniques, he instructs it to generalize on its experience with successful models. With (4) so-called planning methods, he chooses a few from a large number of sub-problems, and instructs the machine to appraise them in several ways. And finally (5) to manage broad problems, he programs the computer to construct models of the world about it; then he tries to program the machine to reason inductively about these models—to discover regularities or "laws of nature," and even to generalize about events beyond its recorded experience.

Minsky has predicted that in thirty years the computer will in many ways be smarter than men, but he concedes that the machine will achieve this high state only after very smart men have worked very long hours. "In ten years," he says, "we may have something with which we can carry on a reasonable conversation. If we work hard, we may have it in five; if we loaf, we may never have it.

What is Susie to Joe?

Merely to describe briefly the artificial-intelligence projects now under way would take volumes. Among the more important are question-answering programs that allow a computer to be interrogated in English: the so-called BASEBALL project at M.I.T.'s Lincoln Laboratory, which answers a variety of queries about American League teams; and SAD SAM (Sentence Appraiser and Diagrammer—Semantic Analyzer Machine) program of Robert Lindsay of the University of Texas, which constructs a model of a family and tells the computer how to reply to queries about family relationships, such as "How is Susie Smith related to Joe and Oscar Brown, and how are the two Browns related?"

There is Oliver Selfridge's pioneering work in pattern recognition, which has led to techniques that have progressed from recognizing simple optical characters like shapes to recognizing voices. Pattern recognition, in turn, leads to the simulation of verbal learning behavior; one program that does this is the Elementary Perceiver and Memorizer (EPAM) of Herbert Simon and Edward A. Feigenbaum of the University of California. EPAM provides a model for information processes that underlie man's power to acquire, differentiate, and relate elementary symbolic material like single syllables; and it has been used to compare computer behavior with human behavior and to construct "adaptive" pattern recognition models. So far, so good. But the human brain forms concepts as well as recognizing digits and words; and although some work is being done on programming a model of human concept formulation, relatively little progress has been made in telling a machine just how to form concepts and to generalize as people do. And as we have already noted, there is much disagreement about the degree of even that progress.

The Impedimenta of the Intellect

Nobody has yet been able to program the machine to imitate what many competent judges would call true creativity, partly because nobody has yet adequately defined creativity. As a working hypothesis, Newell, Shaw, and Simon have described creativity as a special problem-solving ability character- ized by novelty, unconventionality, high persistence, and the power to formu- late very difficult problems. This is fine as far as it goes; a creative person may have and indeed probably needs all these. But he willy-nilly brings to his task much more than these purely intellectual aptitudes. He brings a huge im- pedimentum of basic emotions and aptitudes that were programmed into him congenitally, and have been greatly augmented and modified in a lifetime of conscious and unconscious learning. The terrible temper of his mother's grand- father, his own slightly overactive thyroid and aberrant hypothalamus, his phlegmatic pituitary, his mysterious frustrations, his odd beliefs and preju- dices, his phlogistic gonads, his illusions and superstitions, his chronic consti- pation, his neuroses or psychoses, even the kind of liquor he takes aboard— all these and thousands more combine to color his personality and imagina- tion, and his approach to his work. "No man, within twenty-four hours after having eaten a meal in a Pennsylvania Railroad dining car," H. L. Mencken used to argue, "can conceivably write anything worth reading."

Consider the art of musical composition. "Great music," wrote Paul Elmer More, "is a kind of psychic storm, agitating to fathomless depths the mystery of the past within us." All composers and musicians may not find this a good description—nor may some intelligence research workers—but it sug- gests strikingly what goes into what a qualified judge would identify as a work of musical art. Such a work is not merely the opus of a brain sealed off in a cranium, but the result of a huge inventory of "states" or influences, inherited or acquired, that has colored that brain's way of performing. The dissimilari- ties among Dvořák, Mahler, Richard Strauss, and Sibelius, to mention four late romantic composers, are probably the result mainly of obscure inherited and acquired influences. And what dissimilarities they are! Music critics have spent lifetimes and built towering reputations on expounding and appraising just such differences.

The computer's achievements in creative composition, literary and musi- cal, are remarkable in the sense that Dr. Johnson's dog, which could walk on its hind legs, was remarkable: "It is not done well, but you are surprised to find it done at all." Since the vast bulk of durable music is relatively formal, arranged according to rules, it theoretically should be possible to instruct a computer to compose good music. Some have tried. Perhaps the best example of computer music is the Illiac suite, programmed at the University of Illinois. By common consent, the Illiac suite is no great shakes; one of the moderate remarks about it is that repeated hearings tend to induce exasperation. But so, of course, does some "modern" music composed by humans; and it is possible that one day the computer may be programmed to concoct music that many regard as good.

Research workers have had some luck in turning out popular jingles on the computer without a hard, long, expensive struggle; in twenty years or so the computer will doubtless be mass producing ephemeral tunes of the day more cheaply than Tin Pan Alley's geniuses can turn them out. But it appears that instructing a computer to compose deep, complex, or carefully prepared music is an almost heroic task demanding more talent and time (and money) than simply putting the music down in the first place and not bothering with the computer at all.

Auto-Beatnik Creativity

Similar observations apply to creative writing. Man as a writer is the human counterpart of the data processor, but the inputs he has stored and the mind he processes them with are the result of thousands of kinds of influences. Put in focus and challenged by the job at hand, his brain processes the relevant data and comes up with all manner of output, from majestic imperative sentences like "No Parking North of Here" (official sign in one large city) to epic poems and novels in the grand style.

It should surprise no one that the computer can be programmed easily and cheaply to do simple jobs of gathering and sorting information, and is in fact doing a rough job of abstracting technical works. As more and more of the world's knowledge is stored on tape or drums, and as centralized retrieval systems are developed, the computer will be able to dredge up nearly everything available on a given subject and arrange it in some kind of order. It will doubtless save much research time, and so prove a boon to writers. The machine, moreover, has already been programmed to write simple-minded television whodunits, and some believe it will soon supply soap opera confectors much of their material, or perhaps eventually convert them into programmers. The cost of using the machine will also pace this "progress"; what the computer and programmer can do more cheaply than the solo word merchant, they will do.

That creative writing of more complexity and depth is something else again is suggested by a sample of "Auto-beatnik" verses "generated" by computers programmed under R. M. Worthy at the Advanced Research Department of General Precision, Inc., Glendale, California. Worthy and his experts have arranged several thousand words into groups, set up sentence patterns and even rhyming rules, and directed the machine to pick words from the groups at random, but in pre-specified order. The computer's verses are syntactically correct but semantically empty; as Worthy allows, a machine must have an environment, a perception, an image, or an "experience" to write a significant sentence. "I am involved in all this nonsense," says Worthy, "because I am fascinated with language. And eventually this research will be valuable in many things like translation and information retrieval." Here is one of the machine's verses:

LAMENT FOR A MONGREL

To belch yet not to boast, that is the hug,
The high lullaby's bay discreetly crushes the bug.
Your science was so minute and hilly,
Yes, I am not the jade organ's leather programmer's recipe.
As she is squealing above the cheroot, these obscure toilets
    shall squat,
Moreover, on account of hunger, the room was hot.

A common diversion in computer circles is to speculate on how a programmer would instruct a computer to write as well as Shakespeare. Even the untutored layman can see some of the trees in the forest of problems ahead. Before doing anything, the preceptor must do no less than decide exactly what makes Shakespeare so good, which itself is probably a job for a genius. Then he must write the rules; he must formalize his conclusions about the bard's talents in staggering detail, not omitting even the most trivial implications, so that the machine can proceed logically from one step to another lest it produce elegant gibberish when it starts to "create." The set of instructions or program for such a project would probably run to several times the length of Shakespeare's works. And it might demand more talent than Shakespeare himself possessed.

Creation as Effective Surprise

The indefatigable Newell and Simon, however, are not to be dissuaded. They have speculated on the notion that creativity might not always have to reside in the programmer—that the computer on its own could *match* (not copy) such creations as a Beethoven symphony, *Crime and Punishment,* or a Cezanne landscape. Although they freely admit that no computer has ever come up with an opus approaching any of these, they suggest that none appears to lie beyond computers.

To create, they hold, is to produce effective surprise, not only in others but in the creator; and in principle a computer might do this. "Suppose," they have written, "a computer contains a very large program introduced into it over a long period by different programmers working independently. Suppose that the computer had access to a rich environment of inputs that it has been able to perceive and select from. Suppose—and this is critical—that it is able to make its next step conditional on what it has seen and found, and that it is even able to modify its own program on the basis of past experience, so that it can behave more effectively in the future. At some point, as the complexity of its behavior increases, it may reach a level of richness that produces effective surprise. At that point we shall have to acknowledge that it is creative, or we shall have to change our definition of creativity."

They may. Their definition of creativity, critics feel, may be too narrow because it makes too little allowance for human motivation, or the complex mix of emotions and other drives that compel people to behave as they do. The question is not whether the machine can produce something original; any computer can do something trivial or incomprehensible that nobody has ever done before. The question is whether men can show the machine how to create something that will contain enough human ingredients to meet at least a minimum of approval by perceptive human beings specially qualified to judge the creation. The creations of Newell and Simon's well-educated computer might amount to expensive nonsense unless the computers were fed a vast amount of brilliant instructions on how to handle the sophisticated inputs, and unless these inputs included human motivations.

Human Motivation and the Computer

Human motivations, Ulric Neisser believes, must be considered not only by workers who would instruct the machine to create, but also by those who would increase its power otherwise to simulate human intelligence. Man's intelligence, he points out, is not a faculty independent of the rest of human life, and he identifies three important characteristics of human thought that are conspicuously absent from existing or proposed programs: (1) human thought is part of the cumulative process of the growth of the human organism, to which it contributes and on which it feeds; (2) it is inextricably bound up with feelings and emotions, and (3) unlike the computer's behavior, which is single-minded to the exclusion of everything but the problem assigned it, human activity serves many motives at once, even when a person thinks he is concentrating on a single thing. Recent research by George A. Miller of Harvard, Eugene Galanter of Pennsylvania, and Karl Pribram of Stanford suggests that human behavior is much more "hierarchical" and intricately motivated than hitherto assumed, and Neisser thinks that this multiplicity of motives is not a "supplementary heuristic that can be readily incorporated into a problem solving program."

It is man's complex emotional and other drives, in other words, that give his intelligence depth, breadth, and humanity; nobody has yet found a way of programming them into a computer, and Neisser doubts that anybody soon will. He predicts, however, that programming will become vastly more difficult as the machine is used more and more in solving "human" problems. Pattern recognition, learning, and memory will still be research goals, but a harder job will be to inject a measure of human emotion into the machine.

Some feel Neisser errs in suggesting that anybody will want to imitate man's way of thinking in all its complexities. "The computer can and will be programmed to do and be a lot of things," says one research worker, "including acting just as foolishly as any human being." In other words, nobody may want to stuff a machine full of the useless mental impedimenta lugged around by humans; the great merit of the machine is that it can think accurately and single-mindedly, untainted by irrelevant emotions and obscure and even immoral drives.

Nevertheless, the world is populated by human beings, and their motivations cannot be overlooked. For if the computer is asked to solve a problem in which human motivation is important, it will have to be told exactly what that motivation is, and what to do about it and under what circumstances. That will not be easy.

A Neat, Clean, Consistent Judgment

Meanwhile the prospect for instructing the computer to behave like a real human is remote; and this is precisely why some fear that the machine's role as decision maker will be abused. "If machines really thought as men do," Neisser explains, "there would be no more reason to fear them than to fear men. But computer intelligence is not human, it does not grow, has no emotional basis, and is shallowly motivated. These defects do not matter in technical applications, where the criteria for successful problem solving are relatively simple. They become extremely important if the computer is used to make social, business, economic, military, moral and government decisions, for there our criteria of adequacy are as subtle and as multiply motivated as human thinking itself." A human decision maker, he points out, is supposed to do the right thing even in unexpected circumstances, but the computer can be counted on only to deal with the situation anticipated by the programmer.

In a recent issue of *Science,* David L. Johnson of the University of Washington and Arthur L. Kobler, a Seattle psychologist, plowed through the subject of misusing the computer. The use of the computer, they concede, inevitably will increase. But it is being called on to act in areas where man cannot define his own ability. There is a tendency to let the machine treat special problems as if they were routine calculations; for example, it may be used to plot the route for a new highway by a routine computation of physical factors. But the computation may overlook the importance of locating the highway where it will not create or compound ugliness.

Johnson and Kobler also feel that the "current tendency of men to give up individual identity and escape from responsibility" is enhanced by the computer. It takes man's inputs and turns out a neat, clean, consistent judgment without "obsessive hesitation," commitments, or emotional involvements. In effect, it assumes responsibility; and its neatness and decisiveness can lead men to skip value judgments, to accept unimaginative and partial results as accurate solutions, and to read into its results the ability to solve all problems. Even scientists who are aware of the limitations of machines, the authors reason, can find them so useful in solving narrow and well-defined problems that they may tend to assume the computer can solve all problems. Thus the danger of oversimplifying complex decisions, a danger that has always existed, becomes worse. Another worry is that military computer systems will react so swiftly that the people who nominally make the judgments will not have time to make them. "The need for caution," Johnson and Kobler conclude, "will be greater in the future. Until we can determine more perfectly what we want from the machines, let us not call on mechanized decision systems to act upon human systems without realistic human processing. As we proceed with the inevitable development of computers and artificial intelligence, let us be sure we know what we are doing."

Military and computer experts are already studying the problems raised by the speed of the machines. And to use the warnings to deny the real value of computers would be as foolish as misusing computers. The machines compel men to formulate their problems so much more intelligently and more thoroughly than they ever have that men can hardly be unaware of the shortcomings of their programs. The great majority of computers, as Johnson and Kobler are well aware, are being employed by business. Granted that U.S. business makes mistakes, granted that it has made and will make mistakes with computers, it does not operate in a monopolistic vacuum. Nothing would make a company more vulnerable to smart competitors than to abdicate responsibility to the neat, clean, consistent judgments of a machine.

The computer is here to stay; it cannot be shelved any more than the telescope or the steam engine could have been shelved. Taking everything together, man has a stupendous thing working for him, and one is not being egregiously optimistic to suggest he will make the most of it. Precisely because man is so arduously trying to imitate the behavior of human beings in the computer, he is bound to improve enormously his understanding of both himself and the machine.

Abyss of Progress

E. J. Mishan

In a previous article (*The Nation,* November 27, 1967) I elaborated on the simple proposition that an index which evaluated the annual growth of man-made goods while assiduously ignoring the annual growth of man-made "bads"—the noxious spillover from modern industry and its products—could not be relied upon to plot the course of "real" income over time. Moreover, if the condition of society is such that the annual increment of goods adds a measure of choice that is marginal to people's contentment, while the annual increment of "bads" deprives them of environmental choices that are vital to their sense of well-being, a decline in social welfare may accompany the observed growth of GNP.

In principle, at least, many of these vexatious spillover effects can be measured, though in practice they are only beginning to enter into the cost-benefit studies of a number of public investment projects. The public, however, is becoming increasingly aware of the more blatant spillovers—motorized din on the ground and in the air, pollution of air and water, fume, smog, stench, the destruction of natural beauty and the "slumification" of built-up areas. Its concern has begun to shift these effects gradually toward the center of political discussion. If, eventually, legislation against the more corrosive side effects begins to curb their incidence, the case for economic growth would become somewhat less vulnerable than it is at present.

That case would remain weak, however, so long as we remained alert to the influence of less tangible factors that bear strongly on human welfare. The precondition of sustained economic growth is sustained discontent. Indeed, the success of the advertising agency turns on its ability to implant dissatisfactions in the minds of men; to impel them to earn more and to spend more in the vain endeavor to catch up with a receding horizon of material fulfillment. Again, sustained economic growth entails sustained innovation which, in turn, implies continuous obsolescence, not only of technology itself but of the associated skills and expertise. Specialized knowledge, experience, dexterity, accumulated over the years, the products of hard study and apprenticeship, can become obsolete today in a matter of months. For the scores of thousands that pour annually from our institutions of higher learning there is no assurance about the future, no resting on laurels; they are destined to live on the brink of obsolescence. Inevitably, the initial exhilaration of the younger generation, at the prospect of living in a world of continuous technological change, transforms itself over the years into a state of perpetual anxiety. Finally, the more significant product innovations—the automobile, self-service, television —all act to reduce interdepence, to isolate people from direct communication and direct human experience.

Now intangible consequences of economic growth such as these may be conceded without allowing that they should be decisive. The case for growth could look for support to a number of other developments which, judged at least by 20th-century standards, are socially desirable: the advance of medical knowledge, the conquest of poverty, the spread of education. I shall have nothing to say here of the first—except that no previous generation has depended so heavily on drugs and medical treatment. Of the second, I confine myself to the cynical observation made by Arnold Toynbee that the wealthiest country, the United States, is the worst country in the world in which to be poor. On the question of education, however, I cannot resist more expansive reflections.

In the late 18th century, even in the 19th century, when enlightened opinion came down in favor of nurture rather than nature, and reformers believed in the infinite adaptability of man, the belief that greater leisure for the masses would be used in self-improvement was not altogether implausible. Today it is. Though a smattering of "instant culture" is certainly one of the earmarks of the professional and middle classes, there is no hard evidence to suggest that the mass of ordinary people are interested in culture for its own sake. If the culture pill can be sugared and passed off as entertainment on television screens, it will doubtless be swallowed without a murmur. But the fact remains that it is the sugar, not the culture, that is in demand.

It can be observed, on the other hand, that wherever the link between potential earnings and education is in evidence, the interest in education of people, at all levels of society, distinctly quickens. Thus when economists apply themselves to measure the value of education—in order, generally, to justify the expansion of higher education—it is to vocational education that they turn. But to justify economic growth by reference to the expansion of educational opportunities, which opportunities in turn are valued in terms of their contribution to economic growth, strikes one as being a remarkably circular procedure. It is much like affirming that a man should eat more as this enables him to work harder, so giving him an appetite to eat more.

And certainly the mass education toward which we are moving cannot easily be accepted as a civilizing influence in itself. The overextended mega-universities of the United States, having campuses of 30,000 students or more, is perhaps the only way of dealing with a domestic student population which is expected to pass the 10 million mark within the next five years. But in these vast automated teaching factories one must not look for the virtues once associated with the university—an atmosphere conducive to reflection, discourse and tolerance, leisure to think and to immerse oneself for a while in the disinterested pursuit of learning. In these vast and highly organized teaching corporations, vocational learning (and not vocational learning alone) is increasingly assimilated via closed circuit television and tape recorders. Discussion groups are being superseded by teaching machines. Contact with tutors is giving way to contact with computerized information centers.

I do not argue that this application of technology to teaching methods is less efficient. If students could dehumanize themselves as readily as do the universities, they could well become more knowledgeable than students trained by more traditional methods. But if they cannot; if they resist the new technology, what then? Not surprisingly there is a growing revolt against the Orwellian features of the new automated education. It is not altogether fantastic for a student to feel he is being machine-cut, pressed, molded and trimmed to fit into a complex industrial system that is geared to churn out endless streams of plastic and gadgetry for a mass-consumption society that has no apparent interest in his latent idealism nor cares for his resentments or anxieties—save as they reduce his productivity. Rightly, then, he does not feel proud or privileged. Rightly, then, he feels somehow cheated, and in his resentment or despair makes common cause with others.

If we turn, then, to the professional and academic groups, among which it might be reasonable to suppose that increased leisure would beget increased culture, we find little cause for comfort. Among the professional groups, among doctors, engineers, administrators, scientists, lawyers, accountants and a countless variety of parvenu experts, it is not impossible that official working hours will be reduced over time—though one feels certain that they work a good deal harder now than they ever did before the war. But this trend would provide no clue to the "free" time, the unpre-empted leisure, at the disposal of a member of this group. Since his earnings and, at least as important, his status depend upon his acknowledged proficiency, the painful choice, between using his leisure to add to his enjoyment or to maintain and add to his status and earnings, tends to be resolved in favor of the latter. Indeed, the current "exponential" growth in each branch of knowledge—and the consequent fear of finding himself a victim of the pace of progress, his hard-earned expertise become obsolete—is more likely to increase his anxiety and to reduce his free leisure over the foreseeable future.

A lot may be said about the practical difficulties of coordinating knowledge in the future. But confining ourselves here to the economic compulsion for men to learn more and more about less and less, it need hardly be observed that such learning is the antithesis of the learning of the educated man as traditionally understood—the "man of parts" who could converse and speculate intelligently over a wide range of subjects. Up to the turn of the 19th century it was still possible to be educated in this latter sense. The really tragic consequence of the seemingly irreversible trend toward a more intensive and systematic mining into the surface of knowledge—and this holds true even of the humanities—is the utter impossibility of men ever again being educated men.[1]

But if the educated man, as once conceived, is granted to be a creature of past civilizations, the faithful today seek consolation in the growth of scientific manpower. Applied on a massive scale, scientific research, it is believed, will ultimately unlock the secrets of the abundant life (though for the present it is largely engaged in providing solutions to the tangle of urgent problems brought about by the continued application of science to society). Of course, the possibility that it will do so can be entertained; but so also can the possibility that the excessive love of science is ultimately more subversive of the moral foundations of society than the excessive love of money.

As far back as the ancient civilizations men have recognized the corrupting influence of the single-minded pursuit of money. When this vice becomes an approved part of the ethos of Western society—as it gradually did between the 18th and 19th centuries, though perhaps more so in the United States than in any other country—its corrupting influence is revealed in the character and tastes of the public, as also in the style of living and the shape of towns and cities.

With the growth in popularity of the cinema, followed by the radio, and more recently television, we are becoming conscious of the increased powers of these new media to mold for profit not only public opinion but the character, even the morality, of the young and impressionable. Yet there are some checks even today on the extent to which commerce and commercial entertainment are free to develop a taste for sadism, pornography and violence. There are none, apparently, on propaganda which inculcates an inordinate ambition in a man and encourages unrealizable expectations—obvious prerequisites for a life macerated by discontent and frustration.

I do not for a moment underestimate the corruptive powers of commercial interests in molding the shoddy tastes, the shallow passions, and the mindless opportunism of millions the world over; but I fear far more the unchecked growth of technology and the petulant, parasitic mass society it cannot help producing. The worship of Mammon is damnable, but it is a common failing. The ruthless jostle for a pecuniary gain is debasing, but it is still human. In consequence, a man may feel more comfortable in the market place, among the familiar follies and vices of humanity, than he does contemplating the blind devotions of the priesthood in the temple of science. His declarations to the contrary notwithstanding, the scientist is not at all interested in human welfare. Some may be interested in knowledge "for its own sake." Individually,

---

1.The growing sales of art reprints, of popular histories, of classical records, provides no evidence to the contrary. The taste of the public is guided by "amateur experts," themselves guided by more professional experts, and can be viewed as a response to fashion, or as a purchase of status goods. However "sincere" is their search for learning, such a public is the passive recipient of tidbits of "culture" passed on to them. It has no affinity with the educated elites of other ages.

each is to a greater or lesser extent interested in recognition, in the kudos of published scientific papers. As a fraternity, however, scientists are interested in the search for power—at whatever cost to humanity at large.

Furthermore, today's apotheosis of the intellect, of sheer brain power, has entailed a devaluation of all other qualities that once made up a man. The man of today, ideally, is one who by a new alchemy has transmuted the fullness of his emotional inheritance into a single flame of intellectual concentration. To this new being, to the scientist who is taking over the earth, nothing matters very much save the burning task of ripping out the secrets of the universe. For him, as for his ally, the technocrat, the world of human impulse is an irrelevance. Society is already moving from the traditional world of vice and virtue, comedy and tragedy, into a world where there is only sickness and health, efficiency and inefficiency. Anyone who sincerely believes that from giving the scientists and technocrats their head there will emerge a secure and serene future for the common man is no ordinary optimist.

Again, one asks, have we any choice? If we conclude that we cannot stop the juggernaut of science, let us not conceal our helplessness from ourselves, or take refuge in a bogus pride about the "unquenchable human spirit." Let us have the honesty to affirm that we are not free agents, and at least contemplate the possibility that, like Satan, humanity is predestined to soar onward "with compulsion and laborious flight" to a doom of its own making.

A world in which the great powers stand armed in mortal fear of one another is not one in which a call to halt the advance of technology is likely to make much impression. If we survive the next war, and there comes into being a single world government, it is not impossible that a more humane wisdom will prevail. The habit of empirical research has grown over the centuries, and it cannot be disowned overnight. But in a dispensation wiser than any we can readily visualize—one in which it was finally recognized that no matter how "exciting" or "efficient" was an innovation, its ultimate contribution to human happiness was likely to be slight, whereas the risk of its causing social disharmony or ecological disaster was large—we could, for a while, permit scientists to go on probing the universe if they wished, provided we took measures to insure that they did not change it. This new and wiser society will be wealthy enough, I assume, to finance their activity for some time, and to allay their craving for recognition by the provision of prizes and titles. All their contributions to knowledge would be faithfully recorded for the admiration of their colleagues and for the wonder of future generations. Applications of any of their discoveries would, however, be strictly forbidden. History would not therefore stop. The human drama would continue to unfold. Music, art, literature and politics would, I hope, flourish. Only technological change would be a thing of the past.

Is it possible that, under more favorable circumstances, in a world in which peace was assured, so exacting a wisdom might prevail? If one is old-fashioned enough to believe in free will, the possibility cannot be ruled out. But the likelihood of its coming to pass is not high. The study of history does not abound with examples of ordinary people who are willing to renounce immediate and tangible gains for some ideal conception of the good life.

The Year 2000

R. Buckminster Fuller

This article is a freely edited version of a lecture given at San José State College, California, March 1966.

In relation to increments of time and of prediction, I am confident that I cannot predict for A.D. 2000. Though it is only a little over a generation forward, I do not believe that any human being can foresee with any accuracy as far ahead as that 15 years. What will go on in this next period will be more of a change than has occurred in the whole history of man on the whole earth. All the trend curves which we may examine show rates of acceleration which underline the unprecedented nature of the changes to come. In plotting such trend curves over a period of years, I have sought fundamental information on the experiences of man on earth which might govern the shape of all developments men are experiencing. Charts of inventions, for example, are not satisfactory as the list is open-ended and is difficult to assess in terms of the relative importance of specific inventions. The significant area of information is the rate at which scientists have successfully isolated the chemical elements.

This is the most important pattern of discovery with which man deals, embracing, as it does, all physical phenomena. The rate at which man found chemical elements seems to be the key controlling the development of the application of science to technology, and, following from this, the application and effect of that technology on economics and, ultimately, the effect of the new technology on society itself.

The chart on pages 498 - 499, Profile of the Industrial Revolution, begins with the year A.D. 1200, going up to A.D. 2000. We begin with a list of nine elements: carbon, lead, tin, mercury, silver, copper, sulphur, gold and iron, which were known to man at the opening of our history. We do not know when they were first isolated or knowingly used. The first known isolation of a chemical element is arsenic, in 1200. Following this, there is a 200-year gap and we come to antimony, another 200-year gap and we come to phosphorus. Then the gap narrows to 75 years and we have cobalt. From here on we average an isolation of an element every two years. It is at an extraordinary period in history that the rate begins to accelerate.

If you check the date 1730, not long before the American Revolution, you will notice that there are some separate shoulders, or plateaus, appearing on the chart. These shoulders are slowdowns when we have major wars—the American revolution, various civil wars and World War I. They show that pure science does not prosper at the time of war—which is contrary to all popular notions. Scientists are made to apply science in wartime, rather than look for fundamental information.

We can also see that in 1932, which was thought to be the depth of the depression, man made his 92nd isolation of a chemical element. This completed the element table representing the full family now mastered by man—in the sense of his ability to repeat element isolation and to rearrange elements, as fundamental ingredients of physical environment, in preferred patterns of use. From this point on we may notice something strange. Previous to completion of the table, element isolation occurred irregularly—for example, element number 19 would be the 43rd element isolated; the 45th isolation would number 30, etc. With the post-uraniums the isolations show an absolute regularity of increase—they come in by number. Man begins to control consciously the rate of development of his capability.

We must note this in reviewing the contiguous developments in environ control (as shown at the top of the chart). Just as man is able to go into cold climates by putting on fur skins or into hot by taking off clothes, he enters more hostile environments by having more control devices. The development of these devices is a fundamental measure of man's degree of advantage over his environment. The first time, to our knowledge, that he goes around the world in an invention was, as shown on the chart, in a wooden sailing ship. It comes after the second isolation. Then there is a gap of 350 years and he now goes around the world in a steel steamship. This is an entirely new magnitude of control, no longer dependent on the wind. Upon this, there swiftly follows the world journey in an aluminum airplane, then in the "exotic metals" rocket. There is a very great contraction in time between these developments. The wooden ship takes two years to circumnavigate the earth; the steamship two months, the airplane two days and the orbiting satellite just over an hour. We have at least three accelerations of accelerating accelerations involved here.

The consequence of what we have considered then, in relation to our charting, is that the next point for a significant new chapter would be around 1975, nine years from now. What that will be we can only guess at—sending ourselves around the world by radio?

The key realization is the degree of acceleration of change, and that better than 99 per cent of all important technologies affecting such change are *invisible*. Man cannot see what is going on. He cannot "see" the chemistries, he cannot see the alloys. Most of the important rates and patterns of change

1250 A.D. 1270   1290   1310   1330   1350   1370   1390   1410   1420   1450   1470   1490   1510   1530   1550   1570   1590   1610

SAILING SHIP

LEONARDO DA VINCI

COLUMBUS
COPERNICUS

GALILEO

ALGORISMA INTRODUCES CYPHER INTO EUROPEAN CIVILIZATION FROM
ARABS, THUS PROVIDING SCIENCE WITH PRACTICAL CALCULATING FACILITY

9 ELEMENTS WERE
ACQUIRED BY CIVILIZATION
PRIOR TO HISTORIC RECORD
OF THE EVENTS, PROBABLY
IN ASIA MILLENIUMS AGO

CARBON	#6	C
LEAD	#82	Pb
TIN	#50	Sn
MERCURY	#80	Hg
SILVER	#47	Ag
COPPER	#29	Cu
SULPHUR	#16	S
GOLD	#79	Au
IRON	#26	Fe

10 ARSENIC   #33 As (first recorded discovery) Bavarian

11 ANTIMONY   #51 Sb German

9
8
7
6
5
4
3
2
1
1250 A.D. 1270   1290   1310   1330   1350   1370   1390   1410   1420   1450   1470   1490   1510   1530   1550   1570   1590   1610

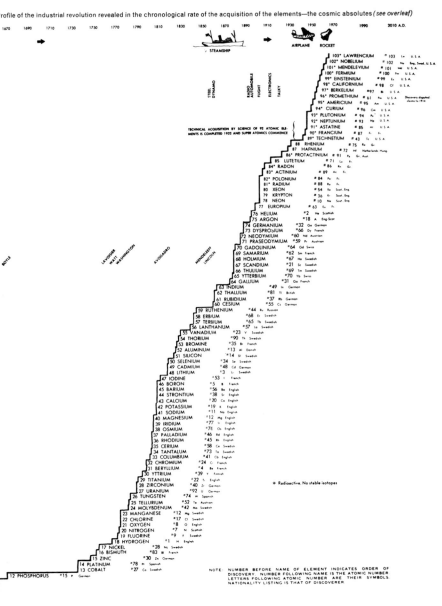

Profile of the industrial revolution revealed in the chronological rate of the acquisition of the elements—the cosmic absolutes (*see overleaf*)

cannot be apprehended by him directly in a sensorial manner. Not only does man have a very narrow range of tunability in the electromagnetic spectrum where he can actually see, but he also has a very narrow spectrum of motion apprehension. He cannot see the hands of the clock moving or the stars or any of the atoms in motion.

In the same way, man has had a limited understanding of the invisible historical factors. Few people are aware of the tremendous power wielded by a relative handful of men, of world masters, in respect to the whole world up to the time of the Great Crash in 1929. World War I was a struggle between one handful, the *outs,* and another, the *ins.* But the old invisible"pirate" masters who ruled the seas and the men on land were not displaced until 1929 when the corresponding developments in science and technology outran their traditional capacities. The J. P. Morgans and the like had operated for centuries through manipulation of the visible, and sensoriable, apprehendable, physical factors influencing trade, economies and "power" balances. Probably 99 per cent of humanity never knew that they existed and didn't know that they had gone. Control passed to their manager/lawyers, the people's politicians, and they are making a mess of it. One finds people around the world feeling extraordinarily well disposed to other people and with none of the emnity and suspicion which is claimed by their political leaders.

We need not be against politicians to realize that their local preoccupations are futile. They have good, convictions and are individually moved as human beings by what they regard a responsibility to "their" side or "our" side. But every political ideology and all extant political systems assume that there is not enough to go around—it is either you or me—there can't be enough for both. So we eventually assume war, and that is the cause of the weapons race. The reasoning was once correct. When there is enough available a healthy human will eat three pounds of dry food a day, drink six pounds of water and breathe fifty-four pounds of air or six pounds net of oxygen. For most of the history of man on earth there has not been enough of that dry food and humanity has fought about this, time and again. Many times there has not been enough water, and humanity has fought over this. There has been no time when there has not been enough air. Humanity has so much air available that no one has even thought of putting meters on air and trying to make money out of it. But, there are times, for example in a great theatre fire, when humanity, completely unused to competing for air, finds itself suffocating and goes mad.

It seems perfectly clear that when there is enough to go around man will not fight any more than he now fights for air. When man is successful in doing so much more with so much less than he can take care of everybody at a higher standard, then there will be no fundamental cause of war.

In the years ahead, as man does become successful, the root cause of war will be eliminated. Scientists assure us over and over again that this is feasible.

There can be enough energy and organized capability for all men to enjoy the whole earth.

This is the most important prediction I can make: in ten years from now we will have changed so completely that no one will say that you have to demonstrate your right to live, that you have to earn a living. Within ten years it will be normal for a man to be successful—just as through all history it has been the norm for more than 99 per cent to be economic and physical failures.

Politics will become obsolete.

At the present moment, we could take all the machinery from all the countries around the world, all the railroad tracks, all the wires, etc., everything we call industrialization—and we could dump all this in the ocean. Within six months, two billion people would die of starvation, having endured great pain. On the other hand, supposing that we take away instead every politician, all the ideologies, all the books on politics—and send them into orbit around the sun. Everybody would keep on eating as before, down will go all the political barriers and we would begin to find ways in which we could send the goods that were in great surplus in one place to another. So people may even begin to eat a little better—in a hurry. This could not be said before.

There are many prognostications about immediate technology. It seems likely, for example, that there will be considerable advances in transportation, but our present ignorance in handling traffic is appalling. At the bottom of our air-ocean world we are still like flounders and crabs travelling in burrows. In surface travel we restrict ourselves to crowded tubes and lines between buildings and trees when we could go omni-directionally. This is in dramatic contrast to air-travel in which as soon as you go any distance at all you lose sight of other airplanes and can go thousands of miles without seeing another human being. It is only when you slow down that you begin to re-establish close contact. The slower we go, the more crowded we get. Finally, as we leave an airport and get out onto a free-way we have the preposterness of running in lines in opposite directions at 65 miles per hour five feet apart—with everybody practicing steering. A decade from now this will look rather silly. With current technological trending in omni-directional transport we will finish our great highway programmes just in time to turn them into some kind of roller skating rink. It may be fun to roller skate from California to New York—and you will have the time to do it if you like.

It is all Buck Rogers and it will happen. But, such speculation is a waste of time, it is more important to consider what will happen to our relationship one with another.

In its broadest aspect this area must be considered under "population." There has been a great debate about the so-called population explosion in recent years. This has been occasioned in part by the fact that we have only recently had accurate census in many countries. Even in Europe population figures only go back a short time.

In the USA, though there was an increase in the post-war birth rate between 1947 and 1954, since then it has declined. This trend is also evident in all the industrialized countries, including Russia. During the last twelve years then the birth rate has been declining in the industrialized countries, yet the main problem is thought to be population increase. The cause of the bulge in census population of these countries is, of course, that more people are living longer. But the underlying reality of the population problem, if there is a problem it is that as we industrialize, the rate of births decrease. We may see this most clearly in, for example, the US, where the early settlers had an average of thirteen children per family and survival rate was very poor. We may then plot the decrease in number of children per family against improvement in technological services, public health, indoor water supply, bathrooms, refrigeration, general improvement in life expectancy, and so on. Clearly, as man industrializes and improves the probability of human survival, whatever the drives or controls of nature are, she does not have to have anywhere as many birth"starts." This is one of the fundamental points about industrialization.

We should also consider the rate at which countries become industrialized. England took two hundred years to get industrialization going and up to the present level. The United States"took off"from England's vantage point and did it in a hundred years. Russia came in and accomplished in fifty years what had taken the USA a hundred years, because it was able to start at a more advantageous point. We find that the new countries come in where others left off, not where they started. Japan did not start flying with the Wright Brothers bi-planes, but with the "Zero" and "Spitfire" types; China has never flown anything but jets. China came into the world of industrialization after the transistors, computers and atomic fission were available—so she will come to industrial parity with the West in about five years. India will probably be even faster. The acceleration of capabilities coming to bear on India and Africa are of the very highest. As far as one can see, industrialization will be world wide by 1985.

By this date, as the world industrial process is completing, and birth rates reducing, every individual human being will still have about ten acres of dry land and approximately twenty acres of ocean averaging half a mile deep. In terms of a family of five that would be fifty acres of land and 100 acres of ocean —150 acres of acres per family. The amount of food supply would be ample.

We may glimpse in such patterning certain total behaviours in universe that we know little about. We noted, for instance, that as survival rate and life sustaining capability increased, fewer birth starts were"required." This may be related to our developing capacities in interchanging our physical parts, of producing mechanical organs, of having progressively fewer human organisms to replenish. The drive in humanity to reproduce as prodigally as possible decreases considerably. This may be reflected in social behaviours—when all the girls begin to look like boys and boys and girls wear the same clothes. This may be part of a discouraging process in the idea of producing more babies.

We shall have to stop looking askance on trends in relation to sex merely as a reproductive capability, i.e. that it is normal to make babies. Society will have to change in its assessment of what the proclivities of humanity may be. Our viewpoints on homosexuality, for example, may have to be reconsidered and more wisely adjusted.

Central to such readjustments will be the concept that man is not alone the physical machine he appears to be. He is not merely the food he consumes, the water he drinks or the air he breathes. His physical processing is only an automated aspect of a total human experience which transcends the physical. As a knot in a series of spliced ropes of manila, cotton, nylon, etc., may be progressively slipped through all the material changes of thickness and texture along the length yet remain as identifiable pattern configuration, so man is an abstract pattern integrity which is sustained through all the physical changes and processing.

We become more aware of this uniqueness of organized and organizing principle in the universe, in science. The long-held myth that science wrests order out of chaos is fast disappearing in due ratio to the extent that all great scientists have found the universe to exhibit *a priori* orderliness. All the various specialities are discovering that their variously remote studies which seemingly 'ordered' local aspects of nature are converging within progressively simpler and more comprehensive patterns. The "ordering" is coming together. When we refer to the computer and automation taking over, we refer really to man's externalization of his internal and organic functions into a total organic system which we call industrialization. This metabolic regenerating automated organism is going to be able to support life in an extraordinary way. The machines will increasingly assume various specialized functions. Man who was born spontaneously comprehensive but was focused by survival needs into specialization is now to be brought back to comprehensivity.

As enormous numbers of men are freed for more education and research and as they become more and more comprehensive in their dealings with nature, there will be engendered a total philosophic awareness of the significance of the whole of human experience. There will be a rediscovery of what Einstein described in 1930, in an article on the "cosmic religious sense"—the intellectual integrity of the universe and an orderliness that was manifestly *a priori* to man.

We are going to have an increasing number of human beings as scientists and philosophers thinking about the total significance of human experience and realizing that there is an intellect far greater and far more powerful than that of man—and anticipatory of the whole trend of his development. An era of extraordinary integrity might ensue.

This would be for me, the most important and exciting aspect of all the trend curves—that in A.D. 2000, to a marked extent, the integrity of humanity will be of an unbelievably high order. What one human being says to another regarding what he thinks or what he has observed will be reliable. There will be play-acting still, but it will very clearly be play-acting. In looking forward to the year 2000, it is not the "Buck Rogers" details which are important but whether the world will be a good place for our children and grandchildren. In the past, man had to do many things shortsightedly and we have wasted a great deal of our natural heritage. We have squandered the fossil fuels which represented an extraordinary "savings" or energy capability account stored up in the earth. The great change now will be in a new type of accounting when we begin to draw more consciously on the fabulous "income" energies of sun, water, and wind and tidal powers—which if not used will not be 'saved' or impounded on the earth. We will adopt new accountancy standards for all wealth. To account our success in terms of gold and various traditional banking practices is irrelevant. Real wealth is organized capability. One of its important characteristics is that it is irreversible—no matter how much wealth you have, you cannot change one iota of yesterday. Wealth can only be used now and in the future. What we really mean by wealth is how many days forward we have energy available and organized for work to keep the machines running, to keep the foods growing, the refrigeration transportation and so on. The basis for our new accounting system will be—"How many forward days of organized capability do we have available to serve how many men." We will be able to make the working assumption that it is normal not only for man to be successful but also normal for him to move as freely as he wishes without interfering with any other man. Our overall accounting assumption will be based on whatever amount of organized energy capability is required so as to make it possible for any man to travel around and enjoy the whole earth, and be completely supported in doing so. There will be no such thing as deficit accounting. You cannot live on deficit accounting. You cannot eat deficitly or drink water deficitly. What is to eat is there—as the water is there.

All such negative accounting procedures went along with the need for exploiting others in the "you or me" phase of man's past struggle for basic survival.

Much of the most exciting and important part about tomorrow is not the technology or the automation at all, but that man is going to come into entirely new relationships with his fellow men. He will retain much more in his everyday relations of what we term the naïvete and idealism of the child. This will be completely justified and not exploited or exploitable in any way. I think then that the way to see what tomorrow is going to look like, is just to look at our children.

Some Psychological Perspectives on the Year 2000

George A. Miller

What plans and problems will concern the leaders of industry, education, and public affairs in the year 2000? Undoubtedly they will be as curious about their future as we are about ours. But what will they be trying to foresee and promote or avoid?

R. J. Herrnstein, a valued friend with whom I have shared many administrative chores, once summarized our experience together in a principle that he called "the conservation of trouble." As each difficulty is solved, a new one takes its place. That, of course, is an optimistic theory of trouble; others would speculate on the exponent to be assigned to its growth. In either case, it is unlikely that the problems our descendants will face in the year 2000 will be simpler than ours. By then, no doubt, they will have developed better methods of extrapolation than we have today—but solving that problem would surely raise others, for how does one maintain an open society when the future is thought to be known?

They will probably be worrying about what to do when the resources of the oceans are exhausted. By then they will have enough information to make eugenics feasible and will be wondering what to do about it. Undoubtedly they will be even more concerned with the population problem, although technology will have changed the focus of their discussions from contraception to persuasion, and from farming to manufacturing as the source of food. Professions will be even more important then than now, and the education of professional men will be a pressing item on their agenda of problems. They will still be wondering how to make life more meaningful and satisfying, especially to those whose talents and skills are of marginal value to society, but they will be discussing this problem in the context of a psycho-pharmacological revolution that will offer possibilities of escape difficult to regulate. War will concern them, but on new terms, in new places, under new conditions. They will try to invent new social institutions to cope with their problems, perhaps joint ventures designed to merge public financing with private initiative. And they will be just as puzzled about the ultimate meaning of it all as we are today.

Is there something that a person trained in the methods and theories of contemporary scientific psychology can do to reduce or alleviate the never decreasing burden of troubles that our children will inherit by the year 2000? Exposure to psychological theory and research does little to prepare one to cope with such questions. In order to enter such a discussion at all, I must first disavow any pretense of scientific privilege; I cannot speak for psychology, but only as an individual who happens, fortuitously, to have had his opinions transformed by that particular kind of professional training.

I believe that two major trends are likely to continue for at least another thirty years: The population will continue to grow, and technology will continue to be a major source of change in the affairs of men. Within that context and limited by those preconceptions, however, there is much to stimulate a psychologist's imagination.

What will life be like when it is shared by seven billion people? Undoubtedly they will need a great deal of self-constraint to tolerate their mutual intrusions, to respect their differences of custom and opinion, to value the individual when there are so many individuals. People will have to be very civilized to get along together, and that requirement raises a host of other questions. Is man really capable of organizing a big civilization? If the church continues to lose its authority in ethics and morality, where will a new impulse toward a more civilized behavior originate? Freud argues that civilization, by curbing natural instincts, places increasing burdens of guilt on the civilized man. Certainly a conscience can be a difficult master; it is always tempting to substitute hypocrisy in its place.

Of course, the human population may not go on expanding at its present rate. According to biologists, in most ecological systems there are forces that limit automatically the number of individuals. When they become too crowded, the animals starve or kill one another, or a debilitative disease suddenly reduces their number dramatically. But will such factors operate automatically in the case of man? I think this is unlikely. Our past wars, horrible as they were, made only a ripple in the population curve. We may be on the verge of a great universal famine, but there are still ways to forestall it. A Malthusian formulation is not adequate, for technology has added dimensions to the human problem that are unprecedented in the evolution of species. It is difficult for me to believe that a contraction of the human population will occur automatically.

If we begin to poison ourselves with pollutants, we will take measures to reduce them. If we are afflicted by debilitative disease, we will subsidize biomedical science to find a cure. If our farms become inadequate, we will develop methods and build factories to produce synthetic proteins. If we invent thermonuclear devices capable of destroying all life, we will find social constraints on their use. No automatic biological principle will take over our destiny, for human intervention is possible, and in emergencies it can be rapid, massive, and effective. Our destiny is in our own hands.

Does it seem unwarranted to call this an optimistic vision? One argument for believing in heaven is that if you are wrong, little is lost, but if you are right, much may be gained. The argument for believing in the survival of rational man is much the same. The alternative vision of a lifeless, radioactive sphere spinning silently through space cannot be dismissed, but I refuse to make plans for it. If we are to take the year 2000 seriously, we had *better* believe that our destiny is in our own hands.

Having said all this, I nevertheless believe that in its own slow and inexorable way the old machinery of evolution continues to work on man side by side with the rapid and (optimistically) controllable evolution that is man's own invention.

Where should we look for evolutionary pressures today? It is a commonplace observation that people differ in their tolerance for crowding. In large measure these differences are a result of social training, of experience with crowded living. But there may also be a biological basis for them, an innate predisposition for some people to thrive better under crowding than others. This possibility is something biologists have tried to study in animals, and about which much has been written. The selective evolutionary pressure that would favor a crowd-adapted organism is reasonably obvious. If there are innate differences, and if we do face a future of living closer and closer together, evolution would inevitably favor certain people over others. By the year 2000, of course, nothing measurable will have had time to happen as a consequence of biological evolution. But we ought to know more about the possibilities and implications of such a change. It might be interesting for psychologists to attempt to measure individual differences in tolerance to crowding, and to try to determine whether there are inheritable traits that could serve as a basis for evolutionary selection. When the day comes that we have the necessary facts to support a realistic eugenics program, such information could be valuable.

The most obvious and foreseeable consequences of population increase and technological innovation are surely the demographic and economic ones, which I have neither desire nor competence to consider. There is, however, one general psychological principle that may have interesting implications. The human mind has a limited capacity for acquiring and storing information. It is a reasonably well-established fact that the number of independent cognitive components a man can cope with at any one time is strictly limited. Evidence could be cited to support this claim, but most people seem willing to accept it as obvious. It is also a plausible speculation, not yet firmly established, that there is some limit to the overall capacity of a man's memory, a limit to the magnitude and complexity of any cognitive system that a man can internalize. More than anything else it is man's capacity to cope with large, interrelated systems of habits and rules that sets him apart from other animals—that enables him to learn a linguistic system, to invent a system of mathematics, or to learn all he must know to be accepted as a member of his society. Gifted as man is in this respect, his capacity is probably not unlimited. Intuitively, we all know this. In our mundane affairs we allocate as much of our memory to a topic as we think its role in our lives justifies. Indeed, by its very nature, the learning process preserves only those facts or skills that recur frequently or are particularly important. As an advertising executive once said, there is only so much that the public wants to know about toothpaste; the advertiser's job is to force his product into that small mental compartment, because he knows that he will thereby almost surely force some competitor's product out.

It would be ridiculous, of course, to insist on some rigid and mechanical limit to human attention or human memory. We can stretch our limited span of attention by carefully organizing information hierarchically and then dealing with our problems at a rather abstract level, relatively secure in the belief that, when necessary, the more detailed information can be reconstructed from the hierarchy stored in memory. But if memory is also limited, this hierarchical strategy must have its limits. Psychologists have not yet demonstrated or measured these limits in their laboratories, but it seems plausible that such limits must exist. And even if memory were an unlimited vessel, the rate at which experience could fill it would still impose limits on the amount it could contain.

As far as the present argument goes, it makes little difference whether the mind is limited in capacity or limited in rate of acquisition. In either case we are faced with an upper bound on the information we can expect an individual to have at his personal disposal. If society itself becomes more complex, the amount of knowledge a child must acquire in the process of socialization increases. What are the implications of this increase for a country whose government is based on the assumption of an informed electorate? If more and more technical competence will be required in order to earn a living, it means that more and more of a man's precious cognitive capacity will have to be devoted to that. We may already be nearing some kind of limit for many of the less gifted among us, and those still able to handle the present level of complexity are in ever increasing demand. The remarkable shift we have seen in the advanced countries from the idle rich and the exploited worker of 1870 to the overworked professional and the unemployed poor of today is at least partly the result of such psychological limits operating in a context of advancing technology. If the meritocracy is already taking shape, it is only a form of scapegoating to blame the purveyors of intelligence tests. The real causes lie far deeper.

At a time when our society is wasting so much of its potential intelligence, it might seem that the solution to this problem would be to improve our educational system and to reduce discrimination based on race, sex, age, color, religion, and nationality. We need every good brain we can train. But such measures, important as they certainly are, would only enable each person to realize his own capacity. If that capacity is limited even under optimal conditions, the problem takes on new dimensions. Even our most intelligent citizens will have to rely increasingly on artificial aids—on such intelligence amplifiers as digests, libraries, computers, special displays, and communication devices. Most important of all, it seems inevitable that even our most intelligent men will have to work increasingly in teams; no single member of any team would be competent to understand all aspects of the shared problem. This change is already beginning in some industries, especially those where technology has advanced too rapidly for management to exploit it effectively without new forms of co-operation among technically trained personnel. These teams are assembled as needed and dissolved when their work is done; their transitory character threatens something of a revolution in managerial practices. This kind of co-operation—experts collecting around each important new problem and then moving on when the problem is solved—has been slowly developing for many years in scientific laboratories. We can expect to see more of this co-operation in the future—in laboratories, universities, industries, government—as our problems become greater and more complex, and our individual mental capacities do not.

No psychologist worthy of the title could close such a topic without at least wondering about its motivational implications. How are these experts to be rewarded? Who gets the credit and recognition when a team solves a problem? All members of the team will be well paid, no doubt, but is that enough? Each expert will be appreciated by his peers, of course, but on the national stage he can be little more than an anonymous consultant moving like a shadow behind the scenes. There are many motives that keep men hard at work at difficult but important tasks: the desire for power, wealth, fame, or knowledge. Which of these will move the technical teams of the future? Co-operating experts will have little personal power over others, and what limited power they do have will be given up when they move on to another task. Dreams of great wealth seem equally unrealistic; we are thinking now of men who, quite literally, live by their wits. Fame will come to few of them; the public can appreciate only a limited number of eminent men. Knowledge they will enjoy, but only under conditions that will force them to recognize that their personal knowledge is inadequate and must serve the purposes of an organization.

In his *Essays in Sociological Theory: Pure and Applied,* Talcott Parsons comments on the misleading stereotypes of the professional man motivated by altruism and the businessman motivated by acquisitiveness. In Parsons's view these stereotypes are more institutional than motivational. In both cases the dominant personal goal of the individual is "success"; the real difference lies in the paths leading to that goal. For both the professions and for business, the social institution must be so organized that objective achievements of value to the institution will bring recognition and "success" in due proportion. But articulating achievement and recognition equitably is a peculiarly thorny problem when it is a collaborative effort that must be rewarded.

An expert collaborator finds satisfaction in doing his job well, and in the sense of affiliation and shared accomplishment that has always motivated small groups of dedicated men. But these are altruistic motives and may not be sufficient. The prospect is disturbing, for only the strongest motivation will drive a man to learn to the limit of his capacity and to go on learning long after his formal education has ended. What alternative rewards can society provide to keep an anonymous genius hard at work? Probably the experts will incorporate themselves and develop their own answers to that question.

These speculations are advanced in the belief that the intellectual elite—both in business and in the professions—are a particularly important segment of the population, and that they will encounter mounting difficulties in generating those phases of the industrial revolution that lie ahead. If the direction of social change that we have learned to call progress is to continue, our best minds will have to find some more effective way to pool their abilities, for many of our most pressing problems are already too large to fit inside any single head.

New techniques for expert collaboration are developing rapidly. Let us assume they will be successful and that we can extrapolate the present trend toward increased complexity and ever more rapidly changing technology. What will be the effect on the masses of people who do not aspire to change society, but merely want to find some meaningful existence in it?

If we are to remain true to our democratic heritage, one of the most obvious implications of the predicted increase in population is that our already crowded educational system will have to be vastly expanded and overhauled. As knowledge increases and work becomes more technical, there will be a corresponding increase in the amount of information that will have to be imparted to a student. And as automation advances and new industries replace old, learning will not be regarded as ending with graduation from school, but will become a way of life for everyone. Put together the increased number of students, the increased knowledge to be communicated, and the increased duration of the educational experience, and then try to imagine what kind of educational system we will need by the year 2000. Can anything short of an educational revolution meet our needs?

I have followed recent innovations in educational practice with considerable interest. Some kind of change is obviously needed. Too often our children's most valuable return for their years in the classroom is a kind of shrewd skill in coping with a large, well-intentioned, but often stultifying social institution. On the theory that any system that is not changed gets worse, there have been valiant attempts to revise curricula, to write better texts, and to provide more teaching aids, while at the same time making the best possible facilities available to every student. All of these excellent improvements and innovations are necessary, but are they sufficient?

I do not wish to sound critical of all that is being done, yet I feel that in their enthusiasm our educational revisionists sometimes forget a basic fact about the learning process. Most of the studies that have made a serious effort to evaluate the effectiveness of these new programs have shown that the method of packaging the information makes relatively little difference. Of course, if the information is wrong or irrelevant, a student cannot learn what is right or relevant; if, however, the same information is presented in alternative ways, the major factor determining how much a child learns is how much time he spends studying. Some learn faster than others, some learn more than others, but on the average, the generalization holds true. The problem, therefore, is how to motivate the students to study.

Obviously, we must see to it that the content of the teaching is clear, accurate, and up-to-date, that the teacher understands it, and that the student, whether he realizes it or not, really needs to know it. All of this is obviously conducive to a profitable educational experience. Yet it would be useless if students refused to study. Conversely, a student who is truly determined to learn something can learn it even under the most impoverished conditions.

I am not putting forth some radically new dogma. The fact that education is the reward for study is so banal that I am embarrassed to mention it, much less emphasize it. If it were not so important, I would prefer to leave it unsaid. But as every educator knows, the central problem of education is to make the students want it. Unfortunately, the problem is just as difficult as it is important. We know how to write better books and print them with three-color illustrations; we know how to shuffle the order of units within a curriculum; we know how to break up a unit into small steps and drill each step separately; we know how to use movies and field trips and special projects; but we do not know how to inject the urge to learn into a student's heart. So we do what we know how to do.

In defense of those who try to improve the packaging of the information that is presented to the student, I must agree that the finest motive of all for studying is love of knowledge for its own sake. Every subject matter has an intrinsic interest of its own, and a student who becomes intrigued with it on its own terms will certainly be the most gratifying to his teacher. My suspicion is that too often this experience is reserved for a fortunate few who are both highly intelligent and protected from more immediate personal distractions in their own lives. For most students study is a painful experience, and the social milieu of the public schools seldom encourages them to bear the pain until they learn to love it. Our schools, I fear, too often illustrate the irony of a self-fulfilling prophecy. A student is labeled as good or bad. First the teachers and then the student himself accept this classification. If he is mislabeled as good, he may become an "over-achiever," but if he is mislabeled bad, he accepts the judgment and fulfills the prophecy. How can we expect every student to acquire a detached love of learning under such conditions?

An essential ingredient in the motivational pattern of a good student is one that David McClelland and his colleagues call "need achievement." Need achievement manifests itself in a desire to do better, to compete against a standard. How our schools are to instill a desire for success in students who have not already acquired it at home is a difficult but important social question. Attempts have been made to teach people to think like achievers, to learn the opinions and behavior patterns that characterize the successful person. The first results have been encouraging; this kind of motivational training may prove both possible and practical. If it does, perhaps we can even reach some social consensus about its use, in which case psychologists would have contributed an important weapon to the educator's arsenal and helped to mobilize our human resources for the social good. But can parents who would refuse their children a relatively innocuous innovation like flouridation be persuaded to embrace such a deliberate public program of personality modification?

To my mind one of the most persuasive answers to this motivational question is that more initiative should be placed in the hands of the learner. If we want to motivate students to study harder, we should enlist their cooperation. This prescription would probably not be a universal panacea, but giving initiative to the student is important in adult education. Under existing conditions, however, it is not easy to give initiative to the student. In order to allow a learner to say more about what he studies, when he studies it, and how far he takes it, a teacher must adapt himself to the student's interests and abilities. When you recall that the teacher is usually outnumbered by thirty or forty to one in most schoolrooms, the impracticality of this solution becomes all too clear. Given realistic economic constraints on the expansion of our present educational system, I do not see how we could relinquish the initiative to students under that system. When it has been tried, as it often has in the beginning grades, the result has usually been to convert the class into a period of supervised recreation. We would have to change the system. I believe there are alternatives open to us that could achieve the desired result. If these alternatives do indeed prove to be better for motivating students to study, our schools may look very different by the year 2000.

The alternative systems I have in mind would exploit the modern, time-shared computer. Imagine a classroom partitioned into semi-isolated booths. In each booth are a pair of headphones, a typewriter keyboard, a screen similar to a television set's, and a photosensitive "light gun." All of these stations (and others in other classrooms) are in communication with a central computer. A student communicates with the computer by typing on the keyboard or by touching his light gun to designated spots on the screen; the computer communicates with a student by playing recorded speech through the student's earphones, or by writing or drawing pictures on the cathode ray tube. Each student can be working on a different lesson, or two on the same lesson can progress at different rates. A teacher walks from booth to booth, answers questions, sees that the stations are operating properly, and supervises requests for new materials.

A science-fiction fantasy? Not at all. Such systems are already operating. The one I have just described is operating in a public school in Palo Alto, California, as a pilot project under the direction of Patrick Suppes and Richard Atkinson of Stanford University. The children are learning about the same amount they would have learned under the regular system, but their attitude toward learning is entirely different. Learning is fun, they are more curious, and they enjoy studying from the computer. The cost—leaving out the cost of development—is only slightly more per student than before.

If the motivational advantages of this system persist when it is no longer a novelty, we can expect to see many more of these systems in the future. There are several reasons to think that a computer-based school makes sense. Students can go at their own pace. One who has trouble can get additional material; one who makes no mistakes can go on to more advanced material. Bright students are not bored while the teacher explains what they already know; dull students are not baffled by being left behind. There is no need for testing; students' records are maintained automatically. A teacher can teach and leave the threatening duty of evaluation to the machine. Within the broad limits set by what materials have been prepared for the computer, the student is free to study those things that are of most interest to him. And a computer treats all children alike, regardless of race, creed, or color.

For many people the computer is synonymous with mechanical depersonalization, and computerized instruction is frequently regarded as a way for the teacher to avoid his personal responsibility to his students. Fears have been expressed that the computer represents an assembly-line approach to the educational process that will increase alienation, identity crises, *anomie,* and so forth. Such attitudes seem overly emotional. The evidence points in the opposite direction. The computer gives the child a measure of individual attention that he could receive in no other way, short of a private tutor. To the extent that initiative can be left in the hands of the learner, rather than given to the machine, I believe these devices can help to solve an important educational problem.

Needless to say, stations do not have to be located in classrooms. They could be in libraries, or factories, or even private houses; all that is required is a telephone line to the computer. It should not be too difficult to make such facilities available for adult education. If economic considerations make it necessary, classroom stations in the public schools could be used for adult education in the evenings. It seems likely that businessmen will develop their own computer-based teaching systems; some of the most enthusiastic proponents of programmed self-instruction are businessmen who have used it to retrain their own personnel.

The shared use of a central computer by many stations at remote locations can be adapted to other purposes than public education. For example, it promises to be one of the more useful tools for enabling teams of experts to collaborate efficiently. I believe that the first time I heard the phrase, "on-line intellectual community with shared data base," it was intended as a summary of the various possibilities that Project INTREX (an M.I.T. adventure in library science headed by Carl Overhage) was considering during its planning conference in the summer of 1965. In fact, a visionary description of the possibilities inherent in making a "data base" (for example, a library or some part of it) accessible "on-line" (direct communication to a computer from a remote location via telephone line) to an "intellectual community" (a group of scholars or scientists working on a common problem) had already been written in 1965 by J.C.R. Licklider in his *Libraries of the Future.*

A number of organizations are presently working toward the introduction of computers into libraries, or vice versa. Not only can a computer provide a wide range of clerical services to its users, but a library of shared references will be available to them, their own data or other materials can be stored there, and the materials of other users can be made available on request—and all of this is accessible by simple requests initiated and fulfilled at the keyboard of a remote teletypewriter. Scholars in widely scattered locations will be able to work closely together without leaving their houses, and they will have the advantage of clerical, stenographic, library, telegraph, and publication services via the system. Something suggestive of such a computer system is already taking shape on a few college campuses; regional, national, or even international networks would be possible if they seemed desirable.

With just a little foresight in the development of these systems, they could turn out to be one of the greatest educational innovations since the invention of printing. If a student were provided with a console of his own, he could, at little or no cost to the intellectual community, have access to the most advanced thinking in his field of interest; a student in small or isolated colleges could be given the same access as the student at a great university. Moreover, the system would be responsive to his requests, so it would satisfy the requirement that initiative remain in the hands of the student.

The computerization of psychology is already well advanced, and the other behavioral and social sciences are not lagging far behind. Larger data bases and more ambitious data analysis are only part of the story. The machines can be programmed to simulate complex psychological and social systems, to conduct experiments, and to provide communication among scientists. The computer could become as important to the behavioral sciences as the microscope is to the biological.

Nevertheless, the application of computers to the study of man raises some difficult problems. Whenever it is proposed to put large quantities of data into a common file—particularly if the data are of the kind that most social scientists are interested in—there is danger that the information may be misused, a danger that has led in some quarters to an emotional resistance to the whole idea. When computer memories become so large that it will be unnecessary to discard information, and when any item can be made available in a few seconds, the temptation for a government to keep complete dossiers on all its citizens, and particularly those who are intellectually most active, will be quite real. It will be necessary to develop and instill a code of professional ethics among the scientists who use such data, and in some cases legislative safeguards may be required to protect the individual from the invasion of his privacy that such technology will make possible. Congress is already concerned with the problem, and we can expect to hear considerable discussion of it in the years ahead. How these safeguards are implemented could have some important consequences for our knowledge of man and society in the year 2000.

Any effort to peer into the future is likely to impress a psychologist with how fortunate economists are in comparison with other social scientists. The modern economist has available an extensive data base of economic statistics that enables him to formulate and test macroeconomic theories of the national economy, and the theories he has developed have given us new ways to control our economic fate. Other social statistics, however, are harder to come by, and the relative lack of solid noneconomic information about the personal and social status of our citizenry is reflected in a corresponding lack of empirically tested macrosocial theories outside the economic sphere. The advent of the on-line intellectual community with shared data base is an open invitation to sociologists, social psychologists, demographers, and others to follow the economists' lead. Let us hope, therefore, that legislative safeguards on the individual's right to privacy will not be so restrictive as to preclude the compilation of large, centralized, integrated data bases in the social sciences. Without them, the planners in the year 2000 will be scarcely better off than we are today.

The computer is here to stay, and, personally, I think that there is more hope than harm in it. There is another area of technological innovation that frightens me far more, an area in which psychologists will certainly be deeply involved, so I must at least mention it. I have in mind the recent developments in pharmacology, biochemistry, and related fields.

We already have drugs that can make us sleep or keep us alert, drugs that control our emotional state, drugs that induce hallucinations; drugs either to improve or to destroy memory are now appearing on the scene. This is just the beginning. In 1966 Dr. Stanley F. Yolles, Director of the National Institute of Mental Health, told the Senate that the next five or ten years would see a hundredfold increase in the number and types of drugs capable of affecting the mind. Before the year 2000 we will have to revise several of our current ideas about what is possible and what is advisable in the use of these new pharmacological agents.

There are many possible applications of our rapidly growing understanding of the mechanism of heredity. It will soon be possible to use fragments of cells to manufacture specific proteins, to control the sex of our offspring, to prevent hereditary defects. What psychological secrets may be locked up inside the cell? Is it too visionary to imagine that direct control of intelligence and personality may be possible? And, if so, what will they decide to do with these possibilities in the year 2000? What social problems would result if geneticists were to announce that they knew how to breed men who would live as long as turtles? We should not forget that we have also created an active program of research in biochemical warfare. Everyone hopes that we will never have occasion to use such weapons, but hoping may not suffice. The problem is too important to leave in the hands of the military, but as yet there is no consensus on what should be done about it. The quality of life in the year 2000 will be profoundly affected by the use we make of this new biotechnology. It is a social problem, not a scientific one, and eventually it must be discussed and decided by all members of our society.

It was in 1895 that the French psychologist Alfred Binet first suggested the use of ink blots for the study of various personality traits. An ink blot is just an ink blot. In order to see more than ink on paper the beholder must contribute something of himself to it, and the way he projects himself into the blot can be quite revealing. The future is no ink blot, but certainly any attempt to describe it must have a large projective component that will tell as much about the describer as about the thing described, and multiple descriptions can only yield a social projection. Nevertheless, the exercise is worth the effort. The future, unlike an ink blot, is still very much at the mercy of what we imagine it to be, and serious efforts to foresee it are less exercises in accurate prediction than they are attempts to reduce the eternal gap between what is humanly desirable and what is humanly possible.

The Test

Arthur Kantrowitz

> *Can we as a society restore our enthusiasm for technology? Part of the answer is a new system for its democratic control, including "advocates" and "judges."*

An outstanding characteristic of our time is the appearance of widespread *fear* of new technology. The fear of nuclear weapons is so deep that most people now regard discussion of nuclear weaponry as obscene. The fear that a computerized society will destroy human values is more widespread than it is articulate. Possibilities just beyond the present capabilities of science, such as genetic surgery, offer opportunities and consequences difficult to imagine. Of all the frightening aspects of technology the most frightening is its unpredictability. Adlai Stevenson once illustrated this unpredictability beautifully in a talk given in 1964. He said, "And I find myself on a par with the greatest scientific minds of the time (1937)—for I, too, failed to foresee nuclear energy, antibiotics, radar, the electronic computer, and rocketry."

In addition to these well-founded and deeply felt fears of the primary effects of technology, there has been in recent years an enormous emphasis on the undesirable secondary (i.e. unintended) effects of technology. For example, there has been a tremendous emphasis on the pollution of our environment. When someone mentions a power plant, many people think of it first as a device for producing pollution, and only secondarily of its role in producing electricity.

Dealing with these secondary effects will typically require some extensions of technology—e.g. environmental technology—and some social and political advances, for example, the creation of incentives or regulations sufficient to prevent pollution of our environment. I think we can gain some perspective on these secondary problems by recalling problems of this class which were considered important some years ago. For example, the exhaustion of our fossil fuel supplies. It was fashionable to calculate in the twenties and thirties that we were very soon going to run out of the energy necessary to drive an industrialized society. This problem has been solved in our time long before it became oppressive. Any difficulty which can obviously be dealt with by more technology should concern us much less in view of our enormous capability for the creation of new technology to meet recognized problems. These are to be distinguished sharply from the primary effects discussed earlier which cannot obviously be helped by more technology.

Nevertheless, these secondary phenomena have frequently served as obvious targets for criticism of the "technological society." We have seen in recent times increasingly vicious attacks directed at these targets. For example, Wilbur Ferry says (*Saturday Review,* March 2, 1968), "The writing [of rules and regulations governing technology] must be done by statesmen and philosophers consciously intent on the general welfare, with the engineers and researchers summoned from their caves to help in the doing when they are needed." Norman Mailer, at a recent symposium, said (*New York Times,* May 26, 1968), "I think American society has become progressively insane because it has become progressively a technological society. A technological society assumes that if it has a logical solution to a problem then that is the entire solution.

"If it decides that the problem, for instance, is to keep food in such a way that it may be eaten six months later, then it proceeds to freeze it, and then it points out to you that six months later when you unfreeze that food you can still eat it. What it does not decide scientifically—although it pretends that this has been a scientific operation—is what portion of that food has been destroyed, what unknown ailments may possibly be inflicted upon the generations of the future." To quote one more assessment of the current scene, Paul

Goodman has written (*Utopian Essays and Practical Proposals*), "And inevitably, given the actual disasters that scientific technology has produced, superstitious respect for the wizards has become tinged with a lust to tear them limb from limb. Calling this anti-scientific bent Luddite, machine-breaking, is to miss the public tone, which is rather a murderousness toward the scientists as persons, more like anti-Semitism." The revolution brought about by the magnificent union of science and technology is now threatened by a massive counterattack driven, I believe, by the very real and justified fears of the consequences of technology but perhaps still more by the unpredictability and the present lack of control of technology.

There can be little doubt that this counterattack has already noticeably decreased the capabilities and the funding of research and development in the United States. This type of counterattack has been powerful for several decades in some of the countries of Western Europe, and it is my opinion that it is to a considerable extent responsible for the "technological gap" which reportedly exists between Western Europe and the United States. On the other hand, enthusiasm for technological progress is deeply implanted in the Soviet Union, and continues to flourish there. I am convinced that only a few years of continuation of our present vigorous counterattack on technology will result in a technological gap between the United States and the Soviet Union which will be reflected in our economic and our strategic posture with catastrophic consequences.

We have come now to a critical point in the development of science-based technology where it has become so powerful as to inspire widespread fear and to provoke widespread reactions. The real question before us now is will we, and can we, take action to meet the very real problems that are raised by this enormous power or will we attempt to escape? Continuation of the vigorous growth of science-based technology requires rekindling of the enthusiasm for its promise which was the main drive, for example, in America before World War II.

This is the test. Will we meet the challenge or will we escape it and slow the progress of our technology?

This test of societies is analogous to many of the tests that individuals face in growing up. Dangerous times occur when children are first entrusted with dangerous devices. The responsibilities that go with increased power are essential parts of the growth process. In some cases these responsibilities are avoided, and the individual escapes the inevitable growing pains. In other cases, the challenge is met and the individual continues to grow. I believe that technological progress has been one of the prime movers in the growth in our society through the centuries. We must not allow those who would escape this challenge to dominate the thinking of young people.

I lack the wisdom to set forth a prescription for meeting this challenge in any generality. However, I am convinced that one of our pressing needs is to achieve a mechanism for the democratic control of a rapidly advancing technology. Note that this doctrine is in direct opposition to the doctrine of the "moral responsibility of scientists," a modern form of "noblesse oblige." (I am not suggesting that individual scientists should be absolved from moral responsibility for their actions; but the scientific community should not aspire to a role as the conscience of our society.) It certainly is not in the tradition of this country to endow an elite with overriding authority and to entrust our future to its sense of moral responsibility. I am convinced, therefore, that in meeting the current challenge we must achieve a mechanism for democratic control of sophisticated technology, in spite of the fact that our elected representatives do not and cannot be expected to have the deep scientific knowledge needed to provide an adequate base for the decisions they must make. Thus communication between the scientific community and the political community is vital to democratic control.

The Decision-Making Process

The governments of all advanced countries have had to make decisions on questions which have an important scientific component—that is, questions involving areas of science so new that no unanimity has been achieved in the scientific community, and so important that the decisions inevitably have important political and perhaps moral implications. I shall refer to these mixed decisions. Historical examples of mixed decisions are: the World War II decision to build an atom bomb; the German decision (a blunder I think) to build ballistic missiles during World War II; the U.S. decision not to use our ballistic missile capabilities to launch a satellite until after the Russians had beat us to it; the current decision to direct our primary space effort toward beating the Russians to the moon.

These decisions all involved technologies new enough so that debatable extrapolation of hard scientific fact was required. All of these decisions were of great political and sometimes moral importance. We now face a variety of mixed decisions, for example, those involved in dealing with the secondary effects discussed earlier, and still more importantly those involved in dealing with the real fears engendered by the great powers of new technology. These decisions must be made before unanimity exists in the scientific community. The problem of communicating with a divided scientific community is and will remain one of the most difficult aspects of making mixed decisions.

The essential input from the scientific community to decision making in the United States is via the scientific advisory committee. Without going into detail about this process, I would like to make several points. First, in seeking scientific advice on questions of great social importance we must recognize that the moral responsibility which many scientists feel very deeply can easily affect their judgment as to the state of scientific fact when the scientific facts are not yet crystal clear. Second, the selection of scientific committees has always been beset by the dilemma that one must choose between those who have gone deeply into the subjects under discussion and accordingly will have preconceived ideas about what the outcome should be, and those who are perhaps unprejudiced but relatively uninformed on the subjects under discussion.

As Warren Weaver has put it (*Science,* Vol. 130, November 20, 1959), "A common procedure is to set up a Special Committee of experts on X in order to find out whether X is a good idea. This committee is, characteristically, national or even international in scope, is formed of external experts of recognized standing (external as regards the agency in question but most emphatically internal as regards X), and always contains a comforting proportion of what might be called right names. These are men intensively interested in X, often with lifelong dedication to X, and sometimes with a recognizably fanatic concentration of interest on X. Quite clearly, they are just the lads to ask if you want to know whether X is a good idea."

Finally, scientific advisory committees have, in many cases, played an influential role in decision making without taking public responsibility for their judgments. In the making of mixed decisions the validity of the scientific input has frequently been brought under question.

Three Proposals for Science Advice

I have three recommendations directed toward institutionalizing the scientific advisory function with a view toward increasing the presumptive validity of the scientific input.

1. Separate the scientific from the political and moral components of a mixed decision.

It has occasionally been maintained that scientific and non-scientific components of a mixed decision are generally inseparable. It is, of course, true that a final political decision cannot be separated from scientific information on which it must be based. The reverse is not true—a scientific question which logically can be phrased as anticipating the results of an experiment can always be separated from any political considerations. (It is true that there are important questions which are best answered by scientists which are not scientific questions according to this definition. An example of this sort of question is the relative competence of scientific groups, which might be important in a decision as to where to locate a major scientific facility. In many cases, however, the essential information which the political community requires from the scientific community is a considered and unbiased statement of the currently available scientific facts. It is to such cases that these suggestions are addressed.)

Thus, the question—should we build a hydrogen bomb?—is not a purely scientific question. A related scientific question—can we build a hydrogen bomb?—could in principle be answered by an experiment.

It is almost inevitable that scientists who have been engaged in research relevant to the scientific side of great mixed decisions should have deeply held political and moral positions on the relationship of their work to society.

Scientific objectivity is very difficult to achieve and is a precious component of wise mixed decisions. I do not believe it is possible for scientists to have deeply held moral and political views about a question and simultaneously maintain complete objectivity concerning its scientific components. In the past, scientific advisory committees have frequently developed close relationships with the officials who have final decisions to make. They have frequently advised political figures about what final decisions to make. They have frequently advised political figures about what final decisions they should reach, not only about the scientific components of a decision, but about the moral and political implications as well. The close relationship may be valuable; however, it does point up a need for an alternative source of scientific judgement which shall forego taking any moral or political stands and seek to achieve the greatest possible objectivity.

2. Separate the judge from the advocate.

To my mind there is no other solution of the problem discussed earlier —of combining the highest level of expertise with lack of prejudice—except the solution arrived at centuries ago in the similar legal problem. If one insists only on expertise in advocates, and expects them to marshall the arguments for one side of a question, one can call on the services of people who have gone most deeply into a particular subject and who have in the course of this work arrived at a point of view. Such advocates, in addition to presenting their side of the case, can be very useful in criticizing the cases made by opposing advocates. The requirement for the judges, on the other hand, is simply that they must clearly understand the rules of scientific evidence, have no intellectual or other commitments regarding matters before them, and finally must have the mature judgment needed to weigh the evidence presented. Thus, it is almost inevitable that a scientific judge would have earned his distinction in areas other than those in which he could qualify as unprejudiced.

It has occasionally been suggested that the advocates should present their points of view directly to the political leaders who have decisions to make. This procedure suffers from the grave difficulty that political leaders will not be able to spare the time necessary to understand scientific debates in sufficient depth to distinguish the relative validity of positions taken by sophisticated advocates. The scientific judge would differ from the political leader sitting in judgment on scientific questions in that his scientific background should enable him to assess more quickly the evidence presented by opposing advocates and to participate in something analogous to cross-examination. He would not, on the other hand, be expected to have the deep acquaintance with the field required of the advocates.

Scientists are traditionally advocates, and in small-scale science judicial functions have never had an importance comparable with that of advocacy. An experiment can always overturn anyone's judgment on a scientific question. However, the judicial function becomes important in large-scale science and technology, when we must anticipate the results of experiments which cannot be performed without the expenditure of great amounts of money or time. This increase in the importance of the judicial function requires the development of a group of distinguished people who will devote themselves to scientific judgment. The point has been frequently made that a scientist needs to keep actively engaged in creative work in order to maintain his expertise. I submit, however, that if a mature scientist is deeply involved in finding the truth between the claims and counter-claims of sophisticated advocates, his education will be continuously improved by the advocates and he will be continuously mentally stretched in the effort to reach wise judgments. Communication from the judges to the scientific community and the public is an essential part of maintaining their expertise and reputation. A provision for publication of judgments, suggested below, will help to accomplish this.

The problems of selecting people to serve as judges and advocates will, of course, be the most difficult matter in reaching wise decisions, under this scheme as under any other. It would be very important that everything possible be done to elevate the positions of advocates, and especially of judges, so as to attract people whose wisdom will match the importance of the judgments they must make.

3. The scientific judgments reached should be published.

In many cases the results of scientific advisory committees have not been made available to the public for reasons other than national security. The existence of such privileged information makes it very difficult for the public to assess the degree to which a mixed decision is based on political grounds.

I would propose that the opinions of scientific judges reached after hearing opposing advocates should be published, within the limits of national security. The publication of these judgments would serve two purposes. First, it would provide the whole political community with a statement of scientific facts as currently seen by unbiased judges after a process in which opposing points of view have been heard and cross-examined. Hopefully these opinions would acquire sufficient presumptive validity to provide an improved base on which political decisions could be reached. Second, the publication of opinions reached by scientific judges would inevitably increase their personal involvement, and thus could help to attract distinguished scientists to serve in the decision-making process.

There is a grave difficulty raised by the traditional conservatism of scientists, even those who have exhibited great imagination and daring in their own work. I have no formula to offer to overcome this bias other than an insistence that the advocates of novel approaches be heard. It is important that they be cross-examined by skeptical experts and that the judges feel a responsibility for not rendering negative judgments on inadequate evidence. It is actually very difficult to offer rigorous proof that something cannot be done, and usually the most that can be said is that "I cannot set how to do it." Scientific judges whose opinions would be published should be more accountable for errors in judgment.

It is very important that this type of formal procedure not be allowed to interfere with the small-scale creative science which must precede any major decision-making. This work has always been pursued with a wide-spread opportunity for initiative, in a kind of private enterprise, laissez-faire system in which I firmly believe. When large-scale funding is required we must restrict the number of pleas for funds that are made. The question must then be asked, would the formalization of institutions for scientific judgment result in harmful restrictions on initiative? However, the scientific advisory procedures which now exist have also been guilty in this respect, and in formalizing of these procedures we could deliberately try to control this narrowing of the number of alternatives pursued simultaneously as a project grows in size.

It is not my intention here to suggest that institutionalizing communication between the scientific and political communities will of itself lead to effective democratic control of technology, or even that this proposed institution would optimize communication. I would suggest, however, that the achievement of superior communication would help toward democratic control, and that mechanisms for achieving improvements in this communication should be vigorously discussed in both the scientific and political communities.

Technology's Audience Appeal

There is another area of communication between the scientific and technological community and the lay public which would promise vast improvements in understanding. I refer now to the sense of beauty which all creative scientists and technologists feel about some of their own work and about the work of some of their colleagues. The words beautiful and elegant are frequently used in discussing creative work in these communities. However, the sense of beauty as distinguished from the sense of power is rarely communicated to laymen. Although one can find writers devoting themselves to communicating the beauty in primitive technology and a huge amount has been written on the power of modern technology, I know of no example of literary efforts to convey the beauty in, for example, an integrated circuit. I am convinced that science and technology represent great art forms in our time and the creative energy that is poured into these areas is immense compared to many better recognized art forms.

Perhaps one of the important difficulties in achieving this audience appeal is the dichotomy of the humanities and science in our universities, one result of which is that people who prepare themselves to communicate with the public do not generally become sufficiently conversant with science or technology to communicate its beauty. On the other hand, the education of a scientist emphasizes proficiency in the efficient but private languages of science, which of itself tends to atrophy his abilities to communicate with the lay public.

However, in view of the traditional academic rivalry between science and the humanities, it seems to me inevitable that the education of students who can better effect this communication will have to be the responsibility of scientists and technologists. We must meet the challenge of communicating to the lay public the beauty of a sophisticated invention and the grandeur of a new view of nature.

Need for a Vigorous Defense

Scientists and especially technologists are quite unprepared to undertake the defense of their professions against the vicious attacks currently being mounted against them, some of which I have quoted previously. Many of us behave in a manner that has been repeated by minorities many times in history who have adopted the views of their detractors. Thus, I know many technologists who are unsure about their profession and who are quite unprepared to defend its beauty and beneficence. Perhaps we need something analogous to the wonderful "black is beautiful" idea.

I have noticed for example many engineers who are very anxious to make a contribution to relieving the pollution problem. In choosing this area they seek to appease the strident voices who are using pollution to persecute technology. Again, one can find many scientists who have devised another way of appeasing the current attack on technology. We have seen in recent decades the advent of purity as a supreme value in academic science. Thus, science with no visible social impact is elevated to a status beyond that accorded activities with visible consequences for mankind. The practitioners of pure science hope to escape persecution by pretending that they are not contributing to "the disaster that scientific technology has produced." Of course, this pretense is pretty thin, but it does provide some shield from their critics.

At the present juncture scientists and technologists must together reaffirm:

—That science-based technology has improved not only mankind's conditions of life, but mankind itself;

—That there is no other choice in this world but to continue to foster a vigorously growing science-based technology;

—That the secondary problems, pollution, side effects, etc., can easily be met, if the will is there, by our enormously powerful technology;

—That science-based technology is one of the great creative forms of our time;

—That real and justified fears of modern technology, problems which cannot obviously be met by more technology, should be treated as powerful stimuli for social growth;

—That instead of facile predictions that nuclear weapons, computerized abolition of routine labor and still more dramatic things to come will inevitably result in disaster for mankind, we should call for the vision of a society which has advanced enough to meet these challenges.

It is my belief that the promise of the application of science and technology to the continuous improvement of mankind is still largely to be fulfilled, that we can see an ever increasing opportunity to create a world limited only by our own imagination. The current reaction against science based technology is destroying this vision and strenuous efforts must be made to unify science and technology and establish communication channels that will again make its promise bright and clear to everyone.

The Future of the Future

John McHale

We are not fundamentally concerned here with a series of predictions about the next hundred or the next thousand years, but rather with the "futures-orientation" itself as an intellectual and social attitude. We are concerned with ways of looking forward and with some of the implications of present scientific and technological developments in our styles of living.

In general, today's modes of confronting the future are vastly different from those of the nineteenth century Utopians. In that period, men were still preoccupied with the inevitability of progress, Western style, via a science and technology which seemed capable of ever greater mastery of man over nature. This was tempered somewhat by the Malthusian feeling that the future was limited to those able to prove their material strength and mastery—a viewpoint which, in its more negative aspects is now largely confined to the military establishments. Today we do not view the future quite in the same way, as a great evolutionary onrush, largely independent of man's intervention and tinged with various premonitions of doom whether or not he chooses to intervene.

We realize that man does not, in the end,"master" nature in the nineteenth century sense, but collaborates with nature—his very existence depends on an intricate balance of forces within which he is also an active agent.

H. G. Wells' *Mind at the End of its Tether* marks the conscious end point of the older intellectual stance towards the future, and one may still see it repeated in those who cannot make the breakthrough to the next period. In essence, there is a kind of intellectual polarization taking place around the mid-twentieth century which separates the intellectual establishment into two —one, those who are still preoccupied with the world as conditioned by its pre-1900 parameters, and those who are attempting to recast and reorient their world view to one which is, in many ways, quite unprecedented in human experience. The watershed of this dichotomy really lies much further back— around the Renaissance. The argument begins there about man's relation to, and conscious control of, his own forward development and reverberates down to our own period. At a particular point in time, the summation of certain discoveries and access to certain technical facilities suddenly invalidates the whole of one side of the debate. From this time forward, which one may locate as recently as World War II, one can isolate the two attitudes in the turn of a phrase, the use of a particular frame of reference.

An important point for the individual is, that once the switch in perspective is accomplished, a good deal of negative baggage drops away. The fundamental realization is that man's future is literally what he chooses to make it —and the conscious degree of control he may exercise in determining his future is quite unprecedented. There are many alternative paths to as many alternate futures. Some we have already begun to take, others await our decision. As man gains more knowledge of the forces operative in, and external to, human society, he is forced to couch his questions about the future in the form of alternative possibilities of present actions in terms of their long-range consequences. The more knowledge, the greater the number of alternative paths and the longer the range of consequences.

This realization has been borne in upon many sectors of society. Governments and industries alike, committed to long range programmes of the most varied nature find that they are increasingly forced to think not of the next ten or twenty years but of the next fifty or a hundred. To launch a manned space vehicle to the moon in 1970 requires that you start work on it about ten years before. Other decisions are of a similar nature. But planning a series of manned rockets is relatively easy in present terms. You can forecast with reasonable accuracy the types of basic research in metal and other alloys which should be initiated this year so that their bulk production may be available in three years to phase with parallel developments in lubricants for near vacuum which you can predict will be available in four years and so on. By compiling the research trends and rates of technological development you can attain to variously workable ten, twenty or even fifty year predictions. Even such apparently straight-forward forecasting, however, is liable to swift alteration, through human serendipity.

The same might be said for much prediction regarding physical resources and their technological exploitation. But even within this area, there are still alternative paths, each with its various contingencies. All are, in varying degree, affected by factors already known or predictable in some form from today's knowledge. When we come to social planning, the situation is very different, but the need to introduce some predictable parameters and concomitant action has become even more urgent. We have viewed the unforeseen consequences of "not predicting"—famine and disease are preventable catastrophes. On the local scale, governments now attempt to predict situations productive of disorder and violence. Industry has become increasingly preoccupied with the markets of the '70s or '80s, the future of this industry or that. Dealing with human futures re-introduces the capacity of human beings to determine their future. This is a central point. Given his present scientific and technological knowledge, man now has an enormously enhanced capacity to choose his future—both collectively and individually.

> Finding out what we want should become a major object of our attention . . . there is a vast difference between letting changes occur and choosing the changes we want to bring about by our technological means.[1]

The outcome of the "futures" chosen will depend on the degree to which we predict them. If we conceive a specific course of action desirable, we will tend to orient ourselves towards it.

The collective aspect of choice of futures is reflected in the growing concern of our local societies, with the allocation of public funds to various programmes. We begin to agree that investments in pre-natal care, child welfare and pre-school education, etc., which may not "pay off" for twenty or thirty years are realistic societal strategies. We attempt to legislate the future pollution of the rivers and the air, the future congestion of cities, on the same basis. The pattern of a desired future based on even the least factual or measurable prediction commits us to consensual action. Our prior "collective" assumption is, increasingly, that the environment and form of our society are within our positive (or negative) control.

The individual's relation to his or her future has become, and is becoming, more flexible. Where a man, even in the advanced countries, would previously feel impelled to prepare himself for one occupation, profession or career, committed more or less to a particular geographic locality and determined for him largely by the circumstance to which he was born, we now have an

---

1. Bertrand de Jouvenel, "Utopia for Practical Purposes," *Daedalus*, Summer, 1965.

emerging situation within which an individual may reasonably expect to change occupation, career role and geographic location many times in his lifetime. The future of the individual is based, again, on whatever expectation of the future he acquires. His paths towards this or that future, though conditioned in part by physical make up, "talents," etc., may be viewed as more largely determined by his particular conceptual mapping. As Dennis Gabor has suggested, we are now "inventing the future." Man's future is most likely that which he may most imaginatively conceive of, which, in turn, will determine his action towards its accomplishment. Life may be viewed as a great number of alternative possibilities—in life style, location, occupation, etc. The so-called "threat" of leisure is no more than a widening of "living" alternatives.

The future of the future becomes, therefore, what we determine it to be both individually and collectively. It is directly related to how we may *conceive* any specific or vague future to be. Such mental "blueprints" are action programmes, whether immediate or not depends on the individual and his collectivity, i.e. society. All actions have consequences and both may be effected on a larger scale, with further reaching contingencies than was ever consciously possible in human history.

Though emphasizing change, we should also note that all change proceeds within a set of regulating patterns. Life on earth has been possible only during the past billion years through the relatively stable interrelationships of the variables of climate, the chemical composition of the atmosphere, the sea, the life-sustaining qualities of the land surface, the natural reservoirs and the water cycles.

Within the relatively thin bio-film of air, earth and waterspace around the planet, all living organisms exist in a delicately balanced ecological relationship. The close tolerances of this symbiosis are presently known to us in only the haziest outline. Apart from the relatively local disturbance of earth cycles through agricultural practices, man until quite recently did not have the developed capacities to interfere seriously with the major life sustaining processes. Since the Industrial Revolution, this has changed abruptly, and from this time forward the "eco-system" also includes man's machines, their products and an incalculable capacity to alter the natural balances.

The first great changes came with the advent of the Industrial Age, based on engines that used energy stored in coal beds, which built cities and navies, wove textiles, and sent steam trains across the widest continents.
Since then, with energy from petroleum and other sources, changes have come more swiftly. Today, radar telescopes scan the universe to record galactic explosions that occurred billions of years ago; oceanographic ships explore the under-sea; electronic devices measure the earth's aura of unused energy and similar equipment traces inputs and outputs of single nerve cells; television cameras orbiting the earth send back photographs of entire sub-continents; electron microscopes photograph a virus; passenger planes fly at almost the speed of sound; and

machines set type in Paris when a key is tapped in New York. These are only a few of the changes that our increasing supply of energy has made possible in the last 60 years.[2]

The word, ecology, is significantly derived from the Greek *Oikos* meaning *house,* so in our references to human ecology, we are really talking about planetary housekeeping.

Writing on the human biosphere[3] G. Borgstrom, points out that the maintenance of three billion humans presently requires a plant yield sufficient to accommodate 14.5 billion other consumers. These other consumers, the animal populations, are an essential element in maintaining the humans by acting as intermediate processors for many plant products indigestible by man. Pigs, for example, consume four times more than America's 400 million people, when measured on a global scale. Despite mechanization, the world horse population still has a protein intake corresponding to that of 653 million humans—the population of China.

Yet in terms of balance, ". . . only one tenth of the caloric intake of the world household consists of animal products." World food consumption is largely vegetarian with 90 per cent of the caloric intake and 60 per cent of protein coming from plants. This underlines the importance of each of the respiration / excretion / decomposition stages in the natural economy, with microbial activity as a key element in the recycling of materials. Amongst the non-human animal population in the food cycles, micro-organisms play a major invisible role.

The dependence of one-sixth of the world's food supply on "artificial" nitrogen produced by the chemical industry, is another factor. To make each million tons of such nitrogen annually, we use a million tons of steel and five million tons of coal. In terms of our methods of crop use and food production, Borgstrom estimates that we will need 50 million tons of such support nitrogen annually by the year 2000. The amounts of other chemicals, e.g. sulphur and key trace elements such as phosphorus, which will require massive support technologies to augment the natural cycles, is only now becoming apparent. But the greatest areas of developing crisis for man in the biosphere are water and air. Approximately 95 per cent of our water is in the ocean and the remaining 5 per cent of fresh "cycling" waters are presently being used at a prodigious rate. Agriculture accounts for 50 per cent, using 400-500 pounds for each one pound of dry plant matter. This water/crop ratio varies as high as 1-1000 and 1-2000, so that agriculture in the lesser developed countries consumes as much water *per capita* as the technologically advanced—where 250 tons of water are used in producing a ton of newsprint and 25 tons for each

2. Boris Pregel,"The Impact of the Nuclear Age," in *America Faces the Nuclear Age.* Edited by D. Landman and J. E. Fairchild, published by Sheridan House, New York, 1961.
3. G. Borgstrom, *The Human Biosphere and its Biological and Chemical Limitations.* Global Impacts of Applied Microbiology Conference, Stockholm, 1963.

ton of steel. When such uses are compounded with mounting waste and sewage disposal, the position is more severe. The increase of pollution in water and air has now become of national concern in many countries. An average industrial city of half a million people disposes of 50 million gallons of sewage a day and produces solid wastes at the rate of about 8 pounds a person each day. Present solid waste disposal even in advanced countries is archaic.

> Pollutants are the residues of things we make use of once and throw away ... As the earth becomes more crowded, there is no longer any "away." One person's trash basket is another person's living space ... our whole economy is based on taking natural resources, converting them into things that are consumer products, selling them to consumers and then forgetting about them. But there are no consumers—only users. The user employs the product, sometimes changes it in form, but does not consume it—he just discards it. Discard creates residues that pollute at an increasing cost to the consumer and his community.[4]

It has been noted that with present waste treatment, by 1980, effluents will be sufficient to consume all the oxygen of all the dry weather flow of 22 river basins in the USA. Within this discharge into rivers and streams goes also detergent materials, industrial wastes and pesticides from the land. Massive fish-kills of around 10 million in the Mississippi basin and the Gulf of Mexico, during 1960–64, were traced to pesticide run-off and other toxic agents from sources thousands of miles away.

With "people kills" the toxic agencies may go unnoticed for much longer. Some 500 new chemical compounds come into industrial use yearly in one country alone, with practically no legislative attention to their long-range deleterious effects which may be nonspecific as to pass for normal deterioration. In the past hundred years the CO_2 concentration in the atmosphere has been increased by about 10 per cent—no small argument in favour of banning *with* the bomb the comparably lethal uses of coal and other fossil fuels as energy sources. Four thousand Londoners died from air pollution in one week in 1952, one thousand in 1956.

> The average person daily eats about two and three-quarter pounds of food, drinks four and a half pints of water and breathes 20 pounds of air. He can postpone eating and drinking, but he cannot postpone breathing ... air pollution affects almost everything in our environment ... from clothing, skin and lungs to metals and paints ... its damage costs are estimated in the US alone between $7 and $9 billion annually.[5]

In addition to fouling the atmosphere, it has been calculated that certain elements, e.g. argon, neon, krypton, etc., indispensable to life are now being "mined" out of the atmosphere by industrial operations at a faster rate than they are being produced by the earth's atmosphere/hydrosphere/lithosphere process.

4. *Waste Management and Control.* US National Academy of Science—National Research Council Publication 1400 (1966).
5. Dr. Karl W. Wolf, *Conference on Space, Science and Urban Life,* Seminar B. NASA 1963.

This cursory overview is not without consequence for the future of architecture and environment planning.

> The town and city planner and public health specialist of tomorrow will have to take a far more comprehensive view of human ecology than most of them yet dream of; and the costs of safeguarding human health, including the psyche, can no longer be put "on the cuff" no matter what they may do to conventional economic progress.[6]

Air pollution is not a "local" problem—the air is not restrained within municipal or national boundaries, nor are the waters of the planet. In terms of any such planning, even the year 2000 is too short range. As a generation of the 'hinge of history' we must accept the challenge of imaginative extrapolation of human requirements beyond 100 or even 1000 years. Where it may be pleaded, for example by special interest groups, that we have enough coal, oil and gas reserves for 500 years, their continued use at the present rate is obviously precluded by their adverse effects on the eco-balance. Leave them in "storage"—until a more evolved society may use them less prodigally and dangerously.

Some of the mandatory requirements for the merely adequate maintenance of the eco-system are already clear. We need to recycle our minerals and metals; increasingly to employ our "income" energies of solar, wind, water and nuclear power, rather than the hazardous, and depletive, "capital" fuels; to draw upon microbiology and its related fields to refashion our food cycle; to reorganize our chaotic industrial undertakings in new symbiotic forms so that the wastes of one may become the raw materials of another; to redesign our urban and other "life style" metabolisms so that they function more easefully.

As we go towards 2000, it will behove us to accept the facts, that the resources of the planet can no more *belong,* by geographical accident, to any individual, corporation, country or national group than the air we breathe. National ownership of a watershed or key mineral deposit is as farcical a proposition as our supposedly national sovereignty of an 'air space.' The situation we face now is analogous to that which was fought over locally in the nineteenth century with regard to pure food legislation, public health and child welfare, etc. The same arguments will, no doubt, be raised again about the rights and privileges of the individual—to poison, swindle and infect his neighbour, or his own descendants.

Our most important discoveries, therefore, may not lie solely with technological innovation—but with social invention. We begin to recognize more clearly that our societal institutions, the ways in which we organize ourselves to live together in human fashion are not immutable, but are as much manmade "invented" forms as television or the car. The city or nation state were comparatively recent inventions of this order which may now, in certain areas

6. Wm. Vogt, "Man's Ecological Dilemma" *Natural History,* December 1965.

of their functioning, be dangerously obsolescent. Just as we have consciously learnt in the past few decades to organize the process of scientific and technological innovation, and its applied development over long time spans, so must we orient ourselves towards more consciously controlled and experimental social innovation. The design of new forms of human organization is already under way in many areas of public and private life and is even more evident at the international level. The UN, for example, is a second generation "bench" prototype of the League of Nations. We now need to initiate new phases of research and development towards more viable forms of this magnitude. Our evolving planetary society must become like a great learning machine in which, "man's intelligence (now) intervenes and directs the process which remains, nonetheless, basically an experimental process."

Without touching upon the more familiar problems of war, hunger and human disease, even a cursory glance at eco-system is sobering. It should be apparent to all, that we now live in such close community, and within such delicate "life" margins, that all our actions are now cast on a planetary scale and that our gross ecological errors may reverberate for centuries.

Epilogue

In the 1970's we find ourselves in the midst of a maelstrom which threatens to destroy age-old institutions and a value system which took centuries to develop. Modern man is forced to live in a new social environment every five to ten years as technology imposes itself upon each and every person. This maelstrom had its auspicious beginning thousands of years ago when man first began to use objects of his own creation to extend his limited capacities, but the mushrooming of technology was not due solely to the efforts of any one man or any particular group. The growth can be attributed to the accumulation of knowledge and experience throughout history in spite of several periods of technological hibernation. Some ask today if this exponential growth will level off or if it will continue its expansion for centuries to come. At times throughout history it has been predicted that this growth would recede but since 1950 technology has pushed forward like a tidal wave. The expansion of man-made material has become a pathological condition disrupting tradition, demanding new priorities, and creating many mutations in the age-old dream of the "good" life.

Was it the launching of Sputnik or the Battle of Berkeley that brought about our first awareness of radical change? Or can we trace this back to the Luddites who raised the questions we seem to avoid today? Henry Adams in 1909 recognized the growth of technology when he stated that it increased 1,000 times between 1800 and 1900. The growth between 1900 and 1971 has been even more phenomenal. Dostoyevsky said, "taking a new step, uttering a new word is what people fear most." He was referring to the proclivity for confusion, revolution, and united action created by drastic change.

In the midst of this avalanche of achievement it is no wonder that reactions are mixed; some people are confused, some anxious, and some frighteningly concerned. Impermanence in an unstable condition exists for all men and all institutions. Yet we talk about change so much that we have become anesthetized to its true meaning. Such activity has caused some philosophers to ask for a moratorium on further research, implying that all future technological growth is innately nega-

533

tive. To suppress change in this era of growth would be ridiculous to say the least. It would be beyond comprehension to deny the need for curbing disease, re-building our cities, and solving the many other social problems that our society possesses. A moratorium is not the answer; more technology is. However these technological answers must be based upon simultaneous technical and social research.

It is necessary to meet mechanical invention with social invention and to prepare people to be flexible and adaptable to new environments. The attainability for man in the midst of this potential is beyond all expectations. But man must desire to determine his future rather than allowing it to rule him or letting the system make irrational decisions based upon profit, greed, ambivalence, or fear.

It is imperative for the student of change to consider that for every technological invention there is a corresponding set of social circumstances. These reveal themselves as second, third, fourth, and even fifth order consequences. Thus technology is not independent of society but rather an integral part of our mores, norms, and traditions. It cuts across the entire fabric of our lives. Social scientists have predicted that during the next ten years there may be as much social change as in the past thirty years. This means that the children of change are being prepared for a society which we cannot comprehend at this time. How can man attempt to perpetuate his culture with change as the only constant? Or is this modern man's basic problem that he attempts to perpetuate a culture which is inconsistent with the dictates of a highly technological environment?

It is of maximum importance that every individual understand himself in a highly technological society. To do this one must also realize that many sub-systems exist to make up the total universe. During the Renaissance people had a reverence for nature but by the 19th century men were attempting to conquer it. There is encouraging evidence that today man realizes he is only part of the system and must not conquer or destroy nature but rather use it constructively, equitably, and rationally. It is the lack of action that occurred in the pre-1970 era that brought man to near ecocide.

Whether we view technology negatively, positively, or consider it unworthy of study, we find ourselves making value decisions about it. Questioning values does not mean destroying them. Today our options are as wide as man's dreams and naturally such breadth threatens the status quo. Technology as a change agent brings to us previously unattainable goals. It is these open-ended potentialities that cause confusion, polarization, and the unstable condition described so vividly by Donald Schon in his book *Technology and Change*.

Technology has positive and negative characteristics. Both must be judged consistently and accurately, for over-optimism can be as threatening to the future as pessimism. However the greatest pitfall for man is sheer ambivalence. Hopefully it is not too late to make up for past stupidity and inept dogma. Apathy is a mechanism of self-defense. It causes man to regress to a fantasy land that appeases his conscience while perpetuating a system that is perhaps doomed to failure. To insure the future, technology's consequences must be met head-on with planning. To circumvent issues invites a lack of social commitment to technical realities. True, the future holds dangers but it holds challenges as well. Outcomes are uncertain but there is a potential for positive direction if we avoid confusion, anxiety, irritability, apathy, and withdrawal.

Positive change can and does take place if the marketplace demands it and can give some sense of direction or structure for nurturing it. But who should be the instigator of change? Who can determine the consequences of every tangible conceived by the managers of technology? The evidence of pollution was apparent decades ago but nothing was done to curb it until the past several years and now it might be too late. Technology can provide the alternatives to a decadent society but this mandates a social response on the part of every responsible citizen. Technology itself is neutral. As a product of man's ingenuity it thrives on his desires and willingness to pay. It is this sort of cost benefit analysis that has left us with a vast reservoir of problems our children will have to solve (e.g., pollution, dehumanization). It is now time to judge man's misuse of his power in terms of the broader society. The few commitments we have had to the better society have been miserable failures as testified by the lack of legislation to curb pollution and repair our decaying cities. Every social system fights to preserve itself, but this is becoming increasingly difficult in a cybernetic society. Change must not be arrested but assessed, revised, and implemented with a guarantee of perpetual evaluation.

Priorities for the Future

It is not difficult to visualize the priorities that man has established for himself in the 1970's. Military needs rank over civilian needs, the affluent take precedence over the needy, and profit justifies ecocide, dehumanization, and social conflict. To list all of the priorities a nation must consider would be an endless task but seven of prime importance are listed below for the deliberation of our youth who will hopefully decide our future by acting upon them.

1. *Control of World Population.* This is an immediate priority since this growth has compounded the problems of disposal, diminishing resources, and living space. The solutions are technically possible but, once again, demand social awareness, education, and action.

2. *Clean Up Our Environment.* The natural filtration system of the earth has become a plugged sewer. Coupled with over-population this becomes a paramount problem. Man needs air to breathe, water to drink, while the population of the 1970's makes these less accessible daily. What is needed is the application of technical solutions, legislation, and a unified effort by industry, government, and the citizen who relies on technology for survival.

3. *Demilitarization of the National Budget.* Technical solutions to social problems are costly and like our natural resources money is not unlimited. We must ask a value question: Is the military machine of a greater priority than re-building our cities, educating the masses, cleaning up our environment, eradicating disease, and on down the list of basic human needs?

4. *Rebuild Our Cities.* Housing is inadequate, employment within the cities is inadequate, and we continue to seem bewildered by a lack of social cohesion within cities built for the early 1900's.

5. *Alteration of the Educational System.* The educational system must become a viable leader in social change. This can be accomplished by utilizing the captive audience to develop individuals who can gather data, synthesize, and make rational decisions. We have not begun to tap the full potential of the human mind. Yet federal funds for education are not a high priority item in the national budget.

6. *Development of a World Community.* The earth's peoples need a unified and cooperative system which will allow for mutual respect. This new ethic would make possible the pooling of resources, the elimination of fear and military priority, and help to ensure the preservation of man.

7. *Alteration of the Governmental Structure.* The present bureaucratic system does not allow for decisions based upon value impact. Until a structure is developed that allows for this our society cannot hope to come to grips with its social problems, many of which have technical solutions.

It is evident that solutions are technical, organizational, economic, and emotional. To deny their existence is to deny a future for our offspring. Who can justly deny any human a sense of purpose and potency in a technological society? This we have done. This we must eliminate.